Flexible and Wearable Sensors

Flexible and Wearable Sensors

Editors

Libo Gao
Zhuoqing Yang

MDPI • Basel • Beijing • Wuhan • Barcelona • Belgrade • Manchester • Tokyo • Cluj • Tianjin

Editors
Libo Gao
Pen-Tung Sah Institute of Micro-Nano Science and Technology
Xiamen University
Xiamen
China

Zhuoqing Yang
National Key Laboratory of Advanced Micro and Nano Manufacture Technology
Shanghai Jiao Tong University
Shang Hai
China

Editorial Office
MDPI
St. Alban-Anlage 66
4052 Basel, Switzerland

This is a reprint of articles from the Special Issue published online in the open access journal *Micromachines* (ISSN 2072-666X) (available at: www.mdpi.com/journal/micromachines/special_issues/Flexible_Wearable_Sensors).

For citation purposes, cite each article independently as indicated on the article page online and as indicated below:

LastName, A.A.; LastName, B.B.; LastName, C.C. Article Title. *Journal Name* **Year**, *Volume Number*, Page Range.

ISBN 978-3-0365-8299-3 (Hbk)
ISBN 978-3-0365-8298-6 (PDF)

Contents

About the Editors

Libo Gao

Libo Gao received a Ph.D. degree from the City University of Hong Kong (CityU) in 2018. He worked as a research assistant at CityU before he joined Xidian University, Xi'an, China, as an Associate Professor. Currently, he is an Associate Professor at the Pen-Tung Sah Institute of Micro-Nano Science and Technology, Xiamen University, Xiamen, China. His research mainly focuses on intelligent micro/nano-electronic devices and microsystems.

Zhuoqing Yang

Zhuoqing Yang received his B.S. and M.S. degrees in electromechanical engineering from Harbin Engineering University, Harbin, China, in 2003 and 2005, respectively, and his Ph.D. degree in microelectronics and solid-state electronics from Shanghai Jiao Tong University (SJTU), Shanghai, China. He worked at the National Institute of Advanced Industrial Science and Technology, Tsukuba, Japan, in 2011. Currently, he is a Full Professor at SJTU, being affiliated with the National Key Laboratory of Science and Technology on Micro/Nano Fabrication. Dr. Yang received a prestigious Japan Society for the Promotion of Science (JSPS) Postdoctoral Fellowship in 2011.

Preface to "Flexible and Wearable Sensors"

In recent years, the field of flexible and wearable electronics has emerged as a promising domain with vast potential in various areas such as human–machine interaction, robotics, and healthcare monitoring. The favorable flexibility and adaptability of these technologies have paved the way for unprecedented advancements, making this field one of the most captivating and rapidly growing areas of interdisciplinary research.

At the heart of flexible electronics lie flexible and wearable sensors, which play a pivotal role in ensuring the exceptional performance of these devices. With their remarkable sensing capabilities, these sensors have become the focal point of both domestic and international research endeavors. Their development is crucial to the realization of truly flexible and wearable electronics.

In light of the significance of flexible and wearable sensors, this Special Issue presents a comprehensive collection of research papers, short communications, and review articles in the field. The focus of these contributions revolves around three main themes:

(1) Novel structural designs, material fabrication, signal processing, and modeling techniques applying a wide range of mechanisms to flexible and wearable sensors. This exploration could lead to groundbreaking advancements in sensor technology.

(2) The utilization of MEMS (Micro-Electro-Mechanical System) techniques in the production process of wearable and flexible sensors, along with simulation analysis with theoretical modeling. These techniques will enable researchers to enhance the performance and functionality of sensors.

(3) The exploration of multiple application scenarios within the realm of multivariable flexible and wearable sensor systems. These applications hold immense potential in revolutionizing various industries and domains.

By bringing together cutting-edge research in these areas, we aim to provide a platform for knowledge exchange and collaboration among researchers, scientists, and practitioners. This Special Issue serves as a valuable resource for readers seeking to delve into the exciting world of flexible and wearable electronics and contribute to their further advancements.

We would like to express our gratitude to all the contributors for their valuable insights and expertise, without which this Special Issue would not have been possible. We hope that this collection of work will inspire new ideas, foster innovation, and push the boundaries of what is possible in the field of flexible and wearable electronics.

Libo Gao and Zhuoqing Yang
Editors

micromachines

Editorial

Editorial for the Special Issue on Flexible and Wearable Sensors

Libo Gao [1,2,*] and Zhuoqing Yang [3]

1 Pen-Tung Sah Institute of Micro-Nano Science and Technology, Xiamen University, Xiamen 361102, China
2 Innovation Laboratory for Sciences and Technologies of Energy Materials of Fujian Province (IKKEM), Xiamen 361005, China
3 National Key Laboratory of Micro/Nano Fabrication Technology, School of Electronic Information and Electrical Engineering, Shanghai Jiao Tong University, Shanghai 200240, China; yzhuoqing@sjtu.edu.cn
* Correspondence: lbgao@xmu.edu.cn

Citation: Gao, L.; Yang, Z. Editorial for the Special Issue on Flexible and Wearable Sensors. *Micromachines* **2023**, *14*, 1400. https://doi.org/10.3390/mi14071400

Received: 4 July 2023
Accepted: 7 July 2023
Published: 9 July 2023

Flexible wearable sensors have garnered significant interest in the fields of human-computer interaction, materials science, and biomedicine. The remarkable progress achieved in combining flexible materials and sensor manufacturing processes has led to substantial growth in flexible wearable sensors, making it a highly attractive and rapidly evolving area of interdisciplinary research. This Special Issue aims to highlight the latest advancements, ongoing challenges, and emerging opportunities in flexible wearable sensors, encompassing both fundamental principles and practical applications. The objective of this Special Issue is to inspire the community by addressing key questions in this field, with the ultimate aim of maximizing the societal impact of flexible wearable sensors.

In this Special Issue on Flexible and Wearable Sensors, we have included 25 papers, including 24 research papers, covering applications in human-computer interaction [1–3], mechanical design [4–8], health monitoring [9–15], manufacturing technology [16–20], algorithms [21–23], and smart cities [24]. Additionally, we feature an intriguing review paper focusing on flexible wearable sensor devices for biomedical applications [25].

In particular, Xue et al. proposed a frictional electromagnetic hybrid harvester with a low starting wind speed and an engineering-practical propeller design approach to achieve output power over a wide range of wind speeds [24]. Vanhala et al. proposed a strategy for long carbon stitched fibers in the form of permeable carbon fiber cloth placed on a stretchable thermoplastic polyurethane matrix to improve the 3D printed matrix's mechanical, electrical, and thermal properties [17]. Gao et al. proposed a flexible skin pressure sensor for monitoring the pulse of the radial artery, which was connected to a flexible processing circuit mount to enable real-time wireless, accurate monitoring of the pulse by a smartphone [12]. Zhang et al. combined 2D transition metal carbides, nitrides, and carbon-nitrides with a honeycomb structure formed by femtosecond filamentary pulses to design and fabricate a high-performance flexible piezoresistive sensor that enables stress-strain detection during human movement [14].

Finally, the review articles in this issue highlight the latest advances in flexible wearable sensor devices for biomedical applications [25]. As the field of flexible wearable electronics continues to grow, the number of relevant articles published on flexible wearable sensors is also increasing, and we look forward to the day when flexible wearable electronics move from the laboratory to full social life.

It is our hope that this Special Issue on Flexible and Wearable Sensors provides readers with valuable insights into the current state of the art in this rapidly evolving research area, presenting some of the latest technologies developed in the field.

Author Contributions: Investigation, L.G. and Z.Y.; resources, L.G. and Z.Y.; writing—original draft preparation, L.G.; writing—review and editing, L.G. and Z.Y.; supervision, L.G.; funding acquisition, L.G. All authors have read and agreed to the published version of the manuscript.

Funding: This research was supported by the National Natural Science Foundation of China (No.62274140), the Fundamental Research Funds for the Central Universities (20720230030).

Data Availability Statement: The data presented in this study are available on request from the corresponding author.

Conflicts of Interest: The authors declare no conflict of interest.

References

1. Nikitina, N.A.; Ryabkin, D.I.; Suchkova, V.V.; Kuksin, A.V.; Pyankov, E.S.; Ichkitidze, L.P.; Maksimkin, A.V.; Kitsyuk, E.P.; Gerasimenko, E.A.; Telyshev, D.V.; et al. Laser-Formed Sensors with Electrically Conductive MWCNT Networks for Gesture Recognition Applications. *Micromachines* **2023**, *14*, 1106. [CrossRef]
2. Yu, Q.; Zhang, J. Flexible Capacitive Pressure Sensor Based on a Double-Sided Microstructure Porous Dielectric Layer. *Micromachines* **2023**, *14*, 111. [CrossRef] [PubMed]
3. Sha, B.; Lü, X.; Jiang, L. High Sensitivity and Wide Range Biomimetic Tactile-Pressure Sensor Based on 2D Graphene Film and 3D Graphene Foam. *Micromachines* **2022**, *13*, 1150. [CrossRef] [PubMed]
4. Hubmann, M.; Bakr, M.; Groten, J.; Pletz, M.; Vanfleteren, J.; Bossuyt, F.; Madadnia, B.; Stadlober, B. Parameter Study on Force Curves of Assembled Electronic Components on Foils during Injection Overmolding Using Simulation. *Micromachines* **2023**, *14*, 876. [CrossRef]
5. Li, B.; Sun, H.; Zhang, H.; Li, Y.; Zang, J.; Cao, X.; Zhu, X.; Zhao, X.; Zhang, Z. Refractive Index Sensor Based on the Fano Resonance in Metal–Insulator–Metal Waveguides Coupled with a Whistle-Shaped Cavity. *Micromachines* **2022**, *13*, 1592. [PubMed]
6. Liao, Z.; Jing, J.; Gao, R.; Guo, Y.; Yao, B.; Zhang, H.; Zhao, Z.; Zhang, W.; Wang, Y.; Zhang, Z.; et al. A Direct-Reading MEMS Conductivity Sensor with a Parallel-Symmetric Four-Electrode Configuration. *Micromachines* **2022**, *13*, 1153. [CrossRef]
7. Gao, R.; Zhang, W.; Jing, J.; Liao, Z.; Zhao, Z.; Yao, B.; Zhang, H.; Guo, Y.; Xu, Y.; Wang, Y.; et al. Design, Fabrication, and Dynamic Environmental Test of a Piezoresistive Pressure Sensor. *Micromachines* **2022**, *13*, 1142. [CrossRef]
8. Liu, M.; Hu, J.; Zeng, Q.; Jian, Z.; Nie, L. Sound Source Localization Based on Multi-Channel Cross-Correlation Weighted Beamforming. *Micromachines* **2022**, *13*, 1010. [CrossRef]
9. Zhang, Z.; Zhang, H.; Zhang, Q.; Zhao, X.; Li, B.; Zang, J.; Zhao, X.; Zhang, T. A Pressure and Temperature Dual-Parameter Sensor Based on a Composite Material for Electronic Wearable Devices. *Micromachines* **2023**, *14*, 690. [CrossRef]
10. Brendgen, R.; Graßmann, C.; Gellner, S.; Schwarz-Pfeiffer, A. Textile One-Component Organic Electrochemical Sensor for Near-Body Applications. *Micromachines* **2022**, *13*, 1980. [CrossRef]
11. Miao, X.; Gao, X.; Su, K.; Li, Y.; Yang, Z. A Flexible Thermocouple Film Sensor for Respiratory Monitoring. *Micromachines* **2022**, *13*, 1873. [CrossRef] [PubMed]
12. Zheng, W.; Xu, H.; Wang, M.; Duan, Q.; Yuan, Y.; Wang, W.; Gao, L. On-Skin Flexible Pressure Sensor with High Sensitivity for Portable Pulse Monitoring. *Micromachines* **2022**, *13*, 1390. [CrossRef]
13. Shi, S.; Liang, J.; Qu, C.; Chen, S.; Sheng, B. Ramie Fabric Treated with Carboxymethylcellulose and Laser Engraved for Strain and Humidity Sensing. *Micromachines* **2022**, *13*, 1309. [CrossRef] [PubMed]
14. Su, Y.; Ma, K.; Yuan, F.; Tang, J.; Liu, M.; Zhang, X. High-Performance Flexible Piezoresistive Sensor Based on Ti_3C_2Tx MXene with a Honeycomb-like Structure for Human Activity Monitoring. *Micromachines* **2022**, *13*, 821. [CrossRef] [PubMed]
15. Xu, D.; Duan, L.; Yan, S.; Wang, Y.; Cao, K.; Wang, W.; Xu, H.; Wang, Y.; Hu, L.; Gao, L. Monolayer MoS_2-Based Flexible and Highly Sensitive Pressure Sensor with Wide Sensing Range. *Micromachines* **2022**, *13*, 660. [CrossRef]
16. Bakr, M.; Hubmann, M.; Bossuyt, F.; Vanfleteren, J. A Study on Over-Molded Copper-Based Flexible Electronic Circuits. *Micromachines* **2022**, *13*, 1751. [CrossRef]
17. Salo, T.; Di Vito, D.; Halme, A.; Vanhala, J. Electromechanical Properties of 3D-Printed Stretchable Carbon Fiber Composites. *Micromachines* **2022**, *13*, 1732. [CrossRef]
18. Pan, X.; Lin, F.; Wu, C.; Zeng, Y.; Chen, G.; Chen, Q.; Sun, D.; Hai, Z. Additive-Manufactured Platinum Thin-Film Strain Gauges for Structural Microstrain Testing at Elevated Temperatures. *Micromachines* **2022**, *13*, 1472. [CrossRef]
19. Lin, F.; Pan, X.; Wu, C.; Zeng, Y.; Chen, G.; Chen, Q.; Sun, D.; Hai, Z. ZrB2/SiCN Thin-Film Strain Gauges for In-Situ Strain Detection of Hot Components. *Micromachines* **2022**, *13*, 1467. [CrossRef]
20. Wu, C.; Lin, F.; Pan, X.; Zeng, Y.; Chen, G.; Xu, L.; He, Y.; Sun, D.; Hai, Z. A SiCN Thin Film Thermistor Based on DVB Modified Polymer-Derived Ceramics. *Micromachines* **2022**, *13*, 1463. [CrossRef]
21. Zhu, X.; Ma, Y.; Guo, D.; Men, J.; Xue, C.; Cao, X.; Zhang, Z. A Framework to Predict Gastric Cancer Based on Tongue Features and Deep Learning. *Micromachines* **2023**, *14*, 53. [CrossRef] [PubMed]
22. Garzón-Posada, A.O.; Paredes-Madrid, L.; Peña, A.; Fontalvo, V.M.; Palacio, C. Enhancing Part-to-Part Repeatability of Force-Sensing Resistors Using a Lean Six Sigma Approach. *Micromachines* **2022**, *13*, 840. [CrossRef] [PubMed]
23. Li, J.; Zhang, Z.; Zhu, X.; Zhao, Y.; Ma, Y.; Zang, J.; Li, B.; Cao, X.; Xue, C. Automatic Classification Framework of Tongue Feature Based on Convolutional Neural Networks. *Micromachines* **2022**, *13*, 501. [CrossRef] [PubMed]

24. Li, G.; Cui, J.; Liu, T.; Zheng, Y.; Hao, C.; Hao, X.; Xue, C. Triboelectric-Electromagnetic Hybrid Wind-Energy Harvester with a Low Startup Wind Speed in Urban Self-Powered Sensing. *Micromachines* **2023**, *14*, 298. [CrossRef] [PubMed]
25. Nan, X.; Wang, X.; Kang, T.; Zhang, J.; Dong, L.; Dong, J.; Xia, P.; Wei, D. Review of Flexible Wearable Sensor Devices for Biomedical Application. *Micromachines* **2022**, *13*, 1395. [CrossRef]

micromachines

Article

Laser-Formed Sensors with Electrically Conductive MWCNT Networks for Gesture Recognition Applications

Natalia A. Nikitina [1,*], Dmitry I. Ryabkin [1,2], Victoria V. Suchkova [1], Artem V. Kuksin [1], Evgeny S. Pyankov [1], Levan P. Ichkitidze [1,2], Aleksey V. Maksimkin [2], Evgeny P. Kitsyuk [3], Ekaterina A. Gerasimenko [1], Dmitry V. Telyshev [1,2], Ivan Bobrinetskiy [4], Sergey V. Selishchev [1] and Alexander Yu. Gerasimenko [1,2,*]

1 Institute of Biomedical Systems, National Research University of Electronic Technology, 124498 Moscow, Russia; ryabkin@bms.zone (D.I.R.); molodykh1999@gmail.com (V.V.S.); nix007@mail.ru (A.V.K.); zugusik@gmail.com (E.S.P.); ichkitidze@bms.zone (L.P.I.); katball@mail.ru (E.A.G.); telyshev@bms.zone (D.V.T.); selishchev@bms.zone (S.V.S.)

2 Institute for Bionic Technologies and Engineering, I.M. Sechenov First Moscow State Medical University, Bolshaya Pirogovskaya Street 2-4, 119991 Moscow, Russia; aleksey_maksimkin@mail.ru

3 Scientific-Manufacturing Complex "Technological Centre", Shokin Square 1, bld. 7 off. 7237, 124498 Moscow, Russia; kitsyuk.e@gmail.com

4 Center for Probe Microscopy and Nanotechnology, National Research University of Electronic Technology, 124498 Moscow, Russia; vkn@nanotube.ru

* Correspondence: demitashal@gmail.com (N.A.N.); gerasimenko@bms.zone (A.Y.G.); Tel.: +7-9857957508 (N.A.N.); +7-9267029778 (A.Y.G.)

Abstract: Currently, an urgent need in the field of wearable electronics is the development of flexible sensors that can be attached to the human body to monitor various physiological indicators and movements. In this work, we propose a method for forming an electrically conductive network of multi-walled carbon nanotubes (MWCNT) in a matrix of silicone elastomer to make stretchable sensors sensitive to mechanical strain. The electrical conductivity and sensitivity characteristics of the sensor were improved by using laser exposure, through the effect of forming strong carbon nanotube (CNT) networks. The initial electrical resistance of the sensors obtained using laser technology was ~3 kOhm (in the absence of deformation) at a low concentration of nanotubes of 3 wt% in composition. For comparison, in a similar manufacturing process, but without laser exposure, the active material had significantly higher values of electrical resistance, which was ~19 kOhm in this case. The laser-fabricated sensors have a high tensile sensitivity (gauge factor ~10), linearity of >0.97, a low hysteresis of 2.4%, tensile strength of 963 kPa, and a fast strain response of 1 ms. The low Young's modulus values of ~47 kPa and the high electrical and sensitivity characteristics of the sensors made it possible to fabricate a smart gesture recognition sensor system based on them, with a recognition accuracy of ~94%. Data reading and visualization were performed using the developed electronic unit based on the ATXMEGA8E5-AU microcontroller and software. The obtained results open great prospects for the application of flexible CNT sensors in intelligent wearable devices (IWDs) for medical and industrial applications.

Keywords: strain sensors; flexible electronics; smart wearable electronics; carbon nanotubes; electrical conductive networks; laser structuring; gesture recognition; sensor glove

Citation: Nikitina, N.A.; Ryabkin, D.I.; Suchkova, V.V.; Kuksin, A.V.; Pyankov, E.S.; Ichkitidze, L.P.; Maksimkin, A.V.; Kitsyuk, E.P.; Gerasimenko, E.A.; Telyshev, D.V.; et al. Laser-Formed Sensors with Electrically Conductive MWCNT Networks for Gesture Recognition Applications. *Micromachines* **2023**, *14*, 1106. https://doi.org/10.3390/mi14061106

Academic Editors: Libo Gao and Zhuoqing Yang

Received: 28 April 2023
Revised: 21 May 2023
Accepted: 22 May 2023
Published: 24 May 2023

1. Introduction

Hand gesture recognition has attracted widespread research interest as an important approach to human–machine interaction. Enabling direct human–computer communication is necessary in many areas: for people with disabilities [1], in virtual reality [2], video games [3], and robotics [4]. Studies have found that regular training of the hands in the form of monotonous, purposeful repetition of specific movements (gestures), leads to a reorganization of the cerebral cortex and the formation of a permanent therapeutic effect on the patients' motor activity recovery [5]. The effectiveness of training depends directly on the accuracy of the gesture recognition. In selecting the best method of gesture detection, it is important to consider that the human hand has a complex, asymmetrical geometric shape and that hand gestures are unpredictable and varied. In this regard, the following limitations should be addressed in constructing a gesture recognition system: segmentation of the gesturing hand, and a selection of specific purposeful gestures in a continuous stream of chaotic body movements.

One of the most popular approaches to recognizing hand gestures is visual. Visual gesture recognition methods use a combination of digital cameras, visible and infrared range to capture finger movement and further interpretation, so gesture reading is performed without the use of additional connected sensors [6]. The visual approach requires the use of a complex, extensive mathematical apparatus to construct a three-dimensional model of the hand [7]. Besides the listed disadvantages, the visual method of gesture recognition is the most expensive because of the high-tech assistive equipment used.

Radar gesture recognition technology is less sensitive to external noise [8,9]. The range resolution for hand gesture recognition is determined by the bandwidth of the radio waves. This approach requires the development of expensive equipment to generate radio waves with a sufficiently large bandwidth to provide the necessary resolution.

Ultrasound is well known for the ability to penetrate several cm under the skin and deliver information about both superficial and deep muscle tissues. This feature allows for the use of ultrasonic sensors for gesture recognition [10]. Recognition is based on an ultrasound image constructed from the propagation parameters of ultrasound waves [11]. This approach allows for a more miniaturized gesture recognition system. However, the accuracy of the ultrasonic gesture recognition system remains insufficient.

One of the most cost-effective approaches to gesture recognition is the use of accelerometers and gyroscopic sensors [12]. Most of these systems use a gyroscope, which reads angular velocity as the hand gestures change and transmits it to a personal computer interfaced to the three-axis gyroscope. Sensor data is processed using a moving average filter to reduce noise [13]. Gyroscopic sensors can only indirectly estimate the position of the finger joints, so the accuracy of movement registration is not high enough.

A promising approach to recording hand movements and gestures is the use of flexible strain sensors. Traditional microelectronic strain sensors, by the rigidity of their construction, cannot provide the necessary response on flat and/or curved surfaces with large deformations, such as the human body [14]. In contrast, the development of flexible strain sensors makes it possible to measure hand gesture movements directly by placing the sensors on the fingers or integrating them into gloves for data transfer [15]. The basic principle of flexible strain sensors is the generation of electrical signals (resistance, capacitance, voltage, etc., depending on the type of active material) in response to applied strain [16]. Capacitive strain sensors demonstrate good linearity, low hysteresis, and high sensitivity [17]. However, the variation in capacitance is typically in the order of pF and requires quite sensitive measuring equipment. In addition, capacitive sensors have some disadvantages, such as high impedance, making them sensitive to the stray capacitance of coaxial cables and leading to corresponding measurement errors [18]. Other types of sensors, such as Bragg fiber array [19], triboelectric [20], and piezoelectric [21] sensors, usually cannot detect slow or static strain due to rapid charge transfer. In general, they cannot detect slow or static deformation due to rapid charge transfer. In addition, their practical application as skin-worn devices remains difficult due to the need for sophisticated measurement equipment. Resistive sensors are the most used [22,23] due to the lack of the need for complex measuring equipment and the ease of manufacturing. Traditionally, due to their simple design, high sensitivity, and linearity, piezoresistive sensors have been the most widely used in commercial microelectromechanical systems [24]. Current focus is placed on the development and improvement of flexible resistive sensors, as well as extending their range of applications.

The active materials of resistive sensors are mostly conductive micro/nano-polymer composites, thin films, or conductive filaments/fabrics [25]. Recent research and development of resistance sensors based on nanostructures and their composites have accelerated following the discovery of materials such as carbon nanotubes (CNT) [26]. CNT based nanocomposites and films are a promising alternative to other smart materials, largely owing to their superior electrical and mechanical properties, ultra-high intrinsic piezoresistivity caused by nanotube chirality [27], unique morphology, ease of manipulation, and chemical versatility. A significant correlation was found between the reversible mechanical strain and the electrical resistivity of single- and multi-walled carbon nanotubes (SWCNT and MWCNT). The correlation demonstrates that CNT can be used as precise nanoscale strain sensors [26]. Moreover, the addition of low concentrations of CNT to polymers provides the materials with piezoresistive properties. The obtained materials show a sufficiently linear and reversible change in electrical resistance under deformation [15,28]. The strain sensitivity of such CNT composites is based on the modification of the percolation network of nanotubes under mechanical action [29]. Recent results have shown that the characteristics of such strain sensors are closely related to the assembly methods and microstructure of the conductive networks [30]. Therefore, to manufacture high-performance flexible strain sensors based on CNT composites, it is crucial to be able to control the formation of the conductive network of nanotubes in the material during production.

Laser exposure is a universal and inexpensive method for modifying and assembling polymer composites [31–35]. Laser exposure of composites with CNT can significantly increase electrical conductivity by increasing the concentration of the carbon filler until the electrical conductivity percolation threshold is reached [36]. Laser exposure causes pyrolysis of the polymer matrix, which leads to the formation of gaseous particles leaving the material, thereby increasing the conductive filler/matrix ratio [37]. This approach makes it possible to make inexpensive materials, since it allows for the use of sufficiently low concentrations of expensive fillers, such as CNT. The morphology and structure of three-dimensional CNT networks are known to be controllable by adjusting the power and laser exposure time [38]. Thus, laser exposure of electrically conductive composites allows for the formation of conductive CNT networks with a specified morphology and improved electrical conductivity. In the process of the exposure of composites, laser energy is absorbed by electrons and transferred to CNT atoms. As a result, branched CNT networks are formed. It has been previously demonstrated that laser exposure can be used to form strong conductive networks of nanotubes and a carbon framework in polymer composites under normal atmospheres, with minimal costs and minimal losses [39–42]. At the same time, laser exposure allows for increasing the electrical conductivity and mechanical hardness of the final material. Laser exposure through improved binding of the nanotubes to each other has resulted in a low electrical resistance value.

In summary, this paper describes the manufacturing of flexible strain sensors consisting of electrically conductive networks of carbon nanotubes in polydimethylsiloxane (PDMS) elastomer. Laser processing has been applied to fabricate various flexible sensors [43–45]. The novelty of the project lies in the use of laser exposure to form electrically conductive networks of carbon nanotubes for the manufacture of gesture-registering strain sensors. This approach significantly increases the conductivity and sensitivity of the strain sensors, allowing for the use of lower MWCNT concentrations. Thus, the formed percolation network of MWCNT is more electrically sensitive to mechanical deformations. Under laser exposure, electrically conductive networks of carbon nanotubes were formed in the PDMS matrix. This has resulted in a high sensitivity of the sensors while maintaining high elongation and strength. Based on the sensors, a touch-sensitive smart system for hand gesture recognition in the form of a glove can be fabricated. Such a smart system is one of the reliable and inexpensive approaches, endowing a person with objective feedback during training of hand motor functions in rehabilitation after severe diseases and during learning.

2. Materials and Methods

2.1. Components Used for Flexible Sensor Manufacturing

The sensors are nanocomposites based on multi-walled carbon nanotubes (MWCNT), where the MWCNT have been encapsulated in a PDMS matrix. MWCNT (NanoTechCenter LLC, Tambov, Russia) in powder form were obtained via CVD synthesis, had an outer diameter of 8–30 nm, inner diameter of 5–15 nm, length ≥20 μm, specific surface area ≥270 m^2/g, and bulk density 0.025–0.06 g/cm^3. The two-component PDMS elastomer Ecoflex 00–10 (Smooth-On Inc., Macungie, PA, USA) had a dynamic viscosity in the mixed state of 14 Pa·s, Young's modulus at 100% elongation 0.06 MPa, hardness (Shore A) 10, density 1.04 g/cm^3, with an operating range of 19 °C to 232 °C. Carbon fiber wires were used to connect the sensor.

The basis of the electronic unit to read and process signals from the sensors was a 10-bit microcontroller ATXMEGA8E5-AU (Microchip Technology Inc., Chandler, AZ, USA), enclosed in ABS plastic.

2.2. Flexible Sensor Fabrication

The sensors were produced by screen-printing with laser-scanning of the surface to form a strong electrically conductive network of MWCNT. The fabrication process is shown schematically in Figure 1. To compare the effect of laser radiation on the electrical properties, sensors without laser exposure were also fabricated. Sensors that were not laser-exposed were also produced. The manufacturing process was similar.

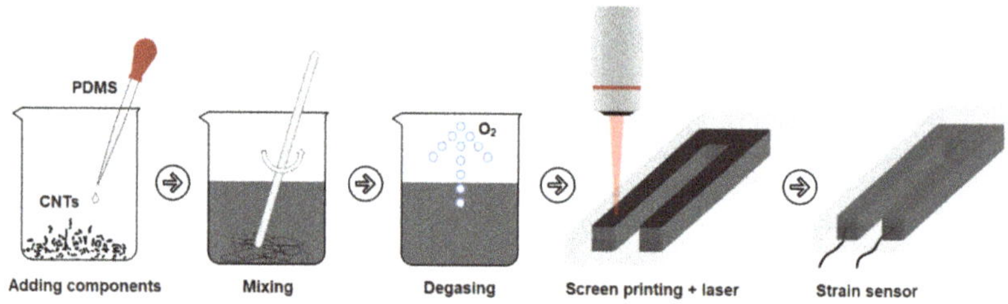

Figure 1. Schematic diagram of sensor manufacturing.

First, components A and B of the PDMS elastomer in the liquid phase were added to the carbon nanotubes in equal proportions and thoroughly mixed mechanically for 10 min. The concentration of MWCNT in the resulting mixture was 3 wt%. This percentage of nanotubes corresponded to the experimentally obtained percolation threshold, above which no significant changes in the electrical conductivity of the MWCNT/PDMS composite were observed. Then, the mixture was transferred to an XFL020 vacuum furnace (France Etuves, Chelles, France) and the degassing process was started (vacuum). After removing the air formed in the mixing process, the mixture MWCNT/PDMS was applied to the substrate via a U-shape screen. The forms printed using a DLP three-dimensional printer were used as a substrate and screen. The sensor dimensions can be adjusted by selecting a new shape of the screen with the required dimensions. The U-shape form of the sensors was chosen for the convenience of the wires which were added to the MWCNT/PDMS composite before the PDMS solidified. In addition, the U-shaped design has a longer active material length than the I-shaped sensor. This provides a greater change in resistance when deformed. The wires were placed on one end of the sensor due to the chosen shape. The thickness of the formed active material layer was 0.5 mm. This thickness is mainly due to the magnitude of the laser radiation drag to form an electrically conductive nanostructured layer. To increase the tensile strength and insulation of the conductive layer, the sensor was coated with 0.25 mm thick layers of PDMS on the top and bottom. The total thickness of the sensor was therefore 1 mm. After the active layer solidified, it was exposed to laser radiation at selected parameters. Pulsed Yb laser was used and a laser exposure pattern was set in the software (Figure 2a). The parameters of the laser exposure are shown in Table 1. Figure 2b shows the pattern under the pilot laser beam. Laser scanning was performed over the entire surface of the MWCNT/PDMS composite in a U-shape. After the MWCNT/PDMS layer was screen-printed and laser-treated, a thin layer of PDMS was poured on top through another rectangular screen. The active material layer with the PDMS-insulating layer was then inverted and the PDMS layer was poured on the other side in the same mold.

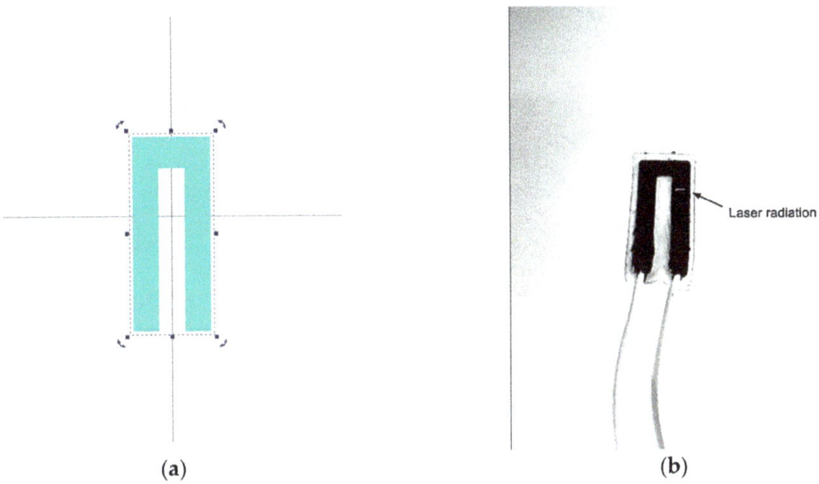

Figure 2. Laser exposure of the active material layer of the sensor: (**a**) template for laser scanning; (**b**) active material layer of the sensor in the process of laser exposure.

Table 1. Parameters of laser exposure.

Parameter	Value
Wavelength	1064 nm
Energy density	0.5 J/cm^2
Pulse duration	100 ns
Frequency	30 kHz
Beam speed	450 mm/s

Laser exposure has been used to improve electrical performance and sensitivity due to the previously demonstrated effect of forming strong MWCNT networks with an increased electrical conductivity of the material [39,40,46]. This includes the proven effect of the enhanced electrical conductivity in various polymer and biopolymer matrices [31–34,41,42]. This approach allows for the use of low MWCNT concentrations (3 wt%) without the loss of sensitivity at large strains.

After laser-processing, the active sensor material (a U-shaped MWCNT/PDMS composite) was coated on both sides with a layer of pure PDMS to insulate it from external damage. The sensors made by this technique consisted of three layers, as demonstrated in Figure 3. The inner layer of the MWCNT/PDMS is the active material of the sensor, ensuring strain sensitivity. The MWCNT/PDMS layer was coated with layers of pure PDMS on both sides.

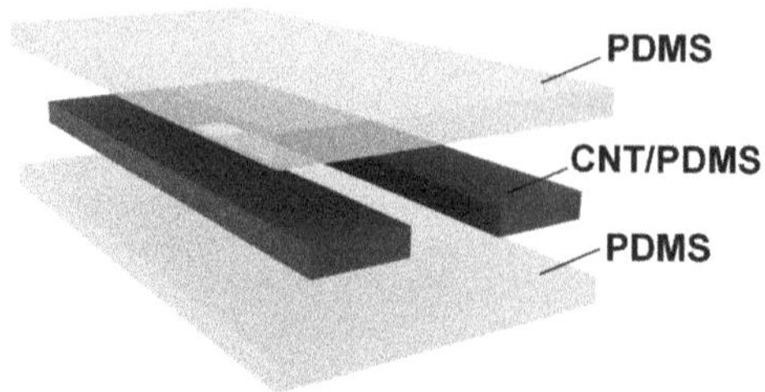

Figure 3. Schematic diagram of the manufactured sensor.

2.3. Sensitivity Study of the Sensors

The study of the sensitivity of the developed sensors included the construction of curves of dependence of electrical resistance on the applied strain and the calculation of the gauge factor (GF).

To plot the curves of electrical resistance dependence on the applied strain, the sensors were connected to a multimeter (UNI-T UT33C+, Rexant International Ltd., Hong Kong, China) and the resistance values were recorded from 0 to 100% stretch (readings were taken every 0.5 cm of elongation).

GF is defined as the ratio of the relative change in electrical resistance to the applied strain. The range of strain for GF determination was 0–100%. This range is related to the fact that the finger bending movement can result in a strain n of less than 100% [47]. The GF of the developed sensors was calculated using the equation [30]:

$$GF = (\Delta R/R_0)/(\Delta l/l_0), \tag{1}$$

where ΔR and Δl are the absolute change in electrical resistance and length, respectively, under strain n, R_0 and l_0, which are the initial value of electrical resistance and length, respectively.

Resistance hysteresis, a typical shortcoming of most resistive sensors, was estimated using the equation [48]:

$$h\,(\%) = ((Rs - Rr)/(R_{max} - R_0)) \times 100\%, \tag{2}$$

where R_s and R_r are the resistance during stretching and releasing, respectively, at the same strains, R_0, which is the initial (minimum) value of resistance, R_{max}, the maximum resistance value.

2.4. Linearity Study of the Sensors

Linearity is the coefficient of determination of GF under stretching up to 100% of the length. Sensor linearity was calculated using the equation:

$$R^2 = \sum(GF_{predicted} - GF_{mean})^2/\sum(GF_{actual} - GF_{mean})^2, \tag{3}$$

where R^2 is the determination coefficient, $GF_{predicted}$ is the predicted GF value, GF_{mean} is the average GF value, and GF_{actual} is the actual calculated GF value. The predicted GF values were determined by constructing a linear regression using the sklearn library in the python programming language.

2.5. Study of Mechanical Characteristics of the Sensors

The sensor elasticity should correspond to the elasticity of human skin in order to accurately replicate the movements during measurements. The elasticity (Young's modulus) was calculated using the equation:

$$E = F \cdot l_0/S \cdot \Delta l, \tag{4}$$

where F is the applied force, l_0 is the initial length, S is the area, and Δl is the absolute change in length because of the applied strain.

2.6. Cyclic Sensor Testing

To confirm the reliability of the developed sensors, cyclic tests were conducted for one month. The sensors were subjected to daily cyclic load: percentage elongation $\varepsilon = 100\%$ and relaxation to the initial length. The number of repetitions per cycle was 10,000. During the test, the resistance values were monitored at the starting point $\varepsilon = 0\%$ and the end point $\varepsilon = 100\%$. The measurements were carried out with a testing machine equipped with a positioning system. For testing, the strain sensor was gripped by the clamps integrated in the moving head of the test machine. To control the resistance, the strain sensor wires were connected to a multimeter. Using a specially developed software, the positioning system has been programmed to stretch the strain sensor for 1 s to $\varepsilon = 100\%$, then return to the initial length of the sensor.

2.7. Manufacturing of an Electronic Signal-Reading Unit

The developed resistive strain sensors convert external deformation into an electrical signal. Their structure contains an electrically conductive active material connected to a flexible substrate. A change in the structure of the active material under applied strain results in a change in electrical resistance. In fact, the conductive network of active materials of the developed sensor acts as a variable resistor under the applied voltage. On this basis, a specialized electronic unit was developed to read the signals of the developed flexible sensors.

The electronic unit was developed as follows: based on the resistance values of the developed sensors, an electrical circuit was simulated using the LTspice software (Analog Devices Inc., Wilmington, NC, USA), and the values of the resistors used in the circuit were selected. The developed electrical circuit is shown in Figure 4a. Next, the circuit for converting the electrical resistance of the sensors into voltage was calculated using a 10-bit ATXMEGA8E5-AU microcontroller (Atmel Corporation, San Jose, CA, USA). The input voltage range applied to the ADC was 0 to 1.14 V. The analog circuit was designed so that the voltage converted by the LM358D operational amplifier (STMicroelectronics, Geneva, Switzerland) from the sensor resistance is in the input voltage range. In this case the sensor acts as a variable resistor. In the electrical circuit, R4 is designated (Figure 4a). In accordance with the developed circuit, an electronic unit was constructed from electronic components (Figure 4b). The electronic circuit (1) was based on the ATXMEGA8E5-AU microcontroller, a Bluetooth module (2) was added to transfer data to a computer with the software installed, a power button (3) allowed for starting the unit, wires (4) provided a sensor connection, and a USB connector (5) was needed to connect to a 5 V power supply or to a computer. Finally, the case was closed with a lid.

2.8. Manufacturing of the Smart System for Gesture Recognition

The flexible sensors made according to the method described above (in Section 2.1) were attached using conventional medical tapes to a soft, thin tissue glove. The sensors were attached to the glove at the middle mobile joints of the fingers and connected to the developed electronic unit. No more than two resistive sensors could be connected to one electronic block. The electronic block was connected to a computer. The software for the electronic unit displayed the ADC values of each connected sensor on the computer screen and plotted the ADC values over time. The ADC values changed with finger movements, and the graph clearly represented the amplitude of the movements. Numerical ADC values were recorded and stored for further processing and machine learning of hand gesture recognition. The data were recorded at intervals of 60 values per second. A schematic representation of the developed smart system for gesture recognition is shown in Figure 5.

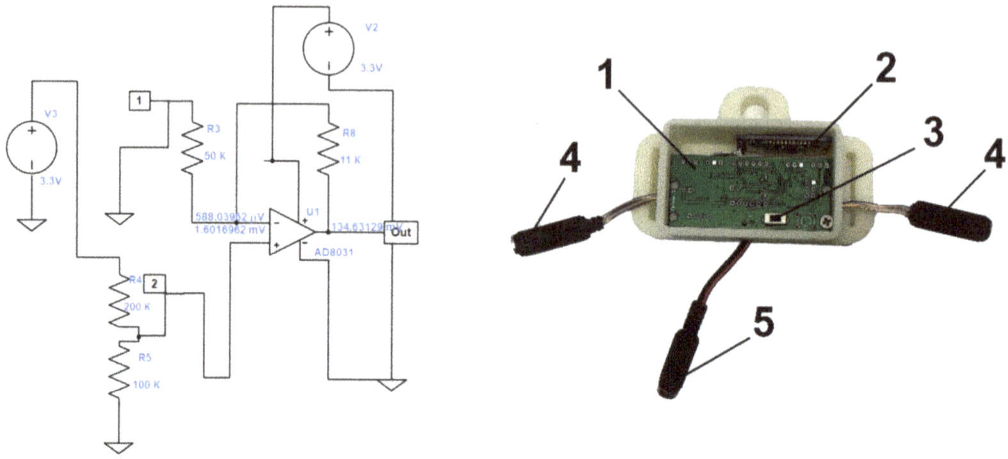

(**a**) (**b**)

Figure 4. (**a**) Developed circuit design of the electronic sensor unit. (**b**) Electronic microcontroller circuit layout, where 1—electronic circuit for reading sensor signals, 2—Bluetooth wireless transmission module, 3—power button, 4—wires for sensor connection, and 5—power connector.

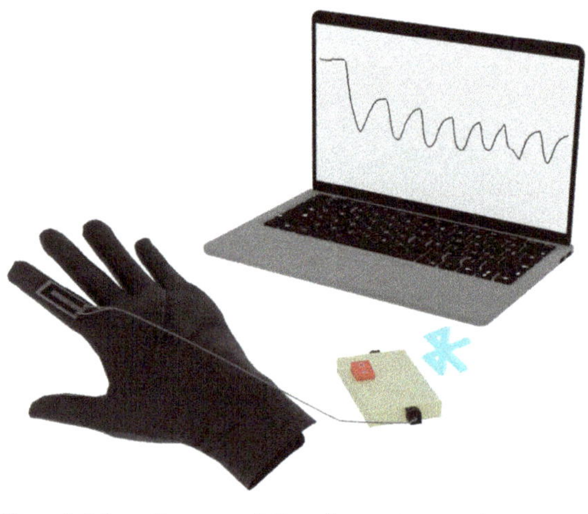

Figure 5. Schematic representation of a smart system for gesture recognition.

2.9. Gesture Recognition Techniques Using a Smart System and Machine Learning

Gesture recognition is a multiclass classification task. The One-vs-Rest approach was used to solve the problem. The One-vs-Rest approach trains n binary classifiers, where n is the number of classes. Each classifier predicts whether an instance belongs to one of n classes or not. The class of which the probability of belonging was the most probable was chosen.

The K nearest neighbor classifier (KNN) and support vector machine (SVM) models, as well as ensembles over decision trees random forest classifier (RF), xgboost classifier (XGB), LightGBM Classifier (LGBM), decision tree classifier (DT), logistic regression (LR) were used for gesture recognition. The ML models KNN, SVM, RF, DT, LR from sklearn (ver. 0.23.2), XGB from xgboost (ver. 1.4.1), and LGBM from LightGB (ver. 3.3.2) were used.

Accuracy was chosen as the quality metric for the gesture recognition task. The accuracy metric for multiclass classification is the ratio of the number of correct predictions to the total number of predictions. With a significant class imbalance, the accuracy metric may show incorrect results. When generating the dataset of resistive sensors, the exposure time of the gestures was equal, thereby achieving a balance in the information for the different gestures.

The formation of the dataset was carried out automatically. Resistive sensors were connected to the electronic unit, which was connected to a personal computer with special software. The software recorded the date, time, resistance values received from the sensors, as well as the demonstrated gesture into a .csv file. The person demonstrated the gesture according to the information from the software. The demonstrated gesture was changed every 5 s. This way, automated data labeling was performed. A maximum of two resistive sensors could be connected to one electronic unit. In this way, the data are written to 3 files. When combining data from files by a sensor from different fingers, the different sampling rate of data recording and the start and end time of the recording are considered.

The optimization of the hyperparameters of models with the best values of quality metrics was performed using the Optuna package (ver. 2.10.0). The data were modified with a moving average filter for better training quality. After filtering, the data were normalized using MinMax Scaler.

3. Results and Discussion

3.1. Structure and Initial Characteristics of the Sensors

Flexible electronic devices are gaining enormous popularity for wearable applications. The flexible sensors manufactured in this paper are the interface component of a smart system being developed to record hand gestures. Essentially, the developed sensors are a flexible resistor that changes its resistance under strain, so the sensors can be useful in many wearable electronics applications.

The sensors manufactured in this work (Figure 6) were a three-layer nanocomposite with dimensions of 35 × 15 × 1 mm. The active layer of sensors, i.e., the MWCNT/PDMS composite had dimensions of 30 × 12 × 0.5 mm (Figure 6d). Flexibility and stretchability of the sensors was provided by the constituent polymer, PDMS elastomer. The deformation sensitivity was provided by the included carbon nanotubes.

In order to compare the electrical and mechanical properties, the sensors were manufactured in two ways: with and without the use of laser exposure.

The structure of the network of conductive particles in polymer composites is known to determine their electrical properties [49]. Considering this, a laser exposure was applied in the sensor fabrication process to form an electrically conductive network of MWCNT by forming bonds between the individual nanotubes. Firstly, laser exposure of carbon nanotube films has previously been found to cause carbon nanotubes to bind and weld together, resulting in strong electrically conductive networks [34,35]. In polymer matrices, laser nanotube network formation not only increases electrical conductivity, but also increases material strength [50].

Figure 7 shows the structure of the electrically conductive network of MWCNT sensors obtained without laser exposure (simple air-drying) and with laser exposure. Images were obtained through scanning electron microscopy (SEM), using an FEI Helios NanoLab 650 microscope (FEI Company, Hillsborough, OR, USA) with an electron column acceleration voltage of 1 kV and an electron probe current of 21 pA. The internal structure was studied for sensors made without laser exposure (Figure 7a) and with laser exposure (Figure 7b). After laser exposure, a redistribution of the MWCNT clusters is observed. Large clusters are separated into smaller clusters, increasing the number of clusters and decreasing their average size. The electrically conductive network of MWCNT after laser exposure has become less complex, with predominantly nanotube bonding areas. A sparser network is preferable for strain sensors because it allows for a lower resistance hysteresis. This is because a sparser network stretches more uniformly and the rearrangement of the conductive network becomes more stable due to the absence of a large number of conductive paths. This means that there is less chance of random contribution to the conductivity of untangled or randomly connected nanotubes. Figure 7c,d shows the areas where the nanotubes are welded together under higher magnification. The arrows show the contact points between the MWCNT.

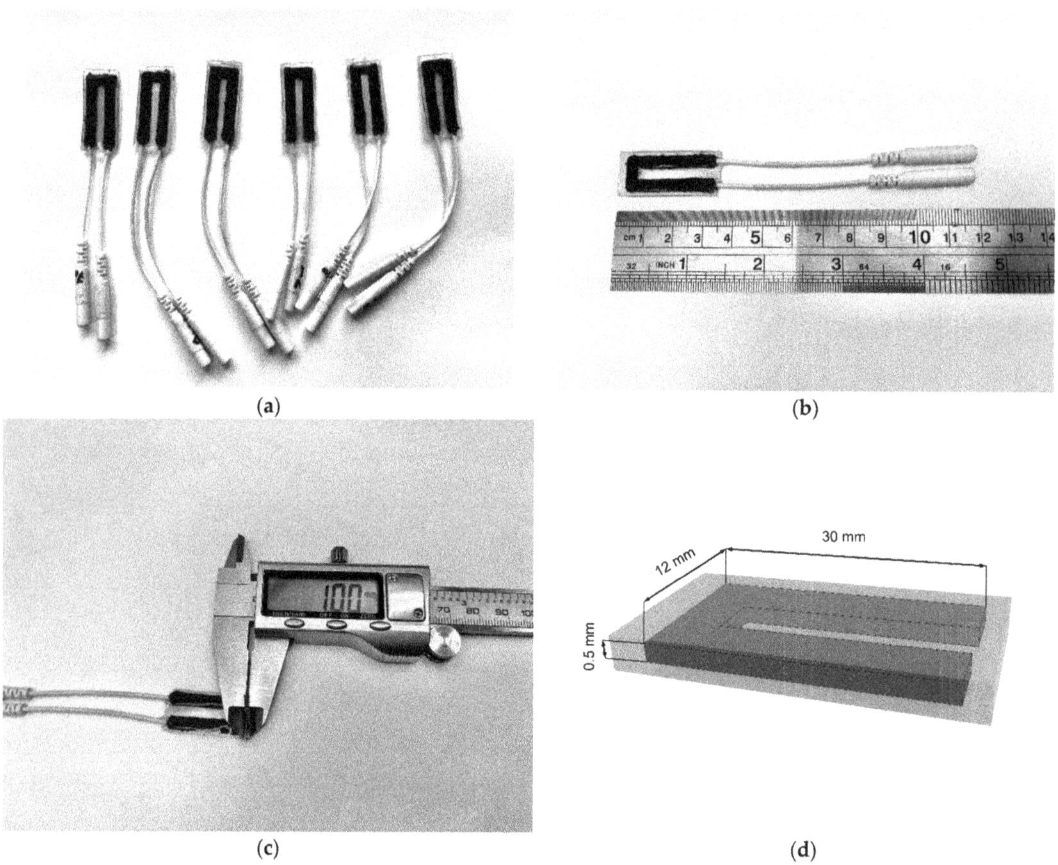

Figure 6. (**a**) Developed sensors based on carbon nanotubes and elastomers. (**b**,**c**) Photo of the sensor with the scale grid. (**d**) Dimensions of the manufactured sensor.

The electrical properties of the sensors obtained with and without laser exposure were compared using different concentrations of 2, 3, 4 wt% of nanotubes (Figure 8). Increasing the MWCNT concentration above 4 wt% significantly increases the viscosity of the MWCNT/PDMS blend. This makes it difficult to homogenize the MWCNT in PDMS during the fabrication of the active material. Concentrations of CNT below 2 wt% will cause the sensor resistance to rise above 1 MOhm. Such high resistance values result in difficulty in detecting changes in resistance and are unacceptable for sensor construction. If 2% is exposed to higher laser power, the silicone in the active sensor layer will overheat and burn out. This will reduce the performance of the sensor.

The initial electrical resistance of the laser-produced sensors was ~3 kOhm (no deformation, Figure 8) at a low concentration of 3 wt% nanotubes in the composition. In comparison, using a similar fabrication process without laser exposure, the active material had significantly higher electrical resistivity values, in this case ~19 kOhm. Increasing the concentration of the nanotubes resulted in a slight increase in strength, which is related to the high strength of the nanotubes themselves, allowing them to be used as a reinforcing material.

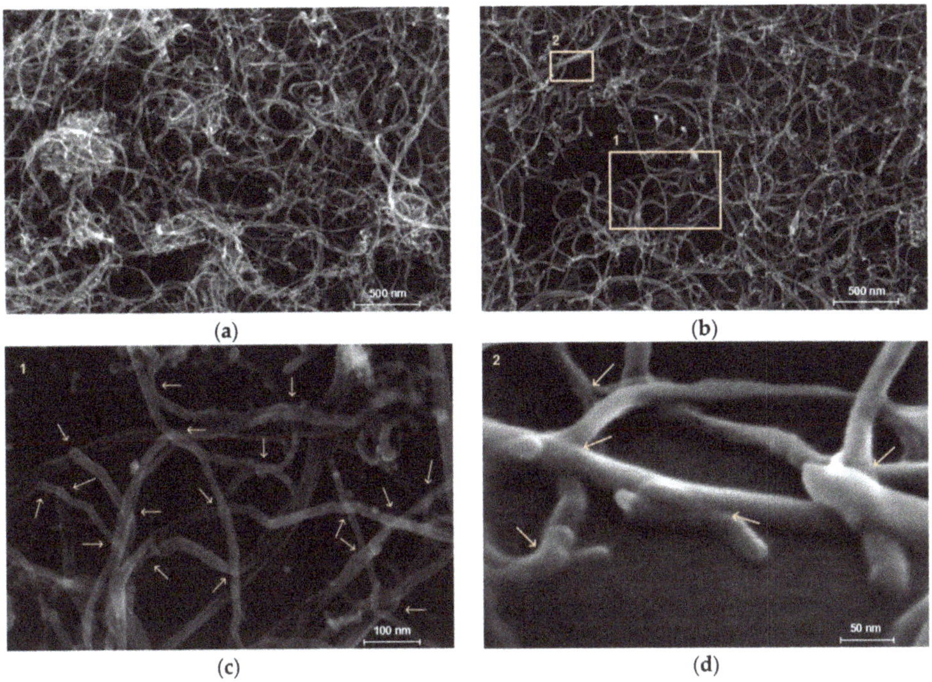

Figure 7. Internal structure of the sensors with enlargement mode ×120,000: (**a**) fabricated without laser exposure and (**b**) with laser exposure. Enlarged areas of laser-irradiated sensors: (**c**) Area 1 with enlargement mode ×500,000, (**d**) Area 2 with enlargement mode ×1,000,000 and at 52° angle. The arrows indicate the welded areas of the nanotubes formed by the laser exposure.

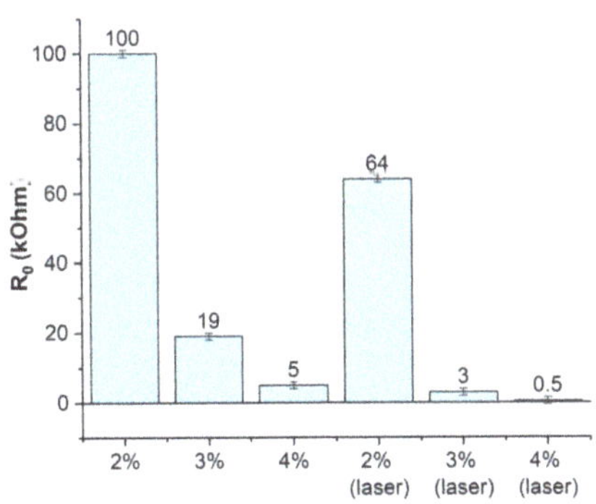

Figure 8. Comparison of the resistance of the active sensor material at different concentrations and fabrication methods.

3.2. *Strain Response*

High strain sensitivity of traditional semiconductor and metal sensors is achieved with a small range of relative strain <3% and only in a certain direction [51]. These types of devices are widely used as "rigid" sensors because they have low flexibility and extensibility, so they are designed to be attached to rigid materials (concrete, metal, plastic) to monitor the condition of the structure or quantify the small strain of the sample. The current demand for wearable electronic devices has changed the structure of resistive type strain sensors from fragile to tensile. For resistive sensors, the working mechanism is the change in electrical resistance in response to tensile (bending) and compressive strains. A common indicator is sensitivity, which is quantified through the GF.

The strain response of the developed sensors obtained with the two methods (for comparison) was evaluated under tensile stress from 0 to 100%. The developed sensors respond to the tensile strain as follows. The electrical resistance of the sensors increases during stretching and decreases during return to the initial state. By recording the changes in electrical resistance, it is possible to judge the magnitude of elongation. The strain response time was 1 ms. Figure 9 shows the plots of the relative change in electrical resistance when the sensors are extended (Figure 9a) and released (Figure 9b). The relative change in resistance is presented as a percentage. The resistance hysteresis is shown in Figure 9c. The sensitivity coefficient of the sensors at different strains is shown in Figure 9d. It should be noted that the sensors exposed to laser exposure showed a higher GF (Figure 9c).

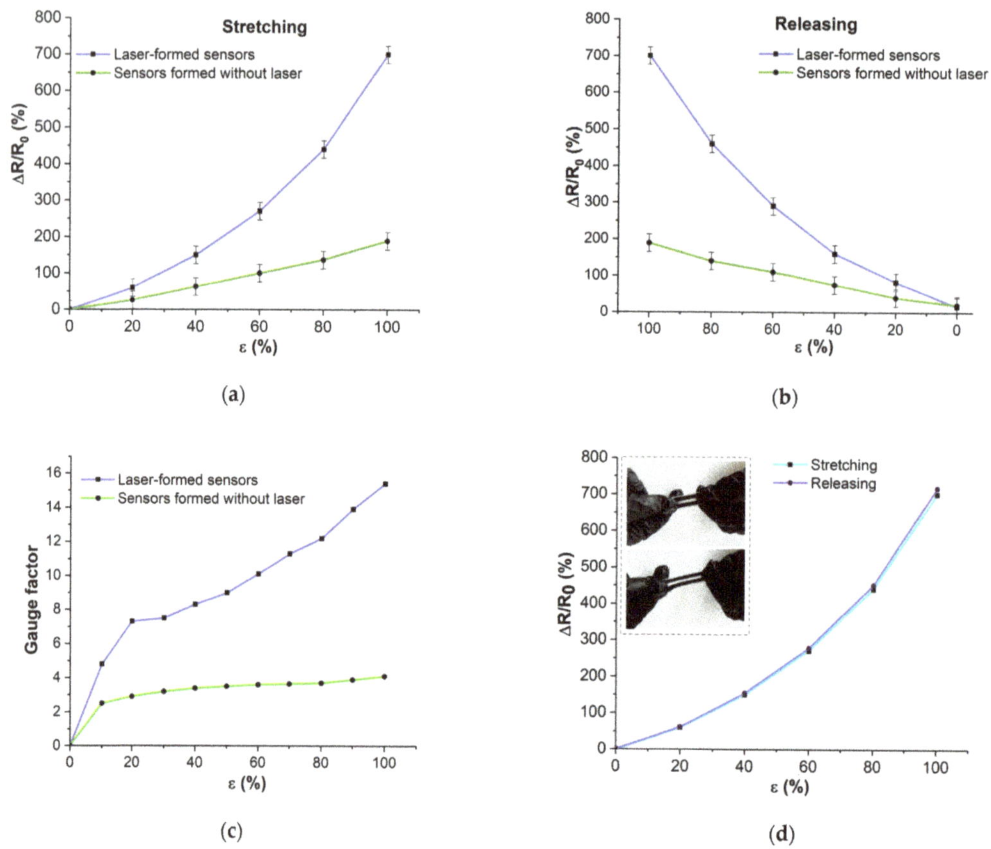

Figure 9. Deformation response of the developed sensors with and without laser exposure: (**a**) dependence of the relative change of resistance during stretching; (**b**) dependence of the relative change of resistance during releasing; (**c**) gauge factor at different strains; (**d**) dependence of the relative change of resistance during tension and release (hysteresis).

It was found that the sensors obtained using the laser-formed conductive network method exhibit a wider variation in electrical resistance under deformation (Figure 9a,b blue lines). This is apparently related to the efficient formation of an electrically conductive network of nanotubes in the sensor material, which conducts an electric current more efficiently. For further research and fabrication of the smart system, laser produced sensors were used, as they showed a higher sensitivity coefficient (Figure 9c). Their average sensitivity coefficient is 10, with a maximum value of 15.4 at 100% strain. Sensitivity is strain dependent and increases with increasing strain range.

The presence of hysteresis is a typical problem of flexible sensors. Hysteresis is related to the viscoelastic nature of flexible polymers and the slip of internal conductive nanomaterials [52,53]. As a result, the conductive network needs time to recover. The calculated hysteresis value of the sensors obtained with laser exposure was 2.4%, while that without laser exposure was 5%. The lower hysteresis of the laser-exposed sensors (Figure 9d) indicates a faster recovery of the conductive network. The lower hysteresis values can be explained by the efficient formation of the conductive network.

Although laser exposure increases the strength of materials, the laser-assisted sensors are flexible enough to follow movements accurately and without constraint. The Young's modulus of the laser-irradiated sensors was 47 kPa, which is comparable to the Young's modulus of human skin, which is 25–250 kPa [54]. Table 2 shows a comparison of the sensor characteristics investigated in this paper.

Table 2. Comparison of the characteristics of sensors.

Method	Initial Resistance, kOhm	Young's Modulus, kPa	Gauge Factor	Strength, kPa	Hysteresis, %
Sensors formed without laser	19	40	2.5–4.1	827	5
Laser-formed sensors	3	47	4.8–15.4	963	3

3.3. *Linearity and Cyclicality*

Laser-formed sensors were selected for the fabrication of an intelligent gesture recognition system. The strain response speed and linearity of these sensors were evaluated. The processing speed of the output signal directly depends on the linearity of the sensor GF. The key feature of using flexible polymer/nanoparticle-based strain sensors for gesture recognition is the presence of two strain linearity areas $\varepsilon = 0–10\%$ and $\varepsilon > 10\%$, caused by hysteresis [25]. Figure 10 compares the actual GF values with the predicted GF using linear regression. The linearity for the 0 to 10% stretch region (Figure 10a) reached 0.998, and for the 10% to 100% stretch region (Figure 10b) 0.97. Since in finger bending, the stretch of the sensors usually does not exceed 10%, the area with the highest linearity is mainly used, which determines the high accuracy and predictability of the results.

Developed sensors, due to their flexibility, present the possibility to measure deformations of body parts caused by anatomical movements. By attaching the sensor to the area under investigation, it is possible to determine strain directly in the area under investigation. The resistance of the developed sensors changes due to the restructuring of the formed network of carbon nanotubes. By recording the resistance changes of the developed sensors, the registration of strain is realized. Resistance changes during strain are due to (1) geometric changes (length and cross-sectional area of the active material) [52], (2) separation of the tunnel gaps due to strain and reconstructing of the conductive network [29], (3) the intrinsic piezoresistive properties of carbon nanotubes [27]. When the deformation is removed, the resistance is restored as the conductive network of nanotubes returns to its original morphology.

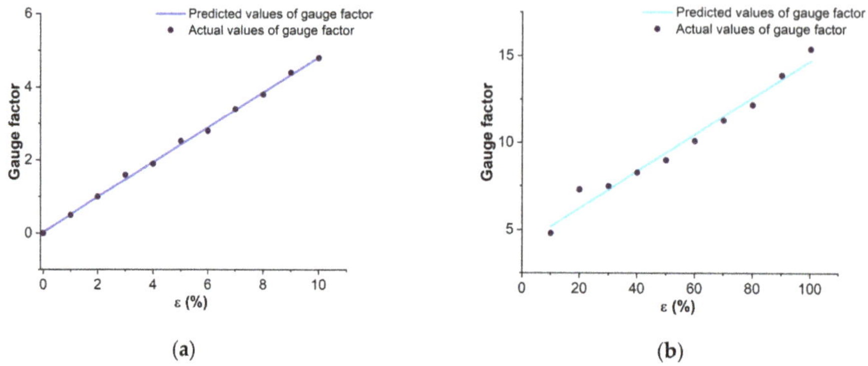

Figure 10. Linearity determination of the GF of the developed sensors where: (**a**) the comparison of predicted GF values with actual values at the area $\varepsilon < 10\%$; (**b**) the comparison of predicted GF values with actual values at the area $\varepsilon > 10\%$.

The results of the study demonstrated that regular loads do not affect the electrical characteristics of the sensors negatively and, accordingly, the sensors can ensure a stable operation of the smart system. The resistance fluctuations remained within the hysteresis range (Figure 11). For 30 days and a total of 300,000 cycles, no degradation of the conductivity and decreased sensitivity of the sensors was observed. This repeatability is caused using the proposed laser processing of the active material of the sensors. The high strength of the created networks of carbon nanotubes makes it possible to keep the conductive properties of the active material for a long period of time.

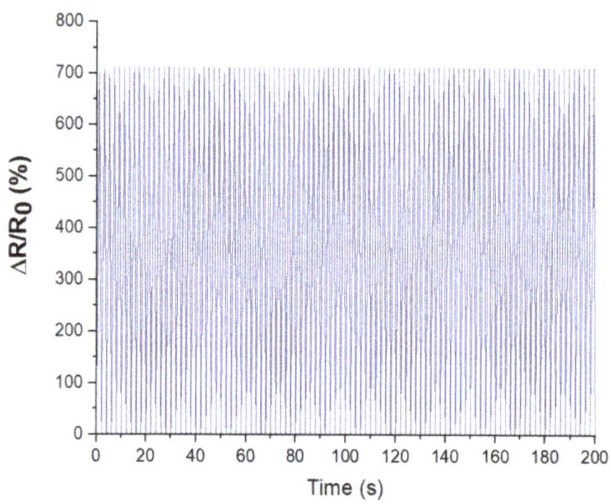

Figure 11. Sensor cyclic load graph: stretching $\varepsilon = 100\%$ and relaxation to the original length within the first 200 s.

3.4. Gesture Recognition

Sensors were manufactured for each finger of the hand (Figure 12a). Sensors were attached to each finger in the moving joints using medical tapes to a soft fabric glove. The glove was comfortable to wear, fit the fingers tightly, and did not restrict movement. The sensors were connected to the developed electronic unit, the appearance of which is shown in Figure 12b. Then, the electronic unit was connected to a personal computer by means of a wireless Bluetooth connection. The appearance of the resulting smart system for gesture recognition is shown in Figure 12c. The program for visualization, reading and storing data from the sensors developed for the electronic unit was run on the computer. The visualization software plotted the amplitude of the ADC values over time. A screenshot of the program window content display by the PC is shown in Figure 12d. The sensors contain wires for connection to the electronic unit. The function of the wires is to transmit the signal from the active sensor layer to the electronic unit. The electronic unit can be powered by a USB cable or a battery charger.

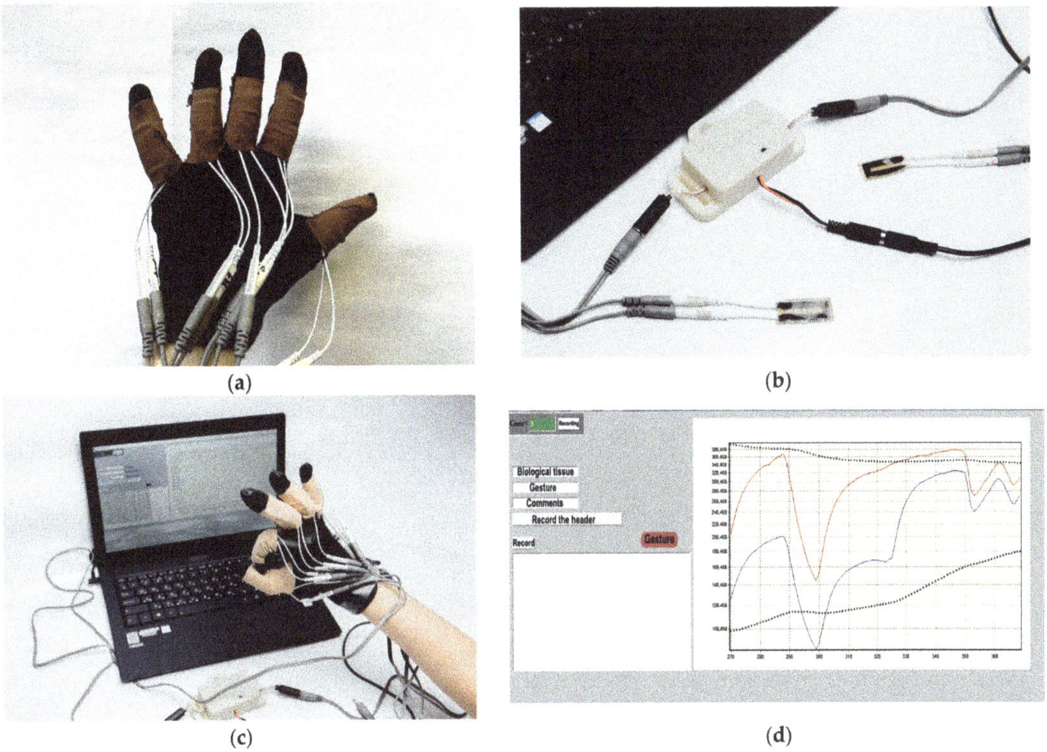

Figure 12. Developed sensors, and a smart system based on them, to recognize hand gestures: (**a**) sensors based on carbon nanotubes and elastomers; (**b**) glove with five installed carbon nanotube and elastomeric sensors for gesture recognition applications; (**c**) a smart gesture recognition system consisting of a glove with sensors and an electronic unit connected to a computer; (**d**) the program for visualization, reading and storing data from the sensors.

To train the ML model for gesture recognition, a dataset was formed. The subject demonstrated in turn the basic and recognizable gesture. The subject wore a glove with sensors. The wires that come from the sensors were connected to the electronic unit to record the signals. The electronic unit was connected to a computer and the developed software run, which recorded the data received from the sensors over time in a separate file. Eleven different gestures were chosen for the gesture recognition experiment, including a basic gesture. The signals produced by the alternation between the base gesture and the studied gesture were recorded. The basic gesture was a hand clenched in a fist. For each gesture, data were collected for 30 min, the basic and recognized gestures changed every 5 s.

Figure 13 shows the plots of dependencies of resistive sensor values during the demonstration of various gestures.

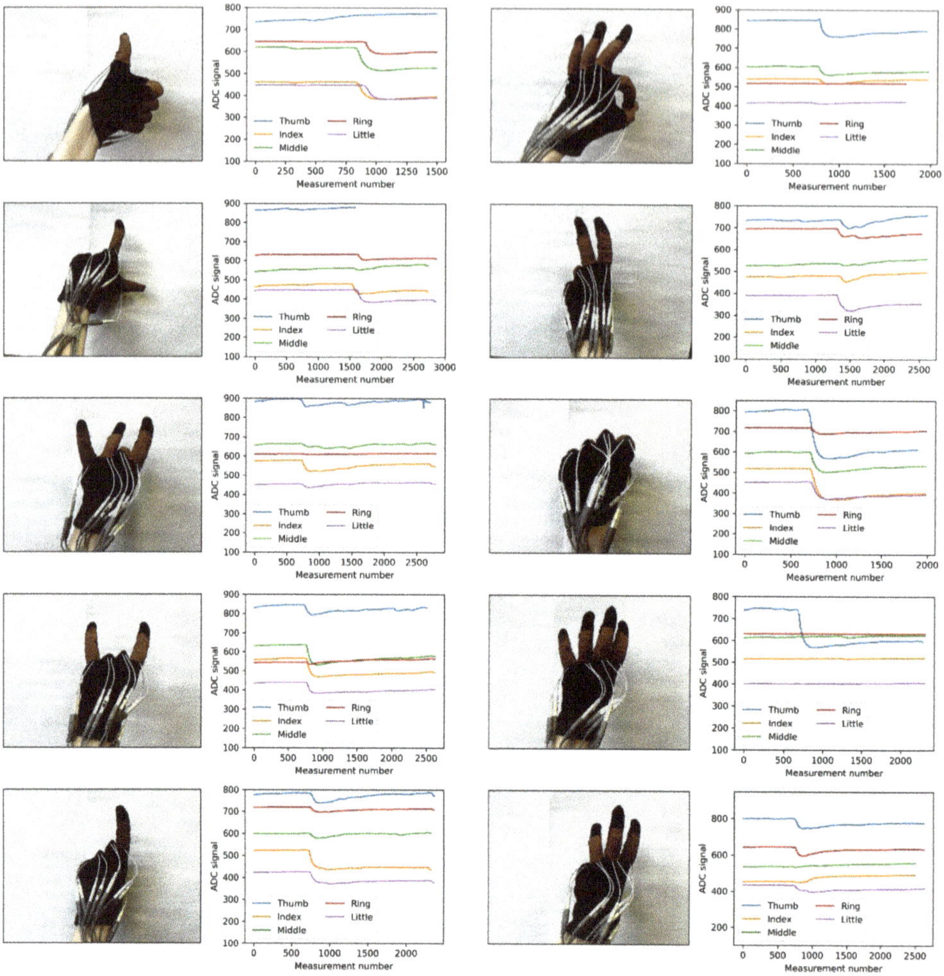

Figure 13. Dependence of resistive sensor readings from five fingers for 10 recognizable gestures.

The resistance value of the sensors is time-varying in long-term measurements. To improve the quality of the ML model predictions it is necessary to apply filtering to the sensor data. Filtering by the moving average was applied:

$$\hat{f} = \frac{\sum_{i=j-n}^{j-1} f_i + \sum_{i=j+1}^{j+n} f_i}{2n}, \tag{5}$$

where f is the value of the feature received from the sensor, $\hat{f}$ is the value after filtering, j is the feature value number in dataset, n is half the width of the interval of values across which the average is calculated. Figure 14 shows graphs of the data as a function of time during the "three" gesture demonstration, before and after filtering. Figure 14 shows the application of filtering to the sensor signal. With prolonged operation, a change in the basic resistance of the sensor occurs. Then there is practically no change in baseline resistance. Filtering is used to correct for changes in the sensor resistance baseline (blue curves before filtering, orange after). Signal fluctuations in the orange curves are the result of changes in sensor resistance and are not noise.

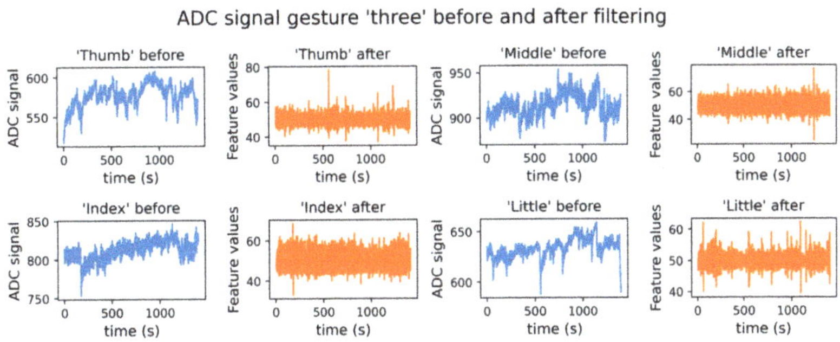

Figure 14. Data in a demonstration gesture "three", before (f) and after filtering ($\hat{f}$) for index, middle, little fingers, and the thumb.

The dataset was collected based on data from sensors detecting the amount of resistance when the subject displayed gestures. Eleven gestures were used in the data collection: 1 basic gesture (the fist), and 10 recognizable gestures. The screen showed information for the subject which gesture to demonstrate, while at the same time, the data from the sensors were written into a .csv with a label value corresponding to the information being shown. This is how the data were labeled. Each gesture was measured for ~30 min, with the displayed gesture changing every 5 s. Data were recorded with a discretization in the range of 20–60 Hz. The final dataset for model training contained over 560,000 lines.

Seven machine learning models were used for gesture recognition: K nearest neighbor classifier (KNN), support vector machine (SVM) random forest classifier (RF), xgboost classifier (XGB), LightGBM Classifier (LGBM), decision tree classifier (DT), logistic regression (LR). The accuracy metric was used as a classification quality metric. Since training several models on the entire dataset is an extremely resource-intensive task, to compare the quality, we used 10^5 rows from the dataset. The following accuracy was obtained for the classifiers: 0.941 ± 0.001 (RF), 0.887 ± 0.001 (KNN), 0.604 ± 0.004 (SVM), 0.734 ± 0.005 (XGB), 0.731 ± 0.003 (LGBM), 0.753 ± 0.003 (DTC), 0.490 ± 0.005 (LR) (Figure 15a). Increasing the dataset size improves the prediction accuracy (Figure 15b). However, if the dataset is larger than 4.5×10^5, the accuracy does not increase significantly.

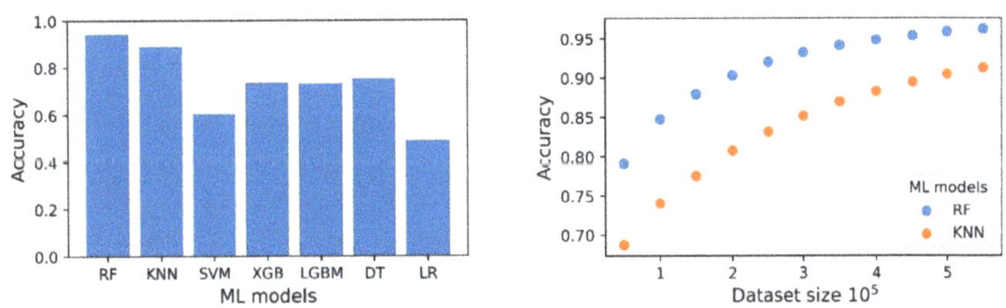

(**a**) (**b**)

Figure 15. Accuracy for the ML models KNN, SVM, RF, XGB, LGBM, DT (**a**) during training on dataset of size 10^5 rows. (**b**) Accuracy for ML models RF, KNN during training on a part of the dataset of size 0.5×10^5 to 5.5×10^5.

For the best performing ML models, the RF and KNN hyperparameters were optimized. After hyperparameter optimization, the KNN model showed an accuracy of 0.965 ± 0.001, and the RF model showed an accuracy of 0.958 ± 0.002.

For wearable applications where significant dynamic loads are often encountered, durability is an important quality indicator. With long load cycles during data collection for machine learning, the flexible sensors built into the smart system ensured the reliable functionality of the developed system.

4. Conclusions

The effect of the formation of electrically conductive networks in the MWCNT/PDMS material under laser exposure is demonstrated. It is found that MWCNT/PDMS-based strain gauges obtained by laser structuring of conductive networks have lower initial resistance (3 kOhm), higher sensitivity (average GF 10) to strain and lower hysteresis (3%) of electrical resistance compared to those obtained without laser exposure. The Young's modulus for the laser-irradiated sensors was 47 kPa, which is comparable to the Young's modulus of human skin.

A sensor-based intelligent hand gesture recognition system based on laser-formed strain sensors has been successfully developed. The numerical values of the change in electrical resistance during finger movements were read and processed using the developed microcontroller electronic unit based on the ATXMEGA8E5-AU microcontroller. The data were transmitted wirelessly to a personal computer and displayed as a graph of the amplitude of movements versus time. The developed software allowed for recording the data in a convenient form for subsequent processing and machine learning for the purpose of machine gesture recognition. The use of a fabricated smart system based on laser-formed sensors with conductive MWCNT networks has achieved a gesture recognition accuracy of ~94%. The developed sensory smart system is one of the reliable and inexpensive approaches to provide objective feedback for training the motor functions of the hand.

Author Contributions: Conceptualization, N.A.N. and A.Y.G.; methodology, L.P.I. and E.P.K.; software, E.S.P. and D.I.R.; validation, D.I.R., A.V.K., V.V.S. and N.A.N.; formal analysis, I.B., A.V.M. and S.V.S.; investigation, N.A.N., V.V.S., A.V.K. and D.I.R.; resources, N.A.N., S.V.S. and A.Y.G.; data curation, D.I.R. and E.S.P.; writing—original draft preparation, N.A.N., D.I.R., V.V.S., A.V.K. and E.A.G.; writing—review and editing, A.Y.G., L.P.I., E.S.P., E.P.K., D.V.T. and A.V.M.; visualization, E.A.G., N.A.N., V.V.S. and D.I.R.; supervision, S.V.S. and A.Y.G.; project administration, A.Y.G.; funding acquisition, A.Y.G. All authors have read and agreed to the published version of the manuscript.

Funding: Structural and electrophysical investigation of sensors were funded by the Russian Science Foundation under grant number 19-19-00401 (https://rscf.ru/en/project/19-19-00401/, accessed on 21 May 2023). The development and research of the sensory system were funded by the Russian Ministry of Science and Higher Education (state assignment No. 075-03-2023-024 from 13 January 2023).

Data Availability Statement: The data taken with the developed smart system, which was used to train the gesture recognition can be found on the following link: https://github.com/DmitriiRybkin/Gesture_recognition (accessed on 28 April 2023).

Acknowledgments: The authors are grateful to colleagues who helped in the SEM investigation, Alexander Dudin and Elena Eganova. This research work was supported by the Academic leadership program Priority 2030 proposed by the Federal State Autonomous Educational Institution of Higher Education I.M. Sechenov First Moscow State Medical University of the Ministry of Health of the Russian Federation (Sechenov University).

Conflicts of Interest: The authors declare no conflict of interest.

References

1. Sathiyanarayanan, M.; Rajan, S. MYO Armband for Physiotherapy Healthcare: A Case Study Using Gesture Recognition Application. In Proceedings of the 2016 8th International Conference on Communication Systems and Networks (COMSNETS), Bangalore, India, 5–10 January 2016; pp. 1–6.
2. Sagayam, K.M.; Hemanth, D.J. Hand Posture and Gesture Recognition Techniques for Virtual Reality Applications: A Survey. *Virtual Real.* **2017**, *21*, 91–107. [CrossRef]
3. Kang, H.; Lee, C.W.; Jung, K. Recognition-Based Gesture Spotting in Video Games. *Pattern Recognit. Lett.* **2004**, *25*, 1701–1714. [CrossRef]
4. Liu, H.; Wang, L. Gesture Recognition for Human-Robot Collaboration: A Review. *Int. J. Ind. Ergon.* **2018**, *68*, 355–367. [CrossRef]

5. Taub, E.; Morris, D.M. Constraint-Induced Movement Therapy to Enhance Recovery after Stroke. *Curr. Atheroscler. Rep.* **2001**, *3*, 279–286. [CrossRef] [PubMed]
6. Pisharady, P.K.; Saerbeck, M. Recent Methods and Databases in Vision-Based Hand Gesture Recognition: A Review. *Comput. Vis. Image Underst.* **2015**, *141*, 152–165. [CrossRef]
7. Sarma, D.; Bhuyan, M.K. Methods, Databases and Recent Advancement of Vision-Based Hand Gesture Recognition for HCI Systems: A Review. *SN Comput. Sci.* **2021**, *2*, 436. [CrossRef]
8. Zhang, Z.; Tian, Z.; Zhou, M. Latern: Dynamic Continuous Hand Gesture Recognition Using FMCW Radar Sensor. *IEEE Sens. J.* **2018**, *18*, 3278–3289. [CrossRef]
9. Kim, S.Y.; Han, H.G.; Kim, J.W.; Lee, S.; Kim, T.W. A Hand Gesture Recognition Sensor Using Reflected Impulses. *IEEE Sens. J.* **2017**, *17*, 2975–2976. [CrossRef]
10. Yang, X.; Sun, X.; Zhou, D.; Li, Y.; Liu, H. Towards Wearable A-Mode Ultrasound Sensing for Real-Time Finger Motion Recognition. *IEEE Trans. Neural Syst. Rehabil. Eng.* **2018**, *26*, 1199–1208. [CrossRef]
11. Akhlaghi, N.; Baker, C.A.; Lahlou, M.; Zafar, H.; Murthy, K.G.; Rangwala, H.S.; Kosecka, J.; Joiner, W.M.; Pancrazio, J.J.; Sikdar, S. Real-Time Classification of Hand Motions Using Ultrasound Imaging of Forearm Muscles. *IEEE Trans. Biomed. Eng.* **2016**, *63*, 1687–1698. [CrossRef]
12. Jiang, S.; Lv, B.; Guo, W.; Zhang, C.; Wang, H.; Sheng, X.; Shull, P.B. Feasibility of Wrist-Worn, Real-Time Hand, and Surface Gesture Recognition via SEMG and IMU Sensing. *IEEE Trans. Industr. Inform.* **2018**, *14*, 3376–3385. [CrossRef]
13. Han, H.; Yoon, S.W. Gyroscope-Based Continuous Human Hand Gesture Recognition for Multi-Modal Wearable Input Device for Human Machine Interaction. *Sensors* **2019**, *19*, 2562. [CrossRef] [PubMed]
14. Liu, X.; Wei, Y.; Qiu, Y. Advanced Flexible Skin-Like Pressure and Strain Sensors for Human Health Monitoring. *Micromachines* **2021**, *12*, 695. [CrossRef] [PubMed]
15. Alamusi; Hu, N.; Fukunaga, H.; Atobe, S.; Liu, Y.; Li, J. Piezoresistive Strain Sensors Made from Carbon Nanotubes Based Polymer Nanocomposites. *Sensors* **2011**, *11*, 10691–10723. [CrossRef] [PubMed]
16. Si, Y.; Chen, S.; Li, M.; Li, S.; Pei, Y.; Guo, X. Flexible Strain Sensors for Wearable Hand Gesture Recognition: From Devices to Systems. *Adv. Intell. Syst.* **2022**, *4*, 2100046. [CrossRef]
17. Nesser, H.; Lubineau, G. Strain Sensing by Electrical Capacitive Variation: From Stretchable Materials to Electronic Interfaces. *Adv. Electron. Mater.* **2021**, *7*, 2100190. [CrossRef]
18. Sha, B.; Lü, X.; Jiang, L. High Sensitivity and Wide Range Biomimetic Tactile-Pressure Sensor Based on 2D Graphene Film and 3D Graphene Foam. *Micromachines* **2022**, *13*, 1150. [CrossRef]
19. Yu, J.; Xu, S.; Jiang, Y.; Chen, H.; Feng, W. Multi-Parameter Sensor Based on the Fiber Bragg Grating Combined with Triangular-Lattice Four-Core Fiber. *Optik* **2020**, *208*, 164094. [CrossRef]
20. He, T.; Shi, Q.; Wang, H.; Wen, F.; Chen, T.; Ouyang, J.; Lee, C. Beyond Energy Harvesting—Multi-Functional Triboelectric Nanosensors on a Textile. *Nano Energy* **2019**, *57*, 338–352. [CrossRef]
21. Panth, M.; Cook, B.; Zhang, Y.; Ewing, D.; Tramble, A.; Wilson, A.; Wu, J. High-Performance Strain Sensors Based on Vertically Aligned Piezoelectric Zinc Oxide Nanowire Array/Graphene Nanohybrids. *ACS Appl. Nano Mater.* **2020**, *3*, 6711–6718. [CrossRef]
22. Li, J.; Fang, L.; Sun, B.; Li, X.; Kang, S.H. Review—Recent Progress in Flexible and Stretchable Piezoresistive Sensors and Their Applications. *J. Electrochem. Soc.* **2020**, *167*, 037561. [CrossRef]
23. Duan, L.; D'hooge, D.R.; Cardon, L. Recent Progress on Flexible and Stretchable Piezoresistive Strain Sensors: From Design to Application. *Prog. Mater. Sci.* **2020**, *114*, 100617. [CrossRef]
24. Liu, X.; Mwangi, M.; Li, X.; O'Brien, M.; Whitesides, G.M. Paper-Based Piezoresistive MEMS Sensors. *Lab Chip* **2011**, *11*, 2189. [CrossRef]
25. Qiu, A.; Li, P.; Yang, Z.; Yao, Y.; Lee, I.; Ma, J. A Path Beyond Metal and Silicon:Polymer/Nanomaterial Composites for Stretchable Strain Sensors. *Adv. Funct. Mater.* **2019**, *29*, 1806306. [CrossRef]
26. Ferrreira, A.; Rocha, J.G.; Ansón-Casaos, A.; Martínez, M.T.; Vaz, F.; Lanceros-Mendez, S. Electromechanical Performance of Poly(Vinylidene Fluoride)/Carbon Nanotube Composites for Strain Sensor Applications. *Sens. Actuators A Phys.* **2012**, *178*, 10–16. [CrossRef]
27. Luo, S.; Liu, T. Structure–Property–Processing Relationships of Single-Wall Carbon Nanotube Thin Film Piezoresistive Sensors. *Carbon* **2013**, *59*, 315–324. [CrossRef]
28. Zhou, J.; Yu, H.; Xu, X.; Han, F.; Lubineau, G. Ultrasensitive, stretchable strain sensors based on fragmented carbon nanotube papers. *ACS Appl. Mater. Interfaces* **2017**, *9*, 4835–4842. [CrossRef]
29. Pandey, G.; Thostenson, E.T. Carbon Nanotube-Based Multifunctional Polymer Nanocomposites. *Polym. Rev.* **2012**, *52*, 355–416. [CrossRef]
30. Wu, Y.-T.; Yan, T.; Pan, Z.-J. Wearable Carbon-Based Resistive Sensors for Strain Detection: A Review. *IEEE Sens. J.* **2021**, *21*, 4030–4043. [CrossRef]
31. Ichkitidze, L.P.; Gerasimenko, A.Y.; Podgaetsky, V.M.; Selishchev, S.V.; Dudin, A.A.; Pavlov, A.A. Electrical Conductivity of the Nanocomposite Layers for Use in Biomedical Systems. Materials Physics and Mechanics. *Mater. Phys. Mech.* **2018**, *37*, 140–145.
32. Ichkitidze, L.P.; Podgaetsky, V.M.; Prihodko, A.S.; Putra, B.M.; Blagov, E.V.; Pavlov, A.A.; Galperin, V.A.; Kitsyuk, E.P.; Shaman, Y.P. A Conductive Composite Nanomaterial with Biocompatible Matrix and Multilayer Carbon Nanotubes. *Biomed. Eng.* **2013**, *47*, 68–72. [CrossRef]

33. Zhurbina, N.N.; Kurilova, U.E.; Ickitidze, L.P.; Podgaetsky, V.M.; Selishchev, S.V.; Suetina, I.A.; Mezentseva, M.V.; Eganova, E.M.; Pavlov, A.A.; Gerasimenko, A.Y. Investigation of Cell Proliferative Activity on the Surface of the Nanocomposite Material Produced by Laser Radiation. In *Saratov Fall Meeting 2015: Third International Symposium on Optics and Biophotonics and Seventh Finnish-Russian Photonics and Laser Symposium (PALS)*; Genina, E.A., Tuchin, V.V., Derbov, V.L., Postnov, D.E., Meglinski, I.V., Larin, K.V., Pravdin, A.B., Eds.; SPIE Press: Bellingham, WA, USA, 2016; p. 991718.
34. Ichkitidze, L.P.; Gerasimenko, A.Y.; Podgaetsky, V.M.; Selishchev, S.V. Layers with the Tensoresistive Properties and Their Possible Applications in Medicine. *Mater. Phys. Mech.* **2018**, *37*, 153–158.
35. Han, T.; Nag, A.; Afsarimanesh, N.; Mukhopadhyay, S.C.; Kundu, S.; Xu, Y. Laser-Assisted Printed Flexible Sensors: A Review. *Sensors* **2019**, *19*, 1462. [CrossRef]
36. El Haber, G.; Noel, L.; Lin, C.-F.; Gree, S.; Vidal, L.; Zan, H.-W.; Hobeika, N.; Lhost, O.; Trolez, Y.; Soppera, O. Near-Infrared Laser Direct Writing of Conductive Patterns on the Surface of Carbon Nanotube Polymer Nanocomposites. *ACS Appl. Mater. Interfaces* **2021**, *13*, 49279–49287. [CrossRef] [PubMed]
37. Caradonna, A.; Badini, C.; Padovano, E.; Veca, A.; De Meo, E.; Pietroluongo, M. Laser Treatments for Improving Electrical Conductivity and Piezoresistive Behavior of Polymer–Carbon Nanofiller Composites. *Micromachines* **2019**, *10*, 63. [CrossRef] [PubMed]
38. Yuan, Y.; Chen, J. Morphology Adjustments of Multi-Walled Carbon Nanotubes by Laser Exposure. *Laser Phys. Lett.* **2016**, *13*, 066001. [CrossRef]
39. Gerasimenko, A.Y.; Kuksin, A.V.; Shaman, Y.P.; Kitsyuk, E.P.; Fedorova, Y.O.; Sysa, A.V.; Pavlov, A.A.; Glukhova, O.E. Electrically Conductive Networks from Hybrids of Carbon Nanotubes and Graphene Created by Laser Radiation. *Nanomaterials* **2021**, *11*, 1875. [CrossRef]
40. Gerasimenko, A.Y.; Kurilova, U.E.; Savelyev, M.S.; Murashko, D.T.; Glukhova, O.E. Laser Fabrication of Composite Layers from Biopolymers with Branched 3D Networks of Single-Walled Carbon Nanotubes for Cardiovascular Implants. *Compos. Struct.* **2021**, *260*, 113517. [CrossRef]
41. Markov, A.; Wördenweber, R.; Ichkitidze, L.; Gerasimenko, A.; Kurilova, U.; Suetina, I.; Mezentseva, M.; Offenhäusser, A.; Telyshev, D. Biocompatible SWCNT Conductive Composites for Biomedical Applications. *Nanomaterials* **2020**, *10*, 2492. [CrossRef]
42. Ichkitidze, L.P.; Podgaetsky, V.M. A Conductive Biocompatible Composite Carbon-Nanotube Material. *Biomed. Eng.* **2012**, *45*, 225–228. [CrossRef]
43. Yu, J.; Wu, J.; Yang, H.; Li, P.; Liu, J.; Wang, M.; Pang, J.; Li, C.; Yang, C.; Xu, K. Extremely Sensitive SERS Sensors Based on a Femtosecond Laser-Fabricated Superhydrophobic/-Philic Microporous Platform. *ACS Appl. Mater. Interfaces* **2022**, *14*, 43877–43885. [CrossRef]
44. Luo, H.; Lu, Y.; Xu, Y.; Yang, G.; Cui, S.; Han, D.; Zhou, Q.; Ouyang, X.; Yang, H.; Cheng, T.; et al. A Fully Soft, Self-Powered Vibration Sensor by Laser Direct Writing. *Nano Energy* **2022**, *103*, 107803. [CrossRef]
45. Cui, S.; Lu, Y.; Kong, D.; Luo, H.; Peng, L.; Yang, G.; Yang, H.; Xu, K. Laser Direct Writing of Ga_2O_3/Liquid Metal-Based Flexible Humidity Sensors. *Opto-Electron. Adv.* **2023**, *6*, 220172-1–220172-10. [CrossRef]
46. Kuksin, A.; Demidenko, N.; Murashko, D.; Osipov, D.; Gerasimenko, A. Creation of 3D Networks of Single-Walled Carbon Nanotubes for Use in Bioelectronics. In Proceedings of the 2020 IEEE Conference of Russian Young Researchers in Electrical and Electronic Engineering (EIConRus), St. Petersburg and Moscow, Russia, 27–30 January 2020; pp. 2503–2506.
47. Webb, R.C.; Bonifas, A.P.; Behnaz, A.; Zhang, Y.; Yu, K.J.; Cheng, H.; Shi, M.; Bian, Z.; Liu, Z.; Kim, Y.-S.; et al. Ultrathin Conformal Devices for Precise and Continuous Thermal Characterization of Human Skin. *Nat. Mater.* **2013**, *12*, 938–944. [CrossRef]
48. Li, Q.; Wang, Y.; Jiang, S.; Li, T.; Ding, X.; Tao, X.; Wang, X. Investigation into Tensile Hysteresis of Polyurethane-Containing Textile Substrates for Coated Strain Sensors. *Mater. Des.* **2020**, *188*, 108451. [CrossRef]
49. Starý, Z. Simultaneous Electrical and Rheological Measurements on Melts of Conductive Polymer Composites under Elongation. *Polymer* **2014**, *55*, 5608–5611. [CrossRef]
50. Gerasimenko, A.Y.; Kitsyuk, E.; Kurilova, U.E.; Suetina, I.A.; Russu, L.; Mezentseva, M.V.; Markov, A.; Narovlyansky, A.N.; Kravchenko, S.; Selishchev, S.; et al. Laser Fabrication of Composite Interfaces Based on Laser-Structured Arrays of Carbon Nanotubes with Albumin for Electrical Stimulation of Heart Cell Growth. *Polymers* **2022**, *14*, 1866. [CrossRef]
51. Yang, S.; Lu, N. Gauge Factor and Stretchability of Silicon-on-Polymer Strain Gauges. *Sensors* **2013**, *13*, 8577–8594. [CrossRef]
52. Li, M.; Pei, Y.; Cao, Y.; Chen, S.; Guo, X. Flexible Strain Sensors: From Devices to Array Integration. *Flex. Print. Electron.* **2021**, *6*, 043002. [CrossRef]
53. Hempel, M.; Nezich, D.; Kong, J.; Hofmann, M. A Novel Class of Strain Gauges Based on Layered Percolative Films of 2D Materials. *Nano Lett.* **2012**, *12*, 5714–5718. [CrossRef]
54. Li, C.; Guan, G.; Reif, R.; Huang, Z.; Wang, R.K. Determining Elastic Properties of Skin by Measuring Surface Waves from an Impulse Mechanical Stimulus Using Phase-Sensitive Optical Coherence Tomography. *J. R. Soc. Interface* **2012**, *9*, 831–841. [CrossRef] [PubMed]

micromachines MDPI

Article

Parameter Study on Force Curves of Assembled Electronic Components on Foils during Injection Overmolding Using Simulation

Martin Hubmann [1,*], Mona Bakr [2], Jonas Groten [3], Martin Pletz [4], Jan Vanfleteren [5,*], Frederick Bossuyt [5], Behnam Madadnia [5] and Barbara Stadlober [3]

[1] Polymer Processing, Department of Polymer Engineering and Science, Montanuniversitaet Leoben, 8700 Leoben, Austria
[2] Center for Microsystems Technology, Ghent University, B-9052 Ghent, Belgium
[3] Joanneum Research Forschungsgesellschaft mbH, Franz-Pichler Str. 30, 8160 Weiz, Austria
[4] Designing Plastics and Composite Materials, Department of Polymer Engineering and Science, Montanuniversitaet Leoben, 8700 Leoben, Austria
[5] Centre for Microsystems Technology, Imec and Ghent University, Technology Park 126, Zwijnaarde, B-9052 Ghent, Belgium
* Correspondence: martin.hubmann@unileoben.ac.at (M.H.); jan.vanfleteren@ugent.be (J.V.)

Abstract: The integration of assembled foils in injection-molded parts is a challenging step. Such assembled foils typically comprise a plastic foil on which a circuit board is printed and electronic components are mounted. Those components can detach during overmolding when high pressures and shear stresses prevail due to the injected viscous thermoplastic melt. Hence, the molding settings significantly impact such parts' successful, damage-free manufacturing. In this paper, a virtual parameter study was performed using injection molding software in which 1206-sized components were overmolded in a plate mold using polycarbonate (PC). In addition, experimental injection molding tests of that design and shear and peel tests were made. The simulated forces increased with decreasing mold thickness and melt temperature and increasing injection speed. The calculated tangential forces in the initial stage of overmolding ranged from 1.3 N to 7.3 N, depending on the setting used. However, the experimental at room temperature-obtained shear forces at break were at least 22 N. Yet, detached components were present in most of the experimentally overmolded foils. Hence, the shear tests performed at room temperature can only provide limited information. In addition, there might be a peel-like load case during overmolding where the flexible foil might bend during overmolding.

Keywords: injection molding; simulation; in-mold electronics; over-molding; Moldflow

Citation: Hubmann, M.; Bakr, M.; Groten, J.; Pletz, M.; Vanfleteren, J.; Bossuyt, F.; Madadnia, B.; Stadlober, B. Parameter Study on Force Curves of Assembled Electronic Components on Foils during Injection Overmolding Using Simulation. *Micromachines* **2023**, *14*, 876. https://doi.org/10.3390/mi14040876

Academic Editors: Zhuoqing Yang and Libo Gao

Received: 20 March 2023
Revised: 11 April 2023
Accepted: 14 April 2023
Published: 19 April 2023

1. Introduction

The need for low-cost, flexible, lightweight electronic devices has risen dramatically. Over-molding electronics is an innovative technique for producing three-dimensional (3D)-shaped smart products. The main concept is to use injection molding to create working circuit carriers with components in two-dimensional (2D) or 3D shapes. The overmolding process has been used in various applications, including the automotive sector, consumer household appliances, and medical equipment [1,2]. First, an electrical circuit is deposited on a flexible substrate, and surface mount devices (SMDs) are attached to the circuit using lead-free solder or conductive adhesives. Then, this electronic circuit is put into an injection molding machine as an insert, and the plastic flows over it. The end product is a plastic part with an integrated circuit and functional electrical components.

However, the prevailing thermal and mechanical loads during the injection molding cycle necessary to shape the molten polymer can lead to failures within the electrical circuits [3–5]. The mechanical forces may be considered separately as shear loads acting in-plane and pressure loads acting normally upon the component surfaces. To bypass the high stresses inherent to the injection molding process, Ott and Drummer [6] proposed using thermoplastic foam injection molding. That way, they could encapsulate epoxy-based printed circuit boards (PCBs) with cavity pressures of approx. 1 MPa only.

Heinle et al. [7] developed an injection mold that allows the forces on small components to be measured during the overmolding cycle. To that end, they added a three-way force sensor with a size of $5 \times 6 \times 1$ mm^3 in the center of a $40 \times 40 \times 2$ mm^3 mold. Hence the sensor could record the force in all directions in space: in flow (F_x), transverse (F_y), and perpendicular (F_z) to the flow direction. Short shots (almost filled part and no packing pressure applied) were made, and sharp peaks of F_x and F_z were recorded during filling that decreased quickly. The centered sensor depicted no force in F_y because of a symmetric flow around it. During cooling, forces F_x and F_z were measured due to shrinkage.

Schirmer et al. [4,8] overmolded assembled foils with components of sizes 0402, 0603, and 0805 with polycarbonate (PC). Larger components showed lower failure rates than smaller components when the conductive adhesive was used. However, the opposite trend was observed when solder was used. Fewer crack formations were observed on components overmolded at higher injection speeds when the conductive adhesive was used, and medium injection speeds were preferred for those with solder. Further, no influence of the orientation of the components, placed in 0° (transverse flow) and 90° (in flow) orientation, was found. Components using conductive adhesive placed close to the film gate were less prone to damage. They reasoned that the melt would become significantly more viscous along the flow path, resulting in more significant damage to the components. Likewise, a faster injection speed would reduce the time the melt would cool during filling and be beneficial [4]. However, the opposite was observed for soldered resistors. Here less damage was visible on components farther away from the gate [8].

Bakr et al. [9] developed an assembled foil design comprising 18 zero-ohm resistors oriented at 0°, 45°, and 90° for overmolding in a flat or corner-shaped mold. The design was also used in this study. Two component sizes, namely, 0805 and 1206, and two copper-based foil substrates with different polymer layers, polyimide (PI) and polyethylene terephthalate (PET), were investigated. Lead-free solder and a conductive adhesive (CA) were used as interconnection materials for the PI foils and components. However, CA and a low melt temperature solder (LTS) were used for the PET foils. Only the PI foil using lead-free solder showed no change in the resistance after overmolding (in the flat mold). In contrast, the PET foils with the used interconnection materials frequently exhibited detached components. No influence of the package size, orientation, or placement within the mold was reported.

Wimmer et al. [10] overmolded assembled polymer films with glued 0805 0 ohm resistors placed in a line in a 4 mm thick test rod with PC. They estimated the shear force on the components using the injection simulation software CADmould (Simcon kunststofftechnische Software GmbH, Würselen, Germany). To that end, the part with components was modeled in the software (mesh size 0.4 mm). Three "sensors" per component with surface area A_{0805} were placed to obtain the local shear rates ($\dot{\gamma}$) and viscosities (η), and hence the shear forces $F_s = \dot{\gamma} \cdot \eta \cdot A_{0805}$ at any time during the cycle [11]. Their simulation showed shear forces $F_s = 0.8 \ldots 1.5$ N. This was substantially lower than the tested shear strength of their components of 15 N to 20 N (measured according to DIN EN 62137-1-2 at room temperature). Of the 42 films overmolded, only one component was washed away. However, components overmolded with lower melt temperatures exhibited larger increases in electrical resistance. This was denoted to crack formations due to the increased viscosity of the colder polymer [10].

State-of-the-art commercial injection molding software such as Autodesk Moldflow Insight (AMI, Autodesk Inc., San Rafael, CA, USA) [12] or Moldex3D (CoreTech System Co., Ltd., Zhubei City, Taiwan) [13] have implemented core shift analyses into their filling simulations. These features estimate possible deformation of (mold) inserts during the injection molding cycle based on the pressure distribution within the melt. Those software extensions, however, were designed with cantilevered core pins in mind that are affected by deformation when a non-uniform flow around them causes pressure differences [12].

In a very recent work [14], we investigated the integration of structural electronics in injection-molded parts. This paper used two thermoplastic polyurethanes (TPUs) as middle layers with different melting temperatures. Parameter studies were performed to investigate the influence of the melt temperature, mold temperature, injection speed, and TPU layer. A shear distortion factor for the TPU layer was derived based on simulations that linked the shear stresses with the injection time and the softening (melting) of the TPU. The distortion of the films was found to reduce with higher melt temperature, lower mold temperature, and faster injection speed. Films using the TPU with the higher melting temperature yielded significantly better results.

This literature overview shows that some studies have investigated the optimization of the injection molding process for assembled foils. However, an in-depth examination of the prevailing forces during filling is still lacking. By analyzing numerically performed parameter studies, this paper aims to gain insight into how the injection molding settings influence the forces on the components.

To that end, the foil design developed by Bakr et al. [9] was utilized (illustrated in Figure 1a), and virtual parameter studies were made using AMI. Here, the influences of the mold thickness, melt temperature, and injection speed on the forces on 1206-sized components when overmolded with PC were investigated. A Python script accessing the AMIs Synergy Application Programming Interface [15] was developed to calculate and visualize the tangential (F_t) and normal forces (F_n) on the components during filling (Figure 1b). The experimental results obtained in [9] were extended with additional injection molding tests, and the number of detached components after overmolding was counted (Figure 1c). In addition, the shear and peel strength values were assessed in mechanical tests (Figure 1d). The findings were then combined to extract guidelines for injection molding of assembled foils (illustrated in Figure 1e).

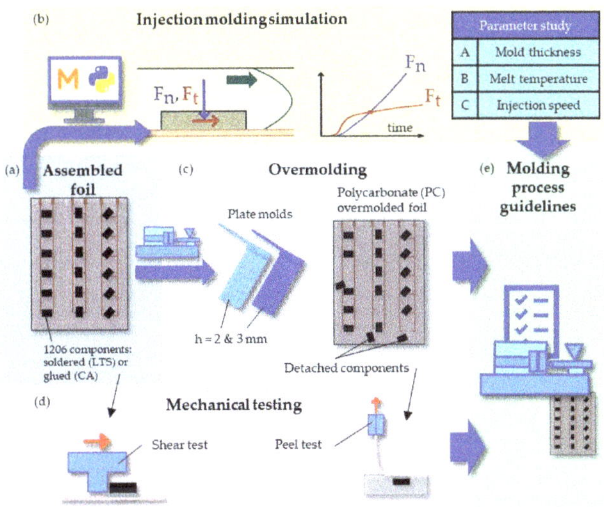

Figure 1. Workflow to study the impact of injection molding parameters on detached components: The foil design with soldered resistor components developed in [9] (**a**) was numerically simulated in parameter studies in which the normal (F_n) and the tangential forces (F_t) on the components during filling were investigated (**b**). Additional injection molding tests were conducted on the results in [9], and the detached components were counted after overmolding (**c**). Further mechanical tests like shear and peel tests investigating the mountings' strength were made (**d**). Finally, the findings concerning simulation, overmolding process, and mechanical testing were combined to extract guidelines for injection molding of assembled foils (**e**).

2. Materials and Methods

2.1. Assembled Foil Design

Figure 2a shows an assembled foil fabricated according to the design developed in [9] comprising 18 components oriented in three directions (0°, 90°, and 45°) with copper tracks used for the connection. A PET-based GTS-MP (Furukawa Electric Co., Ltd., Tokyo, Japan) foil was used with the layer structure and thicknesses given in Figure 2b.

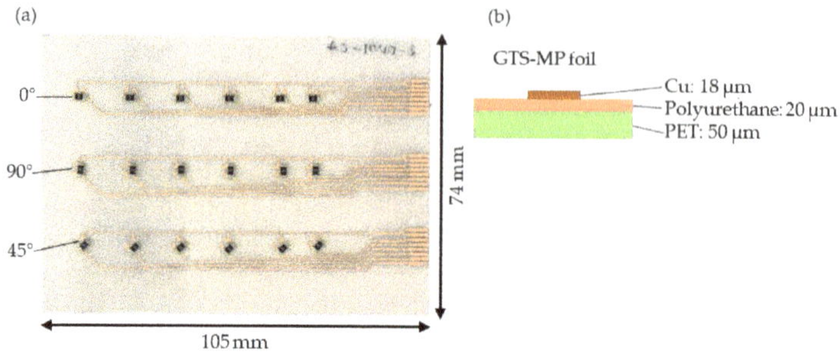

Figure 2. Assembled foil for overmolding (developed in [9]) using 1206 resistor components (**a**) and a layer structure of the foil (**b**).

Zero-ohm resistors of size 1206 (Yageo Corporation, New Taipei City, Taiwan) were used for the foils with dimensions presented in Table 1 [16].

Table 1. Dimensions of the components used [16].

Type	Length (mm)	Width (mm)	Height (mm)
1206	3.10 ± 0.10	1.60 ± 0.10	0.55 ± 0.10

Two materials for manually assembling the components on the foils' copper tracks were used:

- LTS: Low-temperature Sn42Bi57Ag1 Interflux DP 5600 solder alloy (Interflux Electronics N.V., Gent, Belgium) with a melting temperature of 139 °C [17].
- CA: Loctite Ablestik CE 3103WLV thermoset silver-based conductive glue (Henkel AG & Co. KGaA, Düsseldorf, Germany), with a curing temperature of 120 to 150 °C [18].

2.2. Injection Molding

The assembled foils described above were inserted into a plate mold, as depicted in Figure 3, for overmolding with the Makrolon 2805 polycarbonate (PC) (Covestro AG, Leverkusen, Germany) [19]. The cavity thickness was set to either h = 2 mm or 3 mm. A temperature-resistant adhesive tape was used to fix the foils, and the contacts were kept free from overmolding by a retractable insert added on top of the foils (cf. gray part in Figure 3).

Table 2 lists the foils' overmolding settings presented in [9] that were extended by the "harsh" settings of 2-260-CA and 2-260-LTS for this study. A constant injection speed of 70 cm^3/s was used, and n = 3 foils were overmolded per setting.

Table 2. The molding plane of the foils produced in [9] was extended by setting 2-260-CA and 2-260-LTS (mold temperature: 100 °C, injection speed: 70 cm^3/s, n = 3).

Setting No.	Mold Thickness (mm)	Material of Assembly	Melt Temperature (°C)
2-260-CA	2	CA	260
2-260-LTS	2	LTS	260
2-300-CA	2	CA	300
2-300-LTS	2	LTS	300
3-300-CA	3	CA	300
3-300-LTS	3	LTS	300

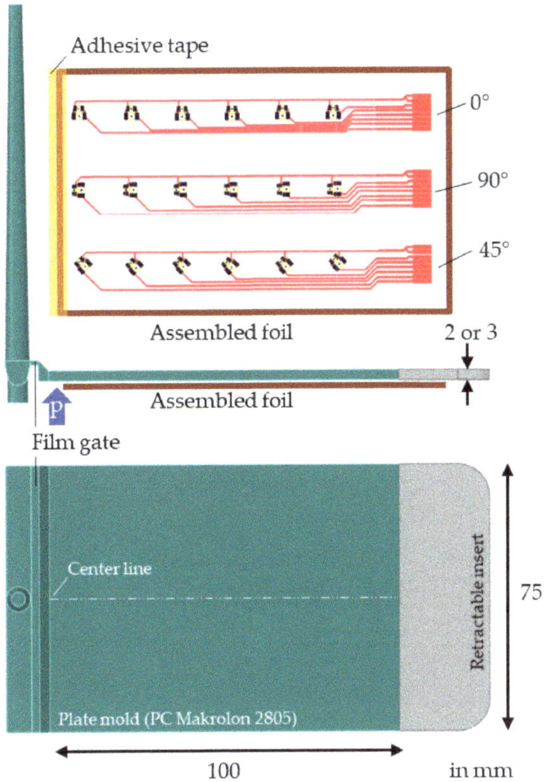

Figure 3. Setup of the plate mold with dimensions, where the foils were placed on the bottom side for overmolding. The insert, drawn in grey, was used to keep the foil contacts free from overmolding. As the blue arrow indicates, the mold was equipped with a pressure sensor close to the gate.

A fully electric Arburg Allrounder 470 A Alldrive (Arburg GmbH + Co KG, Loßburg, Germany) injection molding machine equipped with a 25 mm screw was used. A Wittmann Tempro plus D 160 (WITTMANN Technology GmbH, Wien, Austria) temperature control unit was utilized to heat the mold to 100 °C. The dosing volume was set to either 50 cm^3 for the 3 mm plate or 40 cm^3 for the 2 mm plate. The switchover point (velocity-to pressure-controlled filling) was adapted for each setting to obtain an almost (99%) filled part before the start of packing. The packing pressure was set to 400 bar for 15 s, and a residual cooling time of 50 s was used for all tests.

In addition, the pressure within the cavity close to the gate was recorded using a 4 mm diameter Kistler Type 6157B (Kistler Holding AG, Winterthur, Switzerland) piezoelectric pressure sensor (see Figure 3 for sensor positioning). The retrieved pressure curves were then used to validate the simulation results.

2.3. Simulation

The commercial injection molding simulation software Autodesk Moldflow Insight 2021 (AMI) was used to numerically study the pressures and shear loads on the components during overmolding. To model the filling phase, AMI numerically solves the conservation equations of mass, momentum, and energy using the finite element method (FEM) [20].

The 3D FEM models were created featuring either a 2- or 3-mm plate and 18 components of size 1206. The AMI property part and part insert were assigned to the CADs of the plate and components, respectively. Thus, the AMI solver performed the flow simulation on the region of the plate only while treating the components as fixed pieces added to the cavity. The thin foil (70 μm) and the joints were not modeled. The machine die was modeled as a beam hot runner.

The part region close to the components (<2 mm) and the components were modeled with a mesh size of 0.3 mm (cf. Figure 4a). The global edge length was set to 1 mm with a minimum number of 12 elements through the thickness. The auto sizing scale factor was set to 0.9. Precise match was enabled for the contact interfaces, providing identical surface meshes for contacting bodies.

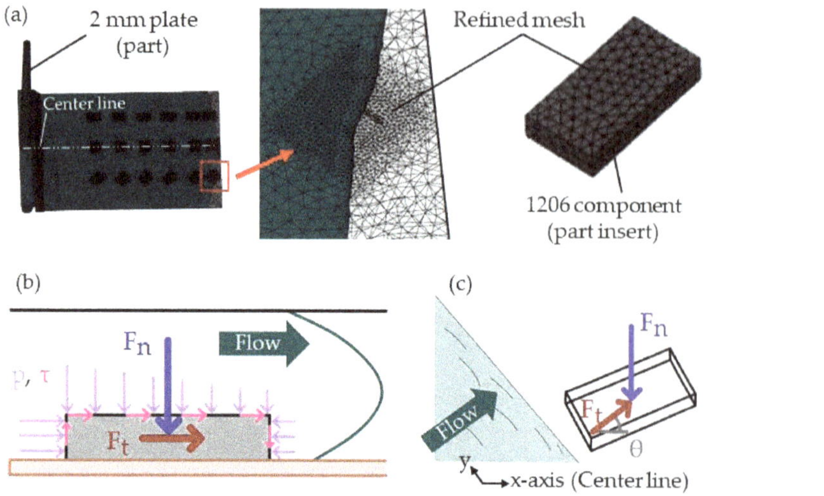

Figure 4. Filling illustration of the 2 mm plate AMI model (**a**). Pressures (p) and shear stresses (τ) accumulate into normal (F_n) and tangential (F_t) forces on the components during filling (**b**). A shear angle (θ) might be used to specify the direction of the tangential force (F_t) with respect to the center line (**c**).

The tetrahedral element count for the 3 mm plate model was 3,198,779/65,832 (part/inserts), and for the 2 mm plate model it was 3,141,193/77,748 (part/inserts). An illustration of the 2 mm model is given in Figure 4a.

All the required simulation material data for the PC Makrolon 2805 overmolding material (such as Cross-WLF viscosity coefficients and Tait pvT coefficients) are available in the Moldflow material database (Autodesk udb-file). The components were simplified and modeled as aluminum oxide (Al_2O_3) ceramic. Here, the values for the density ($\rho = 3.69\ g/cm^3$), the specific heat capacity ($c_p = 880\ J/(kg{\cdot}K)$), and the thermal conductivity ($\lambda = 18.0\ W/(m{\cdot}K)$) were taken from the literature [21].

A uniform heat transfer coefficient (HTC) of 5000 $W/(m^2{\cdot}K)$ (AMI default for the filling phase) was used for the part-component interfaces.

During filling, pressures (p) acting in the normal direction and shear stresses (τ) acting in the flow direction accumulated into forces in normal (F_n) and tangential (F_t) directions on the components, as illustrated in Figure 4b.

The direction of the tangential force (F_t) to the center line of the mold could be specified using a shear angle (θ), as depicted in Figure 4c.

It is not feasible to receive the forces (F_n and F_t with shear angle θ) on a component (viz., part insert) during filling directly within AMI. However, among the results that AMI offers are [22–24]:

- Pressure (scalar nodal part result).
- Shear stress at wall (viz., frozen/molten interface; scalar nodal part surface).
- Velocity (vector nodal part).

A Python script that used the Synergy Application Programming Interface (API) [13] was developed to access the AMI (filling) results and information about the node location and element allocation. The mechanical forces on the inserts were estimated as described in the following and are illustrated in Figure 5:

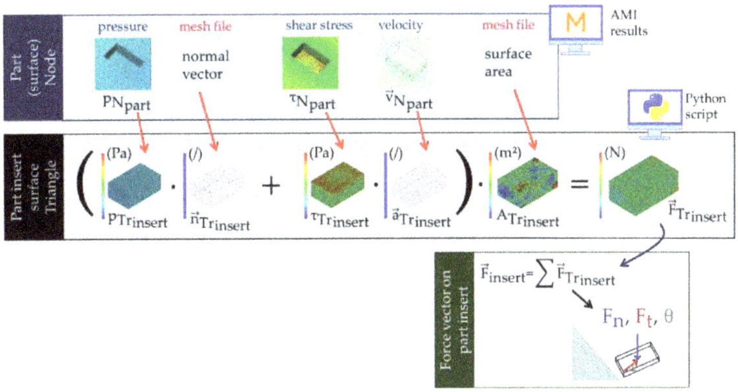

Figure 5. Illustration of the Python API script's functional principle.

The nodal pressure result of the part at the surface ($p_{N_{part}}$) is mapped onto the component insert mesh. This process is made straightforward by the congruent surface meshes between inserts and part (precise match mesh). Following that, a pressure average can be calculated for each insert triangle ($p_{T_{insert}}$). By calculating the normal vector of each insert's triangle ($\vec{n}_{T_{insert}}$), pressures are operating normally upon a surface.

Similarly, the nodal shear stress result of the part ($\tau_{N_{part}}$) can be transformed into an averaged and mapped triangle result of the insert ($\tau_{T_{insert}}$). Shear stresses act in the plane of a surface. This direction is estimated by using the closest nonzero nodal part velocity ($\vec{v}_{N_{part}}$) and mapping it onto the insert mesh. (The melts velocity at the surface is zero due to the assumption of wall adherence. Therefore, the velocity results are based on the closest internal part node.) An averaged vector is computed for each triangle that is projected onto the surface of the triangle and normalized ($\vec{a}_{T_{insert}}$).

The area of each insert triangle is calculated ($A_{T_{insert}}$) and multiplied by the corresponding force contribution of each triangle ($\vec{F}_{T_{insert}} = (p_{T_{insert}} \cdot \vec{n}_{T_{insert}} + \tau_{T_{insert}} \cdot \vec{a}_{T_{insert}}) \cdot A_{T_{insert}}$). The overall mechanical overmolding force of a component is assessed by adding up all the individual triangle forces ($\vec{F}_{insert} = \sum \vec{F}_{T_{insert}}$). This overall force vector can then be separated into the normal force F_n and tangential force F_t with shear angle θ.

Table 3 lists the performed simulated parameter study comprising the 8 simulations termed S01–S08. The parameter mold thickness (A) and melt temperature (B) were investigated in the experiments (listed in Table 2) at two levels. In addition, also a slower injection speed (C) was simulated.

Table 3. Performed injection molding simulations.

Setting No.	A—Mold Thickness (mm)	B—Melt Temperature (°C)	C—Injection Speed (cm^3/s)
S01	3	260	20
S02	3	260	70
S03	3	300	20
S04	3	300	70
S05	2	260	20
S06	2	260	70
S07	2	300	20
S08	2	300	70

Fill was selected as the analysis sequence, and 100 intermediate results were requested. The maximum %volume to fill per time step had to be reduced to 0.5% to ensure that this number of results was obtained. The switchover point was set to 97%, and a constant melt temperature of 100 °C was assumed for all settings.

2.4. Mechanical Test Setups

Shear tests on three 90°-oriented 1206 components using LTS were already performed in [9] and supplemented by equivalent tests for CA. To that end, a Bruker UMT-2 (Bruker Corporation, Billerica, MA, USA) mechanical tester platform with a purpose-built clamping plate fixture for the foils (Figure 6a) was used. The foils were firmly pressed and constrained between two metal plates, as shown in Figure 6b.

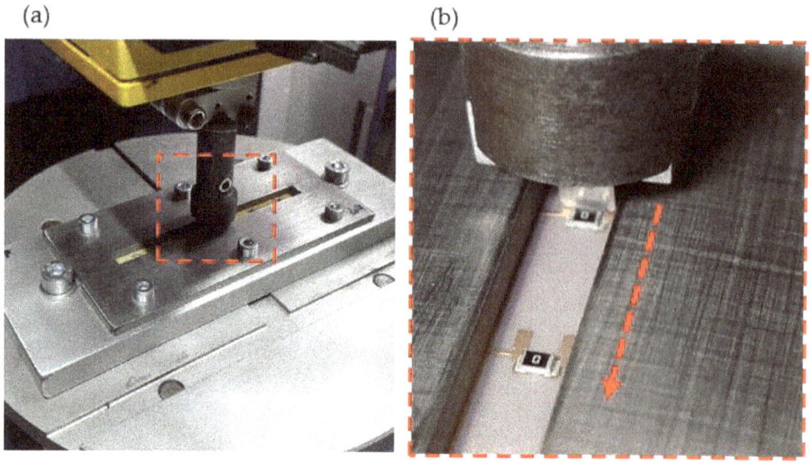

Figure 6. The foils were fixed within a clamping plate fixture (**a**) and sheared off (**b**).

The foils were firmly pressed and constrained between a metal plate and a frame with a 6 mm wide slot, as shown in Figure 6b. Next, a shear head with a width of 4 mm was moved above the foil at 1/4th of the component's height and connected to a 100 N load cell.

The shear tests could only be performed at room temperature at a shear head speed of 6 mm/min (following DIN EN 62137 1-2).

Peel tests were made of already overmolded components to gain insight into the peel strength of the (molded) LTS and CA connections. A 15 mm wide foil strip was laser cut for each row of 0°-, 45°-, and 90°-oriented components. An Instron 5500R (Illinois Tool Works Inc., Glenview, IL, USA) tensile test machine attached with a peel-off fixture was utilized for those tests. First, the molded PC plates were positioned with the foil strip along the centerline of the peel-off fixture. Next, the PC plates were clamped on both sides. Then, one side of the foil strip was loosened from the PC substrate to be clamped by the peel head. The pivoting fixture guaranteed a peel angle of 90°. Finally, a 10 mm/min peel speed and a 100 N static load cell—recording at a frequency of 10 1/s—were chosen. Figure 7a,b) show images of the test setup.

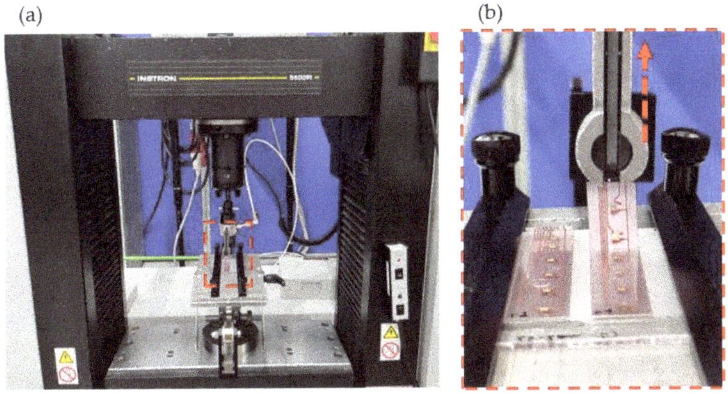

Figure 7. Strips of already overmolded foils with components were laser cut and fixed in a peel-off fixture (**a**) and peeled at an angle of 90° (**b**).

3. Results

3.1. Count of Detached Components

The number of detached components per overmolded foil was counted, as depicted in Figure 8. No pattern in terms of orientation or concerning gate location was noticed.

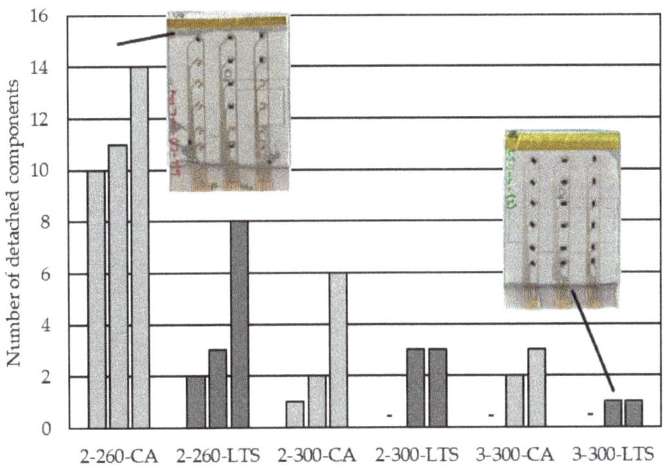

Figure 8. The number of detached components per setting (each foil comprising 18 components total). The labeling reads mold thickness (mm)–melt temperature (°C)–material for assembling (/).

Due to the high variation of detached components within the foil molded settings, comparison across them is limited. However, a lower melt temperature (260 °C vs. 300 °C) clearly provokes detachment. Additionally, there seemed to be more detached components when a lower mold thickness (2 mm compared to 3 mm) was used and when CA was used instead of LTS to mount the components.

3.2. Comparison of Recorded and Simulated Injection Molding Pressures

Figure 9 contrasts the filling pressures for the 2 mm plate mold measured within the cavity using the installed pressure sensor with the simulated filling pressure curves. No packing pressure was applied for those comparisons to emphasize the pressure development during filling.

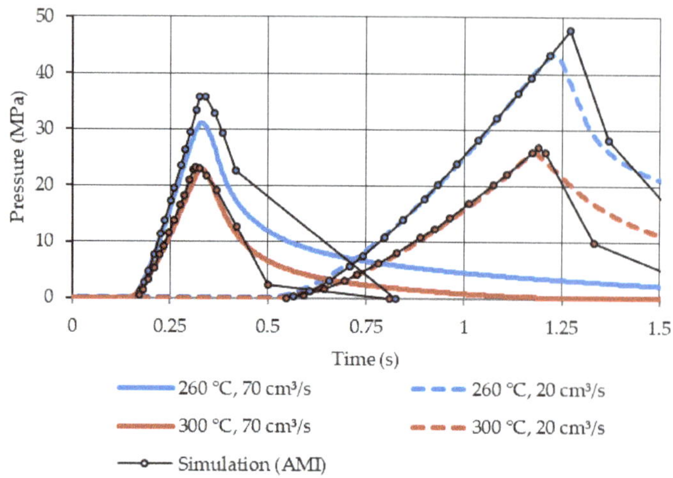

Figure 9. Comparison of the recorded (in blue and red) and simulated (in black) pressure curves for short shots (no packing pressure applied) with the 2 mm plate mold at different melt temperatures and injection speeds. The positioning of the pressure sensor is marked with a blue arrow in Figure 3 (the mold temperature was set to 100 °C).

The measured and simulated pressure developments were in good agreement (similar slopes), and discrepancies in the peak pressures were likely due to differences in the application of the switchover points.

3.3. Simulation of Forces on Components during Filling

It is possible to investigate the force development on each component during filling through the developed Python script. Figure 10a visualizes those forces in the normal (F_n) and the tangential directions (F_t) on the 1206 components of setting S01. As the melt progressed, the normal force (F_n) substantially exceeded the tangential force (Ft). Only in the initial stages of the overmolding was the tangential force (F_t) dominant, as shown in Figure 10b. Thus, we assumed this is also the most critical moment during overmolding—an observation also mentioned in [25].

The shear angles (θ) were almost zero for the 90°-oriented components positioned close to the center line of the plate mold, as depicted in Figure 10c. For the 0°- and 45°-oriented components, however, the shear angles (θ) pointed toward the sidewalls of the mold, which was attributed to the parabolic flow front, as indicated in Figure 10d.

There seemed to be only minor differences in the shape of the force curves regarding orientation (0°, 45°, and 90°) or their position (close or far from the gate).

In Figure 10b, $F_{t=n}$ when $|F_t| = |F_n|$ is marked with a red dot for the 90°-oriented component closest to the gate. This was done to have a clearly defined force at the beginning of the components overmolding. The $F_{t=n}$ value is depicted for all the simulations of the parameter study in Figure 11.

Based on the calculated $F_{t=n}$ forces stated in Figure 11, the polynomial

$$F_{t=n}\ (N) = 19.54 - 2.71 \cdot A - 0.04 \cdot B + 0.04 \cdot C, \quad (1)$$

with A as the mold thickness (mm), B as the melt temperature (°C), and C as the injection speed (cm^3/s) can be derived (coefficient of determination R^2 = 97%). The related effects plot is shown in Figure 12. In such diagrams, the examined factor is varied between the two levels, while the other factors are set to the intermediate value in the regression equation (Equation (1)). Accordingly, as the force $F_{t=n}$ becomes lower, the thicker the cavity (A), the higher the melt temperature (B), and the slower the injection speed (C).

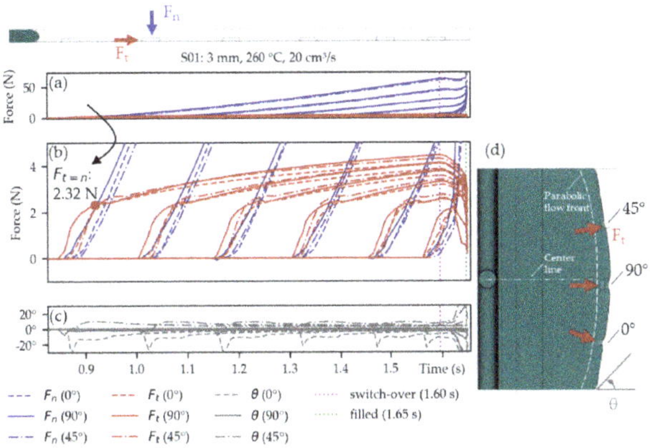

Figure 10. Simulated normal (F_n) and tangential (F_t) forces on the 1206 components during overmolding in (**a**) with augmentation on the tangential forces in (**b**). The 90°-oriented components along the center line show a shear angle (θ) close to zero (**c**). However, the 0°- and 45°-oriented components yield shear angles (θ) pointing toward the mold side walls, which can be attributed to the parabolic flow front (**d**).

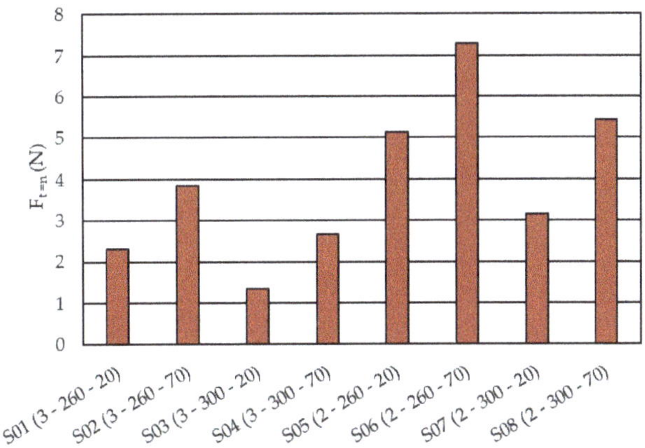

Figure 11. Simulated $F_{t=n}$ forces for the parameter study defined in Table 3. The numbering in brackets reads (mold thickness (mm)–melt temperature (°C)–injection speed (cm^3/s)).

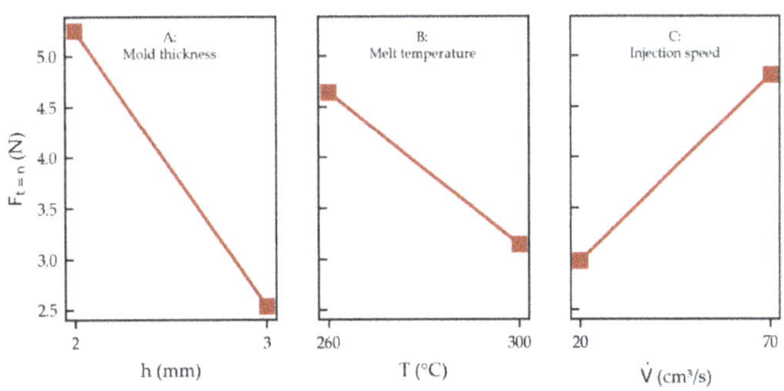

Figure 12. Effect plot for $F_{t=n}$ according to Equation (1).

In the simulations, one can easily set the (filling) pressures to zero ($p_{Npart} = 0$), so that only the shear stresses (τ_{Npart}) remain. Figure 13a shows that in such a hypothetical case, the calculated tangential forces (F_t*) would become significantly lower compared to the previously calculated tangential force (F_t). For instance, the previously defined force $F_{t=n}$ would be halved ($F_{t=n} = F_t \cong 2 \cdot F_t$*). Hence there is a substantial contribution of the pressure gradient during filling, despite the small size of the components (Figure 13b).

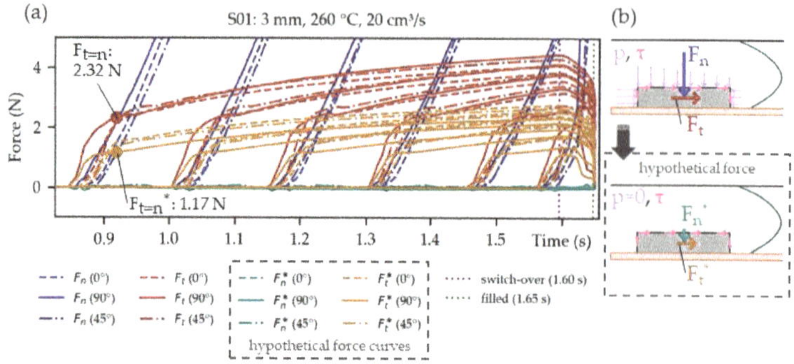

Figure 13. Comparison of the simulated normal (F_n) and tangential (F_t) forces with the hypothetical force curves in the normal (F_n*) and tangential (F_t*) direction (**a**). For the latter, the filling pressures were virtually set to zero ($p_{Npart} = 0$) as illustrated in (**b**). Therefore, the calculated tangential forces (F_t*) were lower compared to the tangential forces (F_t) obtained previously ($F_{t=n}{}^* \cong 50\% \cdot F_{t=n}$), indicating the substantial contribution of the pressure gradient during filling.

3.4. Mechanical Test Evaluation

Figure 14 shows that roughly twice as high shear loads at break can be obtained when LTS (42.8 ± 2.7 N, [9]) is used to mount the 1206 components to the copper pads of the foils compared to when CA (21.9 ± 3.2 N) is used.

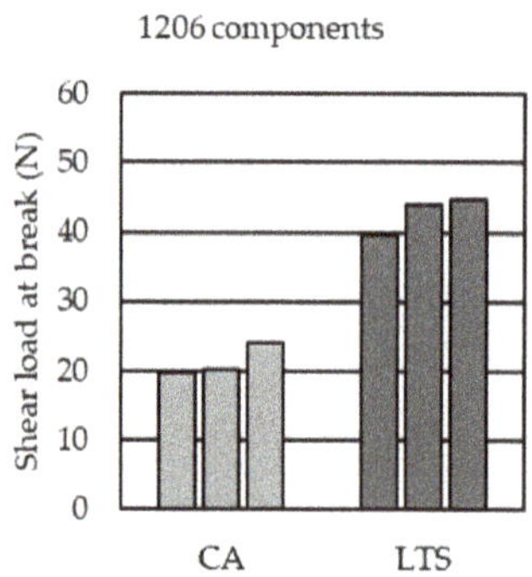

Figure 14. Shear load at break for 1206 components using CA or LTS to mount the copper pads.

Figure 15 shows the peel load vs. peel length plots of the cut 15 mm wide foil strips containing the already overmolded 1206 components, as described in Section 2.4 above.

Figure 15a shows a significant drop in the peel load whenever a component mounted using CA is reached. This was not apparent for LTS-mounted components (Figure 15b), where an initial increase in the peel curve is frequently observed. All peeled components remained in the PC substrate (Figure 15c). Compared to CA, LTS components can be exposed to higher load cases that resemble peel loads. Load cases like this might occur during overmolding when the flexible foil potentially bends during overmolding, as illustrated in Figure 15d.

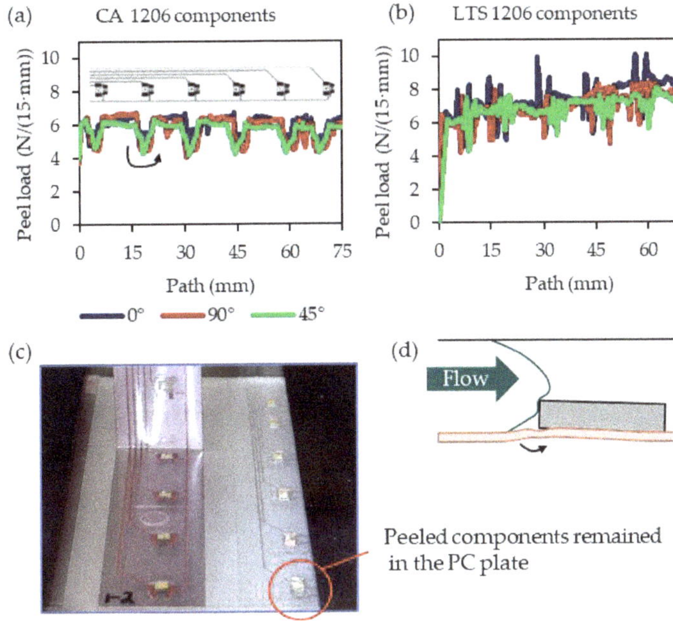

Figure 15. Peel force of the film strips cut from overmolded foils comprising components using low conductive adhesive (**a**) and low-temperature solder (**b**). All peeled components remained in the PC substrate (**c**). Similar load cases might occur during overmolding when the flexible foil potentially bends during overmolding (**d**).

4. Conclusions

The injection molding simulations—in which the force development on 1206-sized components during overmolding with PC in a plate mold was investigated—may be summarized as follows: The tangential forces F_t (acting in the "flow direction") on the components are only dominant in the initial stage of the overmolding. They are quickly surpassed in magnitude by the normal forces F_n (pressing the components to the mold wall). Hence, the potential detachment of components is expected to happen in the initial moments of the overmolding. This observation was also suggested in [25]. The force $F_{t=n}$, when $|F_t| = |F_n|$, seems to be almost independent of the component's orientation (0°, 45°, and 90°) or positioning within the mold (close to or far from the gate). The calculated shear angle θ indicated a push in the direction of the mold sidewalls for components placed at a distance to the center line of the mold caused by the parabolic polymer flow.

According to the simulated parameter study, the force on the components can be reduced by increasing the mold thickness, using a higher melt temperature, and using a slower injection speed. The calculated $F_{t=n}$ ranged from 1.34 N (S03) to 7.23 N (S06) for the 1206 components. Those simulated tangential forces are substantially lower than the experimentally obtained shear loads at breaks of ~22 N and ~43 N for the conductive adhesive (CA, Loctite Ablestik CE 3103WLV) and low-temperature solder (LTS, Interflux DP 5600), respectively. This is a discrepancy to the observation that most of the overmolded assembled foils have detached components. A possible explanation could be that the shear testing was performed at room temperature (following DIN EN 62137 1-2), while injection molding was conducted at elevated temperatures. Further, the flexible foil might cause a more peel-like load case during overmolding. Peel tests of foil strips with overmolded components indicated higher peel loads when LTS was used instead of CA.

The assembled foils overmolded in the 2 mm plate mold at low melt temperature (260 °C) showed more detached components compared to the assembled foils overmolded in the 3 mm plate mold at high melt temperatures (300 °C). This is an expected trend compared to the simulated forces ($F_{t=n}$). Seemingly, when a higher melt temperature is used, the viscosity is lower, and thus lower forces on the components outweigh the potential faster softening of the connections.

Further, no pattern for the detachments could be observed regarding the components' orientations (0°, 45°, or 90°) or distance from the gate. This matches the simulation results, which, as discussed, also do not show substantial differences in the calculated forces (regarding $F_{t=n}$).

The pressure gradients' contribution to the overall component's tangential force F_t is considerable. An examination of the shear stresses alone (as done in [10]) would therefore result in a significant underestimation of the prevailing component's tangential forces ($F_{t=n} = F_t \cong 2 \cdot F_t{}^*$).

Author Contributions: Conceptualization, M.H., M.B., M.P. and B.M.; methodology, M.H. and M.B.; software, M.H.; validation, M.H.; formal analysis, M.H. and M.B.; investigation, M.H. and M.B.; resources, M.H., M.B., J.G. and B.S.; data curation, M.H., M.B. and J.G.; writing—original draft preparation, M.H. and M.B.; writing—review and editing, M.H., M.B., J.G., M.P., J.V., F.B., B.M. and B.S.; visualization, M.H. and M.P.; supervision, J.G. and M.P.; project administration, J.G.; funding acquisition, J.V. and B.S. All authors have read and agreed to the published version of the manuscript.

Funding: This work was conducted by the Austrian COMET program within the project Smart@Surface No. 871495. The project received funding from BMVIT and BMDW, and the provinces of Styria (SFG), Tyrol (Standortagentur Tirol), and Burgenland.

Institutional Review Board Statement: Not applicable.

Data Availability Statement: Not applicable.

Acknowledgments: We especially thank Covestro AG for the overmolding material and the Polymer Competence Center Leoben GmbH (PCCL) for facilitating the shear tests.

Conflicts of Interest: The authors declare no conflict of interest. The funders had no role in the design of the study; in the collection, analyses, or interpretation of data; in the writing of the manuscript; or in the decision to publish the results.

References

1. Pötsch, G.; Michaeli, W. *Injection Molding: An Introduction*, 2nd ed.; Hanser: München, Germany; Cincinnati, OH, USA, 2008; ISBN 3446406352.
2. Singh, G.; Verma, A. A Brief Review on injection moulding manufacturing process. *Mater. Today Proc.* **2017**, *4*, 1423–1433. [CrossRef]
3. Bruk, S. Overmolding of Embedded Electronics. In *Connector Specifier*; PennWell Corporation: Saint Charles, IL, USA, 2001; p. 4.
4. Schirmer, J.; Reichenberger, M.; Wimmer, A.; Reichel, H.; Neermann, S.; Franke, J. Evaluation of Mechanical Stress on Electronic Assemblies during Thermoforming and Injection Molding for Conformable Electronics. In Proceedings of the 2021 14th International Congress Molded Interconnect Devices (MID), Amberg, Germany, 8–11 February 2021; pp. 1–8.
5. Bakr, M.; Su, Y.; Bossuyt, F.; Vanfleteren, J. Effect of Overmolding Process on the Integrity of Electronic Circuits. In Proceedings of the 2019 22nd European Microelectronics and Packaging Conference & Exhibition (EMPC), Technical Papers, Pisa, Italy, 16–19 September 2019; IEEE: Piscataway, NJ, USA, 2019; pp. 1–8, ISBN 978-0-9568086-6-0.
6. Ott, C.; Drummer, D. Low-stress over-molding of media-tight electronics using thermoplastic foam injection molding. *Polym. Eng. Sci.* **2021**, *61*, 1518–1528. [CrossRef]
7. Heinle, M.; Drummer, D. Measuring mechanical stresses on inserts during injection molding. *Adv. Mech. Eng.* **2015**, *7*, 168781401558447. [CrossRef]
8. Schirmer, J.; Reichenberger, M.; Neermann, S.; Franke, J. Conformable Electronics: Thermoforming and Injection Molding of Electronic Components. In Proceedings of the 2021 44th International Spring Seminar on Electronics Technology (ISSE), Bautzen, Germany, 5–9 May 2021; pp. 1–6, ISBN 2161-2536.
9. Bakr, M.; Hubmann, M.; Bossuyt, F.; Vanfleteren, J. A Study on Over-Molded Copper-Based Flexible Electronic Circuits. *Micromachines* **2022**, *13*, 1751. [CrossRef] [PubMed]
10. Wimmer, A.; Schirmer, J.; Reichel, H.; Reichenberger, M. Numerical and Experimental Analysis of High Pressure Forming and Injection Molding Processes with Assembled Film Electronics. In Proceedings of the 2021 14th International Congress Molded Interconnect Devices (MID) Scientific Proceedings, Virtual Event, Amberg, Germany, 8–11 February 2021; Kriebitzsch, I., Ed.; IEEE: Piscataway, NJ, USA, 2021; pp. 1–5, ISBN 978-1-7281-7509-6.

11. Wimmer, A. *Request on: Numerical and Experimental Analysis of High Pressure Forming and Injection Molding Processes with Assembled Film Electronics*; IEEE: Piscataway, NJ, USA, 2021.
12. Bakharev, A.; Fan, Z.; Costa, F.; Han, S.; Kennedy, P. Prediction of Core Shift Effects Using Mold Filling Simulation. In Proceedings of the ANTEC 2004, Chicago, IL, USA, 16–20 May 2004.
13. Chen, Y.-C.; Liao, Y.-J.; Tseng, S.-C.; Giacomin, A.J. Core deflection in injection molding. *J. Non-Newton. Fluid Mech.* **2011**, *166*, 908–914.
14. Hubmann, M.; Madadnia, B.; Groten, J.; Pletz, M.; Vanfleteren, J.; Stadlober, B.; Bossuyt, F.; Kaur, J.; Lucyshyn, T. Process Optimization of Injection Overmolding Structural Electronics with Regard to Film Distortion. *Polymers* **2022**, *14*, 5060. [CrossRef] [PubMed]
15. Autodesk Inc. Moldflow Insight Help | Synergy Application Programming Interface (API) | Autodesk. Available online: https://help.autodesk.com/view/MFIA/2021/ENU/?guid=GUID-853E86EF-7294-424D-81AE-E38130C237A4 (accessed on 9 February 2022).
16. Yageo. Data Sheet: General Purpose Chip Resistors. RC_L Series. Available online: https://www.yageo.com/upload/media/product/productsearch/datasheet/rchip/PYu-RC_Group_51_RoHS_L_11.pdf (accessed on 26 July 2022).
17. S.A. INTERFLUX®ELECTRONICS N.V. DP 5600. Ver: 4.1 28-08-20. Available online: https://cdn.interflux.com/documents/products/DP-5600/TD-DP-5600-EN.pdf (accessed on 1 March 2023).
18. Henkel Corporation. LOCTITE ABLESTIK CE 3103WLV: Technical Data Shee. Available online: https://tds.henkel.com/tds5/Studio/ShowPDF/?pid=ABLESTIK%20CE%203103WLV&format=MTR&subformat=HYS&language=EN&plant=WERCS&authorization=2 (accessed on 1 March 2023).
19. M-Base Engineering + Software GmbH. CAMPUSplastics | Datenblatt Makrolon®2805. Available online: https://www.campusplastics.com/campus/de/datasheet/Makrolon%C2%AE+2805/Covestro+Deutschland+AG/22/7541f4aa (accessed on 2 March 2023).
20. Autodesk Inc. 3D Flow Derivation | Moldflow Insight | Autodesk Knowledge Network. Available online: https://knowledge.autodesk.com/support/moldflow-insight/learn-explore/caas/CloudHelp/cloudhelp/2018/ENU/MoldflowInsight/files/GUID-C94D0362-421F-4EC1-89A6-931BB8D6D365-htm.html (accessed on 13 May 2022).
21. Accuratus Corporation. Aluminum Oxide | Al2O3 Material Properties. Available online: https://accuratus.com/alumox.html (accessed on 2 March 2023).
22. Autodesk Inc. Pressure Result | Moldflow Insight | Autodesk Knowledge Network. Available online: https://knowledge.autodesk.com/support/moldflow-insight/learn-explore/caas/CloudHelp/cloudhelp/2018/ENU/MoldflowInsight/files/GUID-9DF7DB37-2517-4177-A342-D16EC2B636DB-htm.html (accessed on 9 February 2022).
23. Autodesk Inc. Shear Stress at Wall Result | Moldflow Insight 2019 | Autodesk Knowledge Network. Available online: https://knowledge.autodesk.com/support/moldflow-insight/learn-explore/caas/CloudHelp/cloudhelp/2019/ENU/MoldflowInsight-Results/files/GUID-33F77167-AEAE-4E47-B7D8-8BDAFDF4EF37-htm.html (accessed on 9 February 2022).
24. Autodesk Inc. Velocity Result | Moldflow Insight | Autodesk Knowledge Network. Available online: https://knowledge.autodesk.com/support/moldflow-insight/learn-explore/caas/CloudHelp/cloudhelp/2018/ENU/MoldflowInsight-Results/files/GUID-764EE2D5-062F-4814-8C1D-2E4CC38DC2B6-htm.html (accessed on 9 February 2022).
25. IDTechEx Research Asks, In-Mold Electronics: Challenges in Every Step? [Online]. 21 August 2019. Available online: https://www.prnewswire.com/news-releases/idtechex-research-asks-in-mold-electronics-challenges-in-every-step-300905298.html (accessed on 15 July 2022).

micromachines

MDPI

Article

A Pressure and Temperature Dual-Parameter Sensor Based on a Composite Material for Electronic Wearable Devices

Zhidong Zhang [1,†], Huinan Zhang [1,†], Qingchao Zhang [1], Xiaolong Zhao [1], Bo Li [2], Junbin Zang [1], Xuefeng Zhao [3,*] and Tiansheng Zhang [4]

1 Key Laboratory of Instrumentation Science & Dynamic Measurement of Ministry of Education, North University of China, Taiyuan 030051, China; zdzhang@nuc.edu.cn (Z.Z.)
2 School of Software, North University of China, Taiyuan 030051, China
3 State Key Laboratory of ASIC and System, Shanghai Institute of Intelligent Electronics & Systems, School of Microelectronics, Fudan University, Shanghai 200433, China
4 Shanxi Hospital of Acupuncture and Moxibustion, Taiyuan 030006, China; zts910@163.com
* Correspondence: xuefengzhao@fudan.edu.cn
† These authors contributed equally to this work.

Abstract: Wearable sensors integrating multiple functionalities are highly desirable in artificial wearable devices, which are of great significance in the field of biomedical research and for human–computer interactions. However, it is still a great challenge to simultaneously perceive multiple external stimuli such as pressure and temperature with one single sensor. Combining the piezoresistive effect with the negative temperature coefficient of resistance, in this paper, we report on a pressure–temperature dual-parameter sensor composed of a polydimethylsiloxane film, carbon nanotube sponge, and poly(3,4-ethylenedioxythiophene)-poly(styrenesulfonate). The proposed multifunctional sensor can stably monitor pressure signals with a high sensitivity of 16 kPa^{-1}, has a range of up to 2.5 kPa, and also has a fast response time. Meanwhile, the sensor can also respond to temperature changes with an ultrahigh sensitivity rate of 0.84% $°C^{-1}$ in the range of 20 °C to 80 °C. To validate the applicability of our sensor in practical environments, we conducted real-scene tests, which revealed its capability for monitoring = human motion signals while simultaneously sensing external temperature stimuli, reflecting its great application prospects for electronic wearable devices.

Keywords: dual-parameter flexible sensor; PEDOT:PSS; CNTs; external stimuli response; wearable devices

Citation: Zhang, Z.; Zhang, H.; Zhang, Q.; Zhao, X.; Li, B.; Zang, J.; Zhao, X.; Zhang, T. A Pressure and Temperature Dual-Parameter Sensor Based on a Composite Material for Electronic Wearable Devices. *Micromachines* **2023**, *14*, 690. https://doi.org/10.3390/mi14030690

Academic Editors: Libo Gao and Zhuoqing Yang

Received: 15 February 2023
Revised: 1 March 2023
Accepted: 6 March 2023
Published: 21 March 2023

1. Introduction

With the ever-increasing demand for electronic wearable devices, the development of multifunctional tactile sensors has shown significant progress [1–4]. Likewise, many scientific researchers have reported on flexible tactile sensors applied in human motion monitoring, pulse detection, and temperature stimuli perception. The goal of these multifunctional tactile sensors is to accurately identify different external stimuli, and then respond to them with high sensitivity and in a timely manner. Over the years, various well-performing single-function pressure sensors have been reported [4–10], which commonly employ piezoresistive materials whose working mechanism depends on variations in the geometric structure, tunneling resistance, or contact resistance, such as carbon nanotubes [11], carbon black nanoparticles [12], and MXene [13,14]. Nevertheless, as the key for efficient and compact electronic wearable devices, the realization of a multifunctional tactile sensor requires the integration of different types of materials with different physical properties, thereby enabling the ability to simultaneously respond to various stimuli such as pressure and temperature [15,16].

Carbon nanotube sponge, as a high-strain material with a hollow architecture, is an excellent choice for the fabrication of pressure sensors. The typical structural hierarchy enables the CNT sponge to keep its shape, without requiring the support of substrates or electrodes [11,12,17]. Moreover, the microstructure of the CNT sponge consists of CNTs with a large number of micropores uniformly distributed within. This unique internal distribution results in low density and thermal conductivity. Furthermore, poly(3,4-ethylenedioxythiophene)-poly(styrenesulfonate) is an excellent temperature-sensitive material [18–23]. Its advantages are its high conductivity and strong stability, making it a common material for fabricating flexible temperature sensors. However, previous studies have focused on the application of PEDOT:PSS for improved conductivity, while ignoring its own temperature coefficient of resistance. Additionally, by adding certain organic solvents or an organic polyhydroxy, such as DMSO (dimethyl sulfoxide), ethylene glycol, or glycerol, the electrical conductivity of a polymer can be increased dozens or even hundreds of times over. Additionally, the high adhesion strength of PEDOT:PSS allows it to easily form a film on the surface of the CNT sponge.

In this work, we fabricated a multifunctional sensor by first drying the CNT sponge soaked in PEDOT:PSS in the oven. By doing so, the PEDOT:PSS formed a film on the surface of the CNT sponge without destroying its internal structure. The PEDOT:PSS-coated CNT sponge was then sandwiched between two polydimethylsiloxane films to complete the fabrication process. The fabricated pressure–temperature dual-parameter sensor had high sensitivity rates of 16 kPa^{-1} in the low-pressure range and 2.5 kPa^{-1} in the range of 2 kPa to 10 kPa. Even applied to ultrahigh pressure (40 kPa), our sensor showed a sensitivity rate of 0.32 kPa^{-1}. Within a temperature range of 20 °C to 80 °C, the sensor had an ultrahigh sensitivity rate of 0.84% $°C^{-1}$.

2. Experimental Section

2.1. Materials

The PDMS (Sylgard 184 silicone elastomer base and related curing agents) was purchased from Dow Corning Co., Ltd., Midland, MI, USA. The CNT sponge was purchased from Nanjing XFNANO Materials Tech Co., Ltd., China. The PEDOT:PSS was purchased from Shanghai Aladdin Bio-Chem Tech Co., Ltd., China. All materials were used without any further purification.

Parameters:

CNT sponge: Inner diameter: 10–20 nm; outer diameter: 30–50 nm; porosity: 99%;

PEDOT:PSS: Viscosity: 6.0000 mPa.S; solid content: 1.40000%; PH(20 °C): 1.9900.

2.2. Fabrication of PDMS Flexible Substrate

To obtain a flexible PDMS substrate with a thickness of about 1mm, the main and curing agents of the PDMS were mixed at a mass ratio of 10:1 in a plastic cup. Next, the compound reagent was refrigerated for 1 h to remove any bubbles in it. A four-inch silicon wafer was then cleaned with alcohol and dried in the oven. Afterwards, the PDMS mixture was spin-coated onto the silicon wafer at a speed of 300 r/min for 120 s. Then, the wafer was placed on a hot plate at 100 °C for 1 h, where the PDMS flexible substrate was left uncovered. After cooling, the PDMS film was peeled off and used as the flexible substrate.

2.3. Fabrication of CNT Sponge Coated with PEDOT:PSS Film

A 0.5 cm × 0.5 cm CNT sponge was cut and soaked in the PEDOT:PSS for 1 h. The CNT sponge was placed on a clean wafer and dried in the oven for half an hour at 80 °C, after which it was cooled down to complete the fabrication.

2.4. Preparation of the Sensor

One 3 cm × 1 cm PDMS film was cut as the bottom substrate and another 0.5 cm × 0.5 cm film was cut as the top substrate. One 1.5 cm × 0.8 cm electrode was used to receive the current signals. The CNT sponge coated with PEDOT:PSS was laid flat over the interdigital electrode to ensure a close contact, and then the whole structure was sandwiched by two PDMS films. Then, a conductive silver paste with high conductivity was used to connect the pins of interdigital electrodes with the wires for measurements. After drying the silver paste for 1 h at 80 °C in the oven, the preparation of the multifunctional sensor was completed.

2.5. Characterizations and Measurements

The surface and cross-sectional images of the PDMS film, CNT sponge, and CNT sponge wrapped by PEDOT:PSS were taken using field emission scanning electron microscopy (SEM SUPRA-55 purchased from Shanghai Opton Co., Ltd., China). Moreover, the qualitative analysis of the surface element distribution of the CNT sponge coated with PEDOT:PSS was performed via energy-dispersive spectrometry (EDS). Meanwhile, a Kethley 2450 controlled digital source meter was used to measure the real-time I-T and I–V curves.

3. Results and Discussions

3.1. Fabrication and Characterization

As shown in Figure 1a, the complete fabrication process can be mainly divided into two parts. One part involves the preparation of PDMS film, which is a common substrate for flexible sensors. The PDMS mixture was spin-coated onto a silicon wafer uniformly, and after its dried, the PDMS film was peeled off and used as the flexible substrate. The preparation of the composite sponge was relatively simple. After soaking the CNT sponge in PEDOT:PSS solution, a thin film was formed on its surface. The sponge was then dried in the oven while keeping it uncovered. The PEDOT:PSS-coated sponge was sandwiched by two PDMS films, and the composite CNT sponge was closely attached to the interdigital electrode for better conductivity. As shown in Figure 1b,c, the sensor consists of four parts, which from bottom to top are the PDMS film, interdigital electrode, composite CNT sponge, and PDMS film.

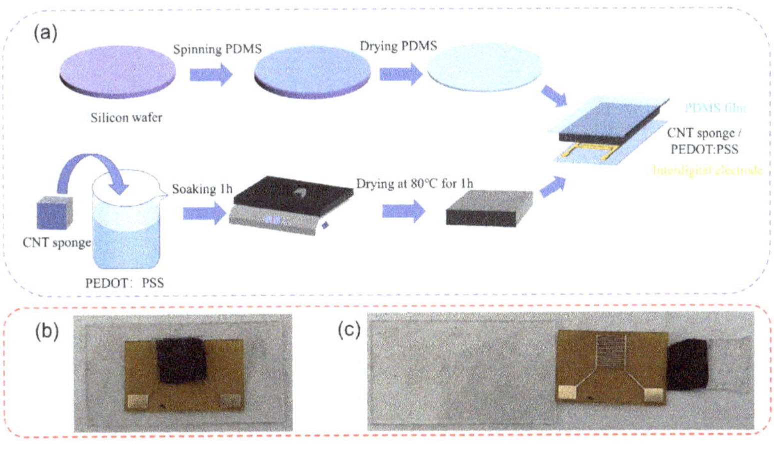

Figure 1. (**a**) Fabrication process of the pressure–temperature dual-parameter sensor. (**b**) Real image of the sensor. (**c**) Real image of each part of the sensor.

PDMS is a commonly used material for the preparation of flexible sensors due to its high dielectric constant and tunable elasticity [17,24–30]. Additionally, its stress–strain relationship can be adjusted by varying the mixing ratio of the main PDMS agent and curing agent in the fabrication. As shown in Figure 2a,d, the cured PDMS film has a flat surface and a thickness of 355.3 um.

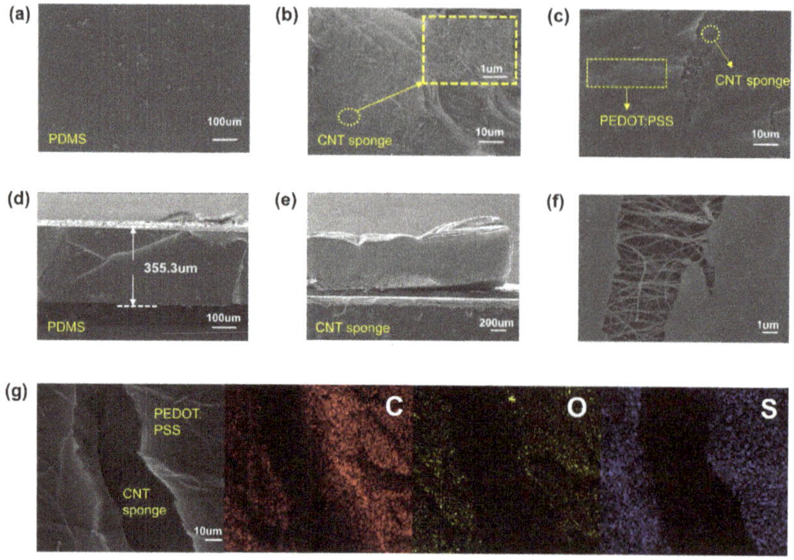

Figure 2. SEM images of the materials used in this work: (**a**) SEM image of the surface of the PDMS substrate; (**b**) SEM image of the surface of the CNT sponge; (**c**) SEM image of the surface of the CNT sponge wrapped in PEDOT:PSS; (**d**) measured thickness of the PDMS substrate; (**e**) measured thickness of the CNT sponge; (**f**) SEM image of the surface of the CNT sponge wrapped in PEDOT:PSS at higher resolution; (**g**) energy-dispersive spectroscopy (EDS) mapping images of the CNT sponge wrapped in PEDOT:PSS.

The CNT sponge is a self-assembled and interconnected conductive material with a sponge-like shape. Due to its light weight and excellent flexibility, they are commonly used as the dielectric layer in the sensor. As shown in Figure 2b, the CNT sponge consists of many carbon nanotubes and has a porosity of up to 99%. Despite the excellent thermal conductivity of single carbon nanotubes, the low density and high thermal resistance at the CNT junctions mean CNT sponge has extremely low thermal conductivity, nearly unaffected by temperature in pressure sensing [27]. Moreover, Figure 2e indicates that the CNT sponge used in this experiment has a thickness of about 1 mm. Progressively, PEDOT:PSS has received a great deal of attention owing to its unique properties, including its transparency, relatively high conductivity, and simple fabrication process. Over recent years, many research studies have focused on its thermoelectric properties and characteristics, such as its electrodes, while its thermistor-type behavior has been less explored [26]. Herein, we fabricate a composite sponge by soaking the CNT sponge in PEDOT:PSS. It can be seen from Figure 2c,f that the PEDOT:PSS is coated on the CNT sponge as an ultrathin film. Notably, very few PEDOT:PSS molecules get into the CNT sponge and most are distributed on the sponge surface. To verify this, we took an EDS mapping image (Figure 2g), where it can be seen that the C and O elements are distributed throughout the crack in the CNT sponge coated with PEDOT:PSS, but there is almost no S element (according to the molecular structure of PEDOT:PSS, the S element is unique compared to the CNT sponge).

The IDE structure is used because of its own good flexibility, bending resistance, and conformal ability. Using the interdigital electrode as the signal response channel of the sensor, the flexible substrate, MXene film, and interdigital electrode can be closely adhered to accurately measure the pressure signal generated by the human body or other carriers. Moreover, the interdigital electrode has the characteristics of multiple conductive channels, so that the breakage or warping of an electrode will not affect the overall output performance of the sensor, thereby greatly improving the stability and reliability of the sensor.

3.2. Pressure Sensing Performance of CNT Sponge

The CNT sponge is a widely used piezoresistive material that has excellent conductivity. To verify its piezoresistive properties, the interdigital electrodes are used to receive the electrical signals and a digital SourceMeter is used to monitor the electrical responses under different pressure values. As shown in Figure 3a, the I-T curves at pressures less than 80 kPa indicate that the current increases with the pressure, reflecting a negative piezoresistive nature. When the deformation occurs, the internal structure of the CNT sponge is compressed, resulting in smaller pores and the exclusion of air. The carbon nanotubes approach each other under pressure and the resistance becomes smaller. The linear relationship between the current and voltage shown in Figure 3b indicate that an ohmic contact forms between the interdigital electrode and the CNT sponge. However, when the applied pressure becomes higher, the amplitude of the current increase starts to decrease. Moreover, as shown in Figure 3c, the CNT sponge exhibits different sensitivity levels in different pressure regions. This is because at low pressure, the amount of contact in the cross-linked network structure inside the sponge is less and the whole structure is relatively loose. When a small amount on pressure is applied to the sponge, the microstructures in the internal cross-linked network structure start to contact each other. Hence, the conductive path expands rapidly, and the resistance change rate increases greatly. In contrast, when the applied pressure is further increased to a larger value, the growth rate of the resistance change becomes smaller, since the cross-linked network structure inside the conductive sponge has already been fully contacted and the conductive path has stabilized. When continuous pressure is applied at an interval of 10 s, the CNT sponge shows good gradient changes, as shown in Figure 3d. Additionally, the CNT sponge has a fast response time of 63 ms, while the recovery time is 71 ms, as shown in Figure 3e. Recently, different piezoresistive materials have been reported, and compared to them [21–25], the sensor from this work can better monitor the pressure and temperature, simultaneously, with high sensitivity.

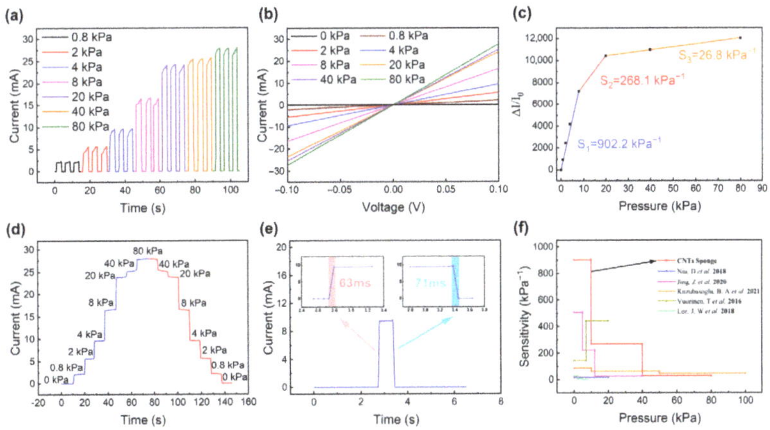

Figure 3. The pressure sensing properties of the CNT sponge: (**a**) the I-T curves at pressures less than 80 kPa; (**b**) the linear relationship of the I–V curves at voltages from −0.1 V to 0.1 V; (**c**) the sensitivity of the CNT sponge, reaching 902.2 kPa^{-1} below 10 kPa, 268.1 kPa^{-1} in the transitional pressure region (10 kPa to 20 kPa), and 26.8 kPa^{-1} in the high pressure region (20 kPa to 80 kPa); (**d**) gradient of the rise and fall of the current in the pressure region from 0 kPa to 80 kPa; (**e**) the response time of the CNT sponge for loading and unloading pressure; (**f**) the performance comparison with other pressure-sensitive materials [21–25].

The pressure sensitivity is determined by the variation of the current and applied pressure. The specific calculation formula is as follows:

$$S = (\frac{I_p - I_0}{I_0})/P \quad (1)$$

where, S (kPa^{-1}) is the sensitivity; I_0 (mA) is the measured current without pressure; I_p (mA) is the measured current when pressure is applied; P (kPa) is the value of applied pressure.

3.3. Temperature-Sensing Performance of CNT Sponge Wrapped by PEDOT:PSS

There have been two types of carbon-based temperature sensors reported recently: positive temperature coefficient (PTC) and negative temperature coefficient (NTC) sensors. PEDOT:PSS is a typical organic semiconductor material with NTC-type temperature-sensitive characteristics, whose sensing mechanism is based on the semiconductor properties of the material. When the ambient temperature is low, the number of carriers in the NTC-sensitive material is relatively small, resulting in higher resistance. However, as the temperature increases, the heat energy increases, thereby activating more carriers and reducing the resistance. To verify this property, we carried out a series of tests. As shown in Figure 4a, the linear relationship of the I–V curves indicates that the PEDOT:PSS film has a close contact with the interdigital electrode. Meanwhile, the rising trend of the current with increasing temperature indicates a negative temperature coefficient of resistance. Additionally, when the temperature rises, the current shows an upward gradient trend, as shown in Figure 4b. Furthermore, as evident from Figure 4c, in the temperature range of 20 °C to 80 °C, the resistance drops by about 210 ohms and the resistance change is about 60%, yielding a sensitivity of 0.84% $°C^{-1}$. In addition, as the temperature rises by about 1 °C, the sensor responds in about 1.1 s (Figure 4d). However, when the temperature drops by 1 °C, the response time of the sensor is larger, around 1.5 s (Figure 4e). In recent years, various thermosensitive materials have been reported, such as PEDOT:PSS/CuPc, PEDOT:PSS/IPA, and PEDOT:PSS/PDMS. Compared to these reported materials [13,14,17,20,26,29–32], the material used in our work offers a wider temperature range and a higher sensitivity level, as shown in Figure 4f.

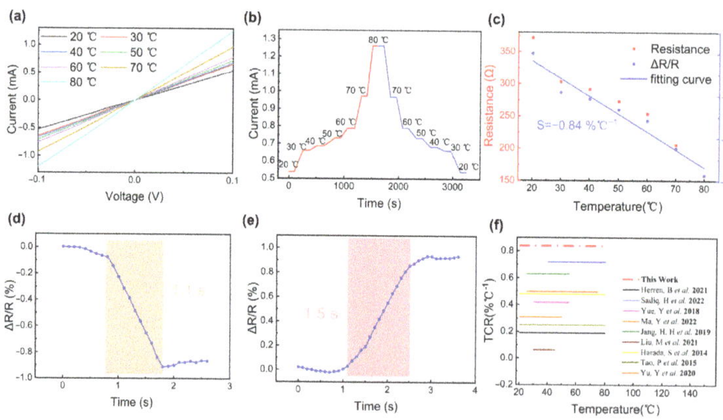

Figure 4. The temperature–sensing properties of the CNT sponge wrapped in PEDOT:PSS: (**a**) the linear relationship of the I–V curves at voltages from −0.1 V to 0.1 V; (**b**) the gradient of the current increase and decrease when under a continuous temperature change; (**c**) the resistance and resistance change rate of the device over a temperature range from 20 °C to 80 °C, exhibiting a sensitivity of 0.84% $°C^{-1}$; (**d**) the response time of the composite sponge when the temperature rises; (**e**) the response time of the composite sponge when temperature drops; (**f**) comparison with other studies [13,14,17,20,26,29–32].

The temperature sensitivity is determined by the variation of resistance and the applied pressure. The specific calculation formula is as follows:

$$S = \left(\frac{R_T - R_0}{R_0}\right) / \Delta T \tag{2}$$

where, S (% °C^{-1}) is the sensitivity; R_0 (Ω) is the initial resistance under 20 °C; R_T (Ω) is the tested resistance when temperature changes; ΔT (°C) is the value of the temperature variation.

3.4. Sensing Performance of the Sensor When Pressure and Temperature Are Applied Simultaneously

To verify the capability of the device to simultaneously handle both pressure and temperature stimuli, we used a digital source meter to detect the I-T curves in a temperature range of 20 °C to 80 °C at 0.8 kPa, and also in a pressure range of 0 kPa to 40 kPa at 30 °C. As shown in Figure 5a, the overall current change shows an increasing trend with the temperature, and the increase rate is small. However, a larger increase rate can be observed with the pressure, as shown in Figure 5b. This is because of the difference in thermal resistance and piezoresistive mechanisms [33], which have different degrees of influence on the resistance. Likewise, the same trend is also reflected in the sensitivity curves. As indicated in Figure 5c,d, the pressure and temperature sensitivity curves both tend to decrease, where the temperature sensitivity curves drop with a faster rate. The reason behind this could be as follows. As shown in Figures 3b and 4a, the device deformation led by pressure has a greater effect on the change in current than the temperature. Thus, when testing the temperature sensitivity curves under a fixed pressure, the current change induced by the pressure is larger, thereby resulting in a stronger decreasing trend. Meanwhile, when testing the pressure sensitivity curves at a fixed temperature, the current change caused by the temperature is smaller, thereby resulting in a weaker decreasing trend. In general, the device has excellent pressure and temperature sensitivity, and is suitable for multifunctional monitoring. Compared with the previous work, the sensor we prepared performs well in the pressure-sensing range, temperature-sensing range, and pressure response time (shown in Table 1). In the next experiment, we plan to complete the decoupling of the pressure and temperature.

Table 1. A comparison with the previous work [34–37].

Materials	Working Range (kPa)	Response Time (ms)	Temperature Range (K)	Independent of P and T Signals	Ref.
PVDF/PANI/PDMS	10	-	0–100	Yes	1
Silk fiber-AgNW & CNTs/ionic liquid	2	250	30–65	Yes	2
Ionnic hydrogel	1	500	30–55	No	3
Metal-organic porous carbon/PDMS	2	60	20–100	No	4
CNTs sponge/PECOT:PSS/PDMS	**40**	**63**	**20–80**	**No**	**This work**

Nowadays, wearable electronic devices have broad application prospects in many fields, the key to which is to meet the requirements for the durability and stability of the sensor. We carried out a series of experiments to verify this. The three consecutive pressure increases and decreases prove that the sensor has a small hysteresis range, as shown in Figure 6a. As shown in Figure 6b,c, we tested the pressure and temperature-sensing performance of the sensor for five consecutive days. The results exhibit good stability. Additionally, with the help of the pressing machine, we performed 500 cycles of repeated experiments of pressure loading and release, the results of which are shown in Figure 6d.

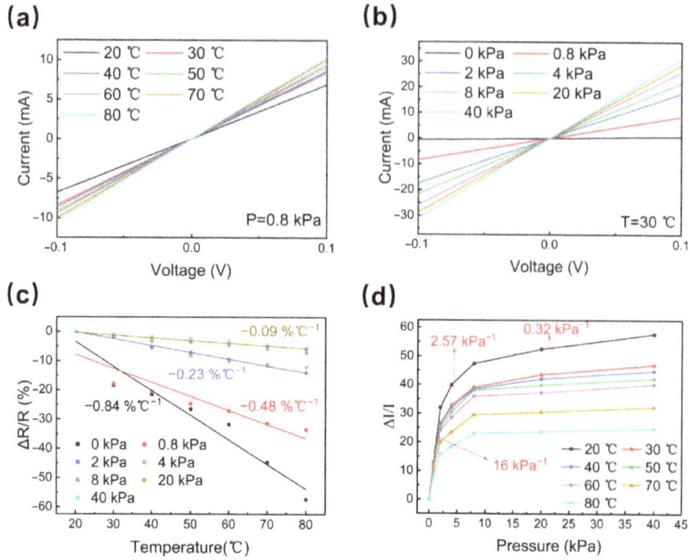

Figure 5. Performance of pressure–temperature co-detection: (**a**) I–V curves in the temperature range of 20–80 °C at 0.8 kPa; (**b**) I–V curves in the pressure range of 0 kPa to 40 kPa at 30 °C; (**c**) resistance change rates and temperature sensitivity curves under different pressures in the range of 0 kPa to 40 kPa; (**d**) current change rates and pressure sensitivity curves at different temperature values in the range of 20–80°C.

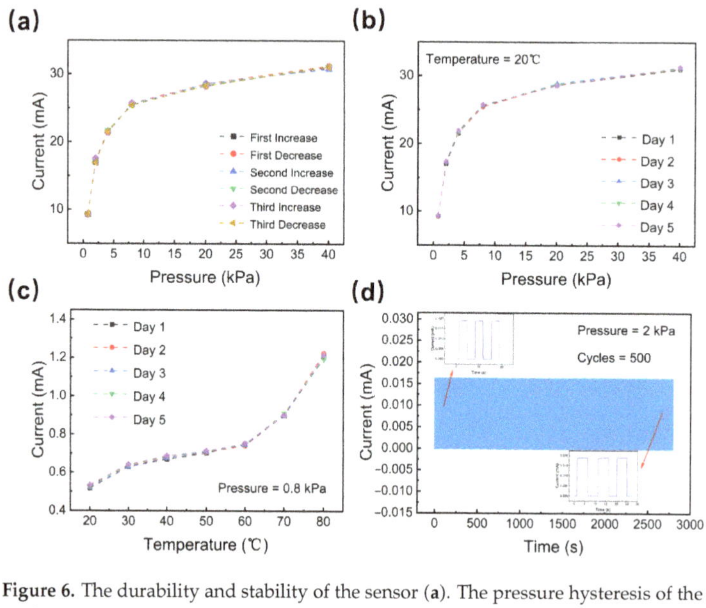

Figure 6. The durability and stability of the sensor (**a**). The pressure hysteresis of the sensor (**b**) and (**c**) the stability of the sensor. (**d**) The durability of the sensor.

3.5. Practical Applications of the Multifunctional Sensor

Ideally, a pressure–temperature multifunctional sensor should have the ability to monitor these two external stimuli simultaneously [38,39]. To verify this property, we conducted experiments involving external contact to the sensor and human motion monitoring. As shown in Figure 7a, we applied a large pressure to the sensor three times through the tip of a pen, and the sensor responded to the stimulus precisely and stably in all three attempts. In Figure 7b, we demonstrate the temperature detection capability of the sensor in response to an approaching finger. It should be noted that because of the PDMS film's thermal insulation, it took about 10 s for the sensor to respond. In order to analyze the pressure–temperature co-interaction of the sensor, we chose two glasses of water of the same mass but at different temperatures for the experiment. As shown in Figure 7c, the increase in temperature resulted in an obvious increase in the current. Meanwhile, some experiments related to human motion monitoring were also carried out. As shown in Figure 7d–f, the I-T curves corresponding to finger bending, knee bending, and throat swallowing experiments all exhibit a continuous and stable signal. Essentially, good monitoring performance for pressure and temperature is the key to achieving efficient wearable devices. As shown in Figure 7g,h, we attached the sensors to the fingers and measured the current of the two sensors when touching the water at different temperatures. The larger current induced when touching the water with higher temperatures indicated that the sensor can precisely respond to a change in temperature. Additionally, when holding the beaker with 80 °C water, the thumb sensor showed a larger current than the rest because of the difference in pressure, which in turn reflects the sensitivity of the sensor to pressure change (shown in Figure 7i). As evident from these results, this multifunctional sensor based on composite materials has great application prospects in wearable devices.

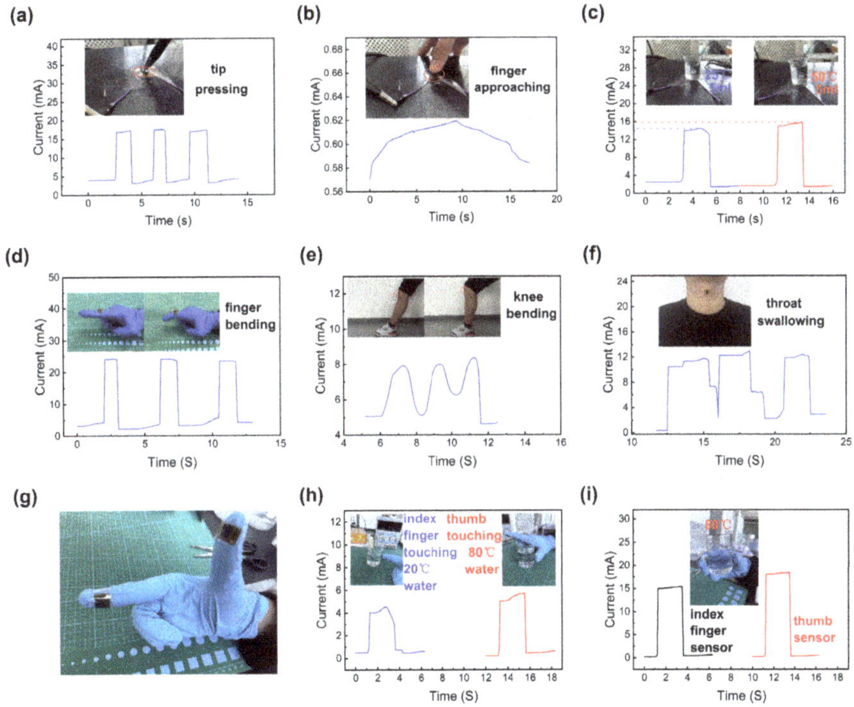

Figure 7. Real-scene tests of the sensor: (**a**) I-T curve when pressing the sensor with a pen tip; (**b**) I-T curve when finger approaches the sensor; (**c**) I-T curve when placing the sensor on a 5 mL glass of 25 °C water and 5 mL of 50 °C water; (**d**) I-T curve when the finger bends; (**e**) I-T curve when the knee bends; (**f**) I-T curve while swallowing; (**g**) the wearable sensors; (**h**) I-T curves of the two sensors when touching the water at different temperatures; (**i**) I-T curves of the two sensors when holding a beaker with 80 °C water.

4. Conclusions

In this work, we reported on a pressure–temperature multifunctional sensor with high sensitivity and a fast response time. The sensor is mainly composed of CNT sponge wrapped by PEDOT:PSS and interdigital electrodes, and sandwiched between two ultrathin PDMS films. In real-scene tests, this sensor showed excellent performance, i.e., a high sensitivity level of 16 kPa^{-1} in the low-pressure region and an ultrahigh sensitivity rate of 0.84% $°C^{-1}$ in the temperature range of 20 °C to 80 °C. Meanwhile, it also exhibited a fast response time to both pressure (63 ms) and temperature (1.1 s) stimuli compared to other studies. Additionally, we also conducted experiments with external stimuli and involving human motion monitoring to verify the applicability of this sensor for use in wearable devices. Owing to the good piezoresistive and thermoresistance effects of the materials, coupled with our unique fabrication scheme, this multifunctional sensor can play a pivotal role in future wearable devices.

Author Contributions: Conceptualization, Z.Z. and H.Z.; methodology, H.Z.; software, Q.Z.; validation, X.Z. (Xuefeng Zhao), T.Z. and H.Z.; formal analysis, H.Z.; investigation, J.Z.; resources, B.L.; data curation, X.Z. (Xiaolong Zhao); writing—original draft preparation, Z.Z.; writing—review and editing, H.Z.; visualization, Q.Z.; supervision, X.Z. (Xuefeng Zhao); project administration, B.L.; funding acquisition, Z.Z. All authors have read and agreed to the published version of the manuscript.

Funding: This research was funded by [the Key Research and Development Program of Shanxi Province] grant number [202102130501011], [Major Scientific and Technological Special Projects of Key Medical Research Projects of Shanxi Province] grant number [2021XM05], [the Fund for the Shanxi "1331 Project" Key Subject Construction] grant number [1331KSC], and [China Postdoctoral Science Foundation] grant number [2022M710722].

Data Availability Statement: The data that support the findings of this study are available upon reasonable request from the authors.

Conflicts of Interest: The authors declare no conflict of interest.

References

1. Lee, W.S.; Jeon, S.; Oh, S.J. Wearable sensors based on colloidal nanocrystals. *Nano Converg.* **2019**, *6*, 10. [CrossRef] [PubMed]
2. Wang, Y.; Yang, B.; Hua, Z.; Zhang, J.; Guo, P.; Hao, D.; Gao, Y.; Huang, J. Recent advancements in flexible and wearable sensors for biomedical and healthcare applications. *J. Phys. D Appl. Phys.* **2021**, *55*, 134001. [CrossRef]
3. Jayathilaka, W.; Qi, K.; Qin, Y.; Chinnappan, A.; Serrano-García, W.; Baskar, C.; Wang, H.; He, J.; Cui, S.; Thomas, S.W.; et al. Significance of Nanomaterials in Wearables: A Review on Wearable Actuators and Sensors. *Adv. Mater.* **2019**, *31*, 1805921. [CrossRef]
4. Wang, X.; Yu, J.; Cui, Y.; Li, W. Research progress of flexible wearable pressure sensors. *Sen. Actuators A-Phys.* **2021**, *330*, 112838. [CrossRef]
5. Lü, X.; Meng, X.; Shi, Y.; Tang, H.; Bao, W. Multilayer microstructured high-sensitive ultrawide-range flexible pressure sensor with modulus gradient. *IEEE Trans. Electron. Devices* **2022**, *69*, 2030–2037. [CrossRef]
6. Shi, Y.; Lü, X.; Wang, W.; Meng, X.; Zhao, J.; Wang, P.; Bai, G. Multilayer flexible pressure sensor with high sensitivity over wide linearity detection range (August 2021). *IEEE Trans. Instrum. Meas.* **2021**, *70*, 1–9. [CrossRef]
7. He, Y.; Zhou, M.; Mahmoud, M.H.H.; Lu, X.; He, G.; Zhang, L.; Huang, M.; Elnaggar, A.Y.; Lei, Q.; Liu, H.; et al. Multifunctional wearable strain/pressure sensor based on conductive carbon nanotubes/silk nonwoven fabric with high durability and low detection limit. *Adv. Compos. Hybrid Mater.* **2022**, *5*, 1939–1950. [CrossRef]
8. He, Y.; Wu, D.; Zhou, M.; Zheng, Y.; Wang, T.; Lu, C.; Zhang, L.; Liu, H.; Liu, C. Wearable Strain Sensors Based on a Porous Polydimethylsiloxane Hybrid with Carbon Nanotubes and Graphene. *ACS Appl. Mater. Interfaces* **2021**, *13*, 15572–15583. [CrossRef]
9. Xu, F.; Li, X.; Shi, Y.; Li, L.; Wang, W.; He, L.; Liu, R. Recent Developments for Flexible Pressure Sensors: A Review. *Micromachines* **2018**, *9*, 580. [CrossRef]
10. Chen, W.; Yan, X. Progress in achieving high-performance piezoresistive and capacitive flexible pressure sensors: A review. *J. Mater. Sci. Technol.* **2020**, *43*, 175. [CrossRef]
11. Zhao, W.; Li, T.; Li, Y.; O'Brien, D.; Terrones, M.; Wei, B.; Suhr, J.; Lu, X. Mechanical properties of nanocomposites reinforced by carbon nanotube sponges. *J. Mater.* **2018**, *4*, 157. [CrossRef]

12. Chen, Y.; Zhang, H.; Yang, Y.; Wang, M.; Cao, A.; Yu, Z. High-Performance Epoxy Nanocomposites Reinforced with Three-Dimensional Carbon Nanotube Sponge for Electromagnetic Interference Shielding. *Adv. Funct. Mater.* **2016**, *26*, 447. [CrossRef]
13. Yue, Y.; Liu, N.; Liu, W.; Li, M.; Ma, Y.; Luo, C.; Wang, S.; Rao, J.; Hu, X.; Su, J.; et al. 3D hybrid porous Mxene-sponge network and its application in piezoresistive sensor. *Nano Energy* **2018**, *50*, 79. [CrossRef]
14. Ma, Y.; Cheng, Y.; Wang, J.; Fu, S.; Zhou, M.; Yang, Y.; Li, B.; Zhang, X.; Nan, C. Flexible and highly-sensitive pressure sensor based on controllably oxidized MXene. *Infomat* **2022**, *4*, e12328. [CrossRef]
15. Liu, R.; He, L.; Cao, M.; Sun, Z.; Zhu, R.; Li, Y. Flexible Temperature Sensors. *Front. Chem.* **2021**, *9*, 539678. [CrossRef]
16. Su, Y.; Ma, C.; Chen, J.; Wu, H.; Luo, W.; Peng, Y.; Luo, Z.; Li, L.; Tan, Y.; Omisore, O.M.; et al. Printable, Highly Sensitive Flexible Temperature Sensors for Human Body Temperature Monitoring: A Review. *Nanoscale Res. Lett.* **2020**, *15*, 200. [CrossRef]
17. Herren, B.; Webster, V.; Davidson, E.; Saha, M.C.; Altan, M.C.; Liu, Y. PDMS Sponges with Embedded Carbon Nanotubes as Piezoresistive Sensors for Human Motion Detection. *Nanomaterials* **2021**, *11*, 1740. [CrossRef]
18. Li, J.; Zhou, G.; Hang, Y.; Chen, C.; He, W.; Wang, S.; Chen, Y.; Wang, C.; Sun, Y.; Wong, C. A Catalytic and Interfacing PEDOT:PSS/CuPc Polymerized on Cloth Fiber to Electro-Metalize Stretchable Copper Conductive Pattern. *Adv. Mater. Interfaces* **2022**, *9*, 2101462. [CrossRef]
19. Liu, Z.; Wu, L.; Qian, J.; Peng, J.; Liu, R.; Xu, Y.; Shi, X.; Qi, C.; Ye, S. Tuned Transport Behavior of the IPA-Treated PEDOT:PSS Flexible Temperature Sensor via Screen Printing. *J. Electron. Mater.* **2021**, *50*, 2356. [CrossRef]
20. Yu, Y.; Peng, S.; Blanloeuil, P.; Wu, S.; Wang, C.H. Wearable Temperature Sensors with Enhanced Sensitivity by Engineering Microcrack Morphology in PEDOT:PSS–PDMS Sensors. *ACS Appl. Mater Interfaces* **2020**, *12*, 36578. [CrossRef]
21. Kuzubasoglu, B.A.; Sayar, E.; Bahadir, S.K. Inkjet-printed CNT/PEDOT:PSS temperature sensor on a textile substrate for wearable intelligent systems. *IEEE Sens. J.* **2021**, *21*, 13090. [CrossRef]
22. Vuorinen, T.; Niittynen, J.; Kankkunen, T.; Kraft, T.M.; Mantysalo, M. Inkjet-Printed Graphene/PEDOT:PSS Temperature Sensors on a Skin-Conformable Polyurethane Substrate. *Sci. Rep.* **2016**, *6*, 35289. [CrossRef] [PubMed]
23. Lee, J.W.; Han, D.C.; Shin, H.J.; Yeom, S.H.; Ju, B.K.; Lee, W. PEDOT:PSS-Based Temperature-Detection Thread for Wearable Devices. *Sensors* **2018**, *18*, 2996. [CrossRef] [PubMed]
24. Niu, D.; Jiang, W.; Ye, G.; Wang, K.; Yin, L.; Shi, Y.; Chen, B.; Luo, F.; Liu, H. Graphene-elastomer nanocomposites based flexible piezoresistive sensors for strain and pressure detection. *Mater. Res. Bull.* **2018**, *102*, 92. [CrossRef]
25. Jing, Z.; Zhang, Q.; Cheng, Y.; Ji, C.; Zhao, D.; Liu, Y.; Jia, W.; Pan, S.; Sang, S. Highly sensitive, reliable and flexible piezoresistive pressure sensors based on graphene-PDMS @ sponge. *J. Micromech. Microeng.* **2020**, *30*, 085012. [CrossRef]
26. Jang, H.H.; Park, J.S.; Choi, B. Flexible piezoresistive pulse sensor using biomimetic PDMS mold replicated negatively from shark skin and PEDOT:PSS thin film. *Sen. Actuators A-Phys.* **2019**, *286*, 107. [CrossRef]
27. Zhang, F.; Feng, Y.; Qin, M.; Ji, T.; Lv, F.; Li, Z.; Gao, L.; Long, P.; Zhao, F.; Feng, W. Stress-sensitive thermally conductive elastic nanocomposite based on interconnected graphite-welded carbon nanotube sponges. *Carbon* **2019**, *145*, 378–388. [CrossRef]
28. Tao, L.; Zhang, K.; Tian, H.; Liu, Y.; Wang, D.; Chen, Y.; Yang, Y.; Ren, T. Graphene-Paper Pressure Sensor for Detecting Human Motions. *ACS Nano* **2017**, *11*, 8790. [CrossRef]
29. Liu, M.; Hang, C.; Zhao, X.; Zhu, L.; Ma, R.; Wang, J.; Lu, H.; Zhang, D. Advance on flexible pressure sensors based on metal and carbonaceous nanomaterial. *Nano Energy* **2021**, *87*, 106181. [CrossRef]
30. Sadiq, H.; Hui, H.; Huang, S.; Mahmood, K.; Sharif, M.H.; Ali, I. A Flexible Pressure Sensor Based on PDMS-CNTs Film for Multiple Applications. *IEEE. Sens. J.* **2022**, *22*, 3033. [CrossRef]
31. Harada, S.; Honda, W.; Arie, T.; Akita, S.; Takei, K.J.A.N. Fully Printed, Highly Sensitive Multifunctional Artificial Electronic Whisker Arrays Integrated with Strain and Temperature Sensors. *ACS Nano* **2014**, *8*, 3921. [CrossRef] [PubMed]
32. Tao, P.; Shang, W.; Song, C.; Shen, Q.; Zhang, F.; Luo, Z.; Yi, N.; Zhang, D.; Deng, T. Bioinspired Engineering of Thermal Materials. *Adv. Mater.* **2015**, *27*, 428. [CrossRef] [PubMed]
33. Harada, S.; Kanao, K.; Yamamoto, Y.; Arie, T.; Akita, S.; Takei, K.J.A.N. Fully Printed Flexible Fingerprint-like Three-Axis Tactile and Slip Force and Temperature Sensors for Artificial Skin. *ACS Nano* **2014**, *8*, 12851. [CrossRef]
34. Zhu, P.; Wang, Y.; Sheng, M.; Wang, Y.; Yu, Y.; Deng, Y. A flexible active dual-parameter sensor for sensitive temperature and physiological signal monitoring via integrating thermoelectric and piezoelectric conversion. *J. Mater. Chem. A* **2019**, *7*, 8258–8267. [CrossRef]
35. Wu, R.; Ma, L.; Hou, C.; Meng, Z.; Guo, W.; Yu, W.; Yu, R.; Hu, F.; Liu, X.Y. Silk Composite Electronic Textile Sensor for High Space Precision 2D Combo Temperature–Pressure Sensing. *Small* **2019**, *15*, 1901558. [CrossRef] [PubMed]
36. Feng, S.; Li, Q.; Wang, S.; Wang, B.; Hou, Y.; Zhang, T. Tunable Dual Temperature–Pressure Sensing and Parameter Self-Separating Based on Ionic Hydrogel via Multisynergistic Network Design. *ACS Appl. Mater. Interfaces* **2019**, *11*, 21049–21057. [CrossRef] [PubMed]
37. Zhao, X.H.; Ma, S.N.; Long, H.; Yuan, H.; Tang, C.Y.; Cheng, P.K.; Tsang, Y.H. Multifunctional Sensor Based on Porous Carbon Derived from Metal–Organic Frameworks for Real Time Health Monitoring. *ACS Appl. Mater. Interfaces* **2018**, *10*, 3986–3993. [CrossRef] [PubMed]

38. Zhang, F.; Zang, Y.; Huang, D.; Di, C.; Zhu, D. Flexible and self-powered temperature–pressure dual-parameter sensors using microstructure-frame-supported organic thermoelectric materials. *Nat. Commun.* **2015**, *6*, 8356. [CrossRef]
39. Wang, Y.; Mao, H.; Wang, Y.; Zhu, P.; Liu, C.; Deng, Y. 3D geometrically structured PANI/CNT-decorated polydimethylsiloxane active pressure and temperature dual-parameter sensors for man–machine interaction applications. *J. Mater. Chem. A* **2020**, *8*, 15167. [CrossRef]

micromachines

Article

Triboelectric-Electromagnetic Hybrid Wind-Energy Harvester with a Low Startup Wind Speed in Urban Self-Powered Sensing

Gang Li, Juan Cui *, Tingshan Liu, Yongqiu Zheng, Congcong Hao, Xiaojian Hao and Chenyang Xue

Key Laboratory of Instrumentation Science & Dynamic Measurement, School of Instrument and Electronics, North University of China, Taiyuan 030051, China
* Correspondence: cuijuan@nuc.edu.cn

Abstract: Wind energy as a renewable energy source is easily available and widely distributed in cities. However, current wind-energy harvesters are inadequate at capturing energy from low-speed winds in urban areas, thereby limiting their application in distributed self-powered sensor networks. A triboelectric–electromagnetic hybrid harvester with a low startup wind speed (LSWS-TEH) is proposed that also provides output power within a wide range of wind speeds. An engineering-implementable propeller design method is developed to reduce the startup wind speed of the harvester. A mechanical analysis of the aerodynamics of the rotating propeller is performed, and optimal propeller parameter settings are found that greatly improved its aerodynamic torque. By combining the high-voltage output of the triboelectric nanogenerator under low-speed winds with the high-power output of the electromagnetic generator under high-speed winds, the harvester can maintain direct current output over a wide wind-speed range after rectification. Experiments show that the harvester activates at wind speeds as low as 1.2 m/s, powers a sensor with multiple integrated components in 1.7 m/s wind speeds, and drives a Bluetooth temperature and humidity sensor in 2.7 m/s wind speeds. The proposed small, effective, inexpensive hybrid energy harvester provides a promising way for self-powered requirements in smart city settings.

Keywords: hybrid nanogenerators; triboelectric nanogenerator; electromagnetic generator; wind-energy harvesting; low startup wind speed; self-powered sensor

Citation: Li, G.; Cui, J.; Liu, T.; Zheng, Y.; Hao, C.; Hao, X.; Xue, C. Triboelectric-Electromagnetic Hybrid Wind-Energy Harvester with a Low Startup Wind Speed in Urban Self-Powered Sensing. *Micromachines* **2023**, *14*, 298. https://doi.org/10.3390/mi14020298

Academic Editor: Mengdi Han

Received: 9 January 2023
Revised: 17 January 2023
Accepted: 18 January 2023
Published: 23 January 2023

1. Introduction

In recent years, the smart city concept has emerged as a solution to address challenges arising from the exponential growth of urban areas and population [1,2]. A key component of the concept is a network composed of distributed wireless sensors [3,4]. Energy efficiencies of such networks are critical factors in establishing the smart city [5]. However, current sensor networks rely mostly on batteries and cables for power supply at huge costs associated with installation and maintenance, drawbacks that potentially have environmental concerns [6]. To achieve continuous wireless power for the network, harvesting clean and renewable energy from the environment is desirable [7]. Wind energy, light energy, mechanical energy, and other environmental energy sources widely exist in cities [8,9]. Among them, wind energy has many advantages, such as its wide distribution in all-weather situations and ease of harvesting [10,11]. Average wind speeds in cities are generally 1.2 m/s to 3.8 m/s in Chinese cities from the National Centers for Environmental Information (NCEI). Such speeds are too low to start most wind-driven micro-harvesters employing electromagnetic generators (EMGs), and rectification loss is very high [12,13]. Therefore, the availability of energy harvesters able to operate in light to moderate breezes is an imperative of the smart city concept.

In 2012, the first triboelectric nanogenerator (TENG) proposed by Wang's team showed great prospects in applications of self-powered systems [14]. With deeper research over the interim, TENGs have offered unique advantages in low-frequency energy collection, specifically, high output voltages from small input excitations [15–17]. They are widely used to harvest energy from environment, such as mechanical energy [18–20], tidal energy [21–23], and wind energy from various environments [24–28]. Currently, wind-energy harvesters operating with TENGs in the low-frequency regime have attained extensive attention because of their low cost, light weight, and high efficiency [29–31]. In cities, wind is characterized by low and broad average wind speeds. Moreover, natural wind patterns are unstable and highly variable. However, current TENGs and EMGs are incapable of achieving high energy conversion efficiencies under both low-speed and high-speed winds [32–34]. For better operations in urban settings, hybrid wind-energy harvesters combining both TENGs and EMGs may be the solution.

Recently, several prototypes of triboelectric–electromagnetic hybrid generators have been developed to harvest wind energy. Fan's group proposed a self-powered wireless transmission sensor for the Internet of things, which uses the synchronous hybrid power generation mode of a TENG and an EMG to power the sensor [35]. Li's group proposed an optimization strategy for flexible collaborative wind energy collection employing asynchronous operation of a TENG and an EMG; the strategy reduces the startup torque and the startup wind speed [36]. Among them, the startup wind speed of these hybrid energy harvesters is 2.2 m/s to 4 m/s [35–39]. However, to meet the energy supply demands of Bluetooth sensors, wind speeds of 4.7 m/s to 9 m/s are required to drive these harvesters [35,36,39]. Realizing the self-power supply needs of the wireless sensor network is difficult in breezes. Because the electromagnetic torque of the EMG has not been reduced, startup wind speeds of the energy harvesters are higher than the wind speeds for breezes, also limiting the output power increase of the TENG. Therefore, EMG designs for hybrid wind-energy harvester must be optimized to reduce the startup wind speed and increase the rotational speed.

With this objective, a triboelectric–electromagnetic hybrid harvester with a low startup wind speed (LSWS-TEH) is fabricated. Developed from the application of fluid mechanics theories in the design of the propeller, this harvester is designed with a low startup wind speed and high-power output over a wide range of wind speeds. In addition, an optimal design structure enables the inertia and electromagnetic torques of the EMG to be reduced, the speed of the harvester to be changed easily, and the resistance torque to be smaller. The harvester can operate and harvest wind energy in urban breezes with greater effectiveness. Experimental results show that its startup wind speed is 1.2 m/s; in wind speeds of 1.7 m/s, it can supply stable power to sensors with multiple integrated components. In addition, under wind speeds of 2.7 m/s, its output energy can steadily power a Bluetooth temperature and humidity sensor. The LSWS-TEH provides a benchmark for effective wind energy collection in breezes and, therefore, broad prospects in the field of the Internet of things connectivity and smart-city power planning.

2. Principles and Methods

2.1. Theoretical Analysis

2.1.1. Design Principle for a Low Startup Wind Speed

During the startup of a wind harvester under low wind speeds, the mechanical model under steady rotor operations is mathematically expressed as

$$\vec{M_a} - \vec{M_f} - \vec{M_e} - \vec{M_m} = \vec{M_1} + \vec{M_2} + \vec{M_3}, \tag{1}$$

where vectors $\vec{M_a}$, $\vec{M_f}$, $\vec{M_e}$, and $\vec{M_m}$ are the aerodynamic torque generated by the propeller, the frictional torque on the rotor, the eccentric torque on the propeller, and the electromagnetic torque applied to the electromagnetic rotor from Ampère forces, respectively. Vectors $\vec{M_1}$, $\vec{M_2}$, and $\vec{M_3}$ are the torques required for the rotation of the propeller, the electromagnetic rotor, and the other parts.

From Equation (1), the main measures to reduce the startup wind speed of the harvester are to reduce the frictional, eccentric, and electromagnetic torques, and to increase the aerodynamic torque of the propeller. Friction and eccentric torques are determined in the machining process and cannot be completely eliminated. Any reduction in electromagnetic torque affects the power-generation capacity of the electromagnetic generator. Therefore, to attain the objective, increasing the aerodynamic torque of the propeller is adopted.

Expanding the right-hand side of Equation (1),

$$\vec{M_1} + \vec{M_2} + \vec{M_3} = \sum_i m_i r_i \vec{\omega} + \sum_j m_j r_j \vec{\omega} + \sum_k m_k r_k \vec{\omega}, \quad (2)$$

where $\vec{\omega}$ is the angular velocity of the rotating part; the pairings (m_a, r_a) with $a = i, j, k$ refer to the weight and radius of a differential element of the propeller, electromagnetic rotor, and other rotating parts, respectively.

From Equation (2), reducing the weight of the rotating part and decreasing the radius of rotation, while meeting structural strength requirements, can also effectively reduce the startup wind speed of harvesters. The design also exploits these two measures.

2.1.2. Propeller Design

Considering the limitations on size, material, and processing technology, a flat blade is selected for the propeller for its ease in design and manufacture. According to fluid dynamics, the relationship linking inflow angle φ, attack angle α, and pitch angle β is as follows:

$$\varphi = \alpha + \beta. \quad (3)$$

The moving fluid produces a lift force that drives the propeller's rotation and offers a more superior startup performance with low-speed winds. A high lift–drag ratio ξ is pursued to promote propeller rotation; ξ is calculated using

$$\xi = \frac{C_{FBL}}{C_{FBD}} = \frac{\sin 2\alpha}{2\sin^2 \alpha}, \quad (4)$$

where C_{FBL} and C_{FBD} are the lift and drag coefficients of the flat blade. The functional relationship between ξ and α shows that, to improve ξ in Figure 1a, α should be as small as possible.

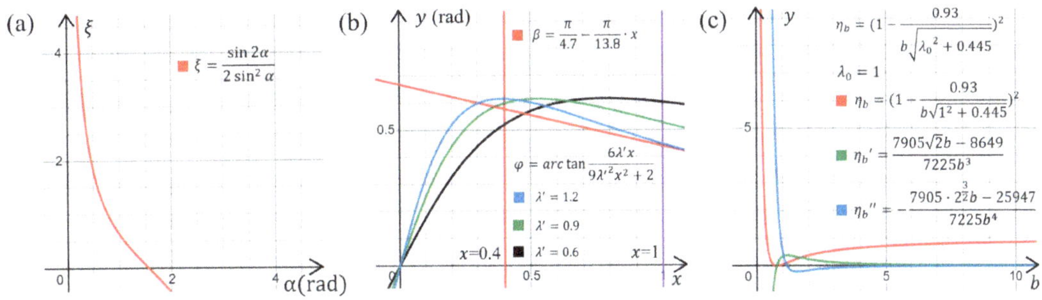

Figure 1. Relationship analysis for propeller parameter selection (see key legends for parameter expressions and settings): (**a**) between ξ and α; (**b**) between φ, α, and β; (**c**) between b and η_b.

Axial and tangential induction factors, m and n, are present during stable operations of the propeller; they are expressed as

$$m = \frac{1}{3}, \quad (5)$$

$$n = \frac{m\,(1-m)}{\lambda^2} = \frac{2}{9\lambda^2}, \tag{6}$$

where λ is the tip speed ratio.

The inflow angle φ is calculated from

$$\tan\varphi = \frac{1-m}{(1+n)\lambda} = \frac{6\lambda}{9\lambda^2+2} = \frac{6\lambda' x}{9\lambda'^2 x^2+2}, \tag{7}$$

$$x = \frac{r}{R}, \tag{8}$$

where λ' is the tip speed ratio of any blade element, r and R are the radii from the center of rotation to any blade element and the propeller edge, respectively, and x is the ratio of r to R, which is used to indicate the position of any blade element relative to the propeller.

Combining Equations (3) and (7) yields

$$\varphi = \arctan\frac{6\lambda' x}{9\lambda'^2 x^2+2} - \beta. \tag{9}$$

Although the blade elements of the propeller should have similar α for different r, the manufacturability of a propeller also needs to be considered. The pitch angle of the propeller is designed as

$$\beta = \beta_2 - (\beta_2 - \beta_1)\,x\,, \tag{10}$$

where β_1 and β_2 are the pitch angles at the tip and root of the blade, respectively.

With increasing wind speeds, the rotor speed is positively correlated with λ. For propellers of a low-speed wind turbine, the power coefficient increases to a peak and then decreases, whereas the torque coefficient monotonically decreases with λ in the range of 0–1.9 [40]. With λ between 0.6 and 1.2, the propeller obtains a sufficient torque while offering high efficiency in energy conversion. In comparing plots with λ equal to 0.6, 0.9, and 1.2 (Figure 1b), the curves of Equations (7) and (10) should generally be parallel in the range of x from 0.6 to 1 when λ is 0.9, thus reducing differences in attack angle α at different blade elements of the propeller. Moreover, as λ increases from 0.6 to 1.2, a family of curves generate from Equation (7) should be higher than the plots of Equation (10). The distance between the lowest point of family of curves and the point from Equation (10) is as small as possible when x is determined. Plotting the propeller parameter design curves (Figure 1b) shows that values of x between 0 and 0.4 easily produce stalling, which is not conducive to propeller rotation. Hence, positions of the propeller with x from 0 to 0.4 are simplified as belonging to the pillar.

The relationship between energy conversion efficiency and blade number can be found from Prandtl's relation,

$$\eta_b = (1 - \frac{0.93}{b\sqrt{\lambda_0{}^2+0.445}})^2\,, \tag{11}$$

where b is the blade number, and λ_0 is the given tip speed ratio.

The first and second derivative functions of Equation (11) (Figure 1c) show that the energy conversion efficiency increases slowly with increasing blade number for certain λ_0. Considering that even-numbered blades are prone to resonance and engineering processing difficulties, a seven-bladed configuration is chosen.

The differential method is used to calculate the aerodynamic torque of the propeller,

$$d\vec{M}_a = m\,\Delta\vec{\omega}\,r^2. \tag{12}$$

The blade should have large chord lengths at large radii, and high rotor solidity contributes to increasing the propeller aerodynamic torque.

In summary, a set of engineering solutions is found that established the main parameter settings of the propeller for the final design in Table 1.

Table 1. Settings of the main parameters of the propeller [1].

Designed Tip Speed Ratio	Blade Root Pitch Angle	Blade Tip Pitch Angle	Number of Blades	Rotor Solidity [2]
0.6–1.2	38.3°	25.3°	7	90%

[1] The propeller is simplified as the pillar with x in the range of 0 to 0.4. [2] The rotor solidity is calculated without simplified blades.

2.2. Design and Working Principle of the Wind-Energy Harvester

The LSWS-TEH consists of a rotor, a stator, and a power management circuit (Figure 2). The rotor comprises a propeller, shaft, bearings, and a magnet. The stator includes housing, a coil, and coil holder. Figure 2a shows the rotor and stator assembly. The power management circuit is composed of a charge pump, rectifier bridge, capacitor bank, and switching regulator circuit (Figure 2b). The circuit schematic of rectification and charging circuit is shown in Figure 2c. The coil is wound onto the coil holder and is fixed to the stator housing (Figure 2d). The designed propeller and shell are made by 3D printing technology of SLA light-curing molding, and the material is white resin. The propeller has an overall size of 80 mm in diameter and 30 mm in cylindrical height; therefore, it is much smaller than other triboelectric–electromagnetic hybrid wind-energy harvesters [35–39].

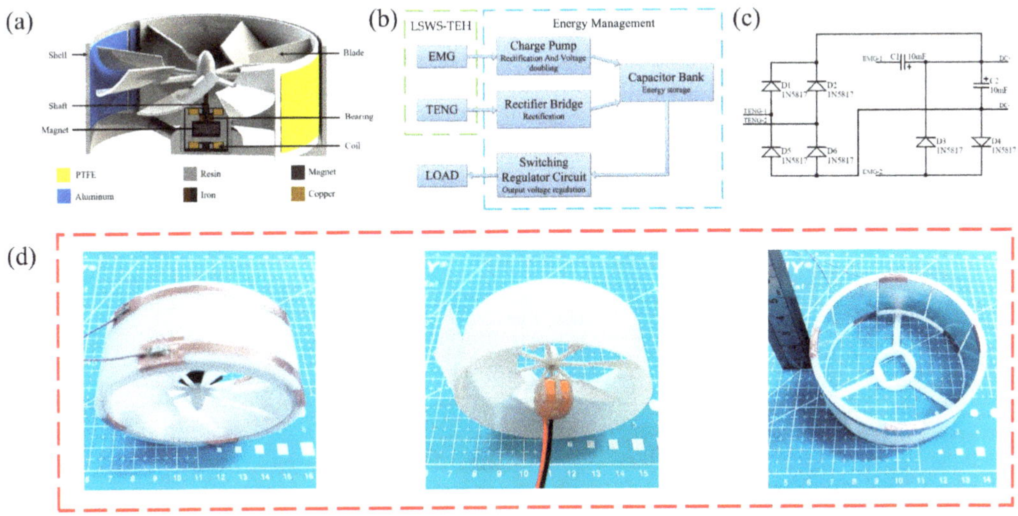

Figure 2. Design and photos of LSWS-TEH: (**a**) structure schematic of the rotor and stator; (**b**) schematic of circuit block diagram; (**c**) rectification and charging circuits; (**d**) three photos show the LSWS-TEH, coils and propeller, housing respectively.

When the ambient wind rotates the propeller, the polytetrafluoroethylene (PTFE) film in the TENG through centrifugation and rotation touches and slides against the aluminum foil. The output induced is an alternating current (AC) that flows between adjacent foils. A rectifier bridge converts AC to direct current (DC), which is then used to charge the capacitor bank. In the EMG, the propeller rotates magnets attached at the bottom of the shaft to produce a rotating electromagnetic field. The coil in the EMG cuts the rotating magnetic field lines to induce an electric potential. The output AC voltage is rectified and boosted by the charge pump to charge the capacitor bank. The switching voltage regulator then converts the unstable voltage of the capacitor bank into a stable voltage, which is a basic requirement of any sensor. If the power of the sensor is lower than the charging power of the capacitor bank, the excess power is stored in the capacitor bank to power the sensor when wind speeds are low.

Under different wind speeds, the TENG and EMG exhibit good synergy. In low-speed winds, the EMG output voltage is smaller than the forward bias voltage of the diode because the rotor rotates slowly, resulting in no voltage output after rectification. However, the TENG has high-voltage output characteristics that can generate a substantial voltage output under low-speed winds. Under high-speed winds, the rotor speed increases and the EMG generates a rectified voltage output. As speeds increase further, the output power and frequency from the TENG also increase.

The power generation principle of the TENG (Figure 3a) is detailed as follows: initially, the PTFE film and the aluminum foil are in contact with each other by centrifugation (Figure 3(aI)) to induce negative and positive electrostatic charges through contact friction. The PTFE film and the aluminum foil are gradually separated under rotation, and the potential difference between adjacent electrodes generates an AC through the external circuit (Figure 3(aII)). The PTFE then makes complete contact with the next electrode (Figure 3(aIII)), meaning that one charge transfer is performed. Next, the adjacent electrodes produce an opposite potential difference to that previously, and the opposite current flows in the external circuit (Figure 3(aIV)). The above cycle is repeated with the TENG outputting a continuous AC. Using COMSOL Multiphysics 5.6 software, the electric potential distribution on the surface of the TENG (Figure 3b) is simulated at four stages during the cycle.

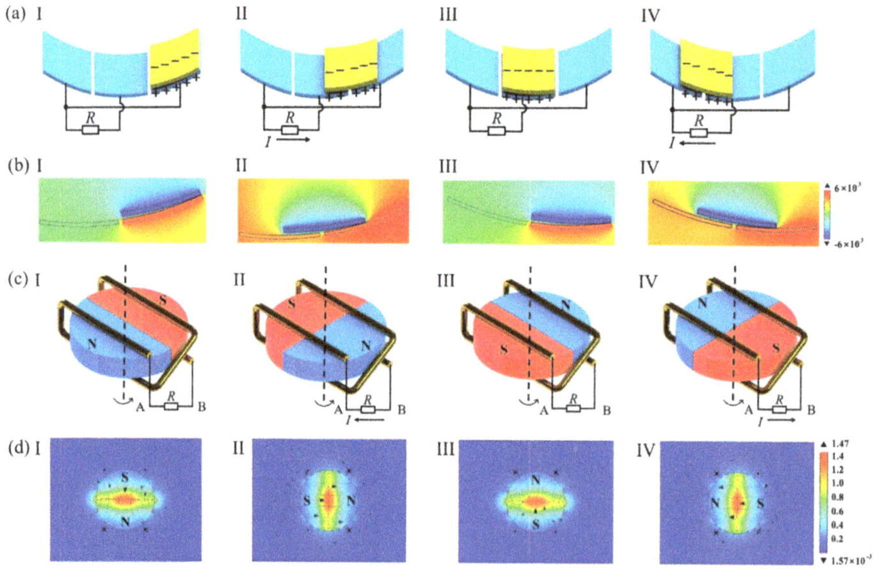

Figure 3. LSWS-TEH operating principle and simulation results: (**aI**) working schematic of the TENG state 1; (**aII**) working schematic of the TENG state 2; (**aIII**) working schematic of the TENG state 3; (**aIV**) working schematic of the TENG state 4; (**bI**) simulation of the surface electric potential distribution of the TENG state 1; (**bII**) simulation of the surface electric potential distribution of the TENG state 2; (**bIII**) simulation of the surface electric potential distribution of the TENG state 3; (**bIV**) simulation of the surface electric potential distribution of the TENG state 4; (**cI**) working schematic of the EMG state 1; (**cII**) working schematic of the EMG state 2; (**cIII**) working schematic of the EMG state 3; (**cIV**) working schematic of the EMG state 4; (**dI**) simulation of the surface flux density of the EMG state 1; (**dII**) simulation of the surface flux density of the EMG state 2; (**dIII**) simulation of the surface flux density of the EMG state 3; (**dIV**) simulation of the surface flux density of the EMG state 4.

The working principle of the EMG is shown in Figure 3c. At the beginning, a zero rate of change in magnetic flux through the coil exists, and the coil provides no voltage output (Figure 3(cI)). At its maximum rate of change (Figure 3(cII)), the induced electric potential in the coil peaks and the current flowing through the external circuit reach a peak. Figure 3(cIII) is similar to Figure 3(cI), but the magnetic field distribution is reversed. Figure 3(cIV) is similar to Figure 3(cII), but the electric potential and current are the opposite of the previous ones. By repeating the cycle, an induced AC is generated through the EMG, and its surface flux density is also simulated using COMSOL software (Figure 3d), with the four stages exhibiting typical fields of the EMG during an AC cycle.

2.3. Experimental Methods

2.3.1. Fabrication of the EMG

One copper wire with diameter of 0.1 mm is wound on the resin holder as two coils. The coils are positioned at a distance of 5 mm. Each coil has 400 turns, and the width and height of the coil are 12.5 mm and 11 mm, respectively. A neodymium (NdFeB) disc magnet with a radius of 4 mm and height of 3 mm is attached to shaft with a diameter of 1.5 mm.

2.3.2. Fabrication of the TENG

Aluminum foil with a thickness of 0.06 mm is used as the metal electrodes. These electrodes are bonded to copper foil of thickness 0.06 mm. The rotating triboelectric electrode is selected as a PTFE film with a height of 20 mm and is glued to the paddle housing.

2.3.3. Testing Systems

The harvester is fixed with a stand in a wind tunnel controlled by an axial flow fan (SFG3-2R, Leifeng Mechatronics Co., Changzhou, China) operating through a frequency converter (SQ1000−2T, Fuci Electromechanical Technology Co., Shanghai, China). The output voltage signal is obtained using an oscilloscope (DSOX3024T, Keysight, Santa Rosa, CA, USA) through a 100 M high-voltage probe (PA5100A, Tektronix, Beaverton, OR, USA), and the current signal is obtained using an electrostatic meter (6514, Keithley, Beaverton, OR, USA). The wind speed is obtained using a thermal anemometer (AR866A, Kexin Measurement & Control Technology Co., Suzhou, China).

The electrical outputs of EMG and TENG are independent. In the process of the output capability test and impedance test of the harvester, the outputs of EMG and TENG are tested separately; that is, only EMG or TENG is connected to the test system during the test. Thus, the measured data reflect the electrical outputs of the single EMG or TENG.

3. Results and Discussion

3.1. Output Performance

The angle between the chord of the propeller and the horizontal plane (pitch angle) was determined, and the effect of propeller curvature on the EMG output was investigated in four different scenarios (Figure 4). In this experiment, the output performance of the EMG was tested with propeller curvatures of 0° (plane), 50.6°, 24.7°, and 16.4°. With a curvature of 50.6°, the arc at the propeller tip was tangent to the horizontal plane; in this instance, the radius of the arc was denoted by r'. Then, the curvature of the propeller was 24.7° at an arc radius of $2r'$ and 16.4° at an arc radius of $3r'$.

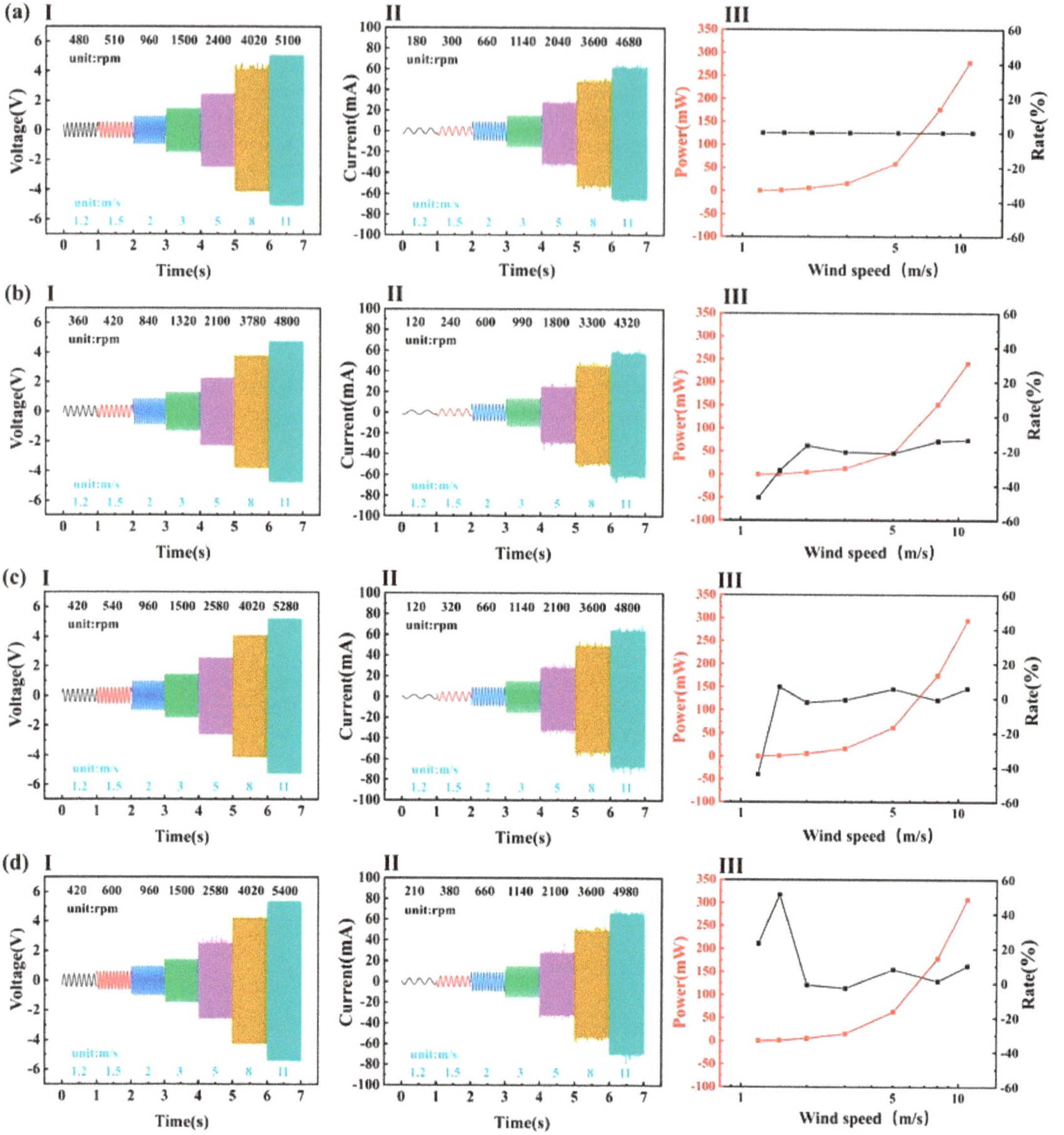

Figure 4. Open-circuit voltage, short-circuit current, power, and rate of change in power of the EMG for different blade curvatures: (**aI**) open-circuit voltage of EMG in flat; (**aII**) short-circuit current of EMG in flat; (**aIII**) power and rate of EMG in flat; (**bI**) open-circuit voltage of EMG in 50.6°; (**bII**) short-circuit current of EMG in 50.6°; (**bIII**) power and rate of EMG in 50.6°; (**cI**) open-circuit voltage of EMG in 24.7°; (**cII**) short-circuit current of EMG in 24.7°; (**cIII**) power and rate of EMG in 24.7°; (**dI**) open-circuit voltage of EMG in 16.4°; (**dII**) short-circuit current of EMG in 16.4°; (**dIII**) power and rate of EMG in 16.4°.

The EMG output decreased more significantly over all wind speeds when the curvature is large (Figure 4a,b). Curvature could improve airflow rate into the propeller, but larger curvatures created vortices that reduced the EMG output. The EMG performance at a wind speed of 1.2 m/s (Figure 4a,c) worsened when the curvature was in the midrange, whereas it showed greater improvements at wind speeds of 1.2 m/s and 1.5 m/s for small curvatures (Figure 4a,d). That is, a suitable curvature could improve the aerodynamic performance of the propeller under low-speed winds. For the application, a propeller curvature of 16.4° was chosen.

Next, the effect of rotating triboelectric electrode length on output performance of the TENG was studied using two metal electrodes in experiments in which the external circuit of the EMG was short-circuited. For the rotating triboelectric electrodes, strips of PTFE film with lengths of 3 cm, 4 cm, and 5 cm and thickness of 0.1 mm were tested (Figure 5). The 3 cm electrode had the lowest startup speed, and the startup wind speed increased with increasing strip length, as shown in Figure 5 (aI–cI). As shown in Figure 5 (aI–cI), the 4 cm long electrode had the highest output voltage because its stiffness produced a suitable curvature, resulting in the highest effective contact area for the TENG. As shown in Figure 5, the 5 cm electrode produced the lowest output power under low-speed winds, due to a reduced effective contact area because of its poor stiffness and easy deformation. It had the highest output current because its stronger centrifugal action increased the effective contact area under high-speed winds.

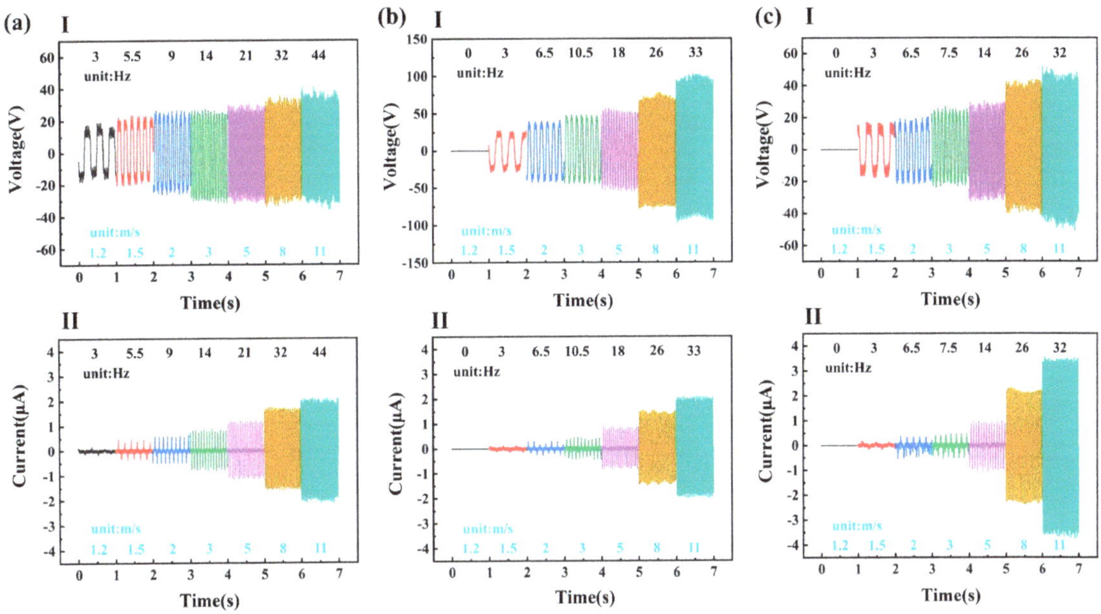

Figure 5. Open-circuit voltage and short-circuit current produced by the TENG for different strip lengths of PTFE film: (**aI**) open-circuit voltage of TENG in 3 cm film; (**aII**) short-circuit current of TENG in 3 cm film; (**bI**) open-circuit voltage of TENG in 4 cm film; (**bII**) short-circuit current of TENG in 4 cm film; (**cI**) open-circuit voltage of TENG in 5 cm film; (**cII**) short-circuit current of TENG in 5 cm film.

By measuring the frequency of the EMG or TENG output voltage, the rotation speed of the harvester can be found by

$$n = 60f_{EMG} = \frac{120f_{TENG}}{p}, \tag{13}$$

where n is rotation speed, f_{EMG} is the frequency of the EMG, f_{TENG} is the frequency of the TENG, and p is the number of TENG metal electrodes.

In addition, calculating the ratio k, i.e., the ratio of the rotating triboelectric electrode length to the number of metal electrodes, the TENG output was found to rise and fall with k. Optimizing k could improve the TENG output performance. The 3 cm electrode had the lowest startup wind speed; hence, it was chosen to balance the startup wind speed with the electrical energy output of TENG.

Lastly, the effect of different numbers of metal electrodes on TENG output performance was investigated without changing the total length of the metal electrodes. The output conditions for the TENG with two metal electrodes are given in Figure 5a. With the same strip length of 3 cm, the TENG electrical outputs obtained with four, six, and eight metal electrodes (Figure 6) show that, with increasing number, the outputs displayed a peak, and the highest output power was achieved with six metal electrodes. Under the same wind speed, higher numbers of metal electrodes generated a higher energy conversion frequency. However, increasing the electrode number further decreased the charge accumulation time that affected the generation of large voltages. Therefore, the harvester design was configured with six metal electrodes.

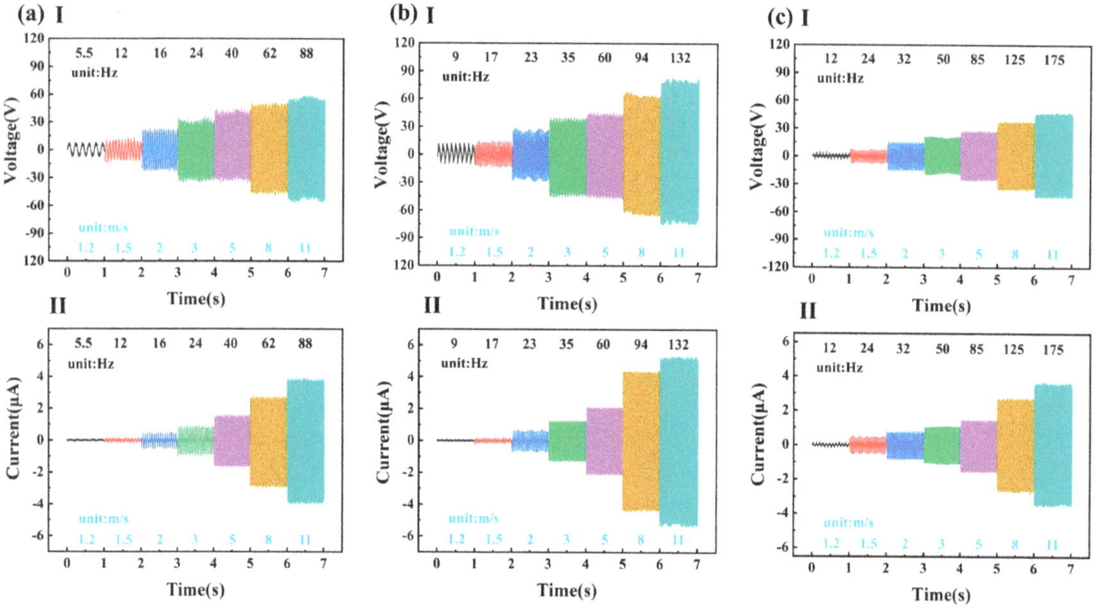

Figure 6. Open-circuit voltage and short-circuit current of the TENG for three different numbers of metal electrodes: (**aI**) open-circuit voltage of TENG in 4 electrodes; (**aII**) short-circuit current of TENG in 4 electrodes; (**bI**) open-circuit voltage of TENG in 6 electrodes; (**bII**) short-circuit current of TENG in 6 electrodes; (**cI**) open-circuit voltage of TENG in 8 electrodes; (**cII**) short-circuit current of TENG in 8 electrodes.

The final design parameters for the harvester were determined according to the results of Section 2.1.2 and the above experiments and analysis.

3.2. Demonstration

The TENG hinders rotor rotation; hence, the changes in peak power and the rates of change in the short-circuit current of the EMG containing the TENG and of the EMG alone were compared for different wind speeds (Figure 7a). At wind speeds less than 3 m/s, the output power for the EMG of LSWS-TEH was slightly reduced under the same wind speeds. The output voltage of the EMG under low-speed winds was too small to reach the forward bias of the diode, resulting in no electrical output after rectification. In contrast, the LSWS-

TEH had a higher electric energy output range compared with that for the EMG because an electric energy output was also present under low-speed winds, attributable to the high-voltage output of the TENG. Although the output power of the harvester decreased slightly in low-wind-speed conditions, it meets the requirements for urban applications in which the average wind speed ranges from 1.2 m/s to 3.8 m/s. Furthermore, high rotation speeds enhance centrifugal forces, which may destroy the balance within the wind-energy harvester and break the propeller. For self-protection, large wind turbines usually lock the propeller. The TENG can protect the harvester during power generation under high-speed winds because friction increases when centrifugal effects increase.

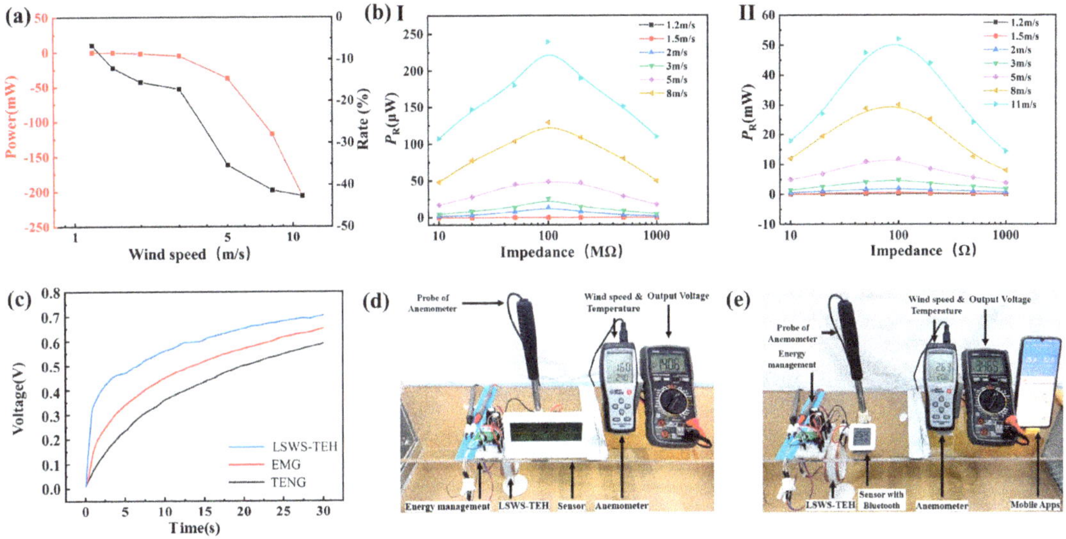

Figure 7. The LSWS-TEH performance demonstration under various wind speeds: (**a**) change in peak power and rate of change for the short-circuit current; (**bI**) peak power outputs for the TENG; (**bII**) peak power outputs for the EMG; (**c**) charging of a 100 μF capacitor with rectifier bridge under a 2 m/s wind speed; (**d**) powering of a sensor with multiple integrated components under wind speeds of 1.7 m/s; (**e**) powering of a Bluetooth temperature and humidity sensor under wind speeds of 2.7 m/s.

The peak output power of the TENG (Figure 7(bI)) and EMG (Figure 7(bII)) under different wind speeds and impedances was investigated. When the harvester was connected to an external load, it could be regarded as a circuit model where the internal resistance and external load divided the open-circuit voltage. Therefore, the output voltage and output current were positively and negatively correlated with the external load, respectively. Furthermore, as the load increased, the output power of TENG and EMG firstly increased and then decreased (Figure 7b). The corresponding optimum resistances for TENG and EMG were about 100 MΩ and 90 Ω, respectively. The charging of a 100 μF capacitor separately by the EMG, TENG, and LSWS-TEH under wind speeds of 2 m/s (Figure 7c) show that the harvester had a higher charging speed and higher capacitance voltage than either EMG or TENG. Moreover, with multiple integrated components, the harvester could power a sensor with a rated power of 150 μW (Figure 7d). In addition, under wind speeds of 2.7 m/s, it could power a Bluetooth temperature and humidity sensor (Figure 7e). With a rated power of 16 mW, the sensor could transfer data to a cell phone APP. Overall, the harvester delivered excellent performances in low-speed winds and had a wide wind speed collection range, meeting the energy-harvesting requirements in urban settings.

The aerodynamic torque, M_a, affects the starting wind speed of the wind-energy harvesters; the torque output of the propeller is

$$M_a = \frac{1}{2}\rho C_m R S V^2 = \frac{1}{2}\pi \rho C_m R^3 V^2, \tag{14}$$

where ρ is the air density, C_m is the moment coefficient, R is the propeller radius, S is the cross-sectional area of the propeller rotating body, and V is the wind speed. The torque output is proportional to the cube of R and the square of V.

The power output, P, can be calculated by

$$P = \frac{1}{2}\rho C_p S V^3 = \frac{1}{2}\pi\rho C_p R^2 V^3, \tag{15}$$

where C_p is the power factor. The power output is proportional to the square of R and the cube of V.

In general, obtaining a high power output is difficult at low wind speeds in small volumes. However, the designed LSWS-TEH achieves a low startup wind speed in a small volume and has a more substantial peak power output, which can drive the sensors at lower wind speed. The performances of different triboelectric–electromagnetic hybrid wind-energy harvesters were compared, as shown in the Supplementary Materials (Table S1). The proposed LSWS-TEH with a smaller size showed a lower startup wind speed and a higher output power.

4. Conclusions

The LSWS-TEH comprising a TENG and an EMG was developed to operate in 1.2 to 11 m/s wind speeds. It has an extremely low startup wind speed and wide operating range. An engineering design strategy was developed for the propeller, which substantially improved the aerodynamic torque of the propeller; accordingly, the harvester was able to meet the requirements for wind-energy harvesting in ultralow and medium–high wind speed in city settings. The TENG resolves the disadvantages inherent in the EMG without output power after rectification at low-speed winds and mitigates propeller damage under high-speed winds. Through specific design features of the structure, both inertial and electromagnetic torques of the EMG were reduced. Experiments demonstrated that the harvester started under wind speeds of 1.2 m/s. A sensor with multiple integrated components was powered using the harvester in wind speeds of 1.7 m/s. Moreover, a Bluetooth temperature and humidity sensor was operational under wind speeds of 2.7 m/s. The harvester can collect wind energy in ultralow and medium–high wind speeds and has wide potential in applications requiring self-powered wireless sensor networks operating in smart city environments.

Supplementary Materials: The following supporting information can be downloaded at https://www.mdpi.com/article/10.3390/mi14020298/s1: Figure S1. Peak power and average power of TENG in LSWS-TEH; Figure S2. Peak power and short-circuit peak current of EMG in LSWS-TEH; Figure S3. Switching regulator circuit; Table S1. The performance comparison of different triboelectric-electromagnetic hybrid wind energy harvesters; Video S1. LSWS-TEH starts in light air; Video S2: LSWS-TEH powers sensors in light breeze; Video S3: LSWS-TEH powers a sensor with Bluetooth in light breeze.

Author Contributions: Conceptualization, C.X. and G.L.; methodology, G.L. and J.C.; software, T.L. and C.H.; validation, G.L. and T.L.; formal analysis, J.C. and C.H.; investigation, X.H. and Y.Z.; resources, C.X.; data curation, G.L.; writing—original draft preparation, G.L. and J.C.; writing—review and editing, J.C. and G.L.; visualization, T.L.; supervision, X.H.; project administration, Y.Z.; funding acquisition, C.X. and Y.Z. All authors have read and agreed to the published version of the manuscript.

Funding: This work was funded by the National Key Research and Development Program of China (Grant No. 2019YFB2004800) and the Fundamental Research Program of Shanxi Province (Grant No. 20210302124033).

Institutional Review Board Statement: Not applicable.

Informed Consent Statement: Not applicable.

Data Availability Statement: The data presented in this study are available on request from the corresponding author.

Conflicts of Interest: The authors declare no conflict of interest.

References

1. Meijer, A.; Bolívar, M.P.R. Governing the smart city: A review of the literature on smart urban governance. *Int. Rev. Adm. Sci.* **2015**, *82*, 392–408. [CrossRef]
2. Silva, B.N.; Khan, M.; Han, K. Towards sustainable smart cities: A review of trends, architectures, components, and open challenges in smart cities. *Sustain. Cities Soc.* **2018**, *38*, 697–713. [CrossRef]
3. Ismagilova, E.; Hughes, L.; Dwivedi, Y.K.; Raman, K.R. Smart cities: Advances in research—An information systems perspective. *Int. J. Inf. Manag.* **2019**, *47*, 88–100. [CrossRef]
4. Ud Din, I.; Guizani, M.; Hassan, S.; Kim, B.-S.; Khurram Khan, M.; Atiquzzaman, M.; Ahmed, S.H. The Internet of Things: A Review of Enabled Technologies and Future Challenges. *IEEE Access* **2019**, *7*, 7606–7640. [CrossRef]
5. Zhang, Z.; He, J.; Wen, T.; Zhai, C.; Han, J.; Mu, J.; Jia, W.; Zhang, B.; Zhang, W.; Chou, X.; et al. Magnetically levitated-triboelectric nanogenerator as a self-powered vibration monitoring sensor. *Nano Energy* **2017**, *33*, 88–97. [CrossRef]
6. Hussain, A.; Wenbi, R.; da Silva, A.L.; Nadher, M.; Mudhish, M. Health and emergency-care platform for the elderly and disabled people in the Smart City. *J. Syst. Softw.* **2015**, *110*, 253–263. [CrossRef]
7. Wang, S.; Wang, X.; Wang, Z.L.; Yang, Y. Efficient Scavenging of Solar and Wind Energies in a Smart City. *ACS Nano* **2016**, *10*, 5696–5700. [CrossRef] [PubMed]
8. Beeby, S.P.; Tudor, M.J.; White, N.M. Energy harvesting vibration sources for microsystems applications. *Meas. Sci. Technol.* **2006**, *17*, R175–R195. [CrossRef]
9. Jettanasen, C.; Songsukthawan, P.; Ngaopitakkul, A. Development of Micro-Mobility Based on Piezoelectric Energy Harvesting for Smart City Applications. *Sustainability* **2020**, *12*, 2933. [CrossRef]
10. Puttichaem, W.; Putivisutisak, S.; Boonyongmaneerat, Y.; Vadhanasindhud, P. Early development of a shaftless horizontal axis wind turbine for generating electricity from air discharged from ventilation systems. *Int. J. Energy Res.* **2020**, *46*, 212–222. [CrossRef]
11. Yu, C.; Niu, S. Development of a Magnetless Flux Switching Machine for Rooftop Wind Power Generation. *IEEE Trans. Energy Convers.* **2015**, *30*, 1703–1711. [CrossRef]
12. Li, J.; Chen, J.; Guo, H. Triboelectric Nanogenerators for Harvesting Wind Energy: Recent Advances and Future Perspectives. *Energies* **2021**, *14*, 6949. [CrossRef]
13. Lu, P.; Pang, H.; Ren, J.; Feng, Y.; An, J.; Liang, X.; Jiang, T.; Wang, Z.L. Swing-Structured Triboelectric–Electromagnetic Hybridized Nanogenerator for Breeze Wind Energy Harvesting. *Adv. Mater. Technol.* **2021**, *6*, 2100496. [CrossRef]
14. Fan, F.-R.; Tian, Z.-Q.; Lin Wang, Z. Flexible triboelectric generator. *Nano Energy* **2012**, *1*, 328–334. [CrossRef]
15. Ko, H.-J.; Kwon, D.-S.; Bae, K.; Kim, J. Self-suspended shell-based triboelectric nanogenerator for omnidirectional wind-energy harvesting. *Nano Energy* **2022**, *96*, 107062. [CrossRef]
16. Liu, X.; Yu, A.; Qin, A.; Zhai, J. Highly Integrated Triboelectric Nanogenerator for Efficiently Harvesting Raindrop Energy. *Adv. Mater. Technol.* **2019**, *4*, 1900608. [CrossRef]
17. Zhang, J.; Lin, S.; Zheng, M.; Wang, Z.L. Triboelectric Nanogenerator as a Probe for Measuring the Charge Transfer between Liquid and Solid Surfaces. *ACS Nano* **2021**, *15*, 14830–14837. [CrossRef]
18. Gao, Q.; Li, Y.; Xie, Z.; Yang, W.; Wang, Z.; Yin, M.; Lu, X.; Cheng, T.; Wang, Z.L. Robust Triboelectric Nanogenerator with Ratchet-like Wheel-Based Design for Harvesting of Environmental Energy. *Adv. Mater. Technol.* **2019**, *5*, 1900801. [CrossRef]
19. García-Casas, X.; Ghaffarinejad, A.; Aparicio, F.J.; Castillo-Seoane, J.; López-Santos, C.; Espinós, J.P.; Cotrino, J.; Sánchez-Valencia, J.R.; Barranco, Á.; Borrás, A. Plasma engineering of microstructured piezo—Triboelectric hybrid nanogenerators for wide bandwidth vibration energy harvesting. *Nano Energy* **2022**, *91*, 106673. [CrossRef]
20. Song, Y.; Wang, N.; Hu, C.; Wang, Z.L.; Yang, Y. Soft triboelectric nanogenerators for mechanical energy scavenging and self-powered sensors. *Nano Energy* **2021**, *84*, 105919. [CrossRef]
21. Matin Nazar, A.; Idala Egbe, K.-J.; Abdollahi, A.; Hariri-Ardebili, M.A. Triboelectric Nanogenerators for Energy Harvesting in Ocean: A Review on Application and Hybridization. *Energies* **2021**, *14*, 5600. [CrossRef]
22. Wang, Z.L.; Jiang, T.; Xu, L. Toward the blue energy dream by triboelectric nanogenerator networks. *Nano Energy* **2017**, *39*, 9–23. [CrossRef]
23. Xu, M.; Zhao, T.; Wang, C.; Zhang, S.L.; Li, Z.; Pan, X.; Wang, Z.L. High Power Density Tower-like Triboelectric Nanogenerator for Harvesting Arbitrary Directional Water Wave Energy. *ACS Nano* **2019**, *13*, 1932–1939. [CrossRef] [PubMed]
24. Chen, P.; An, J.; Shu, S.; Cheng, R.; Nie, J.; Jiang, T.; Wang, Z.L. Super-Durable, Low-Wear, and High-Performance Fur-Brush Triboelectric Nanogenerator for Wind and Water Energy Harvesting for Smart Agriculture. *Adv. Energy Mater.* **2021**, *11*, 2003066. [CrossRef]

25. Gulahmadov, O.; Muradov, M.B.; Kim, J. The wind-driven Scotch yoke-based triboelectric nanogenerator system for energy harvesting. *Int. J. Energy Res.* **2022**, *46*, 10989–10997. [CrossRef]
26. Han, J.; Feng, Y.; Chen, P.; Liang, X.; Pang, H.; Jiang, T.; Wang, Z.L. Wind-Driven Soft-Contact Rotary Triboelectric Nanogenerator Based on Rabbit Fur with High Performance and Durability for Smart Farming. *Adv. Funct. Mater.* **2021**, *32*, 2108580. [CrossRef]
27. Jeon, S.-B.; Kim, S.; Park, S.-J.; Seol, M.-L.; Kim, D.; Chang, Y.K.; Choi, Y.-K. Self-powered electro-coagulation system driven by a wind energy harvesting triboelectric nanogenerator for decentralized water treatment. *Nano Energy* **2016**, *28*, 288–295. [CrossRef]
28. Tcho, I.-W.; Kim, W.-G.; Kim, J.-K.; Kim, D.-W.; Yun, S.-Y.; Son, J.-H.; Choi, Y.-K. A flutter-driven triboelectric nanogenerator for harvesting energy of gentle breezes with a rear-fixed fluttering film. *Nano Energy* **2022**, *98*, 107197. [CrossRef]
29. Cao, H.; Wu, X.; Wu, H.; Pan, Y.; Luo, D.; Azam, A.; Zhang, Z. A Hybrid Self-Powered System Based on Wind Energy Harvesting for Low-Power Sensors on Canyon Bridges. *Int. J. Precis. Eng. Manuf. Green Technol.* **2022**, 1–26. [CrossRef]
30. Ren, Z.; Wu, L.; Pang, Y.; Zhang, W.; Yang, R. Strategies for effectively harvesting wind energy based on triboelectric nanogenerators. *Nano Energy* **2022**, *100*, 107522. [CrossRef]
31. Wu, C.; Wang, A.C.; Ding, W.; Guo, H.; Wang, Z.L. Triboelectric Nanogenerator: A Foundation of the Energy for the New Era. *Adv. Energy Mater.* **2019**, *9*, 1802906. [CrossRef]
32. Han, Q.; Ding, Z.; Sun, W.; Xu, X.; Chu, F. Hybrid triboelectric-electromagnetic generator for self-powered wind speed and direction detection. *Sustain. Energy Technol. Assess.* **2020**, *39*, 100717. [CrossRef]
33. Zhang, B.; Chen, J.; Jin, L.; Deng, W.; Zhang, L.; Zhang, H.; Zhu, M.; Yang, W.; Wang, Z.L. Rotating-Disk-Based Hybridized Electromagnetic-Triboelectric Nanogenerator for Sustainably Powering Wireless Traffic Volume Sensors. *ACS Nano* **2016**, *10*, 6241–6247. [CrossRef] [PubMed]
34. Zhong, Y.; Zhao, H.; Guo, Y.; Rui, P.; Shi, S.; Zhang, W.; Liao, Y.; Wang, P.; Wang, Z.L. An Easily Assembled Electromagnetic-Triboelectric Hybrid Nanogenerator Driven by Magnetic Coupling for Fluid Energy Harvesting and Self-Powered Flow Monitoring in a Smart Home/City. *Adv. Mater. Technol.* **2019**, *4*, 1900741. [CrossRef]
35. Fan, X.; He, J.; Mu, J.; Qian, J.; Zhang, N.; Yang, C.; Hou, X.; Geng, W.; Wang, X.; Chou, X. Triboelectric-electromagnetic hybrid nanogenerator driven by wind for self-powered wireless transmission in Internet of Things and self-powered wind speed sensor. *Nano Energy* **2020**, *68*, 104319. [CrossRef]
36. Li, X.; Gao, Q.; Cao, Y.; Yang, Y.; Liu, S.; Wang, Z.L.; Cheng, T. Optimization strategy of wind energy harvesting via triboelectric-electromagnetic flexible cooperation. *Appl. Energy* **2022**, *307*, 118311. [CrossRef]
37. Wang, P.; Pan, L.; Wang, J.; Xu, M.; Dai, G.; Zou, H.; Dong, K.; Wang, Z.L. An Ultra-Low-Friction Triboelectric-Electromagnetic Hybrid Nanogenerator for Rotation Energy Harvesting and Self-Powered Wind Speed Sensor. *ACS Nano* **2018**, *12*, 9433–9440. [CrossRef]
38. Yong, S.; Wang, J.; Yang, L.; Wang, H.; Luo, H.; Liao, R.; Wang, Z.L. Auto-Switching Self-Powered System for Efficient Broad-Band Wind Energy Harvesting Based on Dual-Rotation Shaft Triboelectric Nanogenerator. *Adv. Energy Mater.* **2021**, *11*, 2101194. [CrossRef]
39. Zhang, B.; Zhang, S.; Li, W.; Gao, Q.; Zhao, D.; Wang, Z.L.; Cheng, T. Self-Powered Sensing for Smart Agriculture by Electromagnetic-Triboelectric Hybrid Generator. *ACS Nano* **2021**, *15*, 20278–20286. [CrossRef]
40. Legourieres, D.; South, P. *Wind Power Plants: Theory and Design*, 1st ed.; Pergamon Press: Oxford, UK, 1982; pp. 50–52.

micromachines

Article

Flexible Capacitive Pressure Sensor Based on a Double-Sided Microstructure Porous Dielectric Layer

Qingyang Yu [1,*] and Jian Zhang [2]

[1] College of Control Science and Engineering, China University of Petroleum, Qingdao 266580, China
[2] Morningcore Holding Co., Ltd., Qingdao 266400, China
* Correspondence: qyyu@upc.edu.cn

Abstract: In the era of intelligent sensing, there is a huge demand for flexible pressure sensors. High sensitivity is the primary requirement for flexible pressure sensors, whereas pressure response range and resolution, which are also key parameters of sensors, are often ignored, resulting in limited applications of flexible pressure sensors. This paper reports a flexible capacitive pressure sensor based on a double-sided microstructure porous dielectric layer. First, a porous structure was developed in the polymer dielectric layer consisting of silicon rubber (SR)/NaCl/carbon black (CB) using the dissolution method, and then hemisphere microstructures were developed on both sides of the layer by adopting the template method. The synergistic effect of the hemispheric surface microstructure and porous internal structure improves the deformability of the dielectric layer, thus achieving high sensitivity (3.15 kPa^{-1}), wide response range (0–200 kPa), and high resolution (i.e., the minimum pressure detected was 27 Pa). The proposed sensing unit and its array have been demonstrated to be effective in large-area pressure sensing and object recognition. The flexible capacitive pressure sensor developed in this paper is highly promising in applications of robot skin and intelligent prosthetic hands.

Keywords: flexible capacitive pressure sensor; double-sided microstructure; hemisphere microstructure; porous structure; pressure sensing; object recognition

Citation: Yu, Q.; Zhang, J. Flexible Capacitive Pressure Sensor Based on a Double-Sided Microstructure Porous Dielectric Layer. *Micromachines* **2023**, *14*, 111. https://doi.org/10.3390/mi14010111

Academic Editor: Vamsi K Yadavalli

Received: 7 November 2022
Revised: 29 November 2022
Accepted: 2 December 2022
Published: 30 December 2022

1. Introduction

In recent years, flexible pressure sensors based on different sensing mechanisms (piezoresistive, capacitive, piezoelectric, triboelectric, and magnetic) have been developed one after another owing to advances in flexible sensing [1–5]. Because of their good flexibility, high sensitivity, and low cost, flexible pressure sensors have been applied in flexible electronic skin, wearable health-monitoring devices, human motion state detection, prosthetic hands, human–computer interaction, etc. [6–10]. Compared with other sensors, flexible capacitive pressure sensors have received more attention due to their simple structure, high sensitivity, rapid dynamic response, and excellent stability [11–13]. Conventional capacitive flexible pressure sensors have sandwiched structures, and they mainly rely on the deformation of the middle dielectric layer to produce a rapid response to pressure, but because they are limited by the poor deformability of the elastic dielectric layer, such sensors still have a poor performance in terms of small-pressure detection and detection range [14,15]. To significantly improve both properties, advanced preparation processes and methods (e.g., the mold method and photolithography [16,17]) are often employed to form regular/irregular microstructure arrays (e.g., pyramids, semicircles, columns, and randomly distributed spine structures, folds [18–22]) on the surface of dielectric-layer elastomers (e.g., polydimethylsiloxane (PDMS), SR/Ecoflex) to improve deformability [23,24]. With this type of dielectric layer, the flat dielectric layer, which has poor compressibility, can produce a relatively large deformation in the range of small pressures, thus significantly improving the initial sensitivity of the sensor and increasing the resolution of the sensed pressure [25]. However, at large applied pressures, the deformation of the surface microstructure reaches its limit, and it is difficult for the sensor to realize a wide detection range. To improve the large deformation capability of the dielectric layer, a porous structured substrate (e.g., foam, sponge) is used [26–28]. This high porosity makes it easy for the sensor to deform under pressure and enables the pressure sensor to have a wide range of capabilities. However, to ensure large deformation under the applied pressure, this existing dielectric layer is often made too large compared to the electrode. The obvious disadvantage of relying on this approach is that it makes sensor packaging difficult, and the potential problem is that the service life is reduced due to structural instability over long-time use, thus affecting the long-term reliability of the sensor. Therefore, to achieve high sensitivity of the sensor while meeting the performance requirement of small-pressure detection and a wide detection range, from the perspective of improving the deformability of the dielectric layer, a double-sided microstructure combining the surface microstructure and internal microstructure is an effective solution.

This study reports a flexible capacitive pressure sensor based on a double-sided microstructure porous dielectric layer. To increase the dielectric constant of the dielectric layer, the dielectric layer was made of a mixture of silicon rubber (SR), salt (NaCl), and carbon black (CB) as a flexible substrate to create an internal porous microstructure by using the characteristics of salt to dissolve water easily. Meanwhile, the hemispherical microstructure was replicated by the template method under the top and bottom of the dielectric layer. Owing to the synergistic effect of the surface hemispheric microstructure and the internal porous structure, the sensor exhibits high sensitivity (3.15 kPa^{-1}), high pressure resolution (i.e., the minimum pressure detected is 27 Pa), and a wide pressure detection range (0–200 kPa). Based on the proposed sensing unit, a large-area pressure sensing array is developed to map the pressure size and position distribution of the object. In addition, by adopting machine learning, object shape recognition is achieved with the dataset collected from the pressure array.

2. Materials and Methods

2.1. Materials

The silicon rubber (SR) and curing agent were purchased from Donghong Craft Material Co., Dongguan, China. The CB (model: BP2000) was purchased from Cabot Corporation (Boston, MA, USA). NaCl was purchased from Shandong Daiyue Salt Co. (Tai'an, China). The conductive silver paste with a volume resistance of 0.2 mΩ was purchased from Guangzhou Kaixiang Electronics Co. (Guangzhou, China). The polyethylene terephthalate film (thickness = 0.025 mm) was purchased from RuiXin Plastic Co. (Jiaxing, China). The PET-based double-sided tape (model: 3M9495LE) and the polyimide single-sided tape (thickness = 50 μm) were purchased online. All materials were used without additional treatment.

2.2. Preparation

2.2.1. Microstructure Mold

An array hemispheric elastic microstructure mold and a surface microstructure-free mold of the same size were fabricated by 3D printing. Specifically, the diameter of hemispheres and the spacing between them were 1.5 mm, the microstructure area was 11 × 11 cm, and the whole mold size was 14 × 14 cm. The microstructure of the mold is shown in Figure 1.

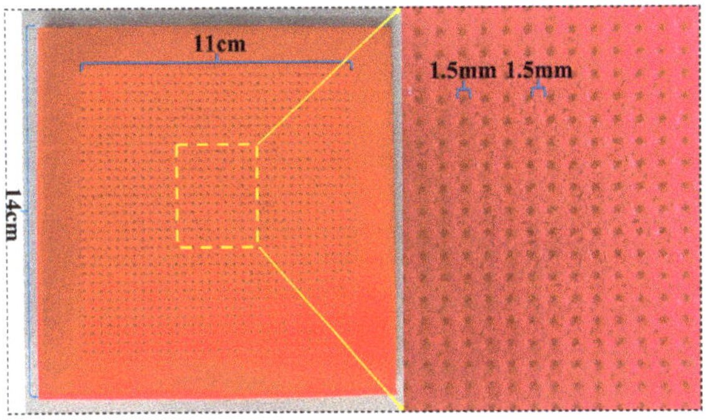

Figure 1. Actual image of microstructure mold.

2.2.2. Composite Material

The mass ratio of SR: NaCl: CB: SR curing agent was set to 50:25:1:1. First, 60 g of SR was added to the beaker, and then 30 g of NaCl powder was added to the beaker with mechanical stirring for 1 h. Then, 1.2 g of CB was added to the beaker and stirred for 40 min. Finally, 1.2 g of curing agent was added to the beaker and stirred for 20 min to mix all the ingredients. The beaker was then placed in a vacuum-drying oven for 10 min to eliminate air bubbles.

2.3. Fabrication of the Dielectric Layer

The fabrication process of the porous double-sided microstructure dielectric layer is shown in Figure 2a. First, the composite material was poured on the surface of mold A with the microstructure adsorbed on the spin coater tray, and then it was spun at 600 rpm for 40 s. Next, the finished mold was placed in a vacuum-drying oven to remove the air (10 min). The above steps were repeated to cover the exposed part of the mold with a uniform layer of composite material, and the spin-coating-vacuuming process was repeated until the entire surface of the microstructure mold was covered with a uniform layer of composite material after vacuuming. This operation of mold A was repeated for mold B. Then, the sides of mold A and mold B containing the composite material (not fully cured yet) were aligned and attached, and the molds were covered with an object of 1 kg that was slightly larger than the molds (to make the two molds fit more tightly). Then, the molds were placed in a vacuum-drying oven for 10 min to remove the air between them to obtain a double-sided microstructure porous dielectric layer. The fabrication process of the porous single-sided microstructure dielectric layer is shown in Figure 2b. The operation of mold A was repeated for mold C to make a single-sided microstructure dielectric layer. Mold D without the microstructure was used to manufacture the porous dielectric layer without the microstructure, as shown in Figure 2c. Afterward, all molds were transferred to an oven for thermal curing (50 °C, 4 h). Subsequently, all the molds were placed in an ultrasonic cleaner with deionized water for 2 h to separate the dielectric layer from the molds. After stripping, all dielectric layers were further shaken for 24 h to remove the internal NaCl. Eventually, the porous double-sided microstructure dielectric layer, the porous single-sided microstructure dielectric layer, and the porous dielectric layer without the microstructure were obtained.

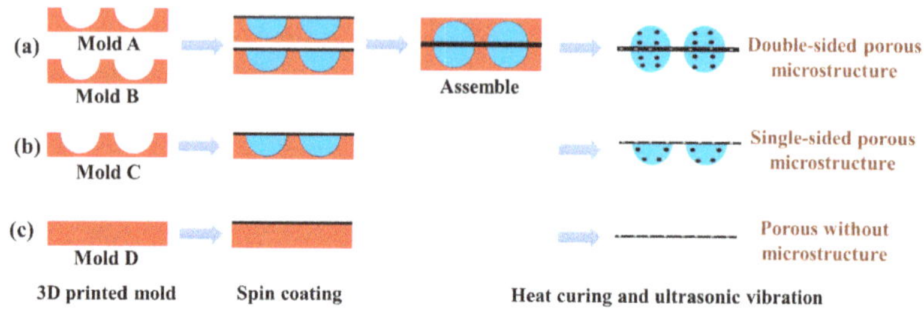

Figure 2. Fabrication process of dielectric layer. (**a**) The porous double-sided microstructure dielectric layer; (**b**) the porous single-sided microstructure dielectric layer; (**c**) the porous dielectric layer without microstructure.

2.4. Fabrication of the Electrode Layer

The electrode layer was prepared by the screen-printing method, and the fabrication process is illustrated in Figure 3a. The flexible substrate PET and the printing screen were fixed on the screen-printing platform, and the conductive silver paste was printed onto the entire electrode pattern surface using a scraper. This process was repeated several times to ensure that the complete electrode pattern was formed on the PET surface, and then the printed PET was placed in a drying oven to cure at 80 °C for 8 min. The actual electrode layer is shown in Figure 3b. The electrode area was a 10 $\times$10 mm square, and a 6 $\times$ 10 mm wire area was led out from one side of the square electrode to facilitate the lead-out for subsequent sensor performance testing.

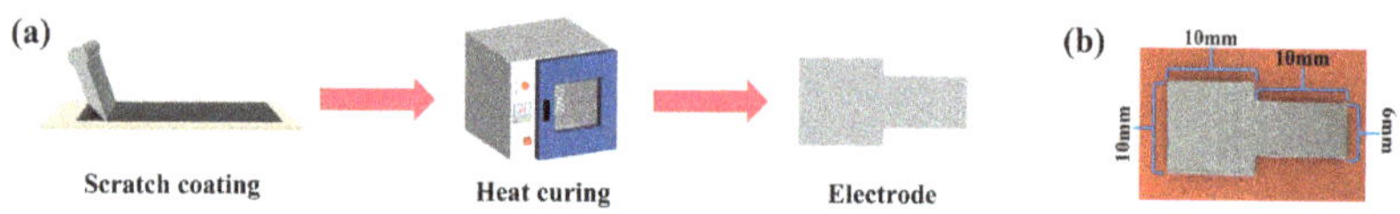

Figure 3. Fabrication of the electrode layer. (**a**) The fabrication process; (**b**) the actual image.

2.5. Sensor Packaging

Figure 4a shows the packaging process of the sensor. The porous double-sided microstructure dielectric layer, the porous single-sided microstructure dielectric layer, and the porous dielectric layer without the microstructure were cut into appropriate sizes and sandwiched between the two printed silver electrodes, respectively. In addition, the PI tape was used to encapsulate the sensor. The conductive aluminum tape connects the conductor area reserved for the electrode to form a pressure sensor. These sensors are labeled as sensor I, sensor II, and sensor III, and their entities are presented in Figure 4b.

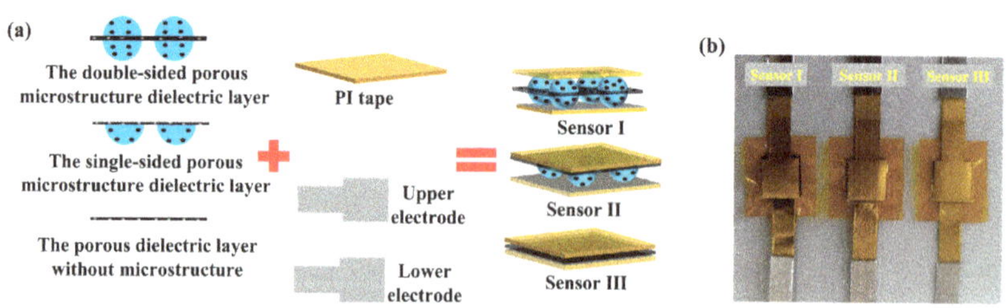

Figure 4. Sensor packaging of the sensor I, the sensor II and the sensor III. (**a**) The packaging process; (**b**) the actual image.

2.6. Fabrication of the Sensor Pressure Array

Following the process of sensor unit fabrication, the electrode array pattern shown in Figure 5a was designed and processed into a printing screen. Then, the conductive silver paste was scraped and cured with heat to obtain the actual electrode array. Meanwhile, conductive silver glue was used to connect the array electrode to the terminal wire. The steps for making the dielectric layer are the same as those for the dielectric layer of sensor I. Afterward, the upper array electrode, middle dielectric layer, and lower array electrode were assembled to enhance the insulation between the upper and lower array electrodes and to improve the performance of the flexible capacitive pressure sensor array. The non-conductive side of the pattern 2 array electrode was fitted together with the dielectric layer, and the conductive side was coated with TPU to make it insulated from the outside world. Finally, the sensor pressure array was encapsulated using 3M tape, and its entirety is shown in Figure 5b.

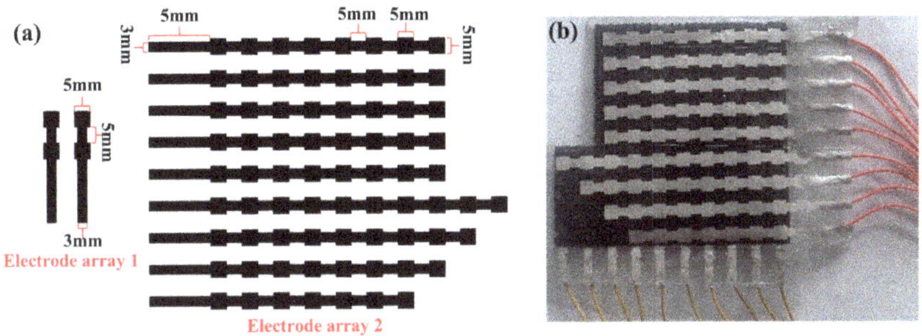

Figure 5. Fabrication of the sensor pressure array. (**a**) The electrode array pattern; (**b**) the actual image.

2.7. Characterization and Measurements

The hemisphere structure of the dielectric layer surface and the surface morphology of the internal porous microstructure were characterized using FESEM. Meanwhile, the capacitance variation was measured at 1 kHz with an LCR meter (TH2826). The external dynamic pressure was applied using a computer-controlled tensile tester (ZQ-990B) (mainly to test the repeatability and response time). In addition, static pressure (test sensitivity) was applied using a manual digital push–pull gauge (HP-50).

3. Results and Discussions

3.1. Finite Element Analysis of Dielectric Layer Sensitivity

The Workbench Platform (ANSYS) was used for the statics simulation of sensor I, sensor II, and sensor III to investigate the reason why the porous double-sided microstructure dielectric layer was most effective for sensitivity enhancement. The three-dimensional model of the sensor created by SolidWorks was imported into the Workbench software. The density of the dielectric layer was set to 0.98 g/cm^3, Young's modulus was 1.2 GPa, and Poisson's ratio was 0.4. During the simulation, the bottom electrode of the sensor was fixed, and the static force was applied to the top electrode surface, ignoring the deformation of the electrode layer. The mechanical simulation results of the sensor at a pressure of 500 Pa, 1 kPa, 5 kPa, and 10 kPa are shown in Figure 6 (the top electrode is hidden). It can be seen that under the same pressure conditions, the deformation of the sensor dielectric layer is ranked as sensor I > sensor II > sensor III. The reason for this difference is that the force on the dielectric layer of sensor I is mainly concentrated at the array elastic hemispheric microstructure on the upper surface and the array elastic hemispheric microstructure on the lower surface; the force on the dielectric layer of sensor II is mainly concentrated at the array elastic hemispheric microstructure on the upper surface and the whole lower surface; the force on the dielectric layer of sensor III is mainly concentrated at the whole upper surface and the whole lower surface. Meanwhile, the force area of the sensor dielectric layer is ranked as sensor I > sensor II > sensor III. In the case of the same force, the smaller the force area, the greater the pressure on the object. Thus, the dielectric layer of sensor A generates the largest deformation, the dielectric layer of sensor B generates the second-largest deformation, and the dielectric layer of sensor C generates the smallest deformation. From the above simulations, the sensor sensitivity is ranked as sensor I > sensor II > sensor III. Theoretically, it is proved that the porous double-sided microstructure dielectric layer has the best effect on improving the sensitivity of the flexible capacitive pressure sensor.

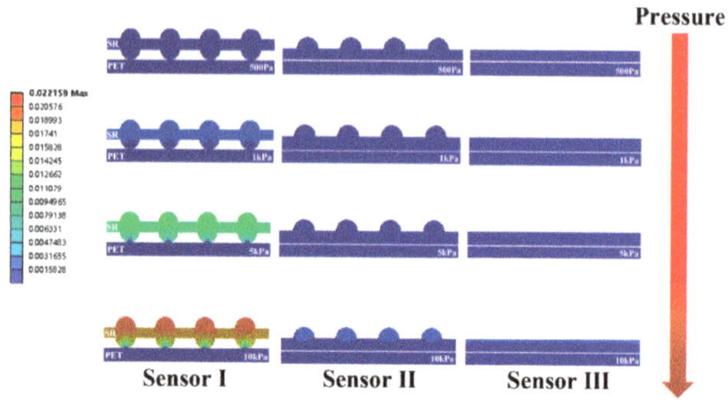

Figure 6. Finite element diagram of sensor I, sensor II, and sensor III under pressures of 0.5 Pa, 1 kPa, 5 kPa, and 10 kPa.

3.2. Sensitivity

A digital push–pull gauge was used to load pressure on the sensor surface, and an LCR meter was employed to collect the corresponding output capacitance. Figure 7 shows the pressure–capacitance curves of the sensor. The sensitivity (*S*) of the capacitance sensor can be defined as [29]:

$$S = \delta(\Delta C/C_0)/\delta P \quad (1)$$

where ΔC is the variation of capacitance ($\Delta C = C - C_0$), C is the capacitance under the corresponding pressure, C_0 is the initial capacitance, and P is the applied pressure. According to Equation (1), the sensitivity of the sensor is the slope of the curve. Then, the sensitivities of the above sensors in different pressure intervals were obtained by the function of segmented linear fitting, and the results are shown in Figure 8. It can be seen that the sensitivity of sensor I is the highest in any pressure range. The experimental results indicate that the porous double-sided microstructure dielectric layer has the best effect in terms of sensitivity.

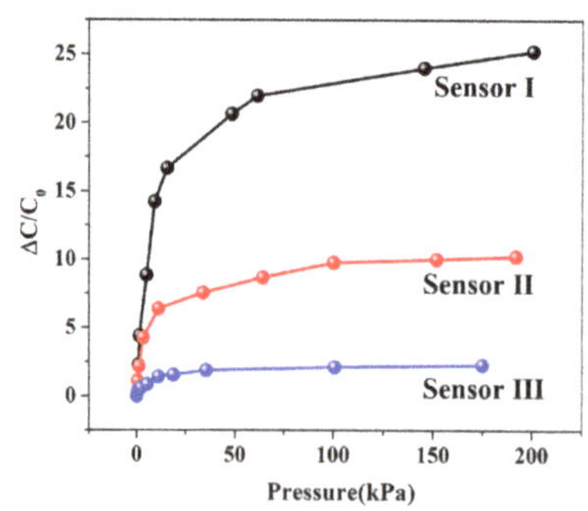

Figure 7. Pressure response curve of the sensor.

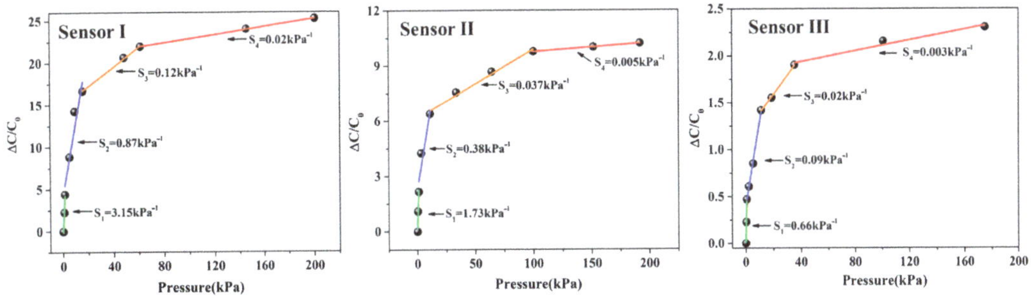

Figure 8. The function of segmented linear fitting.

3.3. Characterization

Figure 9 shows the SEM image of the dielectric layer cross-section of sensor I. It can be observed that there are many holes of different sizes in the interior of the dielectric layer. According to the local magnification, the pore diameter ranges from 10 to 160 µm. Such internal holes greatly improve the elastic performance, which leads to a larger deformation at the same pressure compared to the dielectric layer without holes, thus increasing the sensitivity of the flexible capacitive pressure sensor. In addition, the combination of the porous microstructure inside the dielectric layer and the array elastic microstructures on the surface of the dielectric layer greatly improves the sensitivity of the flexible capacitive pressure sensor. Note that the array elastic hemispheric microstructure (yellow) on the surface of the dielectric layer is not a true hemisphere, and this is caused by the fact that the 3D printing technology produces the mold by printing in layers. As a result, the array elastic hemispheric microstructure mold made is not a true hemisphere, which makes the array elastic microstructure on the surface of the dielectric layer approximate a hemisphere. Overall, among the three types of dielectric layers, the porous double-sided microstructure dielectric layer achieves the best effect in improving the sensitivity of the flexible capacitive pressure sensor, and the sensitivity of the sensor has been greatly improved, which proves that the method is effective in improving the sensitivity of the flexible capacitive pressure sensor.

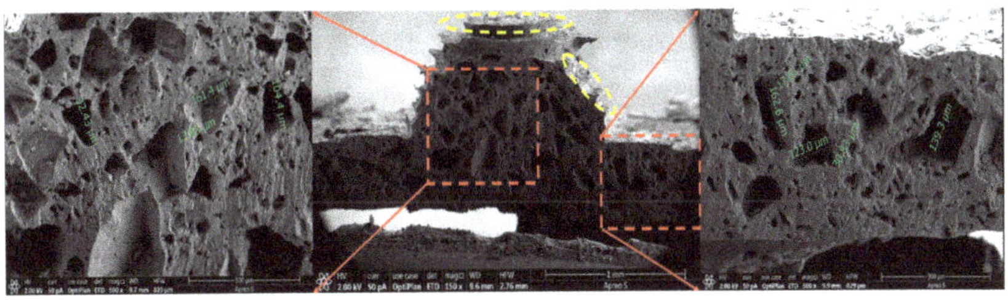

Figure 9. Cross-section SEM image of the dielectric layer of sensor I (the porous microstructure in the red area, and the hemispherical microstructure in the yellow area).

3.4. Dynamic Performance

The pressure response curves are shown in Figure 10a,b, with the square wave pressure and the sharp wave pressure of 0.9 kPa, 6 kPa, 38 kPa, and 52 kPa applied to the sensor surface. It can be seen that when a steady-state force is applied to the sensor, the capacitance change remains stable during the application of the pressure, and there is a clear distinction in the capacitance change for different steady-state forces. When a transient force is applied to the sensor, the capacitance change is almost the same as that of applying a steady-state force of the same size. The above results indicate that the sensor has a stable response to different loading types of pressure and meets the requirement for stability. In addition, we tested the minimum pressure limit by successively loading different weights (100 mg,

200 mg, 500 mg) onto the surface of the sensor. When a 500 mg weight was loaded, the sensor had a significant capacitance output change, as shown in Figure 10c. At this time, the minimum pressure that the sensor can respond to is about 27 Pa. It can be observed that the sensor has a high resolution, which can reflect the small pressure variations well. The sensor pressure response curves for the applied/released 0.9 kPa are shown in Figure 10d–f. The results show that the sensor has a short response time and a recovery time of 120 ms, so it meets the requirement for rapid response to pressure. The sensor resolution refers to the minimum pressure that the sensor can respond to. To test the service life and durability of the sensor, a continuous pressure of 1.5 kPa was applied/released in a time period of 2500 s, and the results are shown in Figure 10g–i. It can be seen that the overall graphs of the sensor's capacitance variation curves are stable without large fluctuations, indicating that the sensor has a reliable service life and durability. Meanwhile, the curve graphs of the local repeatability tests are shown for 220–230 s and 1960–1970 s, respectively. The curves of the sensor capacitance change at the beginning and the end of the applied pressure are almost the same, which is good proof of its reliable service life and durability. The above performance tests indicate that the porous double-sided microstructure dielectric layer enables the flexible capacitive pressure sensor to achieve high sensitivity and excellent performance in terms of detection range and resolution.

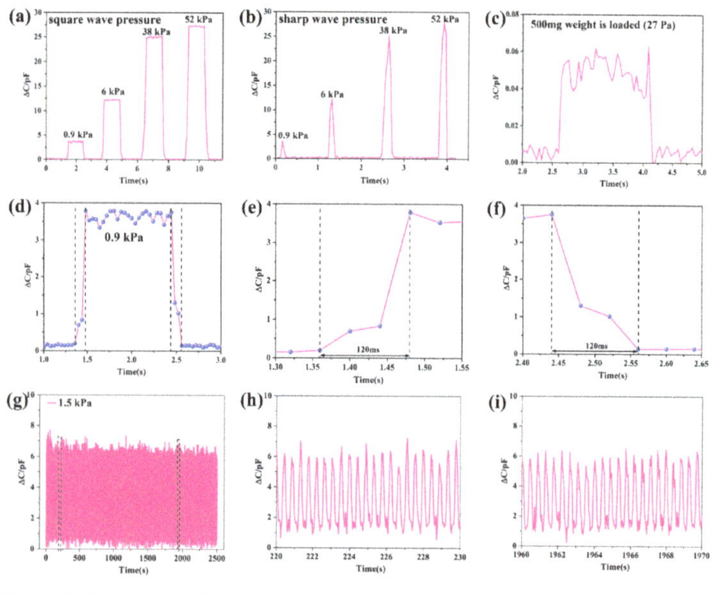

Figure 10. Dynamic performance of the capacitive pressure sensors. (**a**) The output curve under the square wave pressure of 0.9 kPa, 6 kPa, 38 kPa, and 52 kPa; (**b**) the output curve under the sharp wave pressure of 0.9 kPa, 6 kPa, 38 kPa, and 52 kPa; (**c**) minimum response curve of force for 27 Pa; (**d**) the sensor force response curves for 0.9 kPa; (**e**) response time of the sensor; (**f**) recovery time of the sensor; (**g**) repeatability curve of the sensor under the pressure of 1.5 kPa; (**h**) local repeatability curve within 220–230 s; (**i**) local repeatability curve within 1960–1970 s.

3.5. Application

To verify the ability of the flexible capacitive pressure sensor array to sense the real object, two 10 g weights, hexagonal bars, and discs in the form of pressure point, pressure line, and pressure plane were placed on its surface for testing. The results are illustrated in Figure 11. It can be seen that the flexible capacitive pressure sensor array can reflect both the size of the object pressure and the location distribution of the pressure on the surface of the object, with varying response levels and response areas for different objects.

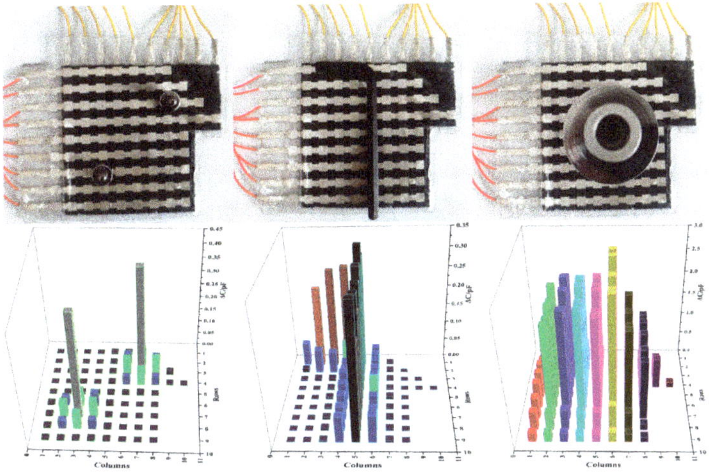

Figure 11. Shape perception of different objects by pressure sensor array.

Object shape recognition plays an important role in robotic hand-grasping perception tasks [30,31]. There are two main methods of object recognition at present: One is an object recognition system based on visual perception, whose performance mainly depends on the quality of the image and the quality of the recognition algorithm. Another method is an object recognition system based on tactile perception. This method is not easily affected by a complex environment, and its direct contact with objects makes the perceived information more abundant and reliable. The flexible sensor array is flexible and easy to integrate. When grasping an irregular object, the flexible sensor array can bend and fold freely to fully contact the irregular object, and the perceived data is more abundant and reliable. Therefore, this study prepared a data glove by integrating a flexible capacitive pressure sensor array into the fingertips of each of the five fingers and the palm of the glove, as shown in Figure 12a. By combining machine learning and pressure sensing to achieve object recognition, the process of the human hand's tactile perception of external information was simulated. The object recognition system was built based on the integration of the data glove, the data collection circuit, and the object recognition model, as shown in Figure 12b. The working principle is as follows: the control chip (model: TMS320F28388) controls the single knife double-throw switch (model: ADG711) to select the pressure sensor array unit one by one; meanwhile, it controls the capacitive signal collection chip (model: PCap02) to collect the signal by the flexible capacitive pressure sensor array and transmits the signal to the host computer through the serial port. The data is preprocessed in Matlab, and the object recognition model is established using a probabilistic neural network (PNN) algorithm. The feature data of six objects (baseball (No. 1), orange (No. 2), mineral water (No. 3), anhydrous ethanol bottle (No. 4), pen (No. 5), and hexagon bar (No. 6)) were collected from the time of grasping and stabilizing to the time of starting to release the objects, as shown in Figure 13. In this approach, a total of 200 sets of data were collected for each object, and a total of 1200 sets of data were collected, each of which contained 94-dimensional features. Then, 80% of the data was randomly selected as the training set of the PNN algorithm to train the object recognition model, and the remaining 20% of the data was used as the test set to test the recognition accuracy of the object recognition model.

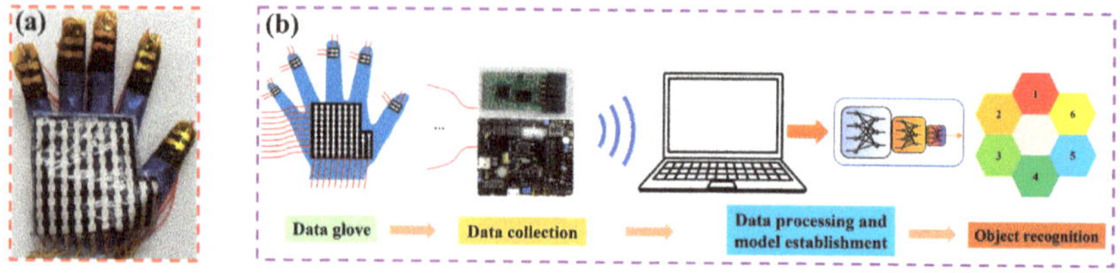

Figure 12. The process of object recognition. (**a**) The data glove with integrated sensors; (**b**) the object recognition system.

Figure 13. The process of grasping six objects (baseball, orange, mineral water, anhydrous ethanol bottle, pen, and hexagon bar).

The overall structure of the PNN algorithm is shown in Figure 14a. The input layer of the model has 71 nodes, which can input 71 eigenvalues. The hidden layer contains 960 neurons. The output layer outputs 1–6, representing six different objects, respectively. The test results are presented in Figure 14b, where "*" represents the real class of the object, and "△" represents the classification result of the object recognition model. It can be seen that Object 1 and Object 2, Object 3 and Object 4, and Object 5 and Object 6 have more recognition errors than other cases. To see more clearly the classification of objects with the object recognition model, the confusion matrix was drawn, as shown in Figure 14c. As shown in the figure, since Object 1 and Object 2, Object 3 and Object 4, and Object 5 and Object 6 are similar in shape, they are easily confused with each other; meanwhile, the probability of confusion between other objects is very low. Table 1 lists the tested recognition accuracy of the object recognition model. It can be seen that the recognition accuracy of all objects exceeds 84%, and the overall recognition accuracy is 87.1%. The results indicate that based on the proposed flexible capacitive pressure sensor array, an adequate and accurate data set is collected to guarantee a high recognition accuracy of the object recognition system, which is expected to provide an accurate tactile perception of the prosthetic hand with the assistance of machine learning.

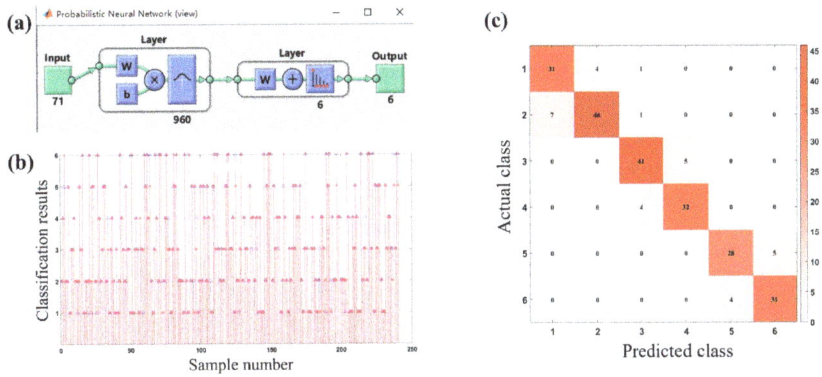

Figure 14. Building and testing of object recognition model. (**a**) Overall structure of PNN algorithm; (**b**) test results of sample data; (**c**) confusion matrix.

Table 1. Recognition accuracy of object recognition model.

Object Number	Recognition Accuracy	Overall Recognition Accuracy
1	86.1%	
2	85.2%	
3	89.1%	87.1%
4	88.9%	
5	84.8%	
6	88.6%	

4. Conclusions

In this work, a flexible capacitive pressure sensor based on a double-sided microstructured porous dielectric layer was developed. Owing to the synergistic effect of the surface hemispherical microstructure and the internal porous structure, the sensor has high sensitivity (3.15 kPa^{-1}), a wide pressure sensing range (0–200 kPa), and high resolution (i.e., the minimum pressure detected is 27 Pa). Meanwhile, the array sensor satisfies the three force-loading forms of pressure perception: "point", "line", and "surface". The sensor array is integrated into the glove to obtain the pressure dataset of the grasped object, and it is combined with machine learning to recognize the shape of the grasped object, achieving an overall accuracy is 87.1%. These results indicate that the sensor has the potential for intelligent prosthetic hand haptic perception. Overall, a reasonable combination of multiple microstructures provides an effective approach for improving sensor performance.

Author Contributions: Conceptualization, Q.Y. and J.Z.; methodology, Q.Y. and J.Z.; software, J.Z.; validation, Q.Y.; formal analysis, Q.Y. and J.Z.; investigation, J.Z.; resources, Q.Y.; data curation, Q.Y. and J.Z.; writing—original draft preparation, Q.Y. and J.Z.; writing—review and editing, Q.Y. and J.Z.; visualization, Q.Y. and J.Z.; supervision, Q.Y.; project administration, Q.Y.; All authors have read and agreed to the published version of the manuscript.

Funding: This research received no external funding.

Data Availability Statement: Not applicable.

Conflicts of Interest: The authors declare no conflict of interest.

References

1. Ji, B.; Zhou, Q.; Wu, J.B.; Gao, Y.B.; Wen, W.J.; Zhou, B.P. Synergistic Optimization toward the Sensitivity and Linearity of Flexible Pressure Sensor via Double Conductive Layer and Porous Microdome Array. *ACS Appl. Mater. Interfaces* **2020**, *12*, 31021–31035. [CrossRef]
2. Lu, P.; Wang, L.; Zhu, P.; Huang, J.; Wang, Y.J.; Bai, N.N.; Wang, Y.; Li, G.; Yang, J.L.; Xie, K.W.; et al. Iontronic Pressure Sensor with High Sensitivity and Linear Response Over a Wide Pressure Range Based on Soft Micropillared Electrodes. *Sci. Bull.* **2021**, *66*, 1091–1100. [CrossRef]
3. Yi, Z.R.; Liu, Z.X.; Li, W.B.; Ruan, T.; Chen, X.; Liu, J.Q.; Yang, B.; Zhang, W.M. Piezoelectric Dynamics of Arterial Pulse for Wearable Continuous Blood Pressure Monitoring. *Adv. Mater.* **2022**, *34*, 2110291. [CrossRef]
4. Lee, S.; Park, J.-W. Fingerprint-Inspired Triboelectric Nanogenerator with a Geometrically Asymmetric Electrode Design for a Self-Powered Dynamic Pressure Sensor. *Nano Energy* **2022**, *101*, 107546. [CrossRef]
5. Yan, Y.C.; Hu, Z.; Yang, Z.B.; Yuan, W.Z.; Song, C.Y.; Pan, J.; Shen, Y.J. Soft magnetic skin for super-resolution tactile sensing with force self-decoupling. *Sci. Robot.* **2021**, *6*, eabc8801. [CrossRef]
6. Wang, P.; Liu, J.; Yu, W.; Li, G.X.; Meng, C.Z.; Guo, S.J. Flexible, Stretchable, Breathable and Sweatproof All-Nanofiber Iontronic Tactile Sensor for Continuous and Comfortable Knee Joint Motion Monitoring. *Nano Energy* **2022**, *103*, 107768. [CrossRef]
7. Abramson, A.; Chan, C.T.; Khan, Y.; Mermin-Bunnell, A.; Matsuhisa, N.; Fong, R.; Shad, R.; Hiesinger, W.; Mallick, P.; Gambhir, S.S.; et al. A Flexible Electronic Strain Sensor Forthe Real-Time Monitoring Oftumor Regression. *Sci. Adv.* **2022**, *8*, 6550. [CrossRef]
8. Guan, J.R.; Zhang, D.Z.; Li, T.T. Flexible Pressure Sensor Based on Molybdenum Diselide/Multi-Walled Carbon Nanotubes for Human Motion Detection. *IEEE Sens. J.* **2021**, *21*, 10491–10497. [CrossRef]
9. Osborn, L.E.; Dragomir, A.; Betthauser, J.L.; Hunt, C.; Nguyen, H.H.; Kaliki, R.R.; Thakor, N.V. Prosthesis with Neuromorphic Multilayered E-dermis Perceives Touch and Pain. *Sci. Robot.* **2018**, *3*, eaat3818. [CrossRef]
10. Huang, J.R.; Wang, H.T.; Li, J.-A.; Zhang, S.M.; Li, H.; Ma, Z.; Xin, M.; Yan, K.; Cheng, W.; He, D.W.; et al. High-Performance Flexible Capacitive Proximity and Pressure Sensors with Spiral Electrodes for Continuous Human-Machine Interaction. *ACS Mater. Lett.* **2022**, *4*, 2261–2272. [CrossRef]
11. Fu, X.; Zhang, J.Q.; Xiao, J.L.; Kang, Y.R.; Yu, L.T.; Jiang, C.P.; Pan, Y.X.; Dong, H.; Gao, S.K.; Wang, Y.C. A High-Resolution, Ultrabroad-Range and Sensitive Capacitive Tactile Sensor Based on a CNT/PDMS Composite for Robotic Handst. *Nanoscale* **2021**, *13*, 18780. [CrossRef]
12. Qin, R.Z.; Hu, M.J.; Li, X.; Liang, T.; Tan, H.Y.; Liu, J.Z.; Shan, G.C. A New Strategy for the Fabrication of a Flexible and Highly Sensitive Capacitive Pressure Sensor. *Microsyst. Nanoeng.* **2021**, *7*, 100. [CrossRef]
13. Li, y.; Cao, Z.G.; Li, T.; Sun, F.C.; Bai, Y.Y.; Lu, Q.F.; Wang, S.Q.; Yang, X.Q.; Hao, M.Z.; Lan, N.; et al. Highly Selective Biomimetic Flexible Tactile Sensor for Neuroprosthetics. *Research* **2020**, *2020*, 8910692. [CrossRef]
14. Wu, J.N.; Yao, Y.G.; Zhang, Y.H.; Shao, T.Y.; Wu, H.; Liu, S.Y.; Li, Z.; Wu, L.M. Rational Design of Flexible Capacitive Sensors with Highly Linear Response Over a Broad Pressure Sensing Range. *Nanoscale* **2020**, *12*, 201198. [CrossRef]
15. Quan, Y.; Wei, X.B.; Xiao, L.; Wu, T.; Pang, H.Y.; Liu, T.F.; Huang, W.; Wu, S.H.; Li, S.B.; Chen, Z. Highly Sensitive and Stable Flexible Pressure Sensors with MicroStructured Electrodes. *J. Alloys Compd.* **2017**, *699*, 824–831. [CrossRef]
16. Bai, N.N.; Wang, L.; Xue, Y.H.; Wang, Y.; Hou, X.Y.; Li, G.; Zhang, Y.; Cai, M.K.; Zhao, L.Y.; Guan, F.Y.; et al. Graded Interlocks for Iontronic Pressure Sensors with High Sensitivity and High Linearity over a Broad Range. *Acs Nano* **2022**, *16*, 4338–4347. [CrossRef]
17. Cheng, W.; Wang, J.; Ma, Z.; Yan, K.; Wang, Y.M.; Wang, H.T.; Li, S.; Li, Y.; Pan, L.J.; Shi, Y. Flexible Pressure Sensor with High Sensitivity and Low Hysteresis Based on a Hierarchically Microstructured Electrode. *IEEE Electron Device Lett.* **2018**, *39*, 288–291. [CrossRef]
18. Ruth, S.R.A.; Beker, L.; Tran, H.; Feig, V.R.; Matsuhisa, N.; Bao, Z.N. Rational Design of Capacitive Pressure Sensors Based on Pyramidal Microstructures for Specialized Monitoring of Biosignals. *Adv. Funct. Mater.* **2020**, *30*, 1903100. [CrossRef]
19. Liang, G.H.; Wang, Y.C.; Mei, D.Q.; Xi, K.L.; Chen, Z.C. Flexible Capacitive Tactile Sensor Array With Truncated Pyramids as Dielectric Layer for Three-Axis Force Measurement. *J. Microelectromech. Syst.* **2015**, *24*, 1510–1519. [CrossRef]
20. Zhuo, B.G.; Chen, S.J.; Zhao, M.G.; Guo, X.J. High Sensitivity Flexible Capacitive Pressure Sensor Using Polydimethylsiloxane Elastomer Dielectric Layer Micro-Structured by 3-D Printed Mold. *IEEE J. Electron Devices Soc.* **2017**, *5*, 219–223. [CrossRef]
21. Bai, N.N.; Wang, L.; Wang, Q.; Deng, J.; Wang, Y.; Lu, P.; Huang, J.; Li, G.; Zhang, Y.; Yang, J.L.; et al. Graded Intrafillable Architecture-Based Iontronic Pressure Sensor with Ultra-Broad-Range High Sensitivity. *Nat. Commun.* **2020**, *11*, 209. [CrossRef]
22. Shuai, X.T.; Zhu, P.L.; Zeng, W.J.; Hu, Y.G.; Liang, X.W.; Zhang, Y.; Sun, R.; Wong, C.-P. Highly Sensitive Flexible Pressure Sensor Based on Silver NanowiresEmbedded Polydimethylsiloxane Electrode with Microarray Structure. *Acs Appl. Mater. Inter.* **2017**, *9*, 26314–26324. [CrossRef]
23. Zeng, X.W.; Wang, Z.X.; Zhang, H.; Yang, W.; Xiang, L.; Zhao, Z.Z.; Peng, L.-M.; Hu, Y.F. Tunable, Ultrasensitive, and Flexible Pressure Sensors Based on Wrinkled Microstructures for Electronic Skins. *ACS Appl. Mater. Inter.* **2019**, *11*, 21218–21226. [CrossRef]
24. Yao, S.J.; Yu, J.P.; Jiang, X.L.; Xu, J.F.; Lan, K.; Yao, Z.H. Fabrication and Experimental Validation of a Sensitive and Robust Tactile Sensing Array with a Micro-Structured Porous Dielectric Layer. *Micromachines* **2022**, *13*, 1724. [CrossRef]
25. Guo, Y.J.; Gao, S.; Yue, W.J.; Zhang, C.W.; Li, Y. AAO-Assisted Low-Cost Flexible Capacitive Pressure Sensors Based on Double-Sided Nanopillars by Facile Fabrication Method. *ACS Appl. Mater. Interfaces* **2019**, *11*, 48594–48603. [CrossRef]

26. Pruvost, M.; Smit, W.J.; Monteux, C.; Poulin, P.; Colin, A. Polymeric Foams for Flexible and Highly Sensitive Low-Pressure Capacitive Sensors. *Npj Flex. Electron.* **2019**, *3*, 7. [CrossRef]
27. Liu, X.Y.; Li, C.L.; Wang, Z.T.; Li, Y.F.; Huang, J.Q.; Yu, H.Y. Wide-Range Flexible Capacitive Pressure Sensors Based on Origami Structure. *IEEE Sens. J.* **2021**, *21*, 9798–9807. [CrossRef]
28. Yu, H.Y.; Guo, C.X.; Ye, X.; Pan, Y.F.; Tu, J.C.; Wu, Z.; Liu, X.Y.; Huang, J.Q.; Ren, Q.Y.; Li, Y.F. Wide-Range Flexible Capacitive Pressure Sensors Based on Dielectrics with Various Porosity. *Micromachines* **2022**, *13*, 1588. [CrossRef]
29. Xiong, Y.X.; Shen, Y.K.; Tian, L.; Hu, Y.G.; Zhu, P.L.; Sun, R.; Wong, C.-P. A Flexible, Ultra-Highly Sensitive and Stable Capacitive Pressure Sensor with Convex Microarrays for Motion and Health Monitoring. *Nano Energy* **2020**, *70*, 104436. [CrossRef]
30. Li, G.Z.; Liu, S.Q.; Wang, L.Q.; Zhu, R. Skin-Inspired Quadruple Tactile Sensors Integrated on a Robot Hand Enable Object Recognition. *Sci. Robot.* **2020**, *5*, 8134. [CrossRef]
31. Sundaram, S.; Kellnhofer, P.; Li, Y.Z.; Zhu, J.-Y.; Torralba, A.; Matusik, W. Learning the Signatures of the Human Grasp Using a Scalable Tactile Glove. *Nature* **2019**, *569*, 698–702. [CrossRef]

micromachines

Article

A Framework to Predict Gastric Cancer Based on Tongue Features and Deep Learning

Xiaolong Zhu [1], Yuhang Ma [1], Dong Guo [2], Jiuzhang Men [2], Chenyang Xue [1], Xiyuan Cao [1,*] and Zhidong Zhang [1,*]

[1] Key Laboratory of Instrumentation Science & Dynamic Measurement, School of Instrument and Electronics, North University of China, Taiyuan 030051, China
[2] Shanxi University of Chinese Medicine, Taiyuan 030051, China
* Correspondence: caoxiyuan@nuc.edu.cn (X.C.); zdzhang@nuc.edu.cn (Z.Z.)

Abstract: Gastric cancer has become a global health issue, severely disrupting daily life. Early detection in gastric cancer patients and immediate treatment contribute significantly to the protection of human health. However, routine gastric cancer examinations carry the risk of complications and are time-consuming. We proposed a framework to predict gastric cancer non-invasively and conveniently. A total of 703 tongue images were acquired using a bespoke tongue image capture instrument, then a dataset containing subjects with and without gastric cancer was created. As the images acquired by this instrument contain non-tongue areas, the Deeplabv3+ network was applied for tongue segmentation to reduce the interference in feature extraction. Nine tongue features were extracted, relationships between tongue features and gastric cancer were explored by using statistical methods and deep learning, finally a prediction framework for gastric cancer was designed. The experimental results showed that the proposed framework had a strong detection ability, with an accuracy of 93.6%. The gastric cancer prediction framework created by combining statistical methods and deep learning proposes a scheme for exploring the relationships between gastric cancer and tongue features. This framework contributes to the effective early diagnosis of patients with gastric cancer.

Keywords: gastric cancer; tongue features; non-invasive; prediction framework; deep learning

Citation: Zhu, X.; Ma, Y.; Guo, D.; Men, J.; Xue, C.; Cao, X.; Zhang, Z. A Framework to Predict Gastric Cancer Based on Tongue Features and Deep Learning. *Micromachines* **2023**, *14*, 53. https://doi.org/10.3390/mi14010053

Academic Editor: Arman Roohi

Received: 8 November 2022
Revised: 5 December 2022
Accepted: 18 December 2022
Published: 25 December 2022

1. Introduction

Gastric cancer is one of the most prevalent malignancies in humans, and the reason for many deaths each year [1]. Early-stage gastric cancer patients are asymptomatic [2], and a few present non-specific symptoms, such as epigastric discomfort and belching. The symptoms tend to be ignored because of their similarity to symptoms of chronic gastric disease. Therefore, identifying gastric cancer patients early, providing immediate treatment, and delaying the deterioration of gastric cancer patients to advanced stages are crucial in the protection of public health.

The most commonly used method for the detection of gastric cancer is gastroscopy [3]. By observing the condition of the gastric mucosa, the physicians assess the severity and location of the lesion. However, extensive expertise and experience are required from physicians to identify suspicious lesions, and the diagnosis results may be affected by the working state of physicians (e.g., alertness) [4]. The advancement of modern technology has enabled researchers to apply artificial intelligence techniques to gastroscopy [5,6]. Compared to specialized endoscopists, artificial intelligence technology is faster in assessing the severity of the lesions and more accurate. However, gastroscopy is an invasive operation that can potentially cause complications, such as perforation and bleeding [7]. Patients with minor symptoms are reluctant to choose gastroscopy to detect gastric cancer.

Tongue diagnosis is a non-invasive diagnostic modality to diagnose the status of patients conveniently. Tongue features not only reveal the physical condition of the organs but also correlate with their functions [8]. Significant changes in tongue features are observed when people develop physical lesions. Practitioners assess the severity of lesions and formulate solutions for the corresponding disease by analyzing the changes in the tongue features of patient [9].

Recently, instead of observing and recording tongue features, research has been focusing on objective tongue feature analysis with the benefit of modern technology. Tongue features have been digitally analyzed to discover convenient ways to detect diseases [10–12]. The extensive application of artificial intelligence techniques [13] in the medical field has increased objectivity in the interpretation of the relationship between tongue features and diseases. Image processing [14] has been utilized to extract information from tongue images and capture physical features that cannot be detected by human eyes. Deep learning [15] has been applied to automatically identify and learn tongue features. Links have been established between tongue features and diseases based on these technologies to predict diseases, such as breast cancer [16], diabetes [17], and gastric cancer [18].

Studies exploring the links between gastric cancer and tongue features have contributed to the detection of gastric cancer. Gastric cancer has been confirmed to be correlated with microbiota in the oral cavity, such as helicobacter pylori [19,20]. The microbiota in the oral cavity of gastric cancer patients has been analyzed by utilizing high-throughput sequencing and the tongue coating thickness has been measured in the literature [21,22]. As such, an association between tongue features and microbiota was established for gastric cancer detection. However, the sensitivity of this method is low and tends to be susceptible to microbial density. Gholami et al. [23] proposed a method that combined artificial intelligence techniques and tongue features to detect gastric cancer. The proposed method had a high accuracy in distinguishing between patients and healthy subjects. Nevertheless, the adaptive reference model that was utilized might change the tongue features, e.g., tongue shape and tongue color.

The main contribution and innovation of this study is to investigate the feasibility of predicting gastric cancer by the tongue. Because no open-source dataset of gastric cancer tongue images exists, a tongue image capture instrument was used to acquire tongue images of gastric cancer subjects and non-gastric cancer subjects. Based on image pre-processing and image enhancement, an original dataset was constructed. Next, the image segmentation algorithm was used for the tongue segmentation task. Then, nine tongue features were extracted with guidance from professional physicians. The association between these tongue features and gastric cancer is explored by using differential analysis, importance analysis and correlation analysis. Four tongue features that contribute to the prediction of gastric cancer are screened and retained. These four features including tongue shape, saliva, tongue coating thickness and tongue coating texture are used as a basis for predicting gastric cancer. Finally, a framework using EfficientNet [24] is established to achieve non-invasive prediction of gastric cancer patients. The details of the entire framework are shown in Figure 1. It provides an approach to study the relationship between tongue features and gastric cancer.

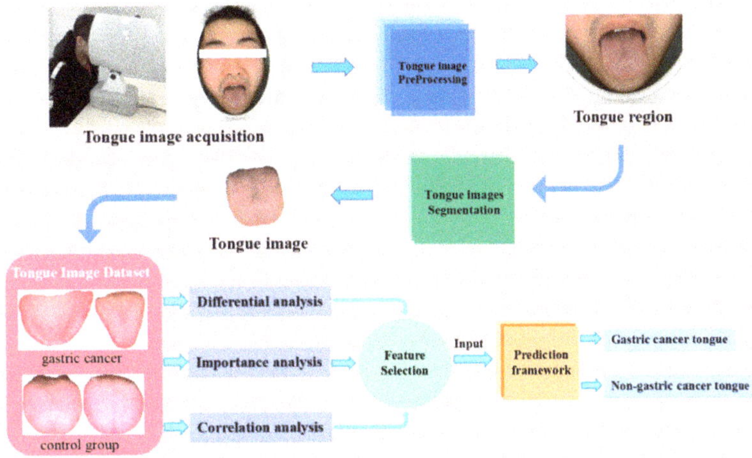

Figure 1. Overview of our framework.

2. Related Works

Gastric cancer is an enormous threat to human health. Early detection of gastric cancer has a positive significance in reducing the mortality rate of gastric cancer patients.

As one of the effective methods to diagnose gastric cancer, the physician can visually observe the changes of gastric mucosa and the size of the lesions. However, the detection results are affected by the work experience and subjective perception of physicians. To overcome these limitations, researchers [25–27] built gastric cancer identification models to detect gastric cancer in endoscopic images by using artificial intelligence techniques. The model was trained based on a large number of endoscopic images of gastric cancer annotated by professional physicians. The experimental results show that the model can accurately identify a large number of gastroscopic images within a short time. However, gastroscopy is an invasive operation that can easily lead to complications. The detection and analysis of volatile organic compounds in exhaled breath is a non-invasive method for diagnosing gastric cancer. Huang et al. [28] used a tubular surface-enhanced Raman scattering sensor to capture volatile organic compounds in human exhaled breath to noninvasively screen patients for gastric cancer, with an accuracy of 89.83%. The results showed that the breath analysis method provides an excellent option for screening gastric cancer. However, the results of breath analysis are susceptible to factors such as the breathing collection method, patient physiological condition, testing environment and analysis method [29].

Jiang et al. [30] constructed a noninvasive, high-precision diagnostic model for gastric cancer, using characteristics such as age, gender, and individual behavioral lifestyle. Compared to other models, the model they built using an extreme gradient-based augmentation algorithm obtained better results. The model achieved an overall accuracy of 85.7% in the test set. Zhu et al. [31] also used machine learning to build a noninvasive prediction model for gastric cancer risk. They used statistical methods to screen for significant features and performed multivariate analysis of significant features to exclude features that were not useful for predicting gastric cancer. The input features of the above models were usually the basic physiological information of the subjects. However, using tongue images to build disease prediction models has been neglected [11,32,33]. As a basic organ of the human body, changes in tongue features reflect physiological and case information. The diagnosis of tongue features can be used as a non-invasive screening method for the detection of gastric cancer. Gholami et al. [23] used tongue color and tongue lint to build a prediction model for gastric cancer, which had an accuracy of 91%. However, they only studied the association between tongue color and tongue lint with gastric cancer without exploring the association between other tongue features and gastric cancer.

Therefore, we extracted nine tongue features and explored the relationship between them and gastric cancer. Through differential analysis, importance analysis, and correlation analysis, tongue shape, saliva, tongue coating thickness and tongue coating texture were finally selected as the input features for predicting gastric cancer. In addition, we used image segmentation to remove non-tongue regions before building the gastric cancer prediction framework to reduce interference and improve the efficiency and accuracy of prediction.

3. Methods

To ensure the established framework had clinical application, we used an instrument to collect the tongue images of gastric cancer patients and non-gastric cancer subjects in hospitals. The images captured by the tongue image acquisition instrument contain non-tongue areas such as the face and the instrument. Non-tongue regions are significantly reduced by image pre-processing, which is helpful to improve the efficiency of image segmentation. The processed tongue images were input into the gastric cancer prediction framework, then the final prediction results were obtained.

3.1. Date Sources

Data was collected from Shanxi Cancer Hospital and the Affiliated Hospital of Shanxi University of Traditional Chinese Medicine from January 2021 to August 2022, with 703 images in total. The dataset consisted of 103 tongue images of gastric cancer patients and 600 tongue images of non-gastric cancer subjects that included patients with other diseases and normal individuals. The diseases of these subjects were definitely diagnosed. Tongue images of the patient were captured under the guidance of specialized physicians. The subjects were arranged to sit in front of the instrument and extend their tongue naturally (Figure 2). During the capture process, the quality of the tongue images was strictly controlled. The light environment was kept stable during image acquisition and blurred images were removed to ensure high-quality tongue acquisition. After image acquisition from each subject, the instrument was disinfected and ventilated to ensure a hygienic and safe acquisition environment.

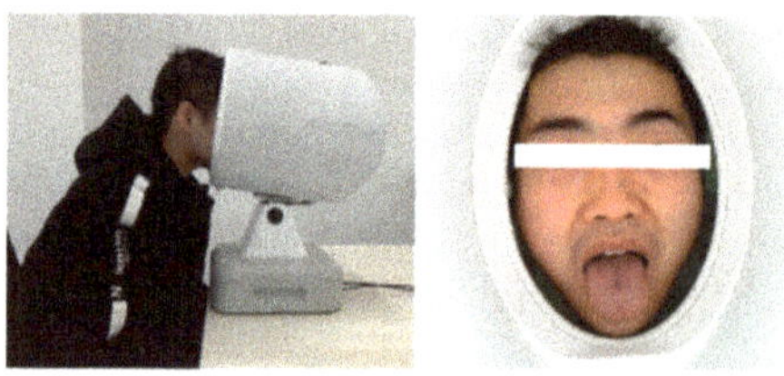

Figure 2. The process of capturing tongue images. During the collection, the subjects place their mandibles on the instrument and extended their tongues naturally.

3.2. Date Preprocess

The resolution of the acquired tongue image was 3264 × 2488 with horizontal and vertical resolution of 96 dpi. High-resolution tongue images containing non-tongue regions consume a lot of computational resources in the deep learning network. The tongue region extraction method proposed by Li [34] was employed to greatly reduce the area of non-tongue regions. This method improved the efficiency of subsequent tongue segmentation, while reducing the computational load of the deep learning network. After cropping the original tongue images, the image annotation software was used to construct a tongue dataset for segmentation model training. To avoid overfitting of the prediction model due to data imbalance, the number of tongue images was expanded to 4375 utilizing data augmentation [35]. The tongue images of the subjects with gastric cancer and non-gastric cancer were expanded in the same ratios, and the image augmentation algorithms used in this study include geometric transformations (e.g., flipping and panning) and the addition of Gaussian noise. The image augmentation algorithm is implemented using the imgaug library in python.

3.3. Segmentation of Tongue Images

In addition to the tongue region, images acquired by the tongue image capture instrument frequently contained areas, such as lips and teeth. The tongue-part of the images was segmented to eliminate the influence of non-tongue areas on the subsequent analysis. Existing automatic tongue segmentation approaches are classified into two categories, namely segmentation models based on traditional methods, such as the region growing method and the thresholding method, and segmentation networks based on deep learning. Traditional segmentation methods are insensitive to the color of the regions close to the tongue and easily mistake these regions for the tongue [36–38]. In this stage, Deeplabv3+ [39] network, which is based on deep learning, was chosen for the tongue segmentation task, and clinically collected tongue images were used as the dataset for segmentation.

The Deeplabv3+ network is based on an encoder-decoder architecture. The encoder part consists of a network for extracting features and multiple parallelly dilated convolution layers. The MobileNetV2 [40] is used as the backbone network. In the encoder part, tongue image features are extracted to generate high-level semantic features. To enhance the learning ability of the network, the low-level semantic features are merged with the high-level semantic. The merged feature map is then processed by the convolution and up-sampling layers to form the last semantic segmentation map. The architecture of the Deeplabv3+ network is shown in Figure 3.

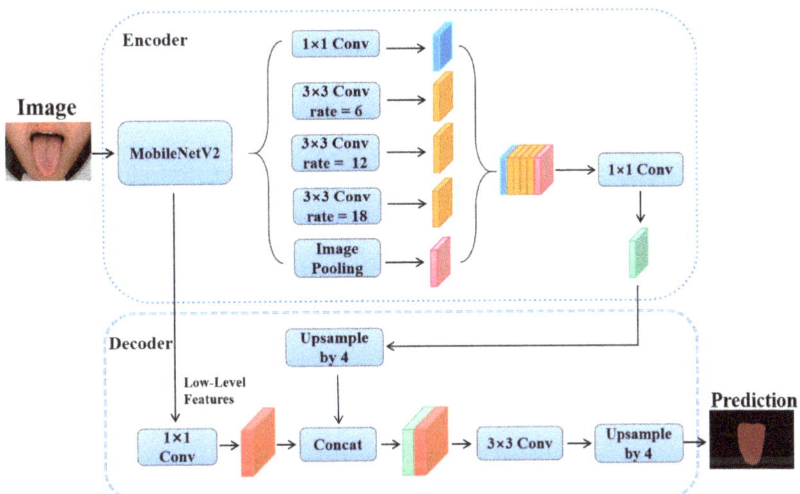

Figure 3. Structure of Deeplabv3+. The high-level semantic features and the low-level semantic features of the tongue image are extracted and fused by the Deeplabv3+ network. The low-level features have high resolution and contain a wealth of detailed information about the image. The high-level features have stronger semantic information. The fusion of features of different scales is an important method to improve the performance of segmentation networks.

The data set used for tongue segmentation was divided in a 7:2:1 ratio into the training set, validation set and test set. The MobileNetV2 network weight trained on ImageNet was used as the beginning weight. The Adam optimizer [41] was used to optimize the model during the training process. The segmentation network was trained with 200 epochs. In the first 100 training epochs, the weights of the MobileNetV2 network were frozen with a learning rate of 1×10^{-2}. In the last 100 epochs of unfrozen training, the learning rate of the global network was 1×10^{-4}.

3.4. Extraction and Analysis of Tongue Features

Combining the professional opinions of several doctors, nine tongue features were extracted to explore the difference between the tongues of gastric cancer patients and non-gastric cancer subjects. The extracted tongue features are listed in the following paragraphs.

1. Tongue shape: fat tongue, thin tongue and normal tongue;
2. Tooth-marked tongue: tooth mark and normal;
3. Spots and prickles tongue: Spots and prickles and normal;
4. Saliva: dry and normal;
5. Tongue coating thickness: thick and thin;
6. Tongue coating texture: greasy and normal;
7. Tongue color: pale white, pale red, red and deep red;
8. Tongue coating color: white and yellow;
9. Tongue Fissure: fissured tongue and normal.

The differences in tongue features between the gastric cancer patients and the control group were contrasted using statistical methods. The cross-plot of tongue features between the two groups is shown in Figure 4. Saliva, tongue coating thickness, tongue fissure, tongue coating texture, and tooth-marked tongue were distinctly different between the two groups. The weight of these tongue features in patients with gastric cancer varied considerably from that in the control group.

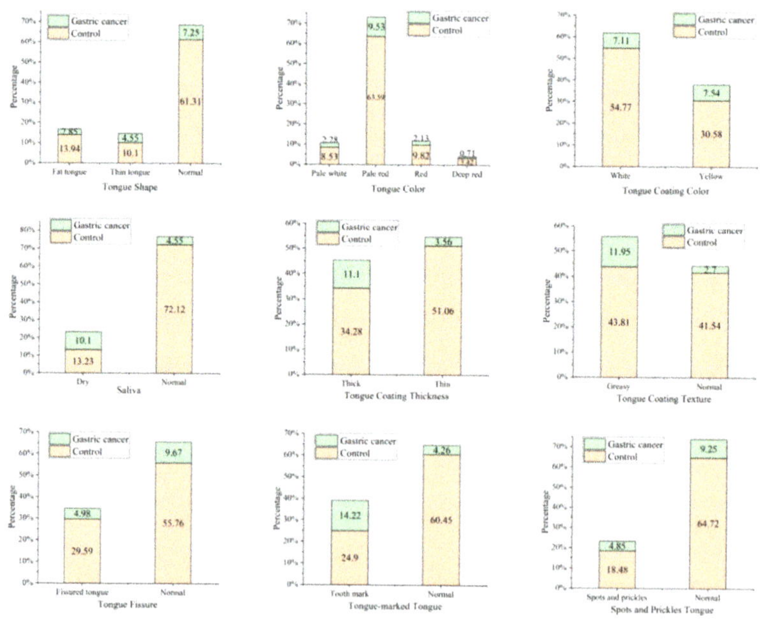

Figure 4. The cross-plot of tongue features between gastric cancer group and non-gastric cancer group. In the cross-plot, the green section represents gastric cancer patients and the orange section represents non-gastric cancer subjects. In the vertical direction, the proportions of different tongue features are clearly shown. In the horizontal direction, the distribution of tongue features in the gastric cancer group and the control group are shown.

The tongue features of the two groups were contrasted using chi-square test and independent t-tests to explore features that were significantly different (p-value < 0.001). Statistics were based on the p-value obtained by the significance test method, with p-value < 0.001 considered as a highly significant difference, 0.001 < p-value < 0.005 considered as a significant difference, and p-value > 0.005 considered as no significant difference.

The differences in tongue features between patients with gastric cancer and the control group are shown in Table 1. Among them, five tongue features, namely tooth-marked tongue, saliva, tongue coating thickness, tongue coating texture, and tongue shape, were significantly different between the gastric cancer patients and the control group (p-value < 0.001). The proportion of yellow tongue coating in gastric cancer patients was significantly higher than in the control group (51.46% vs. 35.83%, 0.001 < p-value < 0.005). In addition, the proportion of spots and prickles tongue in gastric cancer patients was significantly higher than in the control group (36.89% vs. 24.17%, 0.001 < p-value < 0.005). No significant difference was detected between patients with gastric cancer and the control group in tongue color and tongue fissure (p-value > 0.005).

Table 1. Comparison of tongue features between patients with gastric cancer and control subjects.

Tongue Features (n, %)		Control n = 600	Gastric Cancer n = 103	Number of Features	p-Value
Tongue Shape	Fat tongue	98 (83.05%)	20 (16.95%)	118 (16.79%)	<0.001
	Thin tongue	71 (68.93%)	32 (31.07%)	103 (14.65%)	
	Normal tongue	431 (89.42%)	51 (10.58%)	482 (68.56%)	
Tongue Color	Pale white	60 (78.95%)	16 (21.05%)	76 (10.81%)	0.22
	Pale red	447 (86.96%)	67 (13.04%)	514 (73.12%)	
	Red	69 (82.14%)	15 (17.86%)	84 (11.95%)	
	Deep red	24 (82.76%)	5 (17.24%)	29 (4.13%)	
Tongue Coating Color	White	385 (88.51%)	50 (11.49%)	435 (61.88%)	0.0036
	Yellow	215 (80.22%)	53 (19.78%)	268 (38.12%)	
Saliva	Dry	93 (56.71%)	71 (43.29%)	164 (23.33%)	<0.001
	Normal	507 (94.06%)	32 (5.94%)	539 (76.67%)	
Tongue Coating Thickness	Thick	241 (75.55%)	78 (24.45%)	319 (45.38%)	<0.001
	Thin	359 (93.49%)	25 (6.51%)	384 (54.62%)	
Tongue Coating Texture	Greasy	308 (78.57%)	84 (21.43%)	392 (55.76%)	<0.001
	Normal	292 (93.89%)	19 (6.11%)	311 (44.24%)	
Tongue Fissure	Fissured tongue	208 (85.60%)	35 (14.40%)	243 (34.57%)	0.98
	Normal	392 (85.22%)	68 (14.78%)	460 (65.43%)	
Tooth-marked Tongue	Tooth mark	175 (63.64%)	100 (36.36%)	275 (39.12%)	<0.001
	Normal	425 (99.30%)	3 (0.70%)	428 (60.88%)	
Spots and Prickles Tongue	Spots and prickles	145 (79.23%)	38 (20.77%)	183 (23.33%)	0.0096
	Normal	455 (87.50%)	65 (12.50%)	520 (73.97%)	

The XGBT algorithm [42] was utilized to calculate the importance of each tongue feature. Feature importance measures the contribution of each input feature to the prediction result of the model and can highlight the relevance of the feature to the target. Features with lower importance scores were removed, while those with higher importance scores were retained. Filtering tongue features that had an impact on the prediction with the feature importance analysis method reduced the quantity of features input into the neural network model, which contributed to the computational efficiency and accuracy of the model. Figure 5 shows the importance of the tongue features results. The importance scores of features such as dry saliva, fat tongue, and white tongue were high and have a major impact on the model prediction results. The importance scores of pale white tongue, deep red tongue, yellow tongue coating, tooth mark tongue, and spots and prickles tongue were low and are not shown in the figure.

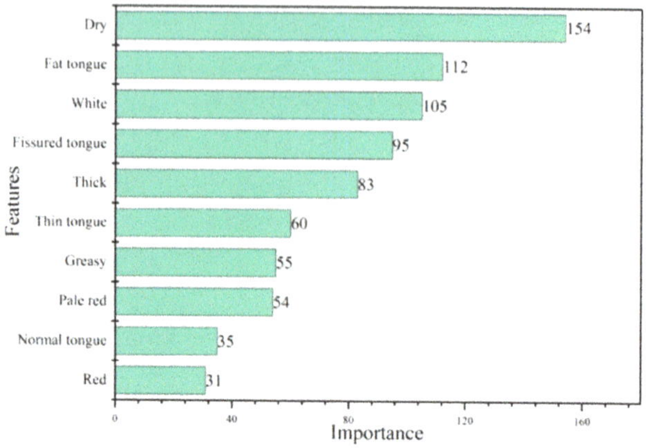

Figure 5. Importance of tongue features.

Subsequently, the association between the nine tongue features of the gastric cancer patients was explored. Correlation analysis was applied to analyze two or more features that were correlated to measure the closeness of the correlation. The correlations between the tongue features of gastric cancer patients are shown in Figure 6. Darker colors in the correlation graph represent a higher correlation between two tongue features. Results show that there is a high probability of association between the thick tongue coating and the greasy tongue coating, while the remaining tongue features were not significantly associated with each other. A negative value for the correlation between two features indicates that an increase in one tongue feature causes a decrease in the other feature.

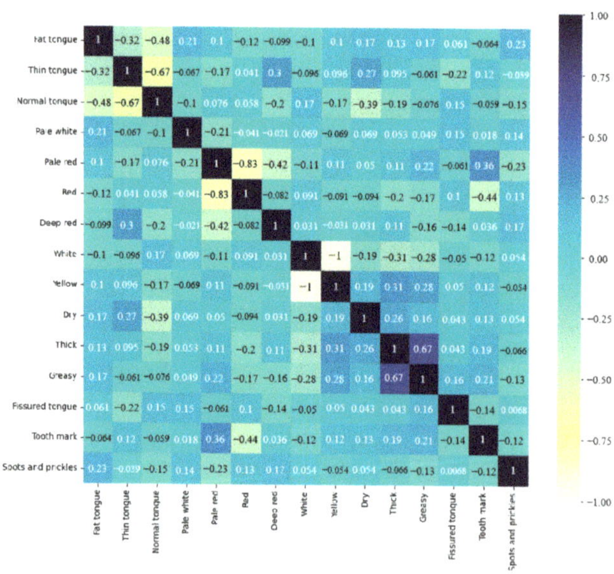

Figure 6. Correlation between tongue features in the gastric cancer group.

Utilizing statistical methods and artificial intelligence techniques for tongue feature analysis, the tongue features with significant differences, high importance scores and high correlations between the gastric cancer patients and the control group were selected as input features for the gastric cancer prediction framework. Finally, tongue shape, saliva, tongue coating thickness and tongue coating texture were also used as features to build a deep learning gastric cancer prediction framework.

To avoid the shortcomings of single feature scale, easy saturation of dimensional scaling and poor classification effect in traditional image classification algorithms, the EfficientNet network was used to classify gastric cancer and non-gastric cancer tongue images. Compared with different CNN models, the EfficientNet model successfully achieved a higher accuracy and efficiency. It used a simple and efficient composite coefficient to scale the network in three dimensions: network depth, network width and image resolution. The MBConv block in MobileNetV2 was used as the backbone of the EfficientNet network. The MBConv block was composed of two convolutional layers, a depth-separable convolutional layer, and a feature extraction module. The basic network architecture of EfficientNet-B0 is shown in Figure 7. Meanwhile, in order to avoid the problem of overfitting and instability due to an insufficient number of image samples, batch normalization and dropout were used to reduce the dependency between convolutional layers, reduce the activity of some neurons in the training process, and suppress the occurrence of overfitting, thus improving the generalization ability of the network and enhancing the robustness of the model classification.

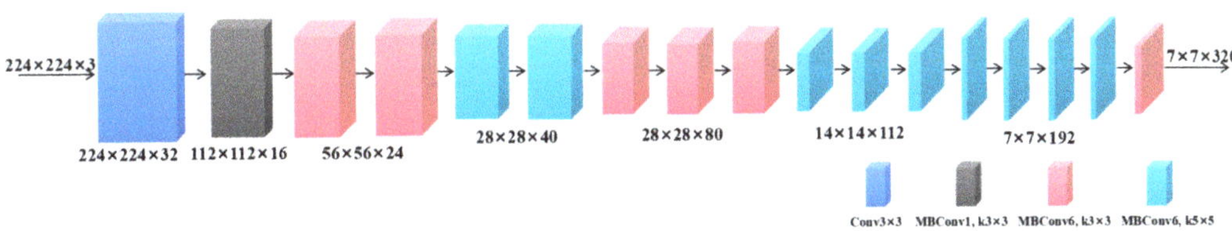

Figure 7. Baseline network for EfficientNet-B0.

The dataset used to build the prediction framework was divided in an 8:1:1 ratio into the training set, validation set and test set. When the EfficientNet network model was trained on the dataset, the official weight was used as the initial weight of the model. After 200 epochs, the training ended with a learning rate of 1×10^{-4}. At the end of the framework, the Softmax classifier normalized the two classification results and the output values were transformed into probabilities. In the final output layer, the output value with the highest probability was selected as the predicted result.

3.5. Performance Metrics

The experiments were performed on a Windows 10 operating system, Intel Core i7-10700, NVIDIA GeForce RTX 2060, based on Pytorch deep learning framework.

In this section, the effectiveness of the segmentation method and gastric cancer prediction framework was assessed with four metrics. The performance of the Deeplabv3+ network was evaluated using the mean intersection over union (MIou) and mean pixel accuracy (MPA) [43]. Accuracy and F1-score were used to evaluate the classification effectiveness of the gastric cancer prediction framework. The definitions and calculations for MIoU, MPA, Accuracy and F1-score are as follows.

MIou is the standard evaluation method for segmentation methods. The specific value of the intersection of the predicted segmentation and ground truth of each class of pixels to the union set is calculated. The ratios of all classes are then summed and averaged to obtain MIou:

$$\mathrm{MIoU} = \frac{1}{k+1}\sum_{i=0}^{k}\frac{p_{ii}}{\sum\limits_{j=0}^{n} p_{ij}+\sum\limits_{j=0}^{n} p_{ji}-p_{ii}} \tag{1}$$

where k is the number of categories in the image except the background, p_{ii} is the number of pixels belonging to category i that are predicted to be in category i, p_{ij} is the number of pixels belonging to category i that are predicted to be in category j.

MPA is based on pixel accuracy (PA). The specific value of the number of pixels correctly predicted in each class to the total number of pixels was calculated to obtain the PA. The PAs of all classes were then summed and averaged to obtain the MPA:

$$\text{MPA} = \frac{1}{k+1}\sum_{i=0}^{k} \frac{p_{ii}}{\sum_{j=0}^{n} p_{ij}} \tag{2}$$

Accuracy is the proportion of samples correctly predicted by the model to the total sample:

$$\text{Accuracy} = \frac{\text{TP} + \text{TN}}{\text{TP} + \text{TN} + \text{FP} + \text{FN}} \tag{3}$$

where TP is the number of positive samples that are correctly identified, TN is the number of negative samples correctly identified, FP is the number of negative samples that are false positives and FN is the number of positive samples that were identified as negatives.

The F1-score is a performance evaluation index of a binary classification model. It is a balance between the precision and recall giving a more balanced metric by calculating the harmonic mean of precision and recall:

$$\text{Precision} = \frac{\text{TP}}{\text{TP} + \text{FP}} \tag{4}$$

$$\text{Rec}all = \frac{\text{TP}}{\text{TP} + \text{FN}} \tag{5}$$

$$\text{F1} - \text{score} = 2 \times \frac{\text{Precision} \times \text{Recall}}{\text{Precision} + \text{Recall}} \tag{6}$$

4. Results

To validate the segmentation effect of the Deeplabv3+ network, other models commonly used for medical image segmentation were built for comparison with the Deeplabv3+ network. The segmentation networks were evaluated with MIou and MPA. The segmentation effect of the three networks evaluated is shown in Table 2. The Deeplabv3+ network made comprehensive use of the MobileNetV2 network and atrous convolution, which minimizes the information loss in the process of image segmentation. The Deeplabv3+ network outperformed PSPNet and U-Net in terms of MIou and MPA, showing that the Deeplabv3+ network model has superior segmentation ability.

Table 2. Results of different segmentation networks.

Method	MPA	MIoU
PSPNet	97.58%	96.6%
U-Net	98.58%	97.14%
Deeplabv3+	98.93%	97.96%

The validity of the gastric cancer prediction framework was tested on a validation set containing tongue images of gastric cancer patients and tongue images without non-gastric cancer. Seventy tongue images in the dataset were used to test the effectiveness of the framework. Representative samples of the tongue images in the test set are shown in Figure 8.

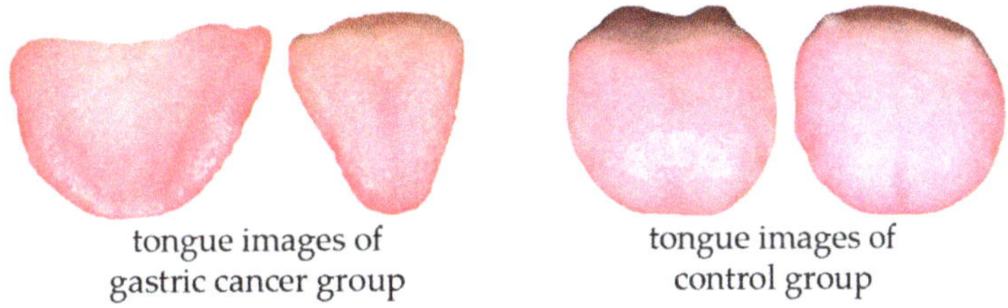

Figure 8. Tongue images in the test set, Image 1 and Image 2 belong in the tongue images of gastric cancer group; Image 3 and Image 4 belong in the tongue images of control group.

The tongue images of gastric cancer patients and the control group in the validation set were input into the framework to assess its prediction effect. The accuracy of the framework was 93.6%, and the F1-score was 92.6%. Table 3 shows the prediction results of the four tongue images in Figure 8. The prediction framework has high accuracy in identifying tongue images of gastric cancer and tongue images of the control group.

Table 3. The final results of the classification.

Result	Image 1	Image 2	Image 3	Image 4
control	16.9%	13.3%	98.6%	99.2%
gastric cancer	83.1%	86.7%	0.14%	0.08%

5. Discussion

A gastric cancer prediction framework was built by statistical analysis of tongue features and deep learning. The framework has high accuracy in detecting gastric cancer patients and the detection process has the advantage of being non-invasive. This study presents a novel approach to exploring the relationship between gastric cancer and tongue features, which has clinical application.

The presented framework has a strong detection ability, but two aspects can be continuously improved. A major part of the time was spent collecting the images in hospitals, but the number of tongue images collected from gastric cancer patients was still low, which is a common problem in the area of medical image studies. Second, tongue images were collected using a standard capture instrument, but the tongue extension posture of each patient has an impact on subsequent tongue feature extraction.

6. Conclusions

Cancer is a major public health problem worldwide. The penetration of artificial intelligence technology into the medical field has become more convenient and accurate to detect gastric cancer. In this paper, we proposed a prediction framework for gastric cancer based on tongue features and deep learning. The framework predicts gastric cancer patients non-invasively and conveniently. Tongue images are acquired by using a standard tongue image capture instrument, and the Deeplabv3+ network is applied to accurately segment the tongue region. Nine tongue features are extracted with guidance from professional doctors, and the relationships between tongue features and gastric cancer are explored by using statistical methods and deep learning. Through differential analysis, importance analysis, and correlation analysis, tongue shape, saliva, tongue coating thickness and tongue coating texture were finally selected as the input features for predicting gastric cancer. Finally, the EfficientNet network was used to build a gastric cancer prediction framework. The experimental results showed that this framework could accurately distinguish between gastric cancer patients and non-gastric cancer subjects with an accuracy of 93.6%. Compared to other similar work, our framework incorporates more tongue features and can more comprehensively predict gastric cancer.

Future work will focus on acquiring more tongue images of gastric cancer patients to extract and further analyze tongue features. Meanwhile, we should improve the framework structure and enhance the robustness of the framework to obtain more accurate gastric cancer prediction results.

Author Contributions: Conceptualization, X.Z. and Z.Z.; methodology, X.Z. and X.C.; software, X.Z.; validation, X.Z. and Y.M.; investigation, D.G.; resources, C.X., Z.Z. and J.M.; data curation, Y.M. and D.G.; writing—original draft preparation, X.Z.; writing—review and editing, X.C. and X.Z.; visualization, X.Z.; supervision, C.X. and Z.Z.; project administration, C.X. and Z.Z. All authors have read and agreed to the published version of the manuscript.

Funding: This work was funded by the National Natural Science Foundation of China [62001430]; the Shanxi Province Traditional Chinese Medicine Strong Province Special Project [204-110457]; the Scientific & Technological Innovation Programs of Higher Education Institutions in Shanxi [2020L0275]; and the Key Research and Development Program of Shanxi Province [202102130501011].

Institutional Review Board Statement: The study was conducted in accordance with the Declaration of Helsinki, and approved by the Biological and Medical Ethics Committee of North University of China (110457).

Informed Consent Statement: Informed consent was obtained from all subjects involved in the study.

Data Availability Statement: The data presented in this study are available on request from the corresponding author. The data are not publicly available due to this data being supplied by Key Laboratory of Instrumentation Science & Dynamic Measurement (North University of China), Ministry of Education and so cannot be made freely available.

Acknowledgments: The authors gratefully acknowledge the medical and technical support of Jiuzhang Men and his team at the Shanxi University of Chinese Medicine's Affiliated Hospital.

Conflicts of Interest: The authors declare no conflict of interest.

References

1. Smyth, E.C.; Nilsson, M.; Grabsch, H.I.; van Grieken, N.C.T.; Lordick, F. Gastric cancer. *Lancet* **2020**, *396*, 635–648. [CrossRef] [PubMed]
2. Lin, L.L.; Huang, H.C.; Juan, H.F. Discovery of biomarkers for gastric cancer: A proteomics approach. *J. Proteom.* **2012**, *75*, 3081–3097. [CrossRef] [PubMed]
3. Choi, J.; Kim, S.G.; Im, J.P.; Kim, J.S.; Jung, H.C.; Song, I.S. Comparison of endoscopic ultrasonography and conventional endoscopy for prediction of depth of tumor invasion in early gastric cancer. *Endoscopy* **2010**, *42*, 705–713. [CrossRef] [PubMed]
4. Yao, K.; Uedo, N.; Kamada, T.; Hirasawa, T.; Nagahama, T.; Yoshinaga, S.; Oka, M.; Inoue, K.; Mabe, K.; Yao, T.; et al. Guidelines for endoscopic diagnosis of early gastric cancer. *Dig. Endosc.* **2020**, *32*, 663–698. [CrossRef] [PubMed]
5. Ueyama, H.; Kato, Y.; Akazawa, Y.; Yatagai, N.; Komori, H.; Takeda, T.; Matsumoto, K.; Ueda, K.; Matsumoto, K.; Hojo, M.; et al. Application of artificial intelligence using a convolutional neural network for diagnosis of early gastric cancer based on magnifying endoscopy with narrow-band imaging. *J. Gastroenterol. Hepatol.* **2021**, *36*, 482–489. [CrossRef]
6. Niikura, R.; Aoki, T.; Shichijo, S.; Yamada, A.; Kawahara, T.; Kato, Y.; Hirata, Y.; Hayakawa, Y.; Suzuki, N.; Ochi, M.; et al. Artificial intelligence versus expert endoscopists for diagnosis of gastric cancer in patients who have undergone upper gastrointestinal endoscopy. *Endoscopy* **2022**, *54*, 780–784. [CrossRef] [PubMed]
7. Emir, T.; Denijal, T. Systemic air embolism as a complication of gastroscopy. *Oxf. Med. Case Reports.* **2019**, *2019*, omz057. [CrossRef]
8. Tania, M.H.; Lwin, K.; Hossain, M.A. Advances in automated tongue diagnosis techniques. *Integr. Med. Res.* **2019**, *8*, 42–56. [CrossRef] [PubMed]
9. Kim, S.R.; Choi, W.; Yeo, I.; Nam, D.H. Comparative Analysis of Tongue Indices between Patients with and without a Self-Reported Yin Deficiency: A Cross-Sectional Study. *Evid.-Based Complement. Altern. Med.* **2017**, *2017*, 1279052. [CrossRef]
10. Kim, J.; Han, G.; Ko, S.J.; Nam, D.H.; Park, J.W.; Ryu, B.; Kim, J. Tongue diagnosis system for quantitative assessment of tongue coating in patients with functional dyspepsia: A clinical trial. *J. Ethnopharmacol.* **2014**, *155*, 709–713. [CrossRef]
11. Zhang, B.; Wang, X.; You, J.; Zhang, D. Tongue color analysis for medical application. *Evid.-Based Complement. Altern. Med.* **2013**, *2013*, 264742. [CrossRef] [PubMed]
12. Cui, Y.; Liao, S.; Wang, H. ROC-Boosting: A Feature Selection Method for Health Identification Using Tongue Image. *Comput. Math. Methods Med.* **2015**, *2015*, 362806. [CrossRef] [PubMed]

13. Ma, J.; Wen, G.; Wang, C.; Jiang, L. Complexity perception classification method for tongue constitution recognition. *Artif. Intell. Med.* **2019**, *96*, 123–133. [CrossRef] [PubMed]
14. Huang, J.; Li, J.; Li, Z.; Zhu, Z.; Shen, C.; Qi, G.; Yu, G. Detection of Diseases Using Machine Learning Image Recognition Technology in Artificial Intelligence. *Comput. Intell. Neurosci.* **2022**, *2022*, 5658641. [CrossRef] [PubMed]
15. LeCun, Y.; Bengio, Y.; Hinton, G. Deep learning. *Nature* **2015**, *521*, 436–444. [CrossRef] [PubMed]
16. Lo, L.C.; Cheng, T.L.; Chen, Y.J.; Natsagdorj, S.; Chiang, J.Y. TCM tongue diagnosis index of early-stage breast cancer. *Complement. Ther. Med.* **2015**, *23*, 705–713. [CrossRef]
17. Li, J.; Yuan, P.; Hu, X.; Huang, J.; Cui, L.; Cui, J.; Ma, X.; Jiang, T.; Yao, X.; Li, J.; et al. A tongue features fusion approach to predicting prediabetes and diabetes with machine learning. *J. Biomed. Inform.* **2021**, *115*, 103693. [CrossRef] [PubMed]
18. Cui, J.; Cui, H.; Yang, M.; Du, S.; Li, J.; Li, Y.; Liu, L.; Zhang, X.; Li, S. Tongue coating microbiome as a potential biomarker for gastritis including precancerous cascade. *Protein Cell* **2019**, *10*, 496–509. [CrossRef]
19. Li, W.Q.; Ma, J.L.; Zhang, L.; Brown, L.M.; Li, J.Y.; Shen, L.; Pan, K.F.; Liu, W.D.; Hu, Y.; Han, Z.X.; et al. Effects of Helicobacter pylori treatment on gastric cancer incidence and mortality in subgroups. *J. Natl. Cancer Inst.* **2014**, 106. [CrossRef]
20. Kainuma, M.; Furusyo, N.; Urita, Y.; Nagata, M.; Ihara, T.; Oji, T.; Nakaguchi, T.; Namiki, T.; Hayashi, J. The association between objective tongue color and endoscopic findings: Results from the Kyushu and Okinawa population study (KOPS). *BMC Complement. Altern. Med.* **2015**, *15*, 372. [CrossRef]
21. Hu, J.; Han, S.; Chen, Y.; Ji, Z. Variations of Tongue Coating Microbiota in Patients with Gastric Cancer. *Biomed. Res. Int.* **2015**, *2015*, 173729. [CrossRef] [PubMed]
22. Xu, J.; Xiang, C.; Zhang, C.; Xu, B.; Wu, J.; Wang, R.; Yang, Y.; Shi, L.; Zhang, J.; Zhan, Z. Microbial biomarkers of common tongue coatings in patients with gastric cancer. *Microb. Pathog.* **2019**, *127*, 97–105. [CrossRef] [PubMed]
23. Gholami, E.; Kamel Tabbakh, S.R.; Kheirabadi, M. Increasing the accuracy in the diagnosis of stomach cancer based on color and lint features of tongue. *Biomed. Signal Process. Control* **2021**, *69*, 102782. [CrossRef]
24. Tan, M.; Le, Q.V. Efficientnet: Rethinking model scaling for convolutional neural networks. *arXiv* **2019**, arXiv:1905.11946.
25. Xiao, Z.; Ji, D.; Li, F.; Li, Z.; Bao, Z. Application of Artificial Intelligence in Early Gastric Cancer Diagnosis. *Digestion* **2022**, *103*, 69–75. [CrossRef] [PubMed]
26. Li, L.; Chen, Y.; Shen, Z.; Zhang, X.; Sang, J.; Ding, Y.; Yang, X.; Li, J.; Chen, M.; Jin, C.; et al. Convolutional neural network for the diagnosis of early gastric cancer based on magnifying narrow band imaging. *Gastric Cancer* **2020**, *23*, 126–132. [CrossRef]
27. Hirasawa, T.; Aoyama, K.; Tanimoto, T.; Ishihara, S.; Shichijo, S.; Ozawa, T.; Ohnishi, T.; Fujishiro, M.; Matsuo, K.; Fujisaki, J.; et al. Application of artificial intelligence using a convolutional neural network for detecting gastric cancer in endoscopic images. *Gastric Cancer* **2018**, *21*, 653–660. [CrossRef]
28. Huang, L.; Zhu, Y.; Xu, C.; Cai, Y.; Yi, Y.; Li, K.; Ren, X.; Jiang, D.; Ge, Y.; Liu, X.; et al. Noninvasive Diagnosis of Gastric Cancer Based on Breath Analysis with a Tubular Surface-Enhanced Raman Scattering Sensor. *ACS. Sens.* **2022**, *7*, 1439–1450. [CrossRef]
29. Hanna, G.B.; Boshier, P.R.; Markar, S.R.; Romano, A. Accuracy and Methodologic Challenges of Volatile Organic Compound-Based Exhaled Breath Tests for Cancer Diagnosis: A Systematic Review and Meta-analysis. *JAMA Oncol.* **2019**, *5*, 182815. [CrossRef] [PubMed]
30. Jiang, S.; Gao, H.; He, J.; Shi, J.; Tong, Y.; Wu, J. Machine learning: A non-invasive prediction method for gastric cancer based on a survey of lifestyle behaviors. *Front. Artif. Intell.* **2022**, *5*, 956385. [CrossRef]
31. Zhu, S.L.; Dong, J.; Zhang, C.; Huang, Y.B.; Pan, W. Application of machine learning in the diagnosis of gastric cancer based on noninvasive characteristics. *PLoS ONE* **2020**, *15*, e0244869. [CrossRef] [PubMed]
32. Hsu, P.C.; Wu, H.K.; Huang, Y.C.; Chang, H.H.; Lee, T.C.; Chen, Y.P.; Chiang, J.Y.; Lo, L.C. The tongue features associated with type 2 diabetes mellitus. *Medicine* **2019**, *98*, e15567. [CrossRef] [PubMed]
33. Li, J.; Chen, Q.; Hu, X.; Yuan, P.; Cui, L.; Tu, L.; Cui, J.; Huang, J.; Jiang, T.; Ma, X.; et al. Establishment of noninvasive diabetes risk prediction model based on tongue features and machine learning techniques. *Int. J. Med. Inform.* **2021**, *149*, 104429. [CrossRef] [PubMed]
34. Li, J.; Zhang, Z.; Zhu, X.; Zhao, Y.; Ma, Y.; Zang, J.; Li, B.; Cao, X.; Xue, C. Automatic Classification Framework of Tongue Feature Based on Convolutional Neural Networks. *Micromachines* **2022**, *13*, 501. [CrossRef]
35. Shorten, C.; Khoshgoftaar, T.M. A survey on Image Data Augmentation for Deep Learning. *J. Big Data* **2019**, *6*, 60. [CrossRef]
36. Mesejo, P.; Ibáñez, Ó.; Cordón, Ó.; Cagnoni, S. A survey on image segmentation using metaheuristic-based deformable models: State of the art and critical analysis. *Appl. Soft Comput.* **2016**, *44*, 1–29. [CrossRef]
37. Pang, B.; Zhang, D.; Wang, K. The bi-elliptical deformable contour and its application to automated tongue segmentation in Chinese medicine. *IEEE Trans. Med. Imaging* **2005**, *24*, 946–956. [CrossRef]
38. Ning, J.; Zhang, D.; Wu, C.; Yue, F. Automatic tongue image segmentation based on gradient vector flow and region merging. *Neural Comput. Appl.* **2010**, *21*, 1819–1826. [CrossRef]
39. Chen, L.C.; Zhu, Y.; Papandreou, G.; Schroff, F.; Adam, H. Encoderdecoder with atrous separable convolution for semantic image segmentation. *arXiv* **2018**, arXiv:1802.02611.
40. Sandler, M.; Howard, A.; Zhu, M.; Zhmoginov, A.; Chen, L. MobileNetV2: Inverted Residuals and Linear Bottlenecks. *arXiv* **2019**, arXiv:1801.04381.
41. Diederik, P.K.; Jimmy, L.B. Adam: A Method for Stochastic Optimization. *arXiv* **2017**, arXiv:1412.6980.

42. Chen, T.; Guestrin, C. XGBoost: A Scalable Tree Boosting System. *arXiv* **2017**, arXiv:1603.02754.
43. Garcia-Garcis, A.; Orts-Escolano, S.; Oprea, S.O.; Villena-Martinez, V.; Garcia-Rodriguez, J. A Review on Deep Learning Techniques Applied to Semantic Segmentation. *arXiv* **2017**, arXiv:1704.06857.

micromachines

Article

Textile One-Component Organic Electrochemical Sensor for Near-Body Applications

Rike Brendgen [1,*], Carsten Graßmann [1], Sandra Gellner [2] and Anne Schwarz-Pfeiffer [3]

[1] Research Institute for Textile and Clothing (FTB), Niederrhein University of Applied Sciences, Webschulstr. 31, 41065 Moenchengladbach, Germany
[2] Faculty Electrical Engineering and Computer Science, Niederrhein University of Applied Sciences, Reinarzstr. 49, 47805 Krefeld, Germany
[3] Faculty of Textile and Clothing Technology, Niederrhein University of Applied Sciences, Webschulstr. 31, 41065 Moenchengladbach, Germany
* Correspondence: rike.brendgen@hs-niederrhein.de; Tel.: +49-216-1186-6099

Abstract: The need for more efficient health services and the trend of a healthy lifestyle pushes the development of smart textiles. Since textiles have always been an object of everyday life, smart textiles promise an extensive user acceptance. Thereby, the manufacture of electrical components based on textile materials is of great interest for applications as biosensors. Organic electrochemical transistors (OECTs) are often used as biosensors for the detection of saline content, adrenaline, glucose, etc., in diverse body fluids. Textile-based OECTs are mostly prepared by combining a liquid electrolyte solution with two separate electro-active yarns that must be precisely arranged in a textile structure. Herein, on the other hand, a biosensor based on a textile single-component organic electrochemical transistor with a hardened electrolyte was developed by common textile technologies such as impregnation and laminating. Its working principle was demonstrated by showing that the herein-produced transistor functions similarly to a switch or an amplifier and that it is able to detect ionic analytes of a saline solution. These findings support the idea of using this new device layout of textile-based OECTs as biosensors in near-body applications, though future work must be carried out to ensure reproducibility and selectivity, and to achieve an increased level of textile integration.

Keywords: textile sensors; biosensing; smart textiles; porous electrode; flexible transistor; OECT

Citation: Brendgen, R.; Graßmann, C.; Gellner, S.; Schwarz-Pfeiffer, A. Textile One-Component Organic Electrochemical Sensor for Near-Body Applications. *Micromachines* **2022**, *13*, 1980. https://doi.org/10.3390/mi13111980

Academic Editors: Libo Gao and Zhuoqing Yang

Received: 6 September 2022
Accepted: 9 November 2022
Published: 15 November 2022

Publisher's Note: MDPI stays neutral with regard to jurisdictional claims in published maps and institutional affiliations.

1. Introduction

"Smart textiles" is a term used to describe textiles that exceed their usual functionalities. As per the definition by the International Organisation for Standardisation (ISO), smart or intelligent textiles are systems consisting of textile and non-textiles components that actively interact with their environment, a user, or an object [1]. Smart textiles have become important as they offer possible applications in numerous industry sectors, such as sports [2], health [3,4], home and living [5], and many more. The growth is especially driven by textile-based sensors used to monitor vital signals for medical and sports purposes. The need for more efficient health services and increasing awareness about a healthy lifestyle contribute to this development. Thereby, textile-based electronics promise extensive user acceptance since textiles have always been an object of everyday life. Furthermore, the growing usage of wireless technology and the miniaturization of electronics contribute to this trend. One challenge hereby is the integration of sensors into the textile material to ensure comfort and wearability [6,7].

Textile-based organic electrochemical transistors (OECTs) can help in this regard. They consist of a gate electrode, an electrolyte, and a source/drain electrode, and often find application in biosensors [8–12]. One big advantage of OECTs is their operation at very low voltages, which is suitable for near-body biosensing applications [13]. By applying a voltage to the gate electrode, the current flow between the source and the drain electrode can be controlled. Thereby, the presence of an electrolyte enables the flow of ions into the channel of the source/drain electrode, and thus changes the current flow [13–15]. For this mechanism, the source/drain electrode needs to be made of an electrically conductive polymer as the basic working principle relies on the doping and dedoping of the conductive polymer, which results in differences in channel conductivity. The changes in the doping state originate from the injection of ions from the electrolyte into the polymer [16]. The electrically conductive polymer poly(3,4-ethylenedioxythiophene) polystyrene sulfonate (PEDOT:PSS) is of interest for applications as electrodes in electrical and chemical transistors, as well as for electrocardiographs, for organic solar cells or organic light-emitting diodes. Next to its high electrical conductivity, PEDOT:PSS also possesses high transparency in visible light, high mechanical flexibility, high thermal stability and good oxidation resistance [6,17–20]. Ultimately, the changes in ion intensity can be detected in the current flow between the source and drain (channel) electrode when the electrolyte is enriched with some sort of ions coming from diverse body fluids. This conduct favors the use as a biosensor [13,14,16]. OECTs based on PEDOT:PSS work in depletion mode, which means that in absence of a positive gate voltage, a hole current flows in the channel (ON-state). When applying a positive gate voltage, ions in form of cations (positively charged) are injected from the electrolyte into the channel and the negatively charged ions (anions) of the PEDOT:PSS become saturated. Consequently, the number of mobile holes in the channel decreases, the conductive polymer is dedoped, the drain current drops and the transistor reaches its off state [16]. Several studies describe textile OECT arrangements, where the gate electrode and source/drain electrode are electro-active yarns that are distinctly separated from each other and only connected by a liquid, gel-like electrolyte solution [21–25]. I Gualandi et al. (2016) presented a printed textile-based OECT, but here again, the electrolyte was externally applied [26], which limited the scope of application to moist environments. This paper, on the other hand, presents an OECT where the three components—gate electrode, electrolyte, and source/drain electrode—are combined in one device so that integration into a garment for future applications as biosensors becomes easier, mostly because the electrolyte is not externally applied. In this way, the herein-developed OECT can easily be integrated into a textile and function as a switch or an amplifier, without additionally applied electrolytes. Therefore, a solid electrolyte layer that distinctly separates, but at the same time permanently connects, the gate and source/drain electrode and still allows the injection of ions into the channel of the source/drain electrode needs to be established. Moreover, a porous layer of PEDOT:PSS was developed that allows diffusion of ionic analytes through the porous source/drain electrode and enriches the electrolyte with ions when an analyte comes in contact with the transistor. At a constant gate and source/drain voltage, this change in ion intensity becomes noticeable by a decrease in current output measured at the source/drain electrode.

2. Materials and Methods

The following sections describe the materials used and the procedure to manufacture and test this textile-based one-device OECT. Firstly, the materials used will be stated, while secondly the manufacturing steps will be described and lastly characterization methods are highlighted.

2.1. Materials

For the different components (gate electrode, source/drain electrode, and electrolyte) of an OECT, various materials were used that will be described.

2.1.1. Gate Electrode

The gate electrode was simply built from a nickel/copper-coated ripstop fabric, provided by LessEMF, USA. According to the datasheet, it had a thickness of 0.08 mm and an aerial weight of 90 g/m^2. The electrical surface resistance is given at 0.03 Ohm/m^2. It was lightweight and flexible, easy to cut and sew, similar to ordinary fabric [27].

2.1.2. Electrolyte

The electrolyte layer was made of an electrode contact gel and hydroxyethyl cellulose (tylose) mixed with glycerol and lithium chloride (LiCl) as an ionic salt. The electrode contact gel was chosen as the base material as it ensures conductivity during ECG and EG examination and contains ions that function as charge carriers. A commercially available electrode contact gel was purchased from P.J. Dahlhausen & Co. GmbH, Germany. Tylose type H 60000 YP2 (SE Tylose GmbH & Co. KG, Wiesbaden, Germany) was used as the binding agent for the electrolyte layer while functioning as a liquid retention agent at the same time. Moreover, glycerol 99.5% (VWR International GmbH, Darmstatd, Germany) was added because of its water-binding and moisturizing properties. Lithium chloride was used as the ionic salt that builds strong hygroscopic crystals and was supplied by Alfa Aesar (Thermo Fisher (Kandel) GmbH, Kande, Germany).

2.1.3. Source/Drain Electrode

The porosity of the source/drain electrode was achieved by impregnating a cellulose-based, biodegradable micro filtering paper (nonwoven) for coffee-making with PEDOT:PSS. Because of its commercial application, it was considered to be porous, which has been proven by the microscopic examination in Section 3.1. The nonwoven textile had a diameter of 6.3 cm but it was cut into strips of 1 cm width so that it could be vertically immersed in the PEDOT:PSS solution. The PEDOT:PSS solution consisted of a 1.3 wt% PEDOT:PSS dispersion (Sigma-Alderich Chemie GmbH, Taufkirchen, Germany) mixed with dimethyl sulfoxide (DMSO)(Sigma-Alderich Chemie GmbH, Taufkirchen, Germany) as the secondary dopant and glycerol 99.5% (VWR International, Darmstadt, Germany) as a water-binding agent.

2.2. Methods

The OECT was manufactured in several steps, whereby only the nickel–copper fabric stayed untreated. The above-mentioned materials for the different components of the OECT (electrolyte and source/drain electrode) were separately mixed before bringing the components together in the one-device OECT.

2.2.1. Electrolyte

The first step for the preparation of the electrolyte was mixing 5% tylose with distilled water and stirring it for 2 h in the dispermat VMA Getzmann GmbH, Germany. Secondly, the diluted tylose was mixed with the commercial ECG-gel in a ratio of 1:1 by hand. Subsequently, 10 wt% glycerol and 1 wt% LiCl were added to this base mixture, and all was stirred for 20 min in the dispermat. As soon as the solution started to form bubbles, one drop of defoamer Lyoprint Air (Huntsman Textile Effects GmbH, Langweid am Lech, Germany) was added to remove the air.

2.2.2. Source/Drain Electrode

For the production of the porous semi-conducting source/drain electrode, firstly the PEDOT:PSS solution needed to be prepared. Therefore, PEDOT:PSS, glycerol and DMSO were mixed in a ratio of: 17.25, 10.12, and 72.63 wt%, respectively. This composition was based on the experimental set-up of Malti et al. [14]. Therein, a nanofibrillated cellulose, glycerol, and DMSO-blended PEDOT:PSS film is presented, and the above-described ratio was derived from the original composition of 16.2 wt% PEDOT:PSS, 6.1 wt% nano cellulose, 68.2 wt% DMSO, and 9.5 wt% glycerol [14]. Instead of using diluted nanocellulose, in the underlying work, a porous nonwoven fabric was applied so that the corresponding weight percentage of nanocellulose was neglected, which resulted in the above-mentioned weight ratio of 17.25, 10.12, and 72.63 wt% (PEDOT:PSS, glycerol, and DMSO). The solution was stirred by a magnetic stirring rod and subsequently placed into an ultrasound bath for 30 min. Next, the filtering paper was vertically immersed into the PEDOT:PSS mixture and left to absorption for at least 4.5 h, then the immersed paper was dried at 50 °C for approximately 2.5 h.

2.2.3. Assembly of the OECT

All three components—the nickel–copper woven textile, PEDOT:PSS-immersed nonwoven fabric, and electrolyte solution—were brought together by laminating with squeegees using the electrolyte as the adhesive material. A hand squeegee with a wet-film thickness of 500 μm was used to apply the electrolyte solution to the nickel–copper fabric, while afterwards the PEDOT:PSS-coated nonwoven fabric was placed in the still liquid electrolyte solution. The laminate was dried in the oven at 60 °C for 10 min so that the electrolyte solution hardened and thereby bonded all three layers. After drying, the single OECTs were cut out with a pair of scissors, whereby the whole dimensions comprised 3 cm in length and 1 cm in width (Figure 1).

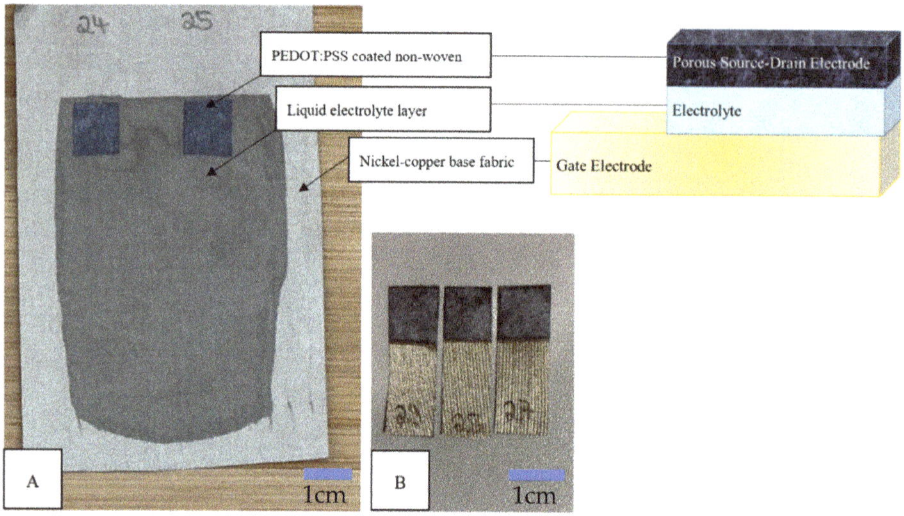

Figure 1. OECT Assembly. (**A**) Coating and assembling of nickel–copper base fabric with electrolyte solution and PEDOT:PSS-coated nonwoven. (**B**) Final OECT composed of the nickel–copper base fabric, hardened electrolyte layer, and PEDOT:PSS coated nonwoven fabric.

2.2.4. Characterization

After the preparation of the one-component OECTs, optical inspections and electrical characterization were performed. The former comprises the analysis of the whole OECT structure as well as of only the uppermost layer—the source/drain electrode—with a scanning electron microscope (SEM) (TM4000Plus, Hitachi High-Tech Corporation, Tokyo, Japan). The source-drain electrode was examined for its porosity and successful coating and was compared with the untreated nonwoven textile. The composed transistor was investigated for its layered structure. In addition, energy dispersive X-ray spectroscopy (EDX) (Bruker Corporation, Billerica, MA, USA) was carried out to identify the chemical elements of the individual components. The porosity of the coated and uncoated nonwoven material was investigated with a PSM 165 capillary fluid pore size meter from Topas (Dresden, Germany). The samples were measured dry and wetted with Topor, which is a special test fluid for this method and mainly contains perfluoro tri *n*-butylamine and isobutyl perfluoro *n*-butylamine. In principle, the test fluid fills all pores of the specimen. When the gas flows through, the specimen increases and the pores become gas permeable at a certain point, which is the bubble point of the material and corresponds to the opening pressure of the largest pore. By further increasing the gas flow rate, it is possible to calculate the pore size distribution and the mean pore size of the material.

For the characterization of the electrical properties of the developed OECT structure, a special testing setup was replicated (Figure 2) per the setup of Gualandi et al. [26]. Thereby, one voltage source controls the voltage of the source electrode whereas the other one controls the gate voltage. The multimeter measures the current output at the drain electrode of the OECT. For the setup, the voltage source PeakTech 6210 and the multimeter PeakTech 4000 were used. Two different measurements were carried out to characterize the electrical properties of the OECT. Firstly, only the source/drain voltage (V_{sd}) was applied while no voltage was exerted on the gate electrode. The source/drain voltage was applied in 0.1 increments from −0.5 up to +0.5 V, and the corresponding source/drain current (I_{sd}) was recorded in μA. Secondly, the gate voltage (V_g) was applied whereas the source/drain voltage was held constant at +0.5 V. Voltage was applied to the gate electrode in steps of 0.5 V from −1.0 up to +1.0 V, and the changing source/drain current was recorded by the multimeter.

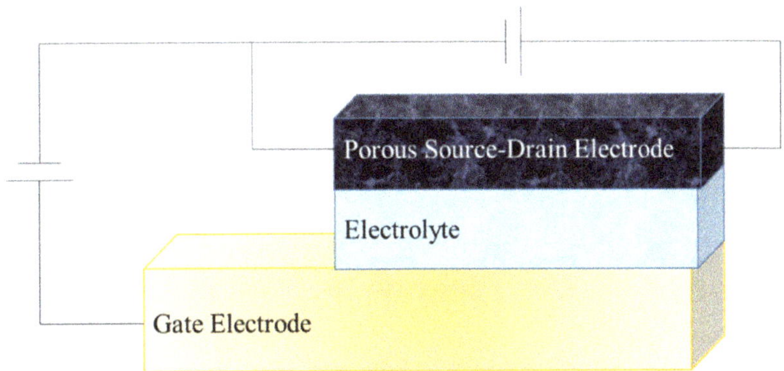

Figure 2. Measurement setup for electrical characterization of textile-based one-component OECT.

3. Results

The optical characterization reveals the successful impregnation of the nonwoven material with PEDOT:PSS. Also, the layered structure of the readily assembled OECT becomes visible. Moreover, the electrical measurements prove the working principle of the one-component OECT, and saline solution was detected. Lastly, also the color changing of the PEDOT:PSS-coated nonwoven textile due to the switching of oxidation states was observed. For the pore size measurement, three samples of uncoated and coated nonwoven materials were measured, and the range of the mean pore size diameter is given. The uncoated nonwoven material had a mean pore size of 18.7 to 21.9 μm, whereas the PEDOT:PSS-coated material had a slightly smaller mean pore size, between 14.5 and 17.4 μm. As expected, the material was still porous and could be used for the construction of textile-based OECTs.

3.1. Optical Characterization of PEDOT:PSS-Coated Nonwoven Material

As the OECT should function not only as a transistor but also as a sensor, the uppermost electrode needs to be porous and permeable for the analytes, which can be proven by optical characterization. Figure 3 shows the comparison of the uncoated nonwoven material to the PEDOT:PSS-coated nonwoven material in different magnifications. The porous structure of the nonwoven material becomes visible at first glance as cavities between the single-fiber strands are visible in all four micrographs. Whereas the single fiber strands are simply entangled in Figure 3A, those fiber strands are glued together over large areas in Figure 2. This finding becomes even more visible in the micrographs with

100 times magnification (Figure 3A,B), where the PEDOT:PSS clearly covers and bonds the single strands. In Figure 3B, large areas covered by PEDOT:PSS are visible, while other parts remain uncoated and reveal the porous structure of the nonwoven fabric. Those open cavities may promote the diffusion of an analyte through the later source/drain electrode, while electrical conductivity is ensured over the whole surface.

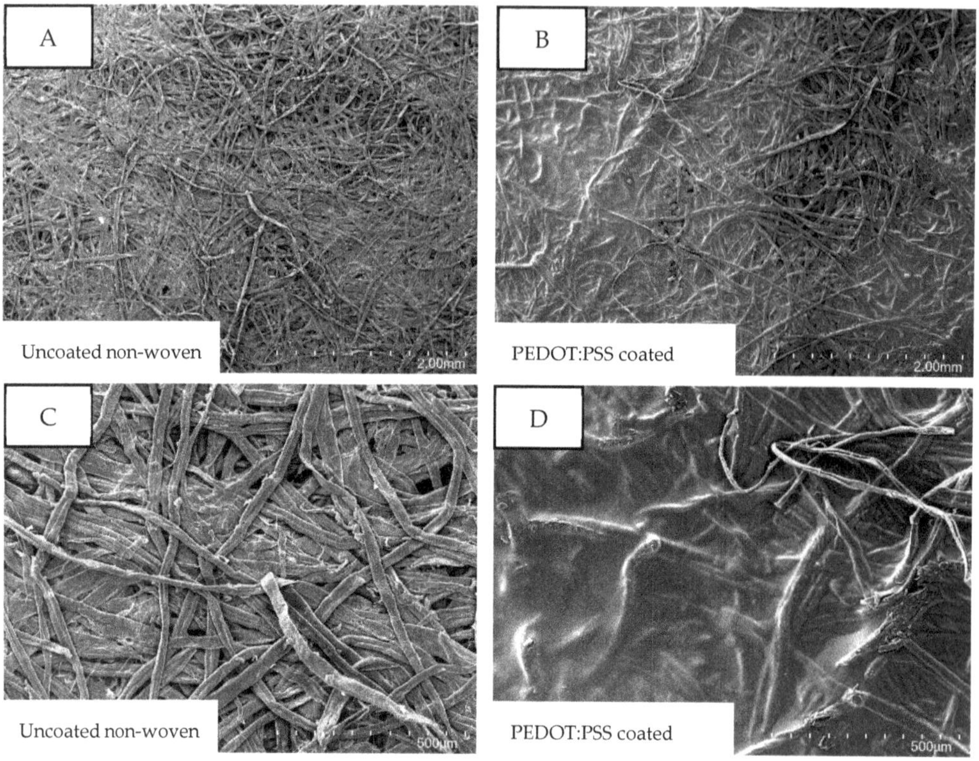

Figure 3. SEM images of (**A**,**C**) uncoated and (**B**,**D**) PEDOT:PSS-coated nonwoven material. (**A**,**B**): 25 times magnification. (**C**,**D**): 100 times magnification.

Figure 4 shows the EDX analysis of the uncoated (Figure 4A,C) and PEDOT:PSS-coated (Figure 4B,D) nonwoven material. Figure 4A,B depict all elements found during the EDX analysis at 200 times magnification of the samples. The elements carbon (C: red), oxygen (O: green), and nitrogen (N: pink) can be explained by the organic origin of the nonwoven material as well as of the PEDOT:PSS. Of interest is the element sulfur (S: blue), which is a characteristic component of PEDOT:PSS, but not of the cellulosic nonwoven material. As expected, the PEDOT:PSS-coated sample (Figure 4B) has a clear blue coloring, while the blue color is nearly non-existing in the uncoated sample (Figure 4A). Figure 4C,D map only the element sulfur and prove the above-described finding; sulfur is detectable in the sample coated with PEDOT:PSS but not in the uncoated sample.

Table 1 shows the atomic percentage of the samples and supports the observation; the atomic percentage of sulfur increases from 0.68 At. -% in the uncoated sample to 4.02 At. -% in the PEDOT:PSS-coated sample. The presence of sodium in the coated sample cannot be explained, but looking at the atomic percentage of 0.66 At. -%, its presence seems to be very little and can be neglected for further consideration.

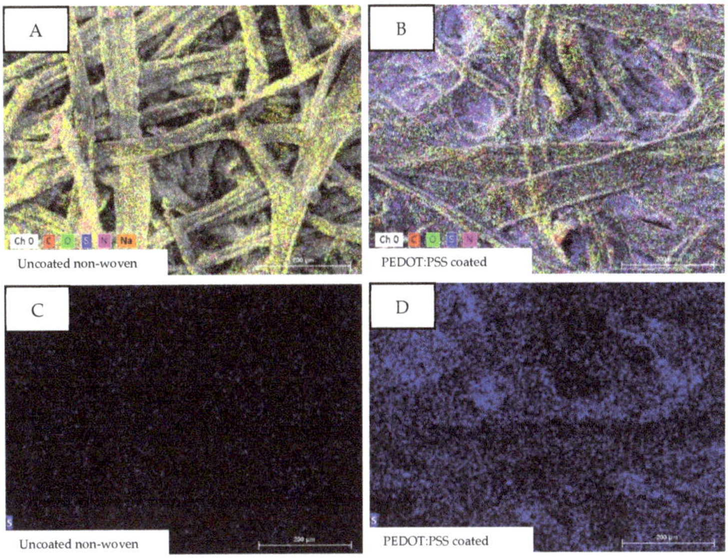

Figure 4. EDX analysis of (**A**) uncoated and (**B**) PEDOT:PSS-coated nonwoven material with special attention to the element sulfur in the uncoated (**C**) and coated sample (**D**).

Table 1. EDX-data of uncoated and PEDOT:PSS-coated nonwoven material, showing an increase of atomic mass of sulfur.

Sample	C [At. -%]	O [At. -%]	N [At. -%]	S [At. -%]	Na [At. -%]
Uncoated	51.57 ± 5.03	46.06 ± 6.01	1.69 ± 0.42	0.68 ± 0.08	/
Coated	59.33 ± 4.51	34.15 ± 3.53	1.84 ± 0.39	4.02 ± 0.27	0.66 ± 0.08

3.2. Optical Characterization of Readily Assembled OECT

Figure 5 shows SEM images of the readily assembled OECT with a focus on the interface, where all three components converge. The interwoven structure on the left side of the images clearly represents the ripstop woven nickel–copper fabric. The PEDOT:PSS-coated nonwoven material is visible on the right side and recognizable by the fibrous structure. In between both components, a small line of electrolyte layer becomes visible that glues together the conductive fabric and the PEDOT:PSS-coated nonwoven material. The electrolyte forms a homogenous layer showing no holes or bubbles that might lead to short circuits in the component.

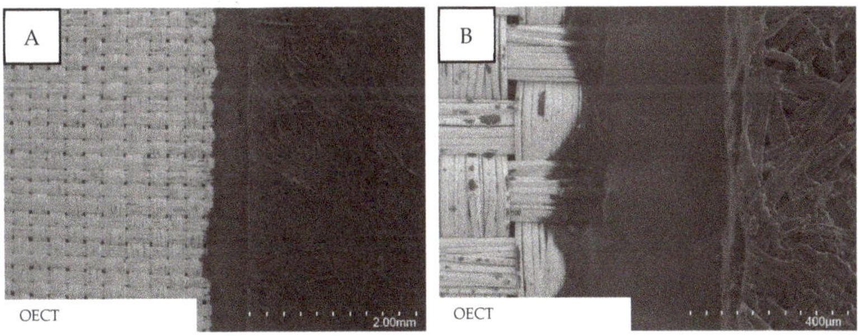

Figure 5. SEM images showing the layered structure of the readily assembled OECT. (**A**): 25 times magnification. (**B**): 120 times magnification.

Figure 6 shows the EDX mapping of the assembled OECT and the following elements are detectable: nickel (Ni: pink), copper (Cu: green), carbon (C: blue), and chlorine (Cl: turquois). Figure 6A shows the mixed mapping of all elements and a distinct separation between nickel and copper (pink–green) on the left side and carbon and chlorine (blue–turquois) on the right side becomes visible. That separation will be clarified even more when looking at the images in Figure 6B–E. Nickel (Figure 6B) and copper (Figure 6C) are primarily present on the left side of the sample, which is indicated by far more colored dots. Blue (Figure 6D) and turquoise (Figure 6E) dots are mainly present on the right side of the sample, which indicates the presence of carbon and chlorine.

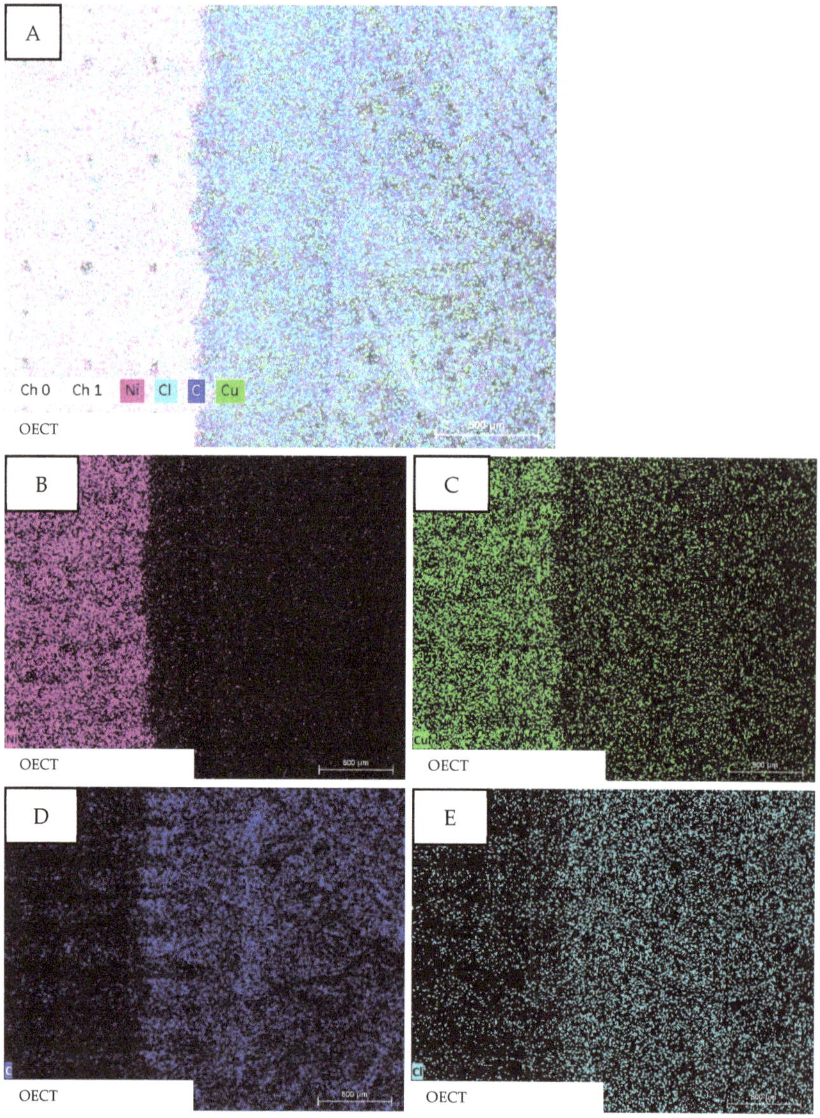

Figure 6. EDX mapping the elementary differences between the gate electrode and source/drain electrode of the readily assembled OECT. (**A**): All elements. (**B**): Nickel. (**C**): Copper. (**D**): Carbon. (**E**): Chloride.

These findings are as expected since nickel–copper-coated ripstop fabric has been used for the gate electrode and thus those elements should be detectable. The presence of carbon, and especially chlorine, can be explained when looking at the formulation of the electrolyte coating. Glycerol and the ECG gel, the main components of the electrolyte coating, are mainly made of carbon, oxygen, and hydrogen, which explains the presence of carbon mostly on the right side of the sample. Additionally, lithium chloride was added to the electrolyte solution and the detected chlorine in the sample can be attributed to that. Lithium is a light element and thus has a low energy of characteristic radiation so it is hardly detectable in EDX mapping. The same applies to hydrogen. Looking at the findings of Figure 4, one would expect sulfur to be present in the EDX mapping of the whole transistor, since the PEDOT:PSS-coated nonwoven material has been integrated into the OECT structure as the source/drain electrode. The non-existence of sulfur can only be explained when assuming that the PEDOT:PSS-coated nonwoven material is completely saturated and covered with the electrolyte solution. Although the PEDOT:PSS-coated nonwoven material was merely placed into the still wet electrolyte solution and not coated with it, the nonwoven material, in a sense, soaked up the solution and thereby the electrolyte solution also penetrated to the upper side of the nonwoven fabric. EDX mapping can only detect the surface of a sample but not the underlying layers, so in this case only carbon and chlorine were detectable for the electrolyte and source/drain structure.

The cross-section (Figure 7) was examined to demonstrate the layered structure of the OECT and prove the successful separation of the gate electrode and source/drain electrode by the hardened electrolyte. Thereby, it is noticeable that the electrolyte settles into the valleys and cavities of the textile electrodes so that the layer thickness of the electrolyte varies. Two measurement points are indicated that visualize the differences in layer thickness. Moreover, it must be stated, that the layer thickness of the hardened electrolyte decreased strongly compared with the wet-film thickness of 500 µm applied with the hand squeegee. This can be attributed to the loss of water during drying.

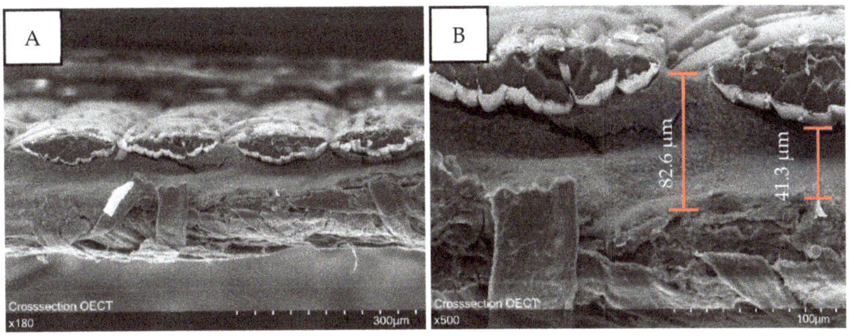

Figure 7. Cross-sectional examination of the layered OECT structure shows how the gate electrode and source/drain electrode are glued together by the electrolyte. (**A**): 180 times magnification. (**B**): 500 times magnification.

3.3. Electrical Characterization

Figure 8 shows the results of the electrical measurement applying source/drain voltage in steps of 0.1 V starting at −0.5 V and increasing up to +0.5 V. Simultaneously, the source/drain current (Isd) was measured as it is the characteristic output for the transistor. The measurement was carried out by recording the data for 30 s before changing the applied voltage. It can be stated that the current (I_{sd}) increased from 2.26 µA at 0.0 V up to −70.66 µA when −0.5 V was applied (modulus). The same tendency can be observed for a positive voltage, though the current varied a little, whereas the current Isd amounts to −70.66 µA at −0.5 V, it was only +66.24 µA at +0.5 V. The measurement proves that the current flowed proportional to the applied voltages, which is as expected since there is an ohmic contact.

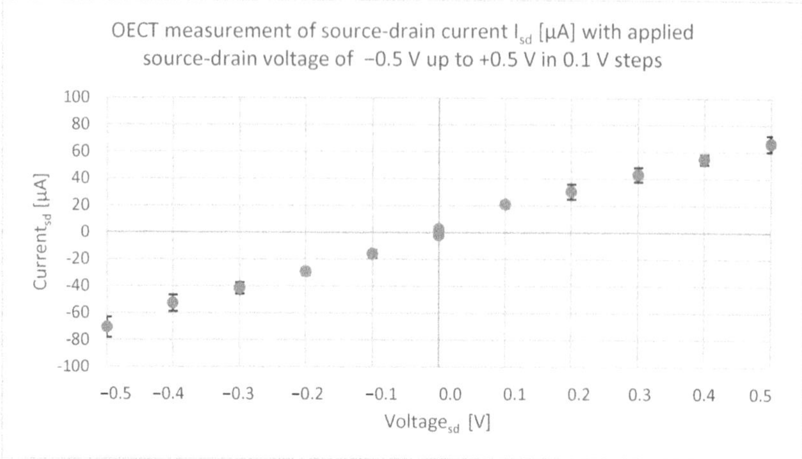

Figure 8. First OECT measurement of source/drain current I_{sd} [μA] showing the characteristic curve of a transistor when the V_{sd} is increased from −0.5 to +0.5 V.

Figure 9 shows the OECT measurement with constant source/drain voltage at +0.5 V and varying gate voltage from −1.0 up to +1.0 V. Again, the source/drain current I_{sd} was measured, and data of 30 s effective testing time was recorded.

Firstly, it can be noted that the current at a gate voltage of 0.0 V complies with the current measured when V_{sd} is varied. For both measurements, the current I_{sd} at source/drain voltage of +0.5 V and no gate voltage lies at +66.24 μA in Figure 8 and +67.97 μA in Figure 9. Moreover, the influence of the applied gate voltage becomes evident. At the same source/drain voltage (V_{sd} = +0.5 V), the current output I_{sd} of the transistor can vary from +93.12 to +30.52 μA only by applying different gate voltages (V_g (1) = −1.0 V, V_g (2) = +1.0 V). It is therefore approximately reduced threefold; less current flowed between the source/drain electrode when a positive gate voltage was applied. The results show that the gate voltage can control the current that flows between the source and the drain electrode, and hence, the device operates as a transistor.

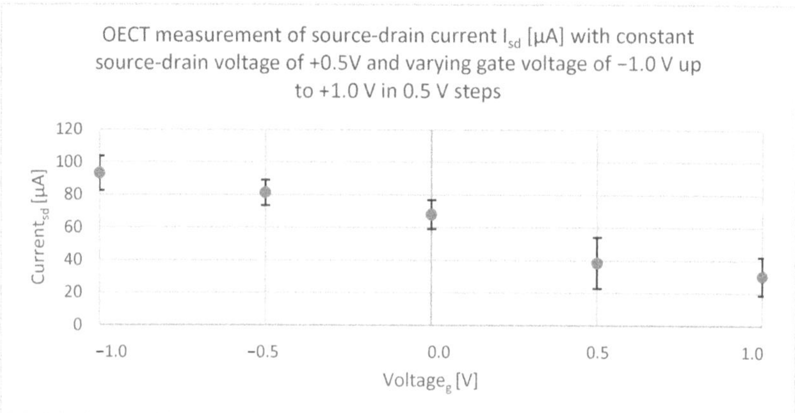

Figure 9. OECT measurement showing that gate voltage (V_g) controls the current that flows between the source and drain electrode (I_{sd}). When V_g is increased from −1.0 V to +1.0 V, the drain current drops.

3.4. Color-Changing Effect of PEDOT:PSS

PEDOT:PSS is known for its color-changing properties depending on the doping state of the conductive polymer. The color palette of PEDOT:PSS ranges from dark blue to light blue to white, or colorless. PEDOT:PSS appears white to transparent in its doped state, which means that it is more conductive. In its dedoped state, PEDOT:PSS appears in a blue color, and its conductive properties are decreased [28–30]. As described above, applying a positive gate voltage changes the doping state of PEDOT:PSS by injecting ions in the form of cations from the electrolyte into the conductive polymer. Thereby, the negatively charged anions of PEDOT:PSS become saturated and the number of mobile holes decreases. PEDOT:PSS gets dedoped, takes on a darker color and the electrical properties deteriorate. This process is reversible and, upon applying a negative gate voltage, PEDOT:PSS gets doped again, which increases the conductivity and changes the color back to light blue or white. This effect has been observable during the electrical measurements as well and became especially visible when switching the gate voltage between −1.0 and +1.0 V.

Figure 10 shows photographs taken during the measurement and it is visible that the color of the PEDOT:PSS-coated nonwoven material slightly changed. The pictures on the left (Figure 10A,C) represent the samples when −1.0 V was applied, and thereby PEDOT:PSS was switched to the doped state. The coated nonwoven material appears brighter. The right side (Figure 10B,D), on the other hand, shows photographs of the dedoped PEDOT:PSS-coated nonwoven material, which was achieved by applying a positive gate voltage of +1.0 V. The nonwoven material took on a darker color.

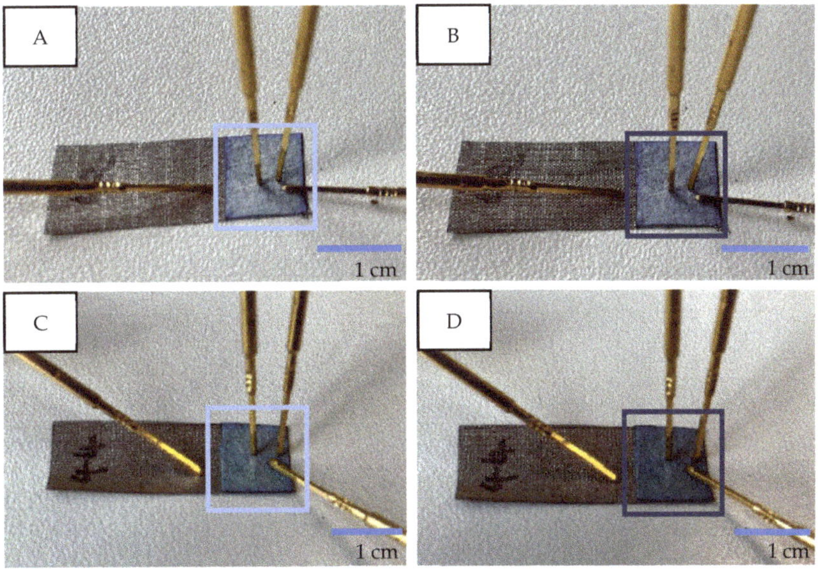

Figure 10. Images of color-changing effect of PEDOT:PSS upon applying different gate voltages (Vg). (**A**,**C**): −1.0 V. (**B**,**D**): +1.0 V.

3.5. Saline Sensing

A conceivable application of OECTs is as biosensors and many studies already discuss the use of OECTs as sensors for the detection of saline content, adrenaline, or glucose in diverse body fluids [24,26,31–38]. Therefore, the herein-developed OECTs were examined for their working behavior in presence of a 0.9% saline solution. For that, 0.09 g sodium chloride (NaCl) was mixed with 10 g distilled water. The mixing ratio is based on the salt content in the human body and hence serves as a simple indicator for use in near-body applications. A 10 μL saline solution was dripped on the source/drain surface (Figure 11), while both the gate and source/drain voltage were set at 0.5 V, respectively.

Figure 11. Detection of salt solution on the surface of source/drain electrode.

The current output (I_{sd}) was simultaneously measured. Figure 12 shows the result of the measurement of one OECT over time. Figure 13, on the other hand, represents the mean value of all data of five OECTs with no saline solution added and 10 μL of 0.9% saline solution added. In Figure 12, a clear and steep drop is visible when the saline solution was added, so that the current output I_{sd} changes from approximately +41 μA to +25 μA. Less current flowed between the source and drain electrode, and the salinity can be detected by that. Also in Figure 13, the sensing mechanism of the produced OECTs is visible as the mean value of all measured OECTs is reduced from approximately 35 to 24 μA when the saline solution was added. Thereby, the standard deviation amounts to 4.7 (no solution added) and 4.76 (0.9% saline solution added) and is hence less than the actual change in current output caused by the addition of saline solution.

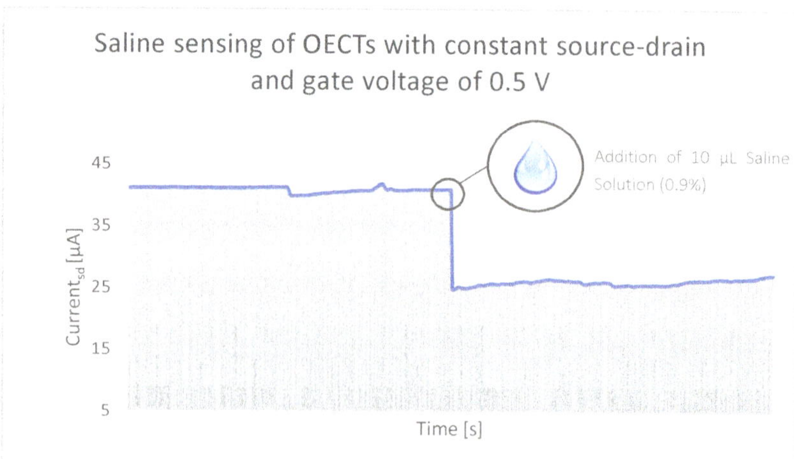

Figure 12. Change in current output under the addition of saline solution over time.

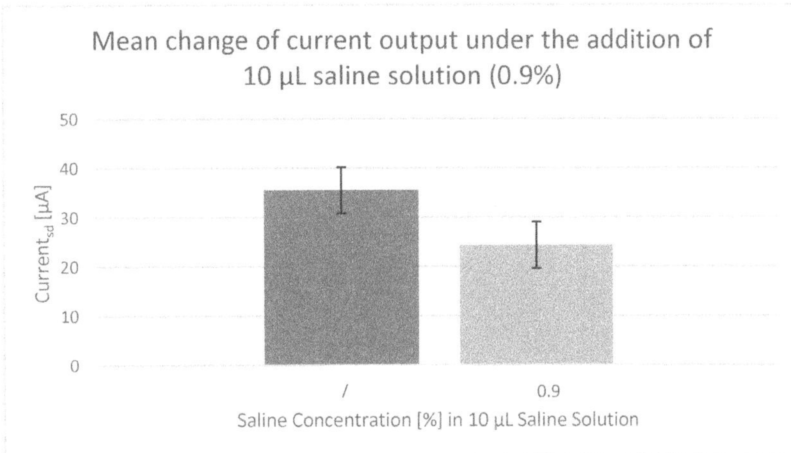

Figure 13. Mean change in current output under the addition of 10 μL saline solution (0.9%).

Next to its detection function, the ability to measure different concentrations is also of importance for a reliable sensor. Therefore, another trial was carried out whereby different concentrations of saline solution (0–2.0%) were pipetted onto the OECT and the current at the source/drain electrode was simultaneously measured (Figure 14). The source/drain and gate voltage were set at 0.5 V as before in the other experiments. To start, the source/drain current was measured when no solution was added at all and lies at 32.48 μA. Subsequently, a solution with no saline content (0.0%), 0.5%, 1.0%, and 2.0% saline content was added and drying of the device was allowed before applying the next solution. The sensor also reacted to the added solution when no saline content was present therein (19.61 μA). However, the reaction of the OECT to a solution containing saline was even stronger and increased further with rising concentration, at 0.5% saline solution current was 16.78 μA. Ultimately, the current at the source/drain electrode was reduced to 11.89 μA when 2.0% saline solution was added.

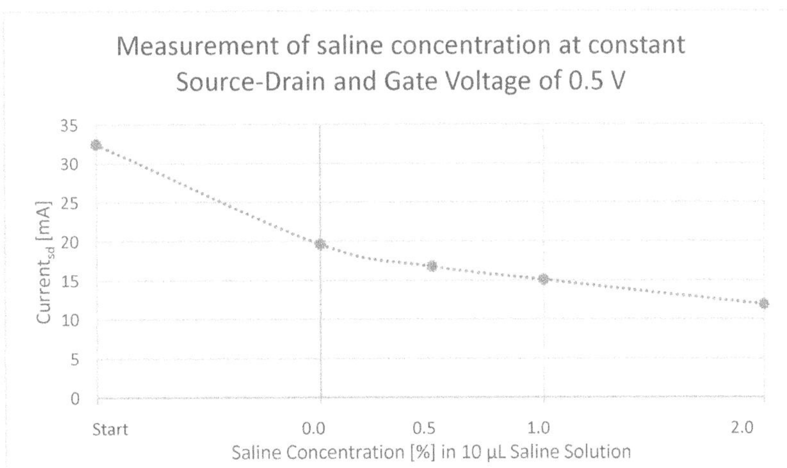

Figure 14. Measurement of different saline concentrations (0–2.0%) at constant source/drain and gate voltage of 0.5 V, showing that the OECT device is not only able to detect but also to measure saline concentration.

4. Discussion

For the production of textile OECTs, common textile technologies (impregnating and laminating) were used. Thereby, a one-component OECT was manufactured from solely textile materials. This simple manufacturing process enables the integration of the device into garments or other smart textiles, where the transistor can function as a sensor, switch, or amplifier.

The optical characterization of the PEDOT:PSS-coated nonwoven material reveals that it is possible to create a porous, yet coated, structure with this rather simple immersion process. This porosity was also proven by determining the mean pore size diameter, which was slightly reduced to a mean pore size range of 14.5 to 17.4 µm after coating. The presence of sulfur in the EDX data can be traced back to PEDOT:PSS. The optical analysis of manufactured OECTs shows a clear layering of the different components of the transistor—source/drain electrode, gate electrode, and electrolyte layer. The EDX mapping identifies characteristic elements of the single components, especially nickel and copper for the gate electrode, and also chloride that has been added as LiCl to the electrolyte.

A measurement setup has been replicated to analyze the transistor characteristics by applying different gate voltages (V_g) and measuring the change in current (I_{sd}) at the source/drain electrode. Measurements were carried out at five different OECTs and all of them functioned as a typical transistor, where the gate voltage controls the current that flows between the source and the gate electrode. Due to that, the current at the source/drain electrode varied from +93.12 µA at a gate voltage of −1.0 V to +30.52 µA at a gate voltage of +1.0 V at a constant source/drain voltage of +0.5 V. The conductivity decreased, which can be proven by Ohm's law: $R = UI$. When inserting the values, the following results are derived:

$$R\left(V_g(1) = -1.0\right) = \frac{0.5\ V_{sd}}{93.12\ \mu A} \approx 5.37\ \mathrm{k}\Omega$$

$$R\left(V_g(1) = +1.0\right) = \frac{0.5\ V_{sd}}{30.52\ \mu A} \approx 16.38\ \mathrm{k}\Omega$$

The calculations prove that a positive gate voltage (V_g) increases the resistance of the transistor and hence reduces its conductivity. As stated in the introduction, OECTs based on PEDOT:PSS work in depletion mode, which means that in the absence of a positive gate voltage or when applying a negative gate voltage, a hole current flows in the channel, and conductivity is increased. However, when applying a positive gate voltage, positively charged cations are injected from the electrolyte into the channel and recombined with the negatively charged anions of the PEDOT:PSS. This recombination reduces the number of mobile holes in the channel and the conductive polymer becomes dedoped, the drain current drops as a result [16]. This mechanism is reinforced even more when adding an ionic substance (0.9% saline solution) as an analyte to the transistor, so that the drain current is reduced from approximately +36 to +24 µA at constant source/drain and gate voltage of +0.5 V. The resistance amounts to 13.89 kΩ when no analyte is added and increases to 18.52 kΩ under the influence of an ionic substance (0.9% saline solution). The analyte gives additional ions to the channel that interacts with the holes in PEDOT:PSS structure, and this ultimately leads to a further reduction of conductivity. Also, the quantifiable measurement of saline concentration was proven as different concentrations changed the current flow at the source/drain electrode. For a reliable sensor, specificity must be conferred on the OECT by modifying the gate electrode, electrolyte, or both. In that way, N. Coppedè et al. managed to carry out real-time measurements of both saline and adrenaline concentration in real human sweat [28]. Other literature demonstrates that OECTs are able to selectively detect glucose by adding an enzyme to the electrolyte or using different gate materials [32–38]. It is conceivable to adapt the OECT in such a way, that it can selectively detect bacteria [39], glucose [9], dopamine [40], lactate [41], and many more in human physiological fluids, such as blood, sweat, or saliva. Moreover, stability and durability need to be ensured, which is why additional experiments must be carried out to investigate the sensory properties of the textile OECT after twisting, bending, folding, etc. comparable to the investigations of A. Yang et al. (2018) [24].

To summarize, the measurements show that, firstly, the gate voltage controls the current output and, secondly, ionic substances can be detected by that OECT structure. Moreover, the doping and dedoping of PEDOT:PSS has not only become visible by the measured current but also by the color changing effect observed during the experiments. Figure 10 visually captures that effect. The given data proves the working principle of the herein-developed textile one-component OECTs.

5. Conclusions

Many textile-based organic electrochemical transistors have been reported that are based on yarn substrates that must be precisely arranged in a textile structure and are connected by a liquid electrolyte. Herein, a one-component OECT was developed that is easily integrable in textile fabrics as a solid electrolyte layer that connects both electrodes. The manufacturing of this device is based on a simple impregnation process to coat a nonwoven textile with PEDOT:PSS and a laminating process to put together the gate electrode, electrolyte, and source/drain electrode. The OECT demonstrates typical transistor characteristics, meaning that the gate voltage controls the current output. At a constant source/drain voltage of +0.5 V and a gate voltage of −1.0 V, the current is at +93.12 μA (I_{sd}). When increasing the gate voltage to +1.0 V, the current drops down to +30.52 μA (I_{sd}), which proves that the herein-developed transistor works in depletion mode. Moreover, saline solution was detected by decreasing the current output when the solution was applied. The easy manufacturing method, the integrability, and the results suggest that the use of these OECTs as biosensors in near-body applications is promising. However, future work must be carried out to ensure reproducibility, selectivity, durability, and full textile integration.

Author Contributions: Conceptualization, R.B.; methodology, R.B., C.G., and S.G.; investigation, R.B.; writing—original draft preparation, R.B.; writing—review and editing, C.G., S.G. and A.S.-P.; visualization, R.B.; supervision, A.S.-P.; project administration, A.S.-P.; funding acquisition, A.S.-P. All authors have read and agreed to the published version of the manuscript.

Funding: This research received no external funding.

Data Availability Statement: Not applicable.

Conflicts of Interest: The authors declare no conflict of interest.

References

1. *ISO/TR 23383:2020*; Textiles and Textile Products-Smart (Intelligent) Textiles-Definitions, Categorisation, Applications and Standardization Needs, Deutsche Fassung CEN/TR_16298:2011. Beuth GmbH: Berlin, Germany, 2020. Available online: https://www.iso.org/obp/ui/#iso:std:iso:tr:23383:ed-1:v1:en (accessed on 5 November 2022).
2. Mazzoldi, A.; Rossi, D.D.; Lorussi, F.; Scilingo, E.P.; Paradiso, R. Smart textiles for wearable motion capture systems. *AUTEX Res. J.* **2022**, *2*, 5.
3. Mečņika, V. Smart Textiles for Healthcare: Applications and Technologies. In *Rural Environment. Education. Personality. (REEP). Proceedings of the International Scientific Conference (Latvia), Jelgava, Latvia, 7–8 February 2014*; Latvia University of Agriculture: Jelgava, Latvia, 2014.
4. Cochrane, C.; Hertleer, C.; Schwarz-Pfeiffer, A. 2-Smart textiles in health: An overview. In *Smart Textiles and their Applications*; Koncar, V., Ed.; Woodhead Publishing: Oxford, UK, 2016; pp. 9–32. [CrossRef]
5. Ojuroye, O.; Torah, R.; Beeby, S.; Wilde, A. Smart Textiles for Smart Home Control and Enriching Future Wireless Sensor Network Data. In *Sensors for Everyday Life: Healthcare Settings*; Postolache, O.A., Mukhopadhyay, S.C., Jayasundera, K.P., Swain, A.K., Eds.; Springer International Publishing: Cham, Switzerland, 2017; pp. 159–183. [CrossRef]
6. Gehrke, I.; Schmelzeisen, D.; Gries, T.; Lutz, V.; Tenner, V. *Smart Textiles Production—Overview of Material, Sensor and Production Technologies for Industrial Smart Textiles*; MDPI: Basel, Switzerland, 2019; p. 35.
7. Cherenack, K.; van Pieterson, L. Smart textiles: Challenges and opportunities. *J. Appl. Phys.* **2012**, *112*, 091301. [CrossRef]
8. Leleux, P.; Rivnay, J.; Lonjaret, T.; Badier, J.M.; Bénar, C.; Hervé, T.; Chauvel, P.; Malliaras, G.G. Organic Electrochemical Transistors for Clinical Applications. *Adv. Healthc. Mater.* **2015**, *4*, 142–147. [CrossRef] [PubMed]

9. Bernards, D.A.; Macaya, D.J.; Nikolou, M.; DeFranco, J.A.; Takamatsu, S.; Malliaras, G.G. Enzymatic sensing with organic electrochemical transistors. *J. Mater. Chem.* **2008**, *18*, 116–120. [CrossRef]
10. Strakosas, X.; Bongo, M.; Owens, R.M. The organic electrochemical transistor for biological applications. *J. Appl. Polym. Sci.* **2015**, *132*. [CrossRef]
11. Wang, N.; Yang, A.; Fu, Y.; Li, Y.; Yan, F. Functionalized Organic Thin Film Transistors for Biosensing. *Acc. Chem. Res.* **2019**, *52*, 277–287. [CrossRef] [PubMed]

12. Marks, A.; Griggs, S.; Gasparini, N.; Moser, M. Organic Electrochemical Transistors: An Emerging Technology for Biosensing. *Adv. Mater. Interfaces* **2022**, *9*, 2102039. [CrossRef]
13. Kergoat, L.; Piro, B.; Berggren, M.; Horowitz, G.; Pham, M.-C. Advances in organic transistor-based biosensors: From organic electrochemical transistors to electrolyte-gated organic field-effect transistors. *Anal. Bioanal. Chem.* **2012**, *402*, 1813–1826. [CrossRef]
14. Malti, A.; Edberg, J.; Granberg, H.; Khan, Z.U.; Andreasen, J.W.; Liu, X.; Zhao, D.; Zhang, H.; Yao, Y.; Brill, J.W.; et al. An Organic Mixed Ion-Electron Conductor for Power Electronics. *Adv. Sci.* **2016**, *3*, 1500305. [CrossRef]
15. Friedlein, J.T.; McLeod, R.R.; Rivnay, J. Device physics of organic electrochemical transistors. *Org. Electron.* **2018**, *63*, 398–414. [CrossRef]
16. Rivnay, J.; Inal, S.; Salleo, A.; Owens, R.M.; Berggren, M.; Malliaras, G.G. Organic electrochemical transistors. *Nat. Rev. Mater.* **2018**, *3*, 2. [CrossRef]
17. Khodakarimi, S.; Hekhmatshoar, M.; Nasiri, M.; Moghaddam, M.; Abbasi, F. Effects of process and post-process treatments on the electrical conductivity of the PEDOT:PSS films. *J. Mater. Sci. Mater. Electron.* **2016**, *27*, 1278–1285. [CrossRef]
18. Mukherjee, S.; Singh, R.; Gopinathan, S.; Murugan, S.; Gawali, S.; Saha, B.; Biswas, J.; Lodha, S.; Kumar, A. Solution-Processed Poly(3,4-ethylenedioxythiophene) Thin Films as Transparent Conductors: Effect of p-Toluenesulfonic Acid in Dimethyl Sulfoxide. *ACS Appl. Mater. Interfaces* **2014**, *6*, 17792–17803. [CrossRef] [PubMed]
19. Sun, K.; Zhang, S.; Li, P.; Xia, Y.; Zhang, X.; Du, D.; Isikgor, F.H.; Ouyang, J. Review on application of PEDOTs and PEDOT:PSS in energy conversion and storage devices. *J. Mater. Sci. Mater. Electron.* **2015**, *26*, 4438–4462. [CrossRef]
20. Leute, U. (Ed.) Wie macht man Kunststoffe leitfähig? In *Elektrisch leitfähige Polymerwerkstoffe: Ein Überblick für Studierende und Praktiker*; Springer Fachmedien: Wiesbaden, Germany, 2015; pp. 3–12. [CrossRef]
21. Mattana, G.; Cosseddu, P.; Fraboni, B.; Malliaras, G.G.; Hinestroza, J.P.; Bonfiglio, A. Organic electronics on natural cotton fibres. *Org. Electron.* **2011**, *12*, 2033–2039. [CrossRef]
22. Müller, C.; Hamedi, M.; Karlsson, R.; Jansson, R.; Marcilla, R.; Hedhammar, M.; Inganäs, O. Woven Electrochemical Transistors on Silk Fibers. *Adv. Mater.* **2011**, *23*, 898–901. [CrossRef]
23. Tarabella, G.; Villani, M.; Calestani, D.; Mosca, R.; Iannotta, S.; Zappettini, A.; Coppedè, N. A single cotton fiber organic electrochemical transistor for liquid electrolyte saline sensing. *J. Mater. Chem.* **2012**, *22*, 23830. [CrossRef]
24. Yang, A.; Li, Y.; Yang, C.; Fu, Y.; Wang, N.; Li, L.; Yan, F. Fabric Organic Electrochemical Transistors for Biosensors. *Adv. Mater.* **2018**, *30*, 1800051. [CrossRef]
25. Tao, X.; Koncar, V.; Dufour, C. Geometry Pattern for the Wire Organic Electrochemical Textile Transistor. *J. Electrochem. Soc.* **2011**, *158*, H572. [CrossRef]
26. Gualandi, I.; Marzocchi, M.; Achilli, A.; Cavedale, D.; Bonfiglio, A.; Fraboni, B. Textile Organic Electrochemical Transistors as a Platform for Wearable Biosensors. *Sci. Rep.* **2016**, *6*, 33637. [CrossRef]
27. Electromagnetic Field Shielding Fabrics. Available online: https://www.lessemf.com/fabric.html (accessed on 25 May 2022).
28. Kawahara, J.; Ersman, P.A.; Engquist, I.; Berggren, M. Improving the color switch contrast in PEDOT:PSS-based electrochromic displays. *Org. Electron.* **2012**, *13*, 469–474. [CrossRef]
29. Ding, Y.; Invernale, M.A.; Sotzing, G.A. Conductivity Trends of PEDOT-PSS Impregnated Fabric and the Effect of Conductivity on Electrochromic Textile. *ACS Appl. Mater. Interfaces* **2010**, *2*, 1588–1593. [CrossRef]
30. Levasseur, D.; Mjejri, I.; Rolland, T.; Rougier, A. Color Tuning by Oxide Addition in PEDOT:PSS-Based Electrochromic Devices. *Polymers* **2019**, *11*, 179. [CrossRef] [PubMed]
31. Coppedè, N.; Tarabella, G.; Villani, M.; Calestani, D.; Iannotta, S.; Zappettini, A. Human stress monitoring through an organic cotton-fiber biosensor. *J. Mater. Chem. B* **2014**, *2*, 5620–5626. [CrossRef] [PubMed]
32. Kanakamedala, S.; Alshakhouri, H.; Agarwal, M.; DeCoster, M. A simple polymer based electrochemical transistor for micromolar glucose sensing. *Sens. Actuators B Chem.* **2011**, *157*, 92–97. [CrossRef]
33. Shim, N.Y.; Bernards, D.A.; Macaya, D.J.; DeFranco, J.A.; Nikolou, M.; Owens, R.M.; Malliaras, G.G. All-Plastic Electrochemical Transistor for Glucose Sensing Using a Ferrocene Mediator. *Sensors* **2009**, *9*, 9896–9902. [CrossRef]
34. Kim, Y.; Do, J.; Kim, J.; Yang, S.Y.; Malliaras, G.G.; Ober, C.K.; Kim, E. A Glucose Sensor Based on an Organic Electrochemical Transistor Structure Using a Vapor Polymerized Poly(3,4-ethylenedioxythiophene) Layer. *Jpn. J. Appl. Phys.* **2010**, *49*, 01AE10. [CrossRef]
35. Macaya, D.J.; Nikolou, M.; Takamatsu, S.; Mabeck, J.T.; Owens, R.M.; Malliaras, G.G. Simple glucose sensors with micromolar sensitivity based on organic electrochemical transistors. *Sens. Actuators B Chem.* **2007**, *1*, 374–378. [CrossRef]
36. Tang, H.; Yan, F.; Lin, P.; Xu, J.; Chan, H.L.W. Highly Sensitive Glucose Biosensors Based on Organic Electrochemical Transistors Using Platinum Gate Electrodes Modified with Enzyme and Nanomaterials. *Adv. Funct. Mater.* **2011**, *21*, 2264–2272. [CrossRef]
37. Liao, J.; Lin, S.; Yang, Y.; Liu, K.; Du, W. Highly selective and sensitive glucose sensors based on organic electrochemical transistors using TiO_2 nanotube arrays-based gate electrodes. *Sens. Actuators B Chem.* **2015**, *208*, 457–463. [CrossRef]
38. Wang, Y.; Qing, X.; Zhou, Q.; Zhang, Y.; Liu, Q.; Liu, K.; Wang, W.; Li, M.; Lu, Z.; Chen, Y.; et al. The woven fiber organic electrochemical transistors based on polypyrrole nanowires/reduced graphene oxide composites for glucose sensing. *Biosens. Bioelectron.* **2017**, *95*, 138–145. [CrossRef] [PubMed]
39. He, R.X.; Zhang, M.; Tan, F.; Leung, P.H.; Zhao, X.Z.; Chan, H.L.; Yang, M.; Yan, F. Detection of bacteria with organic electrochemical transistors. *J. Mater. Chem.* **2012**, *22*, 22072–22076. [CrossRef]

40. Tang, H.; Lin, P.; Chan, H.L.W.; Yan, F. Highly sensitive dopamine biosensors based on organic electrochemical transistors. *Biosens. Bioelectron.* **2011**, *26*, 4559–4563. [CrossRef] [PubMed]
41. Nilsson, D.; Kugler, T.; Svensson, P.-O.; Berggren, M. An all-organic sensor–transistor based on a novel electrochemical transducer concept printed electrochemical sensors on paper. *Sens. Actuators B Chem.* **2002**, *86*, 193–197. [CrossRef]

micromachines

Article

A Flexible Thermocouple Film Sensor for Respiratory Monitoring

Xiaodan Miao [1,*], Xiang Gao [1], Kaiming Su [2], Yahui Li [3] and Zhuoqing Yang [3]

1 School of Mechanical and Automotive Engineering, Shanghai University of Engineering Science, Shanghai 201620, China
2 Department of Otolaryngology-Head and Neck Surgery, Shanghai Jiao Tong University Affiliated Sixth People's Hospital, Shanghai 200233, China
3 National Key Laboratory of Science and Technology on Micro/Nano Fabrication, Shanghai Jiao Tong University, Shanghai 200240, China
* Correspondence: mxd8868@126.com

Abstract: A novel flexible thermocouple film sensor on a polyimide substrate is proposed that is simple and flexible for monitoring the respiratory signal. Several thermocouples were connected in series and patterned on the polyimide substrate, and each one is formed by copper and a constant line connected to each other at two nodes. The respiratory signal was measured by the output voltage, which resulted from the temperature difference between the hot and cold junctions. The sensors were fabricated with surface-microfabrication technology with three sputtering steps. The measurement results showed that the peak voltage decreased by 90% in the case of apnea compared with normal breathing. The sensor has potential application for wearable detection of sleep apnea hypopnea syndrome (OSAHS).

Keywords: respiratory monitoring; thermocouple; flow sensor; heat transfer

Citation: Miao, X.; Gao, X.; Su, K.; Li, Y.; Yang, Z. A Flexible Thermocouple Film Sensor for Respiratory Monitoring. *Micromachines* **2022**, *13*, 1873. https://doi.org/10.3390/mi13111873

Academic Editors: Jianliang Xiao and Aiqun Liu

Received: 17 August 2022
Accepted: 27 October 2022
Published: 31 October 2022

1. Introduction

Obstructive sleep apnea hypopnea syndrome is a serious, potentially fatal sleep respiratory disease that seriously affects the physical and mental health of patients [1]. The respiratory signal is the most important parameter of OSAHS detection. The respiratory sensor can detect sudden infant death syndrome or record a patient's physiological status for sleep studies and sports training [2]. The overall prevalence rate of OSAHS is 9% to 38% in the general adult population, and is much higher in the elderly population [3]. In clinical practice, polysomnography is the standard diagnostic method for OSAHS [4].

According to the classification criteria of the American Academy of Sleep Medicine (AASM), the detection and diagnosis of OSAHS can be divided into four levels: Level 1: standard polysomnography, Level 2: comprehensive portable polysomnography, Level 3: modified portable sleep-apnea testing, and Level 4: continuous single or dual bio-parameter recording. With the expansion of the population and progress of technology, some new classifications were proposed [5,6].

In traditional sleep-monitoring methods, polysomnography is a common device, which is mainly used in hospitals and sleep laboratories. Polysomnography can record a variety of physiological signals of the human body in the process of sleep all night, including electrocardiogram (ECG), electroencephalogram (EEG), electrooculogram (EOG), electromyogram (EMG), respiratory signal, oxygen saturation, and sleep posture [7]. However, due to its high cost and great mental pressure on patients, studies on portable sleep respiration-monitoring devices have become more popular in recent years [8].

Various sensors and sensing methods have been developed to measure respiratory rate and/or lung capacity, including transthoracic impedance, blood O_2/CO_2 concentration, and breathing airflow [9–26]. Breathing air flow is typically detected by sensing pressure or temperature, and the adopted sensors may be resistive, thermoelectrical, pyroelectrical, or piezoelectric [10–26]. Chen designed skin-like hybrid integrated circuits (SHICs) with stretchable sensors that could capture the temperature change of the inhaled and exhaled air. By integrating the flexible devices on human faces, comfort can be enhanced [9]. Jiang proposed a portable sleep respiration-monitoring system including three sensors that could monitor airflow, body posture, and oxygen saturation at the same time. The combination of the three signals could improve the accuracy of diagnosis of OSAHS [10]. Dalola reported a micromachined smart system for flow measurement based on a silicon substrate, consisting of four germanium thermistors embedded in a thin membrane and connected to form a Wheatstone bridge supplied with a constant DC current. It can measure the velocity and transport rate at the same time, which combines calorimetric and hot-wire transduction principles with lower power consumption and thermal loss [22]. Wei et al. represented a novel CMOS process-compatible MEMS sensor for monitoring respiration. This resistive flow sensor was manufactured by the TSMC 0.35 m CMOS/MEMS mixed-signal 2P4M polycide process. The sensor was demonstrated to be sensitive enough to detect the respiratory flow rate, and the relationship between flow rate and sensed voltage was linear [23].

Compared with the work in Refs. [22,23], in which the air-flow sensor was fabricated based on silicon substrate and supplied with a constant DC current, this paper presents a thermocouple thin-film flow sensor that was fabricated based on a polymide substrate with three sputtering steps based on the thermal electromotive force without an extra power supply. As a result, it is more flexible for a wearable diagnosis system outside the hospital with lower cost. It consists of four micromachined thermocouples on a polyimide substrate with a translation circuit. Its working principle is based on the combination of the Seebeck effect and the heat-transfer theorem. The thermocouple is formed by two different metal lines connected to each other at two nodes. One node serves as a hot junction, and its surrounding temperature changes with the respiratory process. The free node serves as cold junction, and its surrounding temperature keeps constant. As a result, the thermal potential is generated from the temperature difference between the hot and cold junctions. The respiratory signal is measured by the output voltage, which is converted from the thermal potential. It was fabricated by surface microfabrication technology and the testing results show that the peak voltage decreased by 90% in the case of apnea compared with normal breathing. The sensor has potential for application to the detection of sleep apnea hypopnea syndrome (OSAHS).

2. Physical Modeling

2.1. Physical Model

The flexible thermocouple sensor consists of four thermocouples, as shown in Figure 1. Each thermocouple is formed by two different metal lines consisting of copper and constantan, which are connected to each other at two nodes. One node serves as a hot junction, and its surrounding temperature changes with the respiratory process. The free node serves as a cold junction, and its surrounding temperature keeps constant. As a result, the thermal potential is generated from the temperature difference between the hot and cold junctions. Under the action of respiration, the temperature of the hot junction increases or decrease and the cold junction remains unchanged, resulting in a temperature difference and thermal electromotive force. Finally, the electromotive-force signal is converted into a respiratory signal.

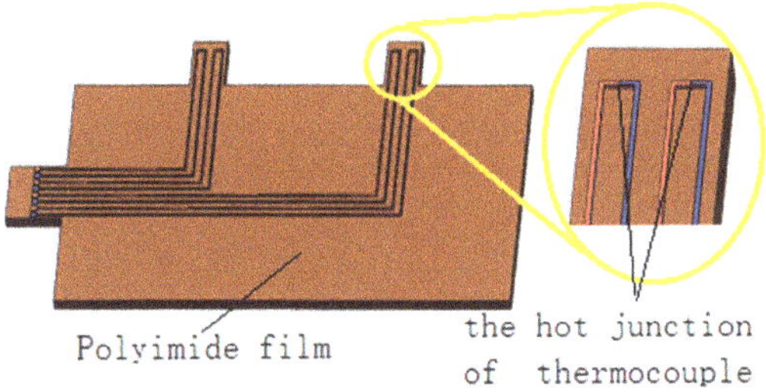

Figure 1. Physical model of the flexible thermocouple film sensor.

2.2. Mathematical Model

The sensor works based on the Seebeck effect, which means that if two different kinds of metals are connected at two nodes with different temperatures, an electric current is generated [17]. The Seebeck effect can be expressed as follows:

$$E_{AB}(T,T_0) = \frac{k}{e}(T-T_0)\ln\frac{n_A}{n_B} + \int_{T_0}^{T}(\sigma_A-\sigma_B)dT \tag{1}$$

where E_{AB} is the thermal electromotive force of the two materials and S_{AB} is the Seebeck coefficient. K is Boltzmann's constant, E is the electron charge, T is the temperature of the hot junction, and T_0 is the temperature of the cold junction. Using S_{AB} as the Seebeck coefficient of the two materials, E_{AB} can be expressed as the integral of the Seebeck coefficient over the temperature difference per the following:

$$E_{AB} = \int_{T_0}^{T} S_{AB}(T)dT \tag{2}$$

Based on different materials, the thermocouples can be divided into many types. The thermocouple type used in this paper was the T type, which is composed of copper and constantan.

The Seebeck coefficient of the thermocouple is related to the temperature of the material and the size effect. According to the research by Yang et al. [21], the Seebeck coefficient of a T-type thermocouple is 43.98~46.47 μV/°C when the thickness of thermocouple is from 0.5 to 2 μm.

The change in respiratory airflow is approximated as a sine function. It is known that the tidal volume of normal people in a sleep state is about 0.6 L [18,19] and a breathing cycle is 3 s. The change in gas volume (Q) of human lungs with time (t) can be expressed as following:

$$Q(t) = 0.3\sin\left(\frac{2\pi t}{3}-\frac{\pi}{2}\right)+0.3 \tag{3}$$

A (A = 100.48 mm^2) is used to represent the area of the nasal entrance of normal people, and the air velocity V at the entrance can be expressed as follows:

$$V(t) = \frac{dQ(t)}{dt}\times\frac{1}{A} = 6.25\cos\left(\frac{2\pi t}{3}-\frac{\pi}{2}\right) \tag{4}$$

The viscosity (ν) and Planck number (Pr) can be obtained. The total length (l) of the heat-transfer part is 5 mm. The Reynolds number (Re_l) can be calculated with Equation (5). It is a laminated flow, since Re_l is less than Re_c.

$$\mathrm{Re}_l = \frac{V(t)l}{\nu} \tag{5}$$

The average convective heat-transfer coefficient when the laminar flow flows along a flat plate can be expressed in Formula (6) as follows:

$$h = 0.664\frac{\lambda}{l}\mathrm{Re}_l^{1/2}\mathrm{Pr}^{1/3} \tag{6}$$

The calculation results of each parameter are shown in Figure 2 and are used as the boundary condition for the finite-element simulation as follows.

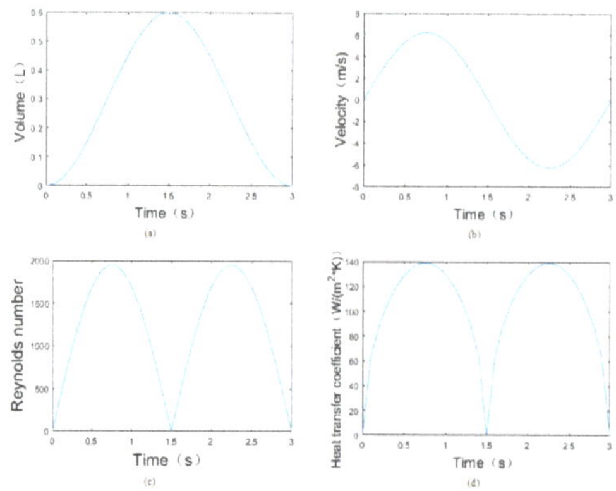

Figure 2. (**a**) The lung gas-volume variations with time; (**b**) the nasal airflow-velocity variations with time; (**c**) the Reynolds-number variations with time; (**d**) the convective heat-transfer coefficient variations with time.

3. Simulation

3.1. Simulation of the Distribution of the Polyimide Substrate

The important parameters of the sensors were simulated, such as the polyimide film thickness and the distribution of the thermocouples under different thermal conditions. Firstly, the static-temperature distribution of the polyimide film was simulated in case of no breathing. The skin temperature was set to 36 °C and the ambient temperature was set to 20 °C. The natural convection heat-transfer coefficient was set to 5 W/(m^2·°C), which decided the heat transfer and the temperature-equilibrium point in steady state. The steady-state results obtained by COMSOL simulation are shown in Figure 3. The temperature on the main part of the polyimide was 36 °C, nearly the temperature in the nostril because of the heat transfer. In comparison, two isosceles-square parts were kept at 20 °C, which were suspended and separated from the skin and kept constant with the air.

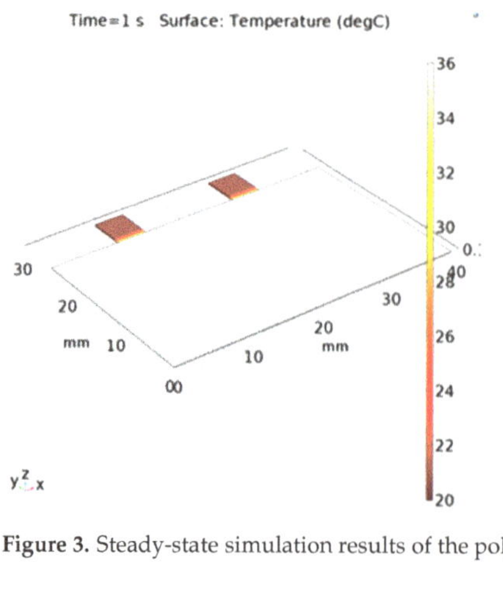

Figure 3. Steady-state simulation results of the polyimide film.

Secondly, the temperature on the center position with different thicknesses of the polyimide film was simulated. The results indicate that the response speed of the temperature decreased slightly with the increase in the thickness, as shown in Figure 4. However, the central steady-state temperature of polyimide film was kept at a stable state. As a result, the thickness of the polyimide film had less influence on the temperature variation of the polyimide film.

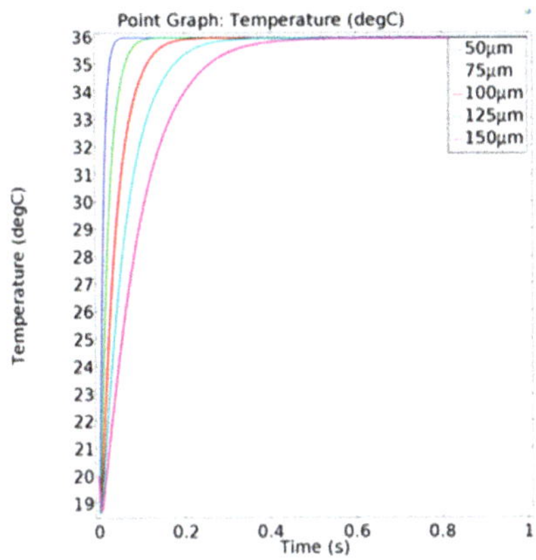

Figure 4. The temperature variation of the film center under different film thicknesses.

Thirdly, the distribution of the cold and hot junctions was simulated. The heat-transfer efficiency of the polyimide film along the thickness was larger because its thickness was far less than the length and width. As a result, the center temperature of the film increased with the skin temperature. As shown in Figure 5, two isosceles-square parts of the thin film were less effected by the skin because they were suspended and separated from the skin in the nostril. The cold conjunctions were located on the two isosceles-square parts to keep them at a constant temperature.

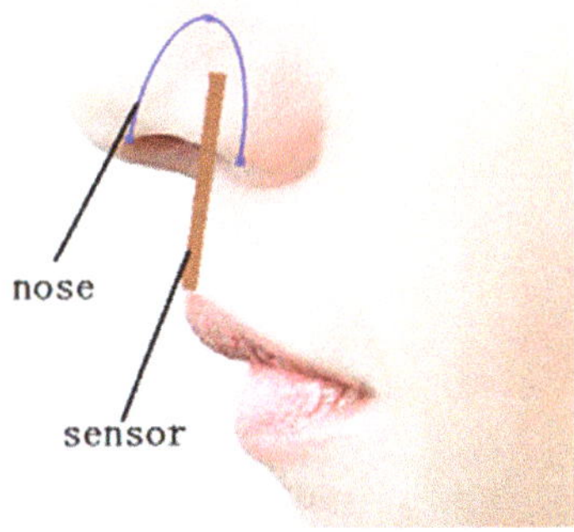

Figure 5. Schematic diagram of sensor position.

The signal of a thermocouple is small, and a more stable and accurate temperature signal can be acquired by connecting several groups of thermocouples in a series, as shown in Figure 6. As shown in Figure 6a, the cold junction was located on the center part of the polyimide film, which was affected by the skin. In order to generate the temperature difference between the cold and hot junctions, the design of the thermocouple was optimized as shown in Figure 6b, where the cold junction was led out with extension wire to avoid the influence of the skin.

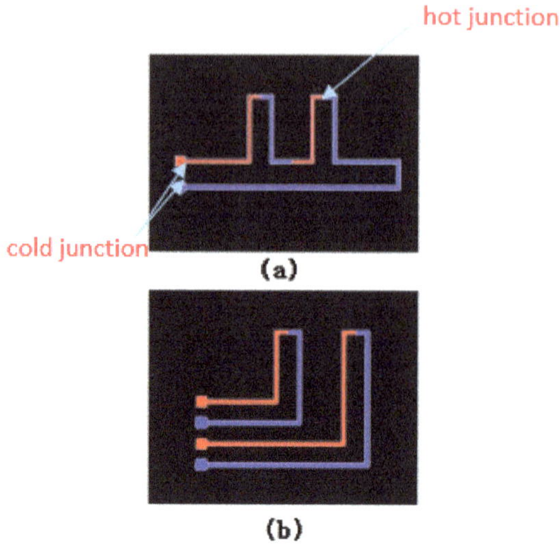

Figure 6. Two structures of the thermocouple. (**a**) The cold junction is located on the center part of the polyimide film; (**b**) the cold junction was led out with extension wire to avoid influence from the skin.

3.2. Simulation of the Single Thermalcouple

The thermoelectric coupling-effect simulation of a single thermocouple was carried out to study its thermal electromotive force. Copper and constantan were selected as the materials, and the Seebeck coefficient S_{AB} was 43 μV/°C. It was assumed that the ambient temperature was 20 °C and the respiratory temperature was 36 °C. Therefore, the temperature difference was 16 °C. Figure 7 shows that the thermal electromotive force was 7×10^{-4} V. According to Formula (2) mentioned above, the thermal electromotive force was 6.88×10^{-4} V. The results of the theoretical simulation and finite-element simulation tended to be consistent.

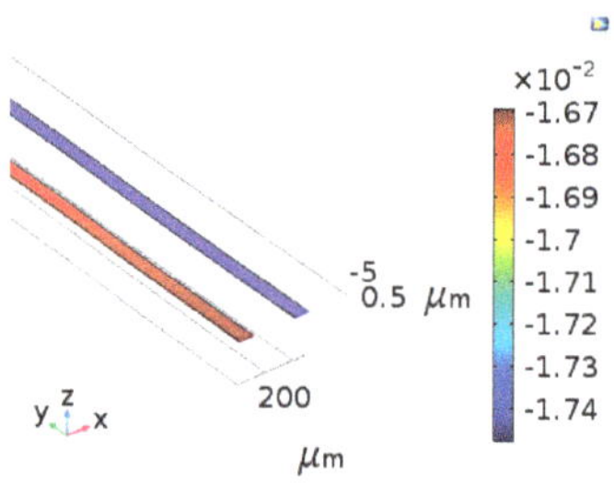

Figure 7. The simulation results of the cold junction of the thermocouple.

The transient simulation was carried out to study the influence of different thicknesses on the response speed of the sensor. On the premise of a fixed Seebeck coefficient, the relationship between the temperature difference and thermal electromotive force is linear. Therefore, the results can be expressed in terms of the temperature of the hot junction. The temperature setting was the same as before and the convective heat-transfer coefficient was set as 100 W/(m^2·°C). In Figure 8, when the temperature reached 90% of the maximum temperature, the corresponding time was defined as the response time, as shown in Table 1.

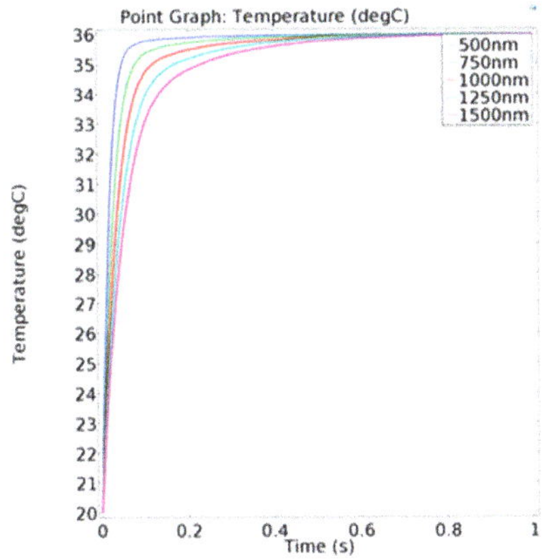

Figure 8. Temperature change of the hot junction at different thermocouple thicknesses.

Table 1. Response of the thermocouple with different thickness.

Thickness of Thermocouple (nm)	Response Time (ms)
500	40
750	57
1000	79
1250	104
1500	134

With the decrease in the thermocouple thickness, the response speed increases. However, the smaller the thickness is, the more difficult the fabrication process. At the same time, the resistance value also increases, which would affect the signal acquisition.

3.3. Simulation of the Sensor under the Working Conditions

The polyimide film, the thermocouple, and the respiratory signal were combined to simulate the temperature variation of the sensor during breathing. The actual ambient temperature at 27 °C was used as a simulation parameter for comparison with the results of the measurement in the environment. In order to verify the ability of the sensor to detect respiration, the respiratory intensity was changed for simulation. If the respiratory intensity decreases by 30%, it can be considered hypopnea, and more than 50% is very serious hypopnea [3–6].

The convective heat-transfer coefficient was used as the simulation condition, and a parametric sweeping was carried out to study the temperature of the hot-junction change under the condition of hypopnea. As shown in Figure 9, the temperature curves of hypopnea and normal respiration were significantly different. The average temperature difference of the hot junction in multiple respiratory cycles under different respiratory flow is shown in Table 2. The decrease range of the average temperature difference relative to 100% respiration intensity can be used as a criterion to estimate whether hypopnea occurs.

Table 2. Average temperature difference of the hot junction under different respiratory flows.

Respiratory Flow (%)	Average Temperature Difference (°C)
100	5.17
90	4.70
80	4.10
70	3.50
60	3.21
50	2.45

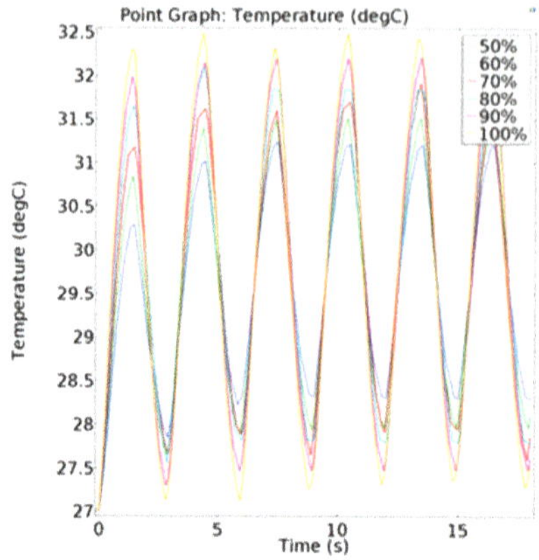

Figure 9. Temperature change of the hot junction under different respiratory flows.

As shown in Figure 9, the maximum temperature of the hot junction of the thermocouple did not exceed 32.5 °C, but human body temperature was at least 36 °C. It was assumed that the PI membrane absorbed part of the heat. In order to verify this hypothesis, a simulation was carried out by changing the PI film thickness while keeping the respiratory flow unchanged. The simulation results are shown in Figure 10. The average temperature difference under different film thicknesses is shown in Table 3.

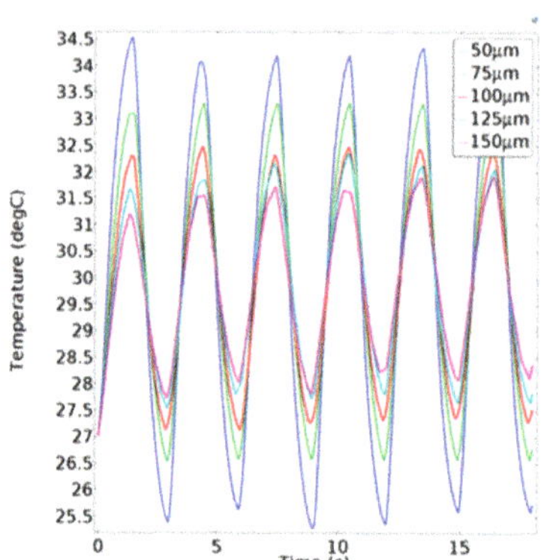

Figure 10. Temperature change of the hot junction under different thin-film thicknesses.

Table 3. Average temperature difference of the hot junction under different thin-film thicknesses.

Thin-Film Thickness (μm)	Average Temperature Difference (°C)
50	8.47
75	6.74
100	5.17
125	4.07
150	3.52

The results show that the thickness of the polyimide film had a great influence on the temperature difference due to the heat absorption by the polyimide film. The larger the thickness of the polyimide film, the smaller the average temperature difference of the thermocouple hot junction in multiple breathing cycles. It can be concluded that an increase in polyimide film thickness leads to a decline in respiratory-monitoring performance of the sensor. Therefore, reducing the thickness can effectively improve the quality of the signal.

4. Microfabrication

In the design and simulation mentioned above, the key parameters were optimized, such as the distribution of the thermocouple, the thickness of the polyimide, and the linewidth of the thermocouple. The flexible thermocouple sensor was fabricated with microfabrication technology, which was mainly composed of sputtering technology. The fabrication process was as follows:

The glass wafer and the polyimide film (PI film) were ultrasonic cleaned in deionized water and finally dried in an oven at 60 °C for 30 min. The thickness of the PI film was 100 μm.

(1) The cleaned PI film was attached to the glass substrate and spin-coated with AZ-4620 positive photoresist with a thickness of 10 μm. Then the glass substrate was heat treated.
(2) The substrate was exposed for 30 s and then developed for 15 s to form the mold of the bottom layer for sputtering.
(3) Then, the chromium seed layer was sputtered at 300 W for 180 s and the copper layer at 300 W for 480 s. The process of sputtering copper was repeated five times.
(4) The photoresist was removed by ultrasonic cleaning in acetone solution for 30 min.

After removing the photoresist, the mask was replaced and the lithography steps were repeated. After sputtering the chromium and constantan layers, the photoresist was removed at last. It should be noted that in order to ensure the adhesion of the two metal layers, the area of the two masks layer need to be partially overlapping. The manufacturing process is shown in Figure 11. The thickness of the sputtering metal measured by step profiler was 500 to 600 nm.

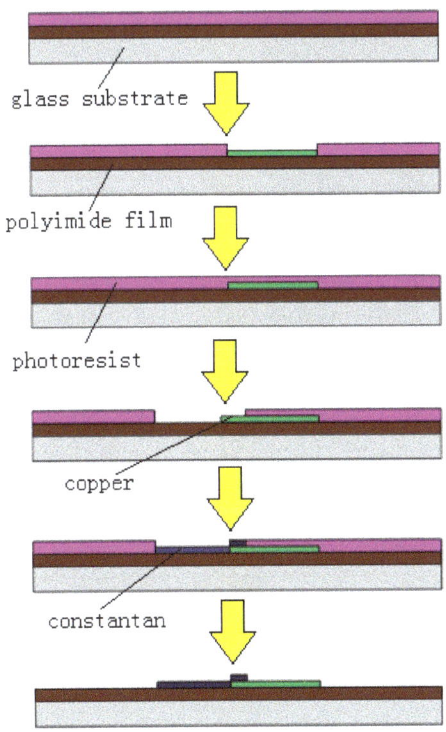

Figure 11. Fabrication-process diagram of the sensor.

5. Experiment and Results

5.1. Microscopic Morphology of the Sensor

The samples were analyzed by ultra-high-resolution FE-SEM, as shown in Figure 12. It can be observed that the upper part of the wire was constantan, which contains both nickel and copper. The lower part was copper wire. In addition, the overlapping of copper and constantan could be clearly observed by magnifying the junction of the two materials.

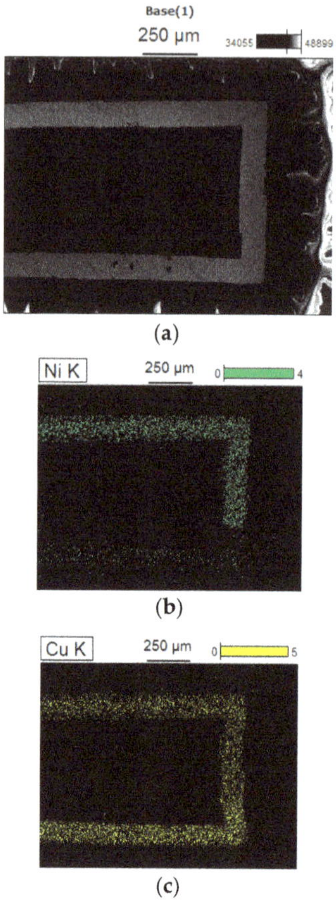

Figure 12. SEM images of thin-film thermocouple. (**a**) Connection point of thermocouple. (**b**) Distribution of Ni. (**c**) Distribution of Cu.

5.2. Sleep Apnea Measurment System and Results

The measurement system was established as shown in Figure 13. The output of the sensor was connected to the oscillographs and an amplifier with a power supply of 1.5 V × 4. In the measurement system, the thermocouple signal was weak and needed to be amplified. A kind of low-power, high-precision amplifier AD620 was applied, which could set the magnification from 1 to 1000 times with only one external resistor. The maximum input offset voltage was 50 μv. The circuit diagram and the print circuit board (PCB) are shown in Figure 14. The ambient temperature was 27 °C, and the multiple of the amplifier was set to 100 times.

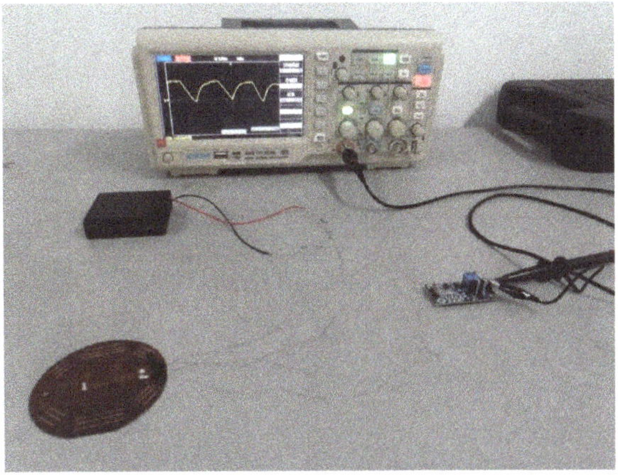

Figure 13. The measurement system of the flexible thermocouple film sensor.

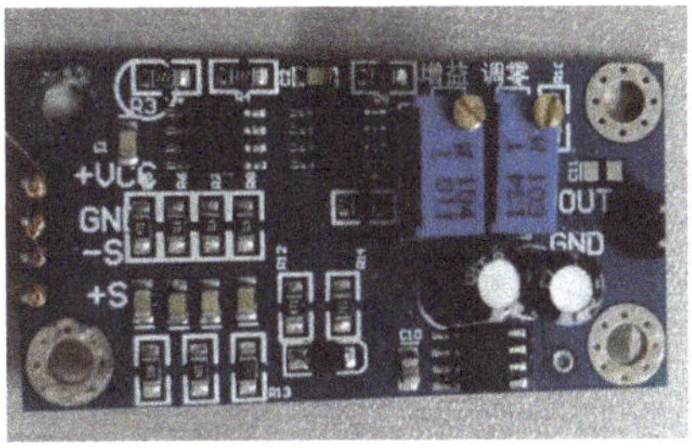

Figure 14. Circuit design of the flexible thermocouple film sensor.

Then, the sensor was attached to the bottom of the nose, with the input part inside the nose and the output part outside the nose, as shown in Figure 15. The different respiratory signal of the sensor was tested by the oscilloscope when the normal and abnormal respirations were modified. The measurement results were obtained as shown in Figures 15–17. As shown in Figure 15, the thermal electromotive-force curve changed periodically with respiration and the peak voltage was 22.4 mv. The temperature of the thermocouple increased in the expiratory phase and decreased in the inspiratory phase. One of the breathing cycles was taken out and converted to the actual temperature, and the maximum temperature was 32.23 °C. Compared with the simulation results, the maximum temperature difference between the test and simulation results was 5.38 °C, as shown in Figures 16 and 17, which indicates that the tested results and the simulation results tended to be consistent.

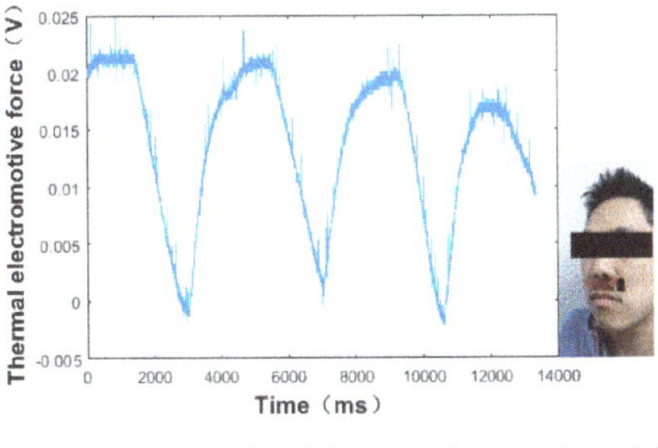

Figure 15. Test results of the normal respiration of the flexible thermocouple film sensor in a cycle.

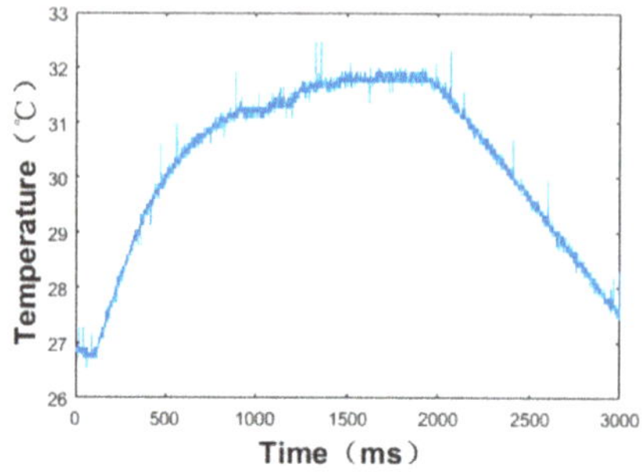

Figure 16. Temperature change of thermocouple hot junction under normal respiration conditions.

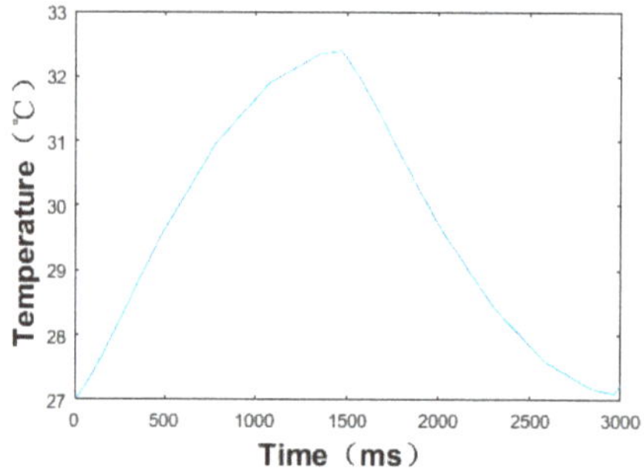

Figure 17. Simulation curve of thermocouple junction end temperature change under normal respiration conditions.

A student attached the flexible sensor on the face beneath the nostril and imitated the breathing of sleep apnea, which can be diagnosed by recognizing a significant decrease in the thermoelectromotive force, as shown in Figure 18. Under normal breathing conditions, there were peaks and troughs in the curve. The peaks indicate the end of exhalation and the troughs indicate the end of inspiration. Because of the noise in the actual detection, there was also a signal in apnea. However, the peak value of apnea decreased more than 90% compared with that of normal breathing, which indicates the occurrence of apnea—at least, very serious apnea.

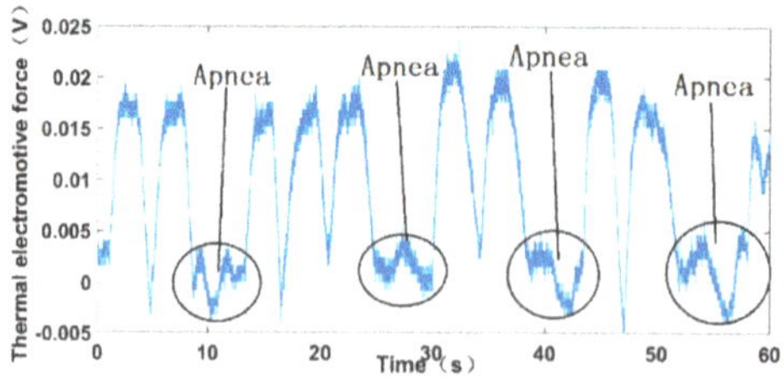

Figure 18. Test results under simulated apnea.

6. Conclusions

This paper presents a flexible thermocouple film sensor for respiratory monitoring, which has potential as an OSAHS-diagnosis system. Since it was fabricated on a flexible polyimide substrate, the sensor could be attached to the skin softly and be portable outside the hospital in the monitoring process. The sensor works based on the electromotive force from the temperature difference between the hot and cold junctions of the polyimide-film thermocouple, based on the combination of the Seebeck effect and the heat-transfer theorem.

In order to acquire the temperature difference between the cold and hot junctions, the cold junction was led out to keep at a constant temperature, whereas the average temperature of the hot junction varied with the breathing air flow. The influence of the thickness of the thermocouple and the polyimide film on the sensors was simulated and optimized. By reducing the thickness of the thermocouple, the response speed could be effectively improved as well as the defective rate, whereas the difficulty of signal measurement was increased. Simulation results also show that reducing the thickness of the polyimide film could effectively improve the signal quality. However, the polyimide film absorbed part of the heat, so the sensor could not reach the maximum temperature.

The sensor was tested in the case of apnea and the peak voltage decreased by 90%, which is in agreement with the simulation results. The results indicate that the flexible thermocouple film sensor could be used in respiratory monitoring for OSAHS detection. In future studies, the stability and wearability, reliability, and intellectualization can be improved and optimized.

Author Contributions: Conceptualization, resources, X.M., Z.Y., and K.S.; methodology, software, formal analysis, and data curation, X.G.; validation, X.G. and Y.L.; investigation, X.M. and X.G.; writing—original draft preparation and writing—review and editing, X.M. and X.G.; project administration and funding acquisition, X.M. All authors have read and agreed to the published version of the manuscript.

Funding: This research was funded by the Natural Science Foundation of China, grant numbers 52172372 and 51605277.

Institutional Review Board Statement: Not applicable.

Informed Consent Statement: Not applicable.

Conflicts of Interest: The authors declare that they have no conflict of interest.

References

1. Buchner, R.; Sosna, C.; Maiwald, M.; Benecke, W.; Lang, W. A high temperature thermopile fabrication process for thermal flow sensors. *Sens. Actuators A Phys.* **2006**, *130–131*, 262–266. [CrossRef]
2. Tang, Z.; Smith, B.; Schild, J.H.; Peckham, P.H. Data transmission from an implantable biotelemeter by load-shift keying using circuit confifiguration modulator. *IEEE Trans. Biomed. Eng.* **1995**, *42*, 524–528. [CrossRef]
3. Yaggi, H.K.; Concato, J.; Kernan, W.N.; Lichtman, J.H.; Brass, L.M.; Mohsenin, V. Obstructive sleep apnea as a risk factor for stroke and death. *N. Eng. J. Med.* **2005**, *353*, 2034–2041. [CrossRef] [PubMed]
4. Senaratna, C.V.; Perret, J.L.; Lodge, C.J.; Lowe, A.J.; Campbell, B.E.; Matheson, M.C.; Hamilton, G.S.; Dharmage, S.C. Prevalence of obstructive sleep apnea in the general population: A systematic review. *Sleep Med. Rev.* **2017**, *34*, 70–81. [CrossRef] [PubMed]
5. Kapur, V.K.; Auckley, D.H.; Chowdhuri, S.; Kuhlmann, D.C.; Mehra, R.; Ramar, K.; Harrod, C. Clinical practice guideline for diagnostic testing for adult obstructive sleep apnea: An american academy of sleep medicine clinical practice guideline. *J. Clin. Sleep Med.* **2017**, *13*, 479–504. [CrossRef] [PubMed]
6. Collop, N.A.; Anderson, W.M.; Boehlecke, B.; Claman, D.; Goldberg, R.; Gottlieb, D.J.; Hudgel, D.; Sateia, M.; Schwab, R. Clinical guidelines for the use of unattended portable monitors in the diagnosis of obstructive sleep apnea in adult patients. *J. Clin. Sleep Med.* **2007**, *3*, 737–747.
7. Collop, N.A.; Tracy, S.L.; Kapur, V.; Mehra, R.; Kuhlmann, D.; Fleishman, S.A.; Ojile, J.M. Obstructive sleep apnea devices for out-of-center (OOC) testing: Technology evaluation. *J. Clin. Sleep. Med.* **2011**, *7*, 531–548. [CrossRef]
8. Marino, M.; Li, Y.; Rueschman, M.N.; Winkelman, J.W.; Ellenbogen, J.M.; Solet, J.M.; Dulin, H.; Berkman, L.F.; Buxton, O.M. Measuring sleep: Accuracy, sensitivity, and specificity of wrist actigraphy compared to polysomnography. *Sleep* **2013**, *36*, 1747–1755. [CrossRef]
9. Chen, Y.; Liu, F.; Lu, B.; Zhang, Y.; Feng, X. Skin-like hybrid integrated circuits conformal to face for continuous respiratory monitoring. *Adv. Electron. Mater.* **2020**, *6*, 2000145. [CrossRef]
10. Jiang, P.; Zhu, R.; Dong, X.; Chang, Y. Combination mode of physiological signals for diagnosis of OSAS using portable monitor. *Sleep Breath* **2018**, *22*, 123–129. [CrossRef]
11. Gupta, H.; Arumuru, V.; Jha, R. Industrial fluid flow measurement using optical fiber sensors: A review. *IEEE Sens. J.* **2021**, *21*, 7130–7142. [CrossRef]

12. Jin, H.; Tao, X.; Dong, S.R. Flexible surface acoustic wave respiration sensor for monitoring obstructive sleep apnea syndrome. *J. Micromech. Microeng.* **2017**, *27*, 115006. [CrossRef]
13. Lin, H.C.; Su, C.L.; Ong, J.H.; Chen, Y.W.; Lin, C.Y.; Chiang, L.L.; Hung, C.H. Pulse oximetry monitor feasible for early screening of Obstructive Sleep Apnea (OSA). *J. Med. Biol. Eng.* **2020**, *40*, 62–70. [CrossRef]
14. Que, R.-Y.; Zhu, R. A compact flexible thermal flow sensor for detecting two-dimensional flow vector. *IEEE Sens. J.* **2015**, *15*, 1931–1936. [CrossRef]
15. Wang, D.F.; Isagawa, K.; Kobayashi, T.; Itoh, T.; Maeda, R. Developing passive piezoelectric MEMS sensor applicable to two-wire DC appliances with current switching. *Micro Nano Lett.* **2012**, *7*, 68–71. [CrossRef]
16. Kottapalli, A.G.P.; Asadnia, M.; Miao, J.M. Polymer MEMS pressure sensor arrays for fish-like underwater sensing applications. *Micro Nano Lett.* **2012**, *7*, 1189–1192. [CrossRef]
17. Kuo, J.T.W.; Yu, L.; Meng, E. Micromachined thermal flow sensors—A review. *Micromachines* **2012**, *3*, 550–573. [CrossRef]
18. Feng, H.L.; Miao, X.D.; Yang, Z.Q. Design, simulation and experimental study of the linear magnetic microactuator. *Micromachines* **2018**, *9*, 454. [CrossRef]
19. Elsheikh, M.H.; Shnawah, D.A.; Sabri, M.F.M.; Said, S.B.M.; Hassan, M.H.; Bashir, M.B.A.; Mohamad, M. A review on thermoelectric renewable energy: Principle parameters that affect their performance. *Renew. Sust. Energ. Rev.* **2014**, *30*, 337–355. [CrossRef]
20. Zhang, X.; Han, Y.; Li, J.; Zhang, L.; Jia, X.; Wang, E. Portable, universal, and visual ion sensing platform based on the light emitting diode-based self-referencing-ion selective field-effect transistor. *Anal. Chem.* **2014**, *86*, 1380–1384. [CrossRef]
21. Lihong, Y.; Yuanshen, Z.; Cenming, F. The influence of size effect on sensitivity of Cu/CuNi thin-film thermocouple. *Phys. Procedia* **2011**, *22*, 95–100.
22. Dalola, S.; Cerimovic, S.; Kohl, F.; Beigelbeck, R.; Schalko, J.; Ferrari, V.; Marioli, D.; Keplinger, F.; Sauter, T. MEMS thermal flow sensor with smart electronic interface circuit. *IEEE Sens. J.* **2012**, *12*, 3318–3328. [CrossRef]
23. Wei, C.-L.; Lin, C.-F.; Tseng, I.-T. A novel MEMS respiratory flow sensor. *IEEE Sens. J.* **2010**, *10*, 16–18. [CrossRef]
24. Wang, X.; Ke, Z.; Liao, G.; Pan, X.; Yang, Y.; Xu, W. Fast-response breathing monitoring system for human respiration disease detection. *IEEE Sens. J.* **2022**, *22*, 10411–10419. [CrossRef]
25. Jiang, T.; Deng, L.R.; Liang, J.; Wu, Y.; Shao, Z.; Wang, D. Wearable breath monitoring via a hot-film/calorimetric airflow sensing system. *Biosens. Bielectronics* **2020**, *163*, 112288. [CrossRef]
26. Hwang, B.; Han, Y.; Matteini, P. Bending fatigue behavior of Ag nanowire/Cu thin-film hybrid interconnects for wearable electronics. *Facta Univ. Ser. Mech. Eng.* **2022**. [CrossRef]

micromachines

MDPI

Article

A Study on Over-Molded Copper-Based Flexible Electronic Circuits

Mona Bakr [1,*], Martin Hubmann [2], Frederick Bossuyt [1] and Jan Vanfleteren [1]

[1] Center for Microsystems Technology, Ghent University, 9052 Ghent, Belgium
[2] Department of Polymer Engineering and Science, Polymer Processing, Montanuniversitaet Leoben, 8700 Leoben, Austria
* Correspondence: mona.bakr@ugent.be; Tel.: +32-92646606

Abstract: Over-molding has been proposed in recent years as an integrated functional flexible circuit board in a plastic part. This method uses the conventional process for film insert technology. Over-molding has attracted significant attention across many industries due to its potential to deliver different electrical functions in a variety of different part geometries, especially in automotive interiors and home appliances. While it has great application potential, manufacturing challenges continue throughout foil fabrication and injection molding. This raises challenges for designers and researchers responsible for maintaining the reliability of such electronic flexible circuits. Therefore, the purpose of this research paper is to improve some of the over-molding process parameters. On 0805 and 1206 over-molded zero-ohm resistors, electrical, mechanical, and failure characterization was performed. Those components were mounted in parallel, perpendicular, and 45° angled arrangements on two different polymer substrates, polyimide (PI) and polyethylene terephthalate (PET) using lead-free solder, low-melt solder, and conductive adhesive paste. Moreover, as an over-molding material, polycarbonate (PC) with medium viscosity was used. The effect of using different mold shapes (corner mold, 2 mm flat mold, and 3 mm flat mold) and injection molding process parameters (injection speeds and melt temperature) was studied.

Keywords: flexible printed circuit boards; injection molding; film insert-based technology

Citation: Bakr, M.; Hubmann, M.; Bossuyt, F.; Vanfleteren, J. A Study on Over-Molded Copper-Based Flexible Electronic Circuits. *Micromachines* **2022**, *13*, 1751. https://doi.org/10.3390/mi13101751

Academic Editors: Libo Gao and Zhuoqing Yang

Received: 30 September 2022
Accepted: 14 October 2022
Published: 16 October 2022

1. Introduction

Plastic products are becoming very popular for mass production since they are cheap, versatile, and easy to manufacture in a variety of forms. Plastic products arefabricated using different manufacturing processes including, injection molding, thermoforming, blow molding, and extrusion. Since the 1990s, a special injection molding process based on insert film technology has gained great importance from researchers and manufacturers. This process is called the in-mold or over-molding process. It is based on a foil insert that could be an electrical foil with assembled components which is inserted into an injection mold that defines the shape of the part that is to be produced. An injection molding machine then rapidly injects the molten thermoplastic into the mold and then the temperature is controlled to quickly cool the produced part. A packing pressure is applied that pushes additional melt into the cavity during the initial stage of this cooling phase to compensate for the volumetric shrinkage of the solidifying melt. The in-mold process is greatly preferred in the automotive industry because it can offer lightweight plastic products with good mechanical strength together with functional circuits [1,2]. This elevates user expectations since this technology can generate three-dimensional structures. Therefore, it is sufficient that the electronic systems fit a convenient form factor of the shape into which they will be integrated. Many researchers worked intensively to integrate different electrical functionalities like light emitting diodes (LEDs), light guides, numerous touch switches or slides, and near field communication (NFC) antennas into plastic products using the in-mold process [3,4]. By fabricating optical touch panels, organic light emitting diodes (OLEDs), foils, and disposable healthcare sensors, T. Alajoki et al. [5] demonstrated the feasibility of hybrid in-mold integration. O. Rusanen et al. [6] developed the IMSE technology (injection-molded structural electronics) by integrating pre-formed silver-based printed electronics and standard electronics components into 3D plastics. Moreover, Juntunen et al. [7], and Kololuoma et al. [8] realized wearables. A novel concept that could serve as a platform for the integration of photovoltaic (PV) cells into plastic products was developed by M. Bakr et al. [9]. Moreover, adhesion mechanics and a demonstration of how to power LEDs wirelessly using an NFC antenna and a chip were discussed in [10]. Wimmer et al. [11] concluded that printed conductive structures must be optimized for the forming process to achieve and maintain the highest electrical conductivity possible. O. M. Tuomikoski et al. [12] demonstrated indoor air quality monitoring and red-green-blue (RGB) luminaire devices with both printed and injection-molded parts. In addition to this, some researchers have expressed interest in examining the impact of the process on different electronic packages, flexible foil materials, metallization, assembly methods, and mold shapes to determine the effect of plastic materials on microsystems [13–17].

The most critical factor in over-molding technology is the compatibility of the various materials to achieve sufficient adhesion between the insert foil and the polymer injected by the injection molding machine. Many research studies were conducted to determine the impact of material combination and injection molding processing parameters on the interface adhesion, which, in turn, determines the product's properties. These experiments did not include any electronics, instead, only foils were over-molded with the injected polymer. Yamaguchi, S. [18] studied the influence of crystallization on the adhesion of polypropylene (PP) foil to PP-injected polymer at the interface. Additionally, Chen et al. examined the characterization of temperature profiles while using PET and PC foils with PC [19] and PP [20] as an injected polymer, observing that both combinations yielded satisfactory results. Moreover, other studies have defined fundamental concepts of interface adhesion using one or more of the following surface treatment mechanisms: interface interlocking, molecular diffusion and entanglement, crystallization, and adsorption theory [21–24]. However, surface treatments are not suggested for large-scale production because they increase the overall cost.

The primary purpose of this research paper is to demonstrate the use of different copper-based foils as substrate foil. In contrast to our work in [10], this paper examines the influence of various mold shapes and the influence of used component assembly materials like lead-free solder, conductive adhesive, and low-melt solder on the electrical performance after over-molding. The first section introduces the over-molding process for integrating electronic components, followed by the experimental procedures which include mold and foil design, adhesion test, process simulation, component assembly, and over-molding. Finally, the final section evaluates the shear test and electrical characterization results.

2. Experimental Procedures

The process flow begins with the mold design and ends with the over-molded plastic part.

2.1. Mold Design

The mold was designed to enable the over-molding in three different cavities. A flat plate mold whose wall thickness can be adjusted to 2 or 3 mm and a corner-shaped mold with an inner radius of 4 mm and wall thickness of 3 mm as depicted in Figure 1.

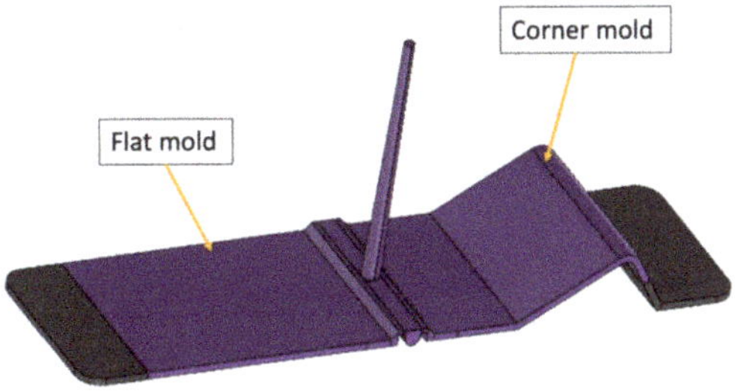

Figure 1. Mold design.

2.2. Foil Design

According to the mold design, we designed the circuit foils to fit in the three mold cavities with dimensions 105 × 74 mm as illustrated in Figure 2, where the connection pads are properly positioned on the right side; each pad is connected to a resistor location and then to the ground. The pads are connected to 18 components and are enabled to take the measurements before and after over-molding. Additionally, it enables the assembly of 0-ohm resistors in a variety of packages (0805 and 1206) and orientations (0°, 90°, and 45°).

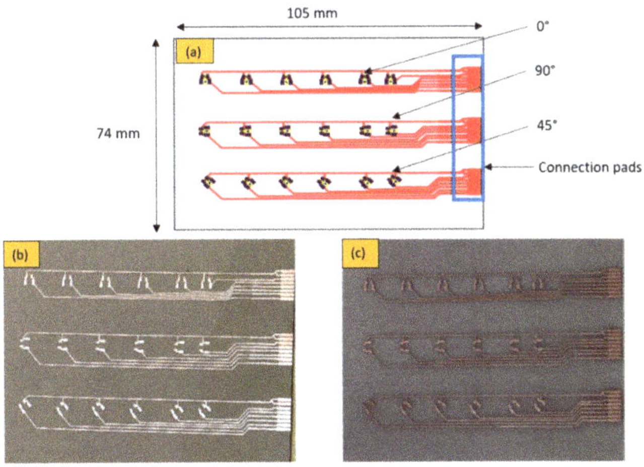

Figure 2. (**a**) Schematic foil design, (**b**) PI-Cu foil, and (**c**) PET-Cu foil.

2.3. Adhesion and Peel Tests

In this paper, the medium viscosity PC Makrolon 2805 (Covestro AG, Leverkusen, Germany) [25] is used as an injection material. The amorphous and transparent PC is known for its mechanical strength as well as high heat and flame resistance [26]. A durable over-molded flexible electronic product should always be fabricated with suitable material combinations with good adhesion between the flexible circuit foil (base plastic substrate) and the over-molded material (injected plastic) as depicted in Figure 3. In this research, the adhesion performance between PC and two commercial copper cladding foils was studied. The foils were composed of three layers: a polymer layer, a copper layer, and an adhesive layer. The composition of the used foils is given in Table 1.

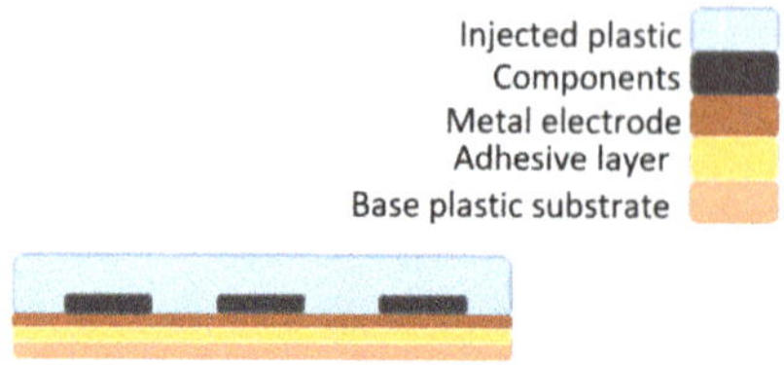

Figure 3. Typical materials stack.

Table 1. Composition of different commercial copper cladding foils.

Foil Layers	FR9210 [27]	GTS5500 [28]
Polymer layer	50 µm PI layer	50 µm PET layer
Adhesive layer	25 µm acrylic-based adhesive layer	20 µm polyurethane-based adhesive layer
Copper (Cu) layer	35 µm	18 µm

The peel strength between the foil adhesive layers shown in Table 1 and the PC molding material was evaluated using a 90° peel test. To that end, the copper was fully removed from some foils that were subsequently over-molded within a 2 mm plate mold with 280 °C melt temperature and injection speed of 70 cm^3/s. Table 2 presents the over-molded samples for the peel test.

Table 2. Listing of the over-molded parts for peel test.

No.	Foil	Mold Temperature (°C)	Residual Cooling Time (s)	Surface Treatment
PI-I	FR9120	80	50	-
PI-II	FR9120	100	50	-
PI-III	FR9120	100	50	-
PI-IV	FR9120	100	50	Corona
PI-V	FR9120	100	20	Oxygen plasma
PI-VI	FR9120	120	50	-
PET-I	GTS	80	50	-
PET-II	GTS	100	50	-
PET-III	GTS	100	50	Corona
PET-IV	GTS	100	50	Oxygen plasma

The molded foils were then laser cut into 15 mm wide foils strips as depicted in Figure 4a (n = 4) and peeled using an Instron 5500R (Illinois Tool Works Inc., Glenview, IL, US) tensile test machine with an attached peel-off fixture as shown in Figure 4b. A peel speed of 10 mm/min was chosen and static load cells were 10 N and 100 N used for PI and PET foils, respectively.

The mean peel strength was recorded for a peel length of 40 mm for each foil strip. The averaged peel strength per investigated foil with standard deviation indicating the variation between the foil strips is given in Figure 5 for PI foils with an acrylic adhesive layer and in Figure 6 for PET foils with polyurethane as an adhesive layer. The polyurethane-based adhesive in PET foils showed about 10 times larger adhesion with the injected PC compared to the acrylic-based layer in PI foils.

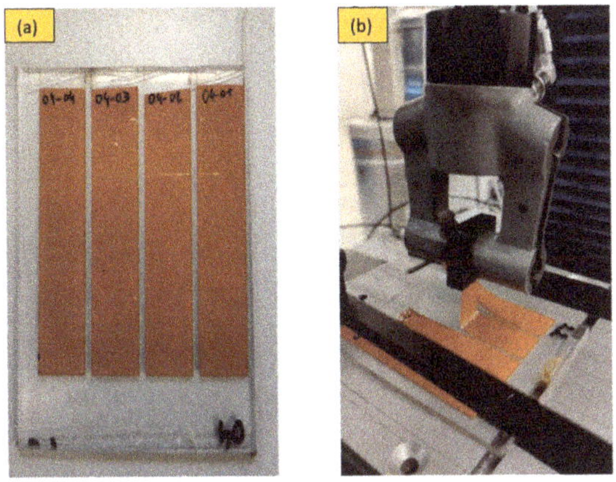

Figure 4. Peel-off tests setup.15 mm wide foils strips (**a**), Instron 5500R tensile test machine (**b**).

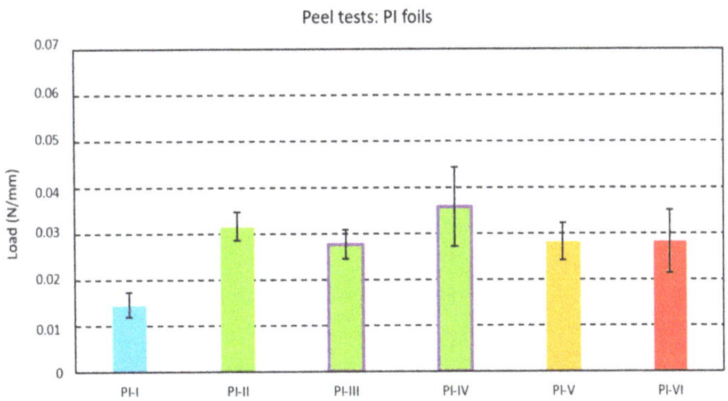

Figure 5. Peel strength for the PI foils with an acrylic-based adhesive layer.

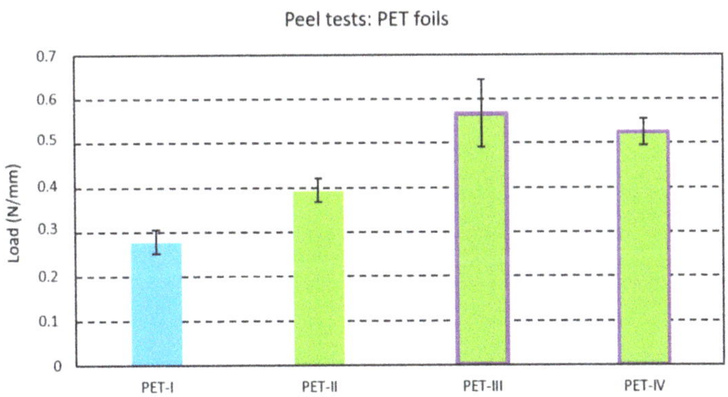

Figure 6. Peel strength for the PET foils with a polyurethane-based adhesive layer.

One-Way ANOVAs with Tukey tests (with 95% confidence interval) for the PI foils (from PI-I to P-VI) yielded a significantly lower adhesion for the foil molded at the low mold temperature (PI-I). While increasing the mold temperature from 80 °C (PI-I) to 100 °C (PI-II) increased the adhesion, a further increase to 120 °C (PI-VI) could not be confirmed to aid bonding. The long residual cooling time of 50 s (PI-II)—effectively tempering the acrylic close to its glass transition temperature [29]—could not be confirmed to aid adhesion compared to a shorter cooling time of 20 s (PI-V). No effect on the bonding could be confirmed when surface-treating the acrylic layer using corona activation (PI-III) or using oxygen plasma (PI-IV).

Likewise, One-Way ANOVAs with Tukey tests (with a 95% confidence interval) for PET foils were made. Increasing the mold temperature from 80 °C (PET-I) to 100 °C (PET-II) was again found to yield a significantly higher peel strength. A further increase could be noted when applying corona or oxygen plasma treatment to the polyurethane-based adhesive layer (PET-III and PET-IV).

2.4. Simulation

The temperature profile of the resistors during the injection molding cycle was numerically investigated using the commercial injection molding simulation software Autodesk Moldflow Insight (AMI) as depicted in Figure 7. An AMI model was created exemplarily for the 2 mm plate (3,141,193 elements), comprising the eighteen 1206 components (in total 77,748 elements) molded as solid aluminum oxide blocks. The two layers of the PI films were modeled separately as acrylic and polyimide parts (188,900 elements). Moreover, the mold (3,116,641 elements) with cooling channels was included too. Figure 7 shows the temperature development on the 1206 components during over-molding with the PC when a melt temperature of 300 °C, a mold temperature of 100 °C, and an injection speed of 70 cm^3/s is used. The grey dashed line indicates the end of filling (~0.4 s) with the inset focusing on this stage of the injection molding cycle. The highest temperatures of approximately 215 °C are reached later during packing. The used temperatures are lower than the melting temperature of the soldering material which is 260 °C. Therefore, it can be assumed that the resistors will remain fully functional and reliable after over-molding.

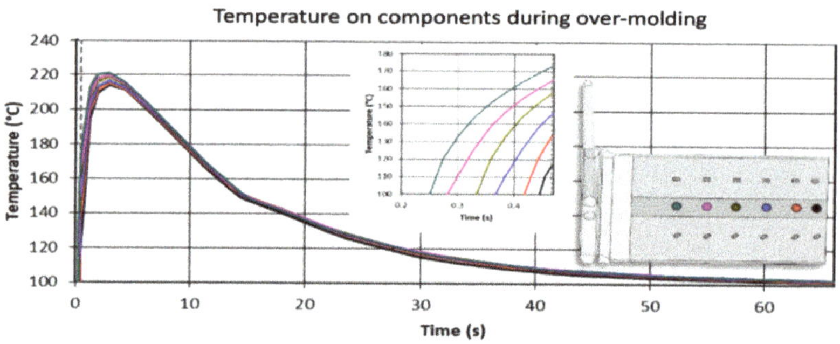

Figure 7. The temperature profile of the components during over-molding with PC.

2.5. Foil Fabrication and Assembly

For all used copper-PI and copper-PET foils, the copper was patterned using lithography and etching. The foil fabrication includes the cleaning of the surface from any foreign substances and particles, photoresist coating, UV exposure and photoresist development, the etching of the exposed metal film, and removal of the photoresist. To protect the patterned copper from oxidation, the copper is treated with an organic solder preservative (OSP). Finally, laser ablation is occasionally used to cut the foil into the desired insert shape and to create an opening that acts as a gate for the polymer to flow during the over-molding process.

The mold is roughly 98.5 mm in length, and we ensured that components were assembled 22 mm away from the mold side wall. The assembly step involves mounting SMD resistors (zero-ohm resistors) acquired from Yageo to the PI-Cu and PET-Cu foils. The assembly plan for the PI-Cu foils included 18 foils assembled using a standard lead-free solder paste [30] and 18 foils assembled using thermoset silver-based conductive glue

(Henkel CE 3103) [31]. Lead-free solder and conductive glue were applied to the copper pads of the copper traces, followed by the resistor assembly. Following that, the foils are heated until the solder particles reflow and harden in a reflow oven. On the other hand, in the case of conductive glue, it is cured for 20 min at 120 °C in a convection oven. Figure 8 depicts the used foils assembly plan for the PI-Cu foils and Figure 9 shows a real-life image of a foil with assembled components.

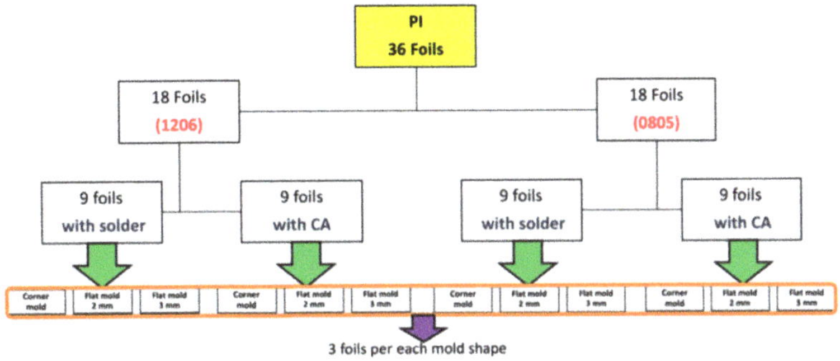

Figure 8. Overview of PI-Cu foils.

Figure 9. Real-life PI-Cu foil with components.

Moreover, the assembly plan for the PET-Cu foils included 9 foils assembled using a low-temperature melt solder paste (Interflux DP5600, Interflux Electronics, Gent, Belgium) [32] and 9 foils assembled using thermoset silver-based conductive glue (Henkel CE 3103). The same steps were taken for the PI-Cu foils. The only difference is the use of the reflow oven for the low-melt solder paste and low-melting temperature PET foils. The reflow oven was configured for preheating for 5 min at 160 °C, followed by 15 min of reflow at the same temperature at 160 °C, then cooling for 6 min. Figure 10 depicts the used foils assembly plan for the PET-Cu foils and Figure 11 shows a real-life image of a foil with assembled components.

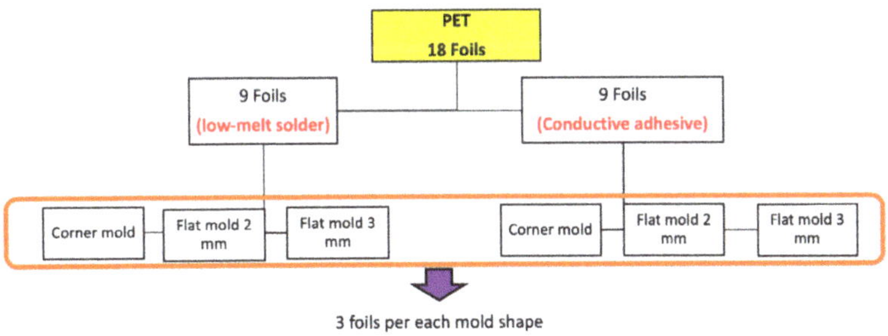

Figure 10. Overview of used PET-Cu foils.

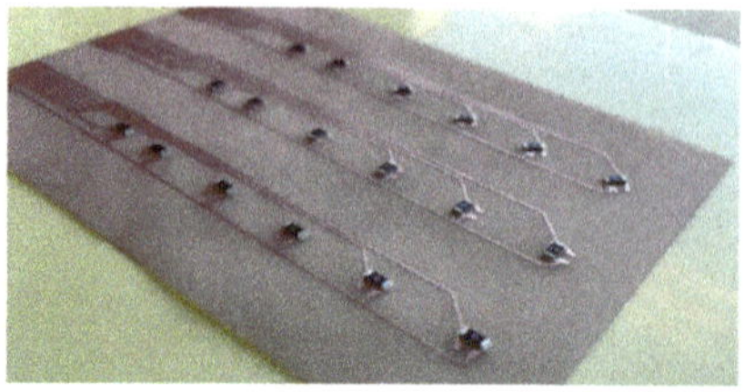

Figure 11. Real-life PET-Cu foil with components.

2.6. Over-Molding

The foils were placed in the mold as indicated in Figure 12 and fixed using temperature-resistant adhesive tape. The detachable inserts were then added to keep the contacts free from being over-molded, they were located as depicted in Figure 13. As a drawback of this simple approach, the films occasionally showed wrinkles after over-molding at the transition to the insert.

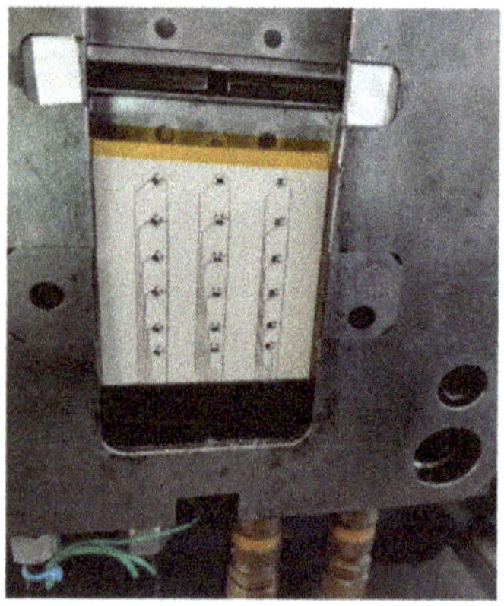

Figure 12. Foil fixed in a flat mold.

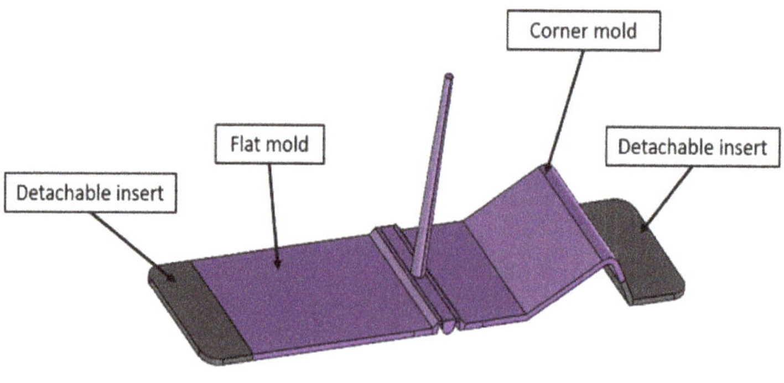

Figure 13. Location of the detachable inserts.

An Arburg Allrounder 470A (Arburg GmbH + Co. KG, Loßburg, Germany) injection molding machine equipped with a 25 mm screw was used for performing the over-molding. The two cycles of a Wittmann Tempro plus D 160 (WITTMANN Technology GmbH, Austria) temperature control unit were used to regulate the temperatures of the two mold halves to 100 °C. This mold temperature was chosen based on the performed peel test results outlined in Section 2.3 above.

The used PC (Makrolon 2805, Coversto AG, Leverkusen, Germany) was dried for 3 h at 120 °C before molding. The packing pressure was set to 400 bar for 15 s and the residual cooling time to 50 s for all tests. The dosing volume was set either to 50 cm^3 for the corner and the 3 mm plate or to 40 cm^3 for the 2 mm plate. The switch-over point (velocity-to-pressure-controlled filling) was adapted for each injection speed and melt.

Figure 14 lists the 36 produced over-molded PI-Cu foils with their corresponding settings. Those foils yielded a broad molding window in which (visually) undamaged parts could be produced. The aim was to find those boundaries of the injection molding process by changing the melt temperature between 240, 260, 280, and 300 °C and the injection speed to either 20 or 70 cm^3/s. In preceding tests with the PET foils (without components) it was shown that they were significantly more sensitive to the formation of wrinkles during over-molding. This issue could be reduced when selecting a higher melt temperature and faster injection speed setting of 300 °C and 70 cm^3/s, respectively.

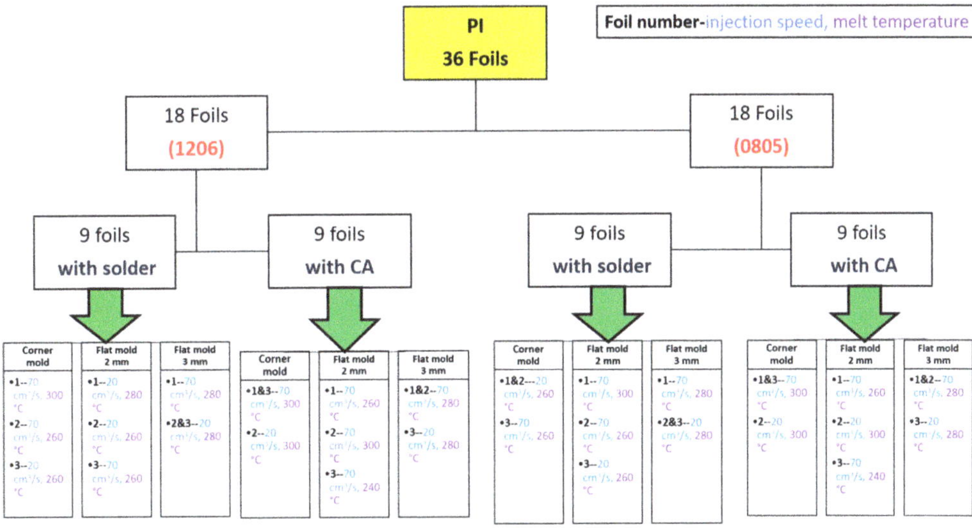

Figure 14. Overview of over-molded foils.

3. Results

3.1. Shear Test

The strength of the joints was assessed using shear tests resembling the DIN EN 62137 1–2 norm. The shear tests were performed on a Bruker UMT-2 (Bruker Corporation, Billerica, MA, US) mechanical tester platform with a 100 N load cell at the Polymer Competence Center Leoben GmbH (PCCL). A schematic illustration of the setup is given in Figure 15. The components were aligned in 180° orientation and sheared using a rectangular 4 mm width chisel that moved at 1/4th of the component's height at a speed of 6 mm/min.

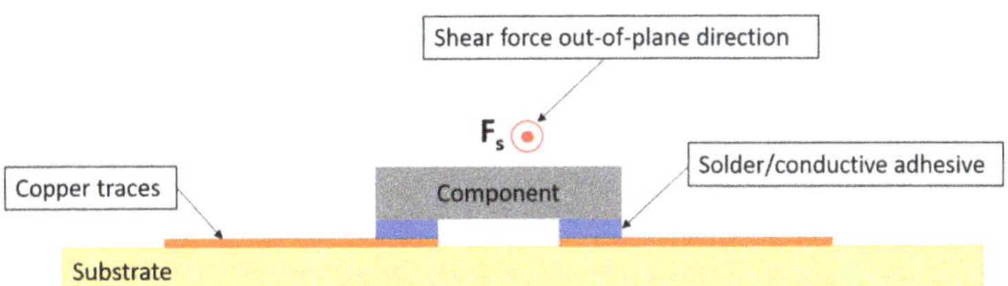

Figure 15. Cross-sectional view for shear off component.

Table 3 displays the shear load at the break for the investigated 0805 and 1206 components assembled using solder, low-melt solder, and conductive adhesive on PI and PET foils. As a general observation, the load at break increased by increasing the component size. Moreover, higher shear loads at break were recorded when the components were soldered not glued using the conductive adhesive.

Table 3. Shear force for assembled components.

Shear Force and Standard Deviation	1206	0805
Solder on PI foils	62 ± 4 N	41 ± 5 N
Low-melt solder on PET foils	43 ± 3 N	31 ± 1 N
CA on PI foils	39 ± 8 N	29 ± 6 N

3.2. Foils Evaluation

Foils tested were assembled using 0805 and 1206 SMD 0-ohm resistors. These resistors were selected in our study because they can be easily assembled manually in a reproducible way. To know the actual resistance of the resistors, we used a multimeter to check the value of ten 0 Ω resistors from both packages to act as the $R_{initial}$ of the component, which was 0.2 Ω before assembling the resistors on the foil. Each 0 Ω resistor mounted on the flex has a different track length (L) between the component and the connection pad as depicted in Figure 16.

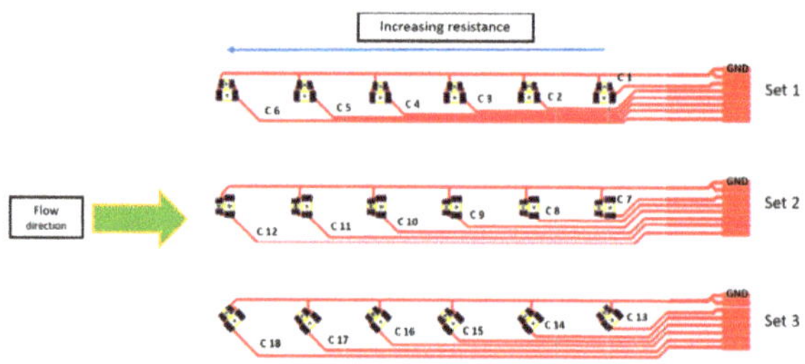

Figure 16. Labeling of components on the foil.

The resistance of the assembled 0 Ω resistor increases proportionately to the length of the copper track. The measured resistance is the sum of the resistance of the used component, the resistance of the copper track, the resistance of the used connection (solder, low-temperature solder, and conductive adhesive), and the probe's contact resistance. In other words:

$$R\ measured = R\ initial + R\ track\ length + R\ connection + R\ contact \quad (1)$$

Assembly using lead-free solder proved its applicability when using PA6 as the injected molding material and PI-Cu foils as presented in our previous work in [10]. However, in this paper, assembly using low-temperature solder and conductive adhesive is also studied. Moreover, different foils, PET-Cu-based foils are evaluated to broaden the research area on the influence of over-molding parameters on electrical characterization. The following sections are divided into two main subsections, PI-Cu foils and PET-Cu foils including a detailed explanation of the measurements.

3.2.1. Measurements on PI-Cu Foils

Measurements on Samples Assembled with Lead-Free Solder

Using lead-free solder, the resistance before over-molding was the same for the 18 components. Therefore, we can check the effect of track length on the measurements using an average value of six resistance values (R) from component one (C1) to component six (C6) as presented in Table 4.

Table 4. Average resistance values of the assembled resistors before OVM.

Components	Resistance Values ($R_{measured}$)					
C1 to C6	R1 = 0.2 Ω	R2 = 0.3 Ω	R3 = 0.3 Ω	R4 = 0.4 Ω	R5 = 0.4 Ω	R6 = 0.5 Ω

To compare $R_{measured}$ and $R_{theoretical}$ we had to consider $R_{initial}$ = 0.2 Ω, the ρ of copper = 1.7×10^{-8} mΩ, the track thickness = 35 µm, and the track width = 300 µm. We calculated R $_{theoretical}$ for 18 resistors using Equation (1), and Table 5 presents the data for set 1 of the resistors. No significant difference in resistance was found between the measured and calculated values. Moreover, this comparison demonstrates that the primary contributor to the measured resistance is the change in resistance of the copper track length; both solder connections and probe contacts contribute insignificantly to the measured resistance.

Table 5. Difference between $R_{measured}$ and $R_{theoretical}$.

	Length between Resistor and Contact Pad	$R_{measured}$	$R_{theoretical}$
R1	32.5 mm	0.2 Ω	0.305 Ω
R2	41.24 mm	0.3 Ω	0.33 Ω
R3	54.78 mm	0.3 Ω	0.37 Ω
R4	68.45 mm	0.4 Ω	0.42 Ω
R5	82.3 mm	0.4 Ω	0.46 Ω
R6	97.4 mm	0.5 Ω	0.52 Ω

The next step is to clamp the foils into the mold and perform an over-molding cycle. Figure 17 depicts the resistance measurements for the corner mold for the 1206 and 0805 components. The measurements were taken for three foils before (blue bars) and after (red bars) over-molding had almost the same readings except for foil 1 after over-molding which had an increase in resistance of 0.1 Ω.

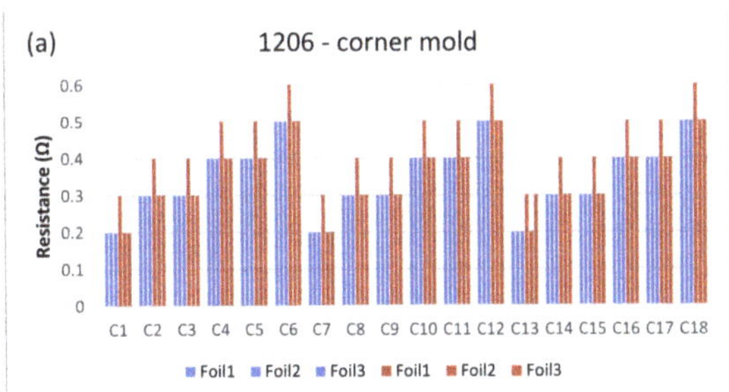

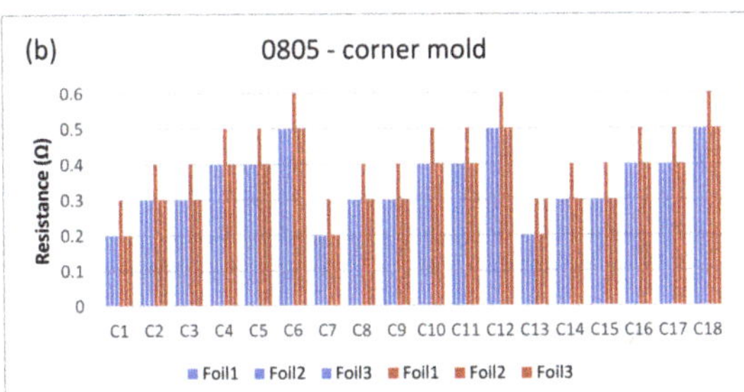

Figure 17. Resistance measurements for 1206 resistors (**a**) and 0805 resistors (**b**) molded in corner mold. Blue bars (before over-molding) and red bars (after over-molding).

Figures 18 and 19 show the measurements for the three foils in the flat mold in 2-mm and 3-mm cavities, respectively. The over-molding did not affect the measured resistance, indicating that solder-based components can withstand the over-molding process and also that the internal stresses in the flat molds are lower than in the corner mold.

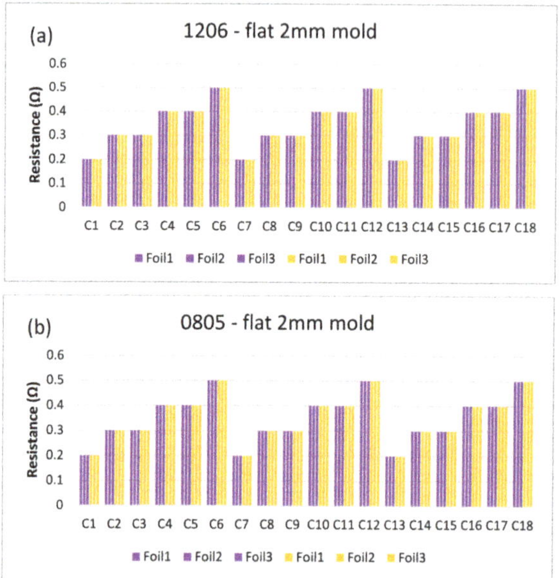

Figure 18. Resistance measurements for 1206 resistors (**a**) and 0805 resistors (**b**) molded in the 2 mm flat mold. Purple bars (before over-molding) and yellow bars (after over-molding).

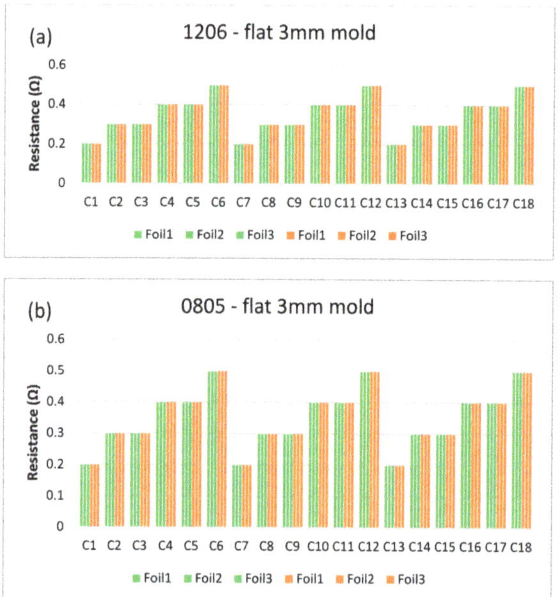

Figure 19. Resistance measurements for 1206 resistors (**a**) and 0805 resistors (**b**) molded in the 3 mm flat mold. Green bars (before over-molding) and orange bars (after over-molding).

Moreover, a visual examination was conducted to verify that the solder connections between the components and the copper pad were intact. The cross-section images were obtained in two-component positions: on a flat surface and on a curved slightly formed surface. Figure 20 illustrates a component constructed on a flat surface, whereas Figure 21 depicts a component assembled on a curved surface.

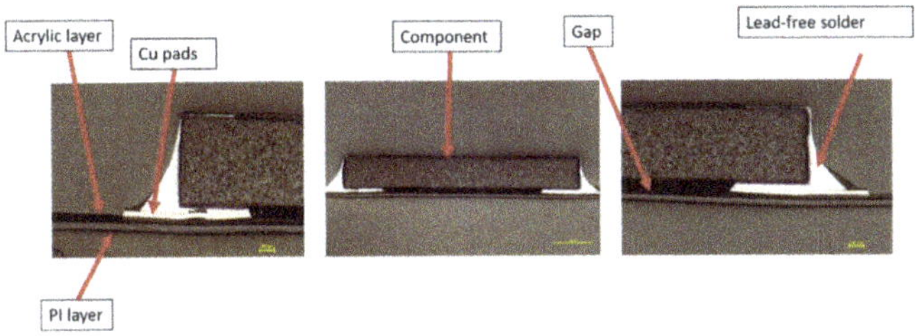

Figure 20. Cross-section images for flat surface.

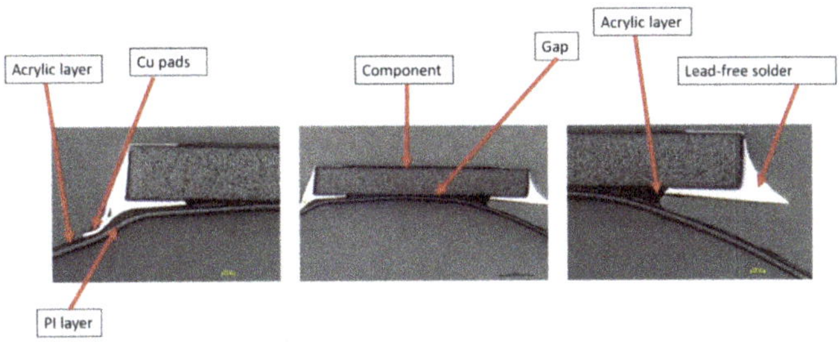

Figure 21. Cross-section images for curved surface.

Cross-section images of both positions, curved and flat surfaces, demonstrate the strong connection formed by assembled components and copper tracks when lead-free solder is used. Even in the case when the component is near the curvature's location, the solder is attached to the acrylic adhesive layer. These images demonstrate our work's novelty on why solder joints are preferred in the over-molding process when using acrylic-based flexible foils. Additionally, it may be used in curvatures with a small radius, such as the 4 mm in our case.

Measurements on Samples Assembled with Conductive Adhesive

The resistance of the assembled component was tested before and after over-molding and calculated again using Eq.1. As indicated before in the shear force section (Table 3), components bonded with conductive glue yielded lower shear force at break than soldered components. This affected the electrical measurements in some way.

Figure 22 depicts the resistance for the foils assembled with 1206-size components. Some components had very high resistance values that led to an open loop (OL) reading on the multimeter. Such reading means that resistance is opposition to the free flow of current within a circuit and the higher the resistance, the harder it is for the current to flow from one point to another. Other components were out of place after over-molding, and, for both cases, their resistance was eliminated from the calculations in order to have more accurate data as is the case for C1, C13, and C17 in Figure 22.

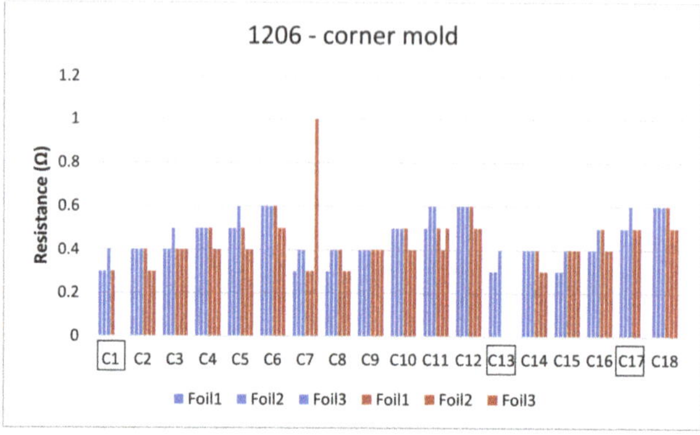

Figure 22. Resistance measurements for 1206 resistors molded in corner mold.

After over-molding, components number 13 (for two foils) had no data after over-molding. Therefore, we investigated the foils and found that the three components of C13 were out of place after over-molding as depicted in Figure 23.

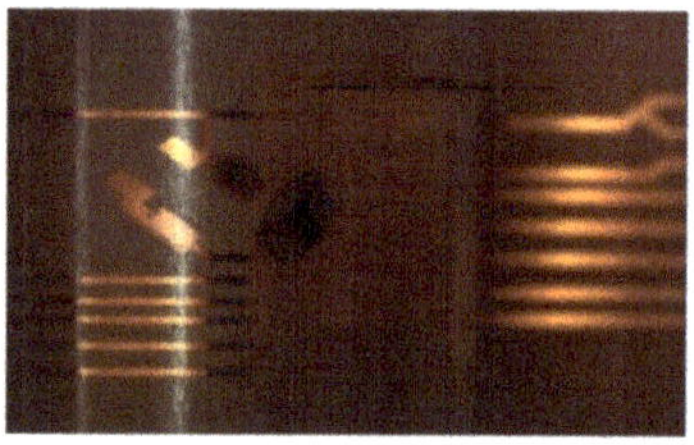

Figure 23. An out-of-place component at 2000 µm.

However, in the case of C1, the components were on the foil but were not conducting. Therefore, cross-section images were taken to check the conductive adhesive connection reliability between copper pads and components' conductive adhesive joints, as depicted in Figure 24. This figure shows how a curved surface may disconnect the copper and the acrylic layer from the conductive adhesive. The same failure was also detected for C17 and C13 on foil 3.

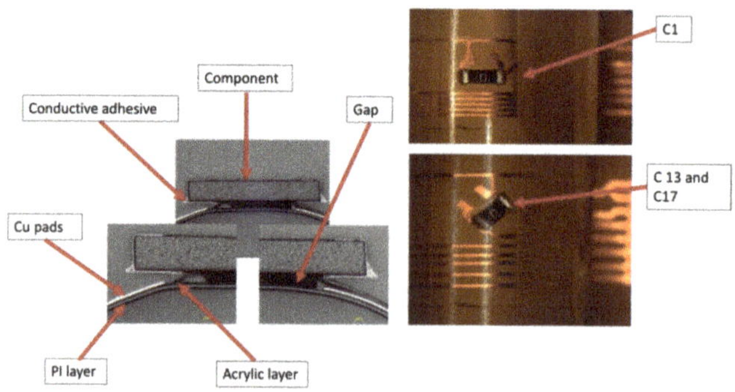

Figure 24. Cross-sectional images at 100 µm for C1, C13, and C17 components (**Left**). Top view of the component at 3000 µm (**Right**).

On the other hand, Figure 25 depicts the resistance measurements for 0805-size components. Four components had OL readings after over-molding, these components are C1, C6, C13, and C14.

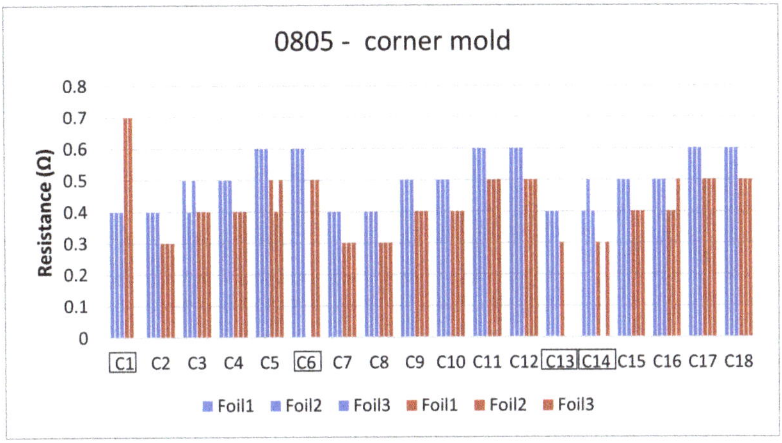

Figure 25. Resistance measurements for 0805 resistors molded in corner mold.

Regarding components C1, C13, and C14, the same concept of the weak bond between the conductive adhesive and the joints apply as presented in Figure 24. However, for C6, the conductive adhesive joints were on the component, but due to poor adhesion between the PC and PI-Cu-based foil, the copper traces were delaminated from the conductive adhesive joints and stayed on the components, thus, the connection was broken as depicted in Figure 26.

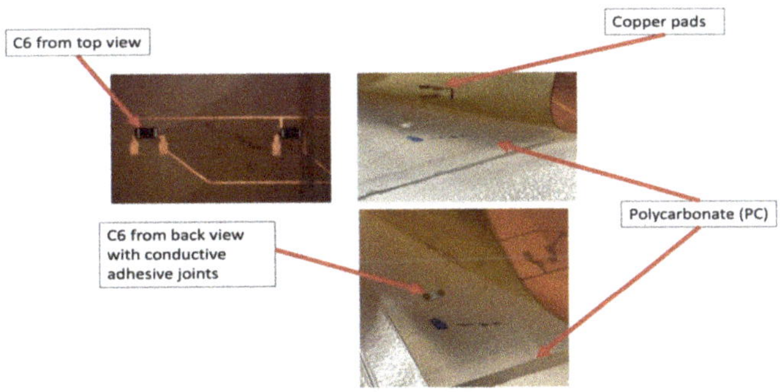

Figure 26. Poor adhesion between PI-Cu and PC.

Generally, components cannot be incorporated or present in curvatures as part of the in-mold technology design requirements. However, the components in the lead-free solder-based foils were functional after over-molding despite the curvature. Moreover, according to our tests, the use of the lead-free solder on copper pads gives a stronger connection between the component and the pad as proven in our shear tests. We can conclude that a corner shape mold and weaker bonds between components and the used foil could lead to several failures. Figure 27 depicts the failure locations. Components C1, C7, C14, and C13 could have a high change in resistance due to their position in the curved area. Moreover, components C6 and C17 which are close to the flow entry point may become displaced or out of place due to the polymer flow.

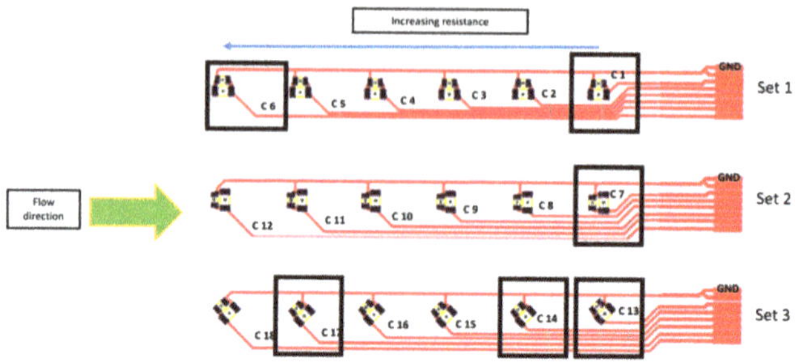

Figure 27. Typical failure locations for the corner mold.

Regarding the 2 mm flat mold, Figure 28 depicts the resistance measurements for 1206 resistors before (purple bars) and after (yellow bars) over-molding. In foil 2, component C5 was not assembled in a reliable way due to manual assembly and, therefore, there is no data present for it before and after over-molding.

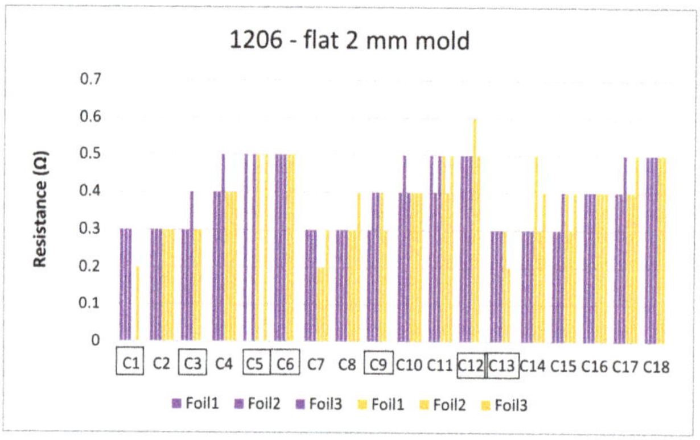

Figure 28. Resistance measurements for 1206 resistors molded in the 2 mm flat mold.

In this group of foils, we had failures for foil 1 and foil 3. For foil 1, C1 was out of place after over-molding as depicted in Figure 29. On the other hand, for foil 3, many components were out of place. The main reason for such an observation is the improper alignment of the foil within the mold, resulting in a wrinkled foil with removed components as presented in Figure 30.

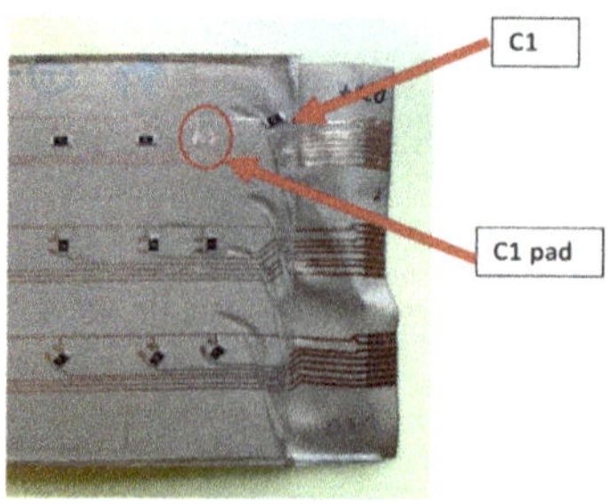

Figure 29. An out-of-place component.

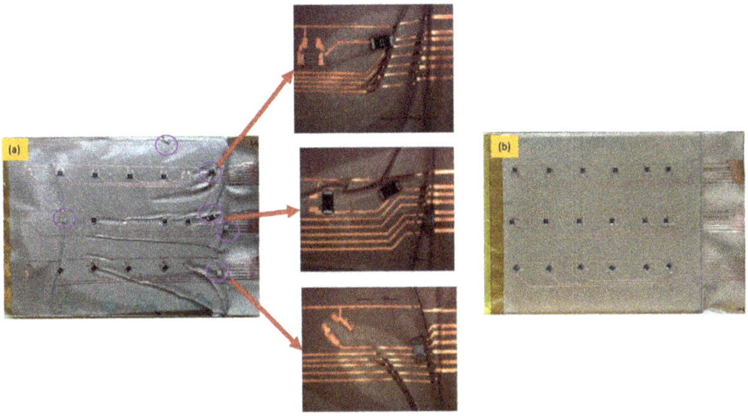

Figure 30. Bad over-molding with misplaced foil, removed components and wrinkles (**a**), good over-molded foil with all components and without wrinkles (**b**).

Most foils that had problems in terms of detached components and wrinkled foil appearance were over-molded at low-melt temperatures of 240 °C and 260 °C. At melt temperatures of 280 °C and 300 °C, no detachment was observed on the foils using CA regardless of the used injection speed or mold thickness.

Figure 31 shows the resistance for 0805 components in a 2 mm flat mold. As depicted, one of the C1 components gave an open loop reading and was still on the foil. Therefore, cross-section images were taken to check the conductive adhesive joints on a flat surface as depicted in Figure 32.

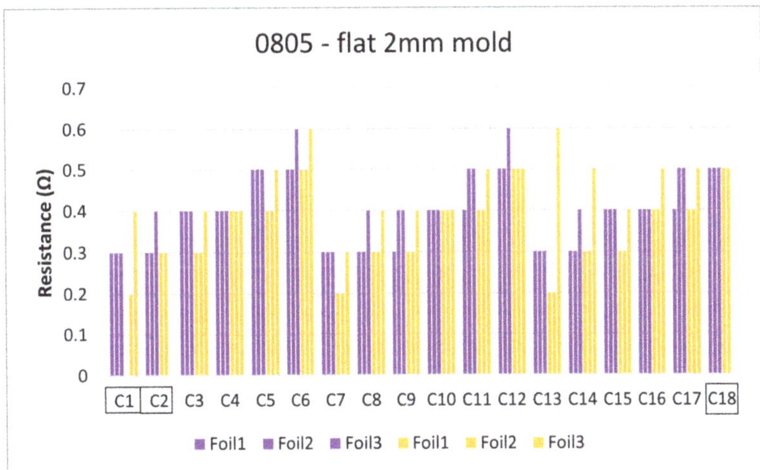

Figure 31. Resistance measurements for 0805 resistors molded in the 2 mm flat mold.

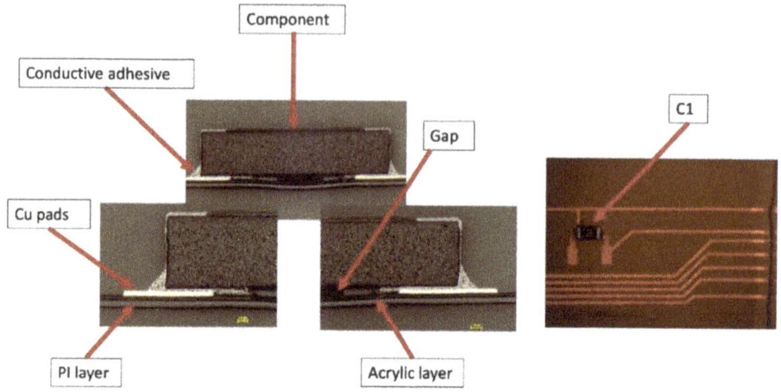

Figure 32. Cross-sectional images at 100 μm for C1 component (**Left**). Top view of the component at 3000 μm (**Right**).

Additionally, as indicated before in Section 2.3, the detachable inserts were added to protect the contacts and to facilitate the resistance measurements after over-molding. However, the films occasionally developed wrinkles as a result of over-molding during the transition to the insert and some of the foils had compressed copper tracks, as illustrated in Figure 33. This wrinkling effect caused a discontinuity in the copper tracks and, therefore, some components also had very high resistance as in the case of components C2 and C18.

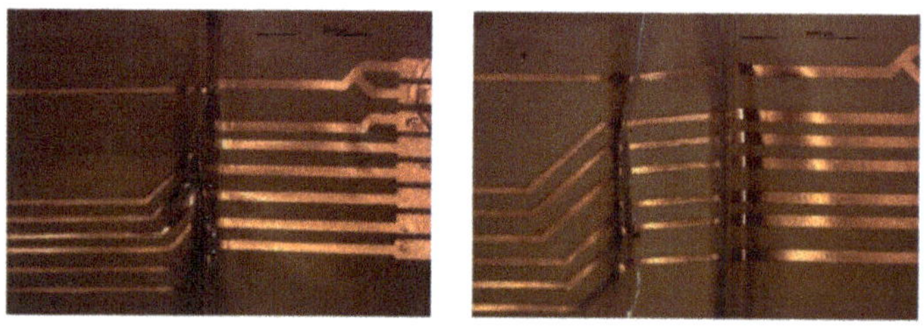

Figure 33. Compressed copper tracks at 2000 μm magnification.

In general, such wrinkle failure may also occur in solder-based foils as well since it is a failure due to clamping the foils into the mold using the inserts, not the material of assembly used.

Finally, the resistance measurements were taken for the 3 foils over-molded in the 3 mm plate mold, green bars represent resistance before over-molding tests and orange bars represent after over-molding as depicted in Figure 34.

With the used settings, foils over-molded in the 3 mm mold showed fewer failures compared to those in the 2 mm and the corner molds. This is due to the lower prevailing shear stresses and pressure during filling. Moreover, the components assembled using conductive adhesive varied significantly from those assembled with solder. This could be because of the nature of the conductive adhesive material that makes it difficult to manually apply it in equal joints during the assembly step.

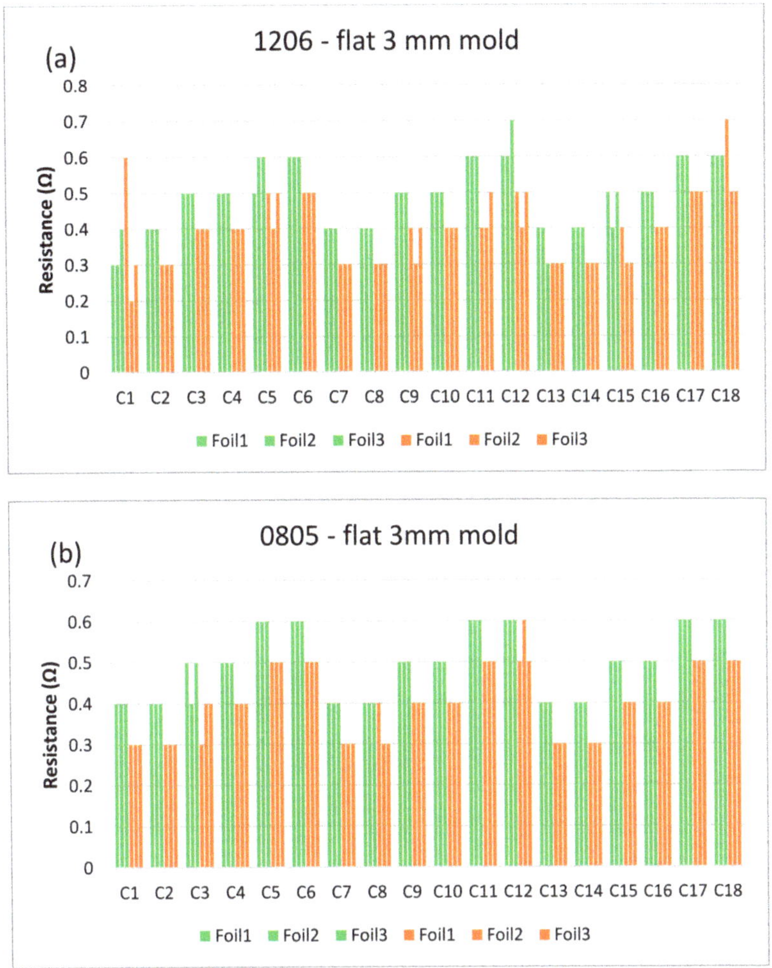

Figure 34. Resistance measurements for 1206 resistors (**a**) and 0805 resistors (**b**) molded in the 3 mm flat mold.

3.2.2. Measurements on PET-Cu Foils

Measurements on Samples Assembled with Low-Temperature Solder

In this section, we used only 1206 components for assembly. This is because, according to our study, component size is independent of the electrical measurements and also because they are easier when manually soldered. Figure 35 depicts the resistance values for 1206 resistors over-molded in the corner mold taken before (blue bars) and after (red bars) over-molding. Components bonded with low-melt solder yielded less shear force at break lower than those bonded with solder and higher than those using conductive adhesive as discussed in Section 3.1.

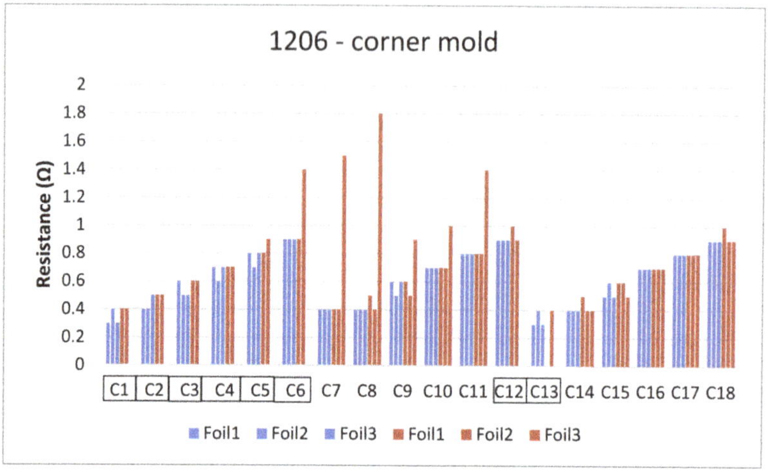

Figure 35. Resistance measurements for the corner mold.

Foil 3 had a ground break as depicted in Figure 36, affecting the measurements between C1-C6. Some components were out-of-place after over-molding like in the case of C12 and C13.

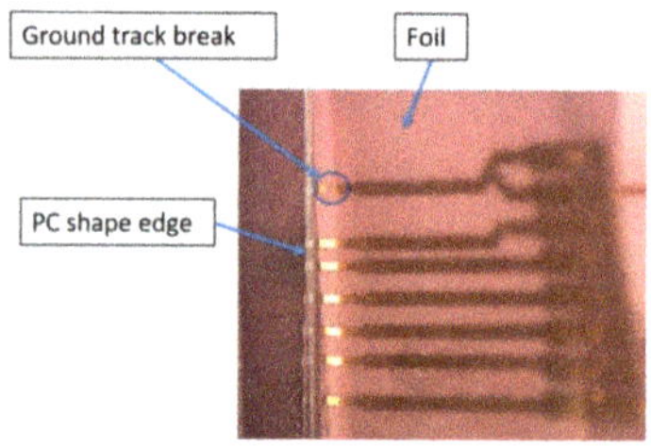

Figure 36. Ground track break in foil 3.

Cross-section images were taken for one of the components, C7, which is in the curved area to check the connection reliability for the low-melt solder on such a surface as depicted in Figure 37. These images show that the low-melt solder was not fully connected to the copper pad, yet component C7 was functional after over-molding.

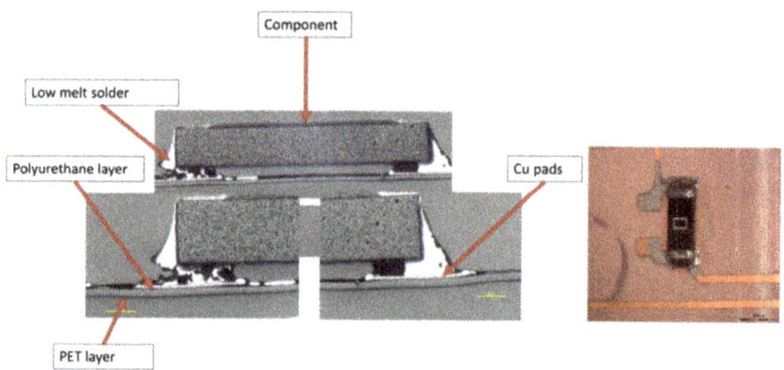

Figure 37. Cross-sectional images at 500 μm for component C7 (**Left**). Top view of the component at 2000 μm (**Right**).

Figure 38 displays the resistance measurements of the components over-molded in the 2 mm plate mold before (purple bars) and after (yellow bars).

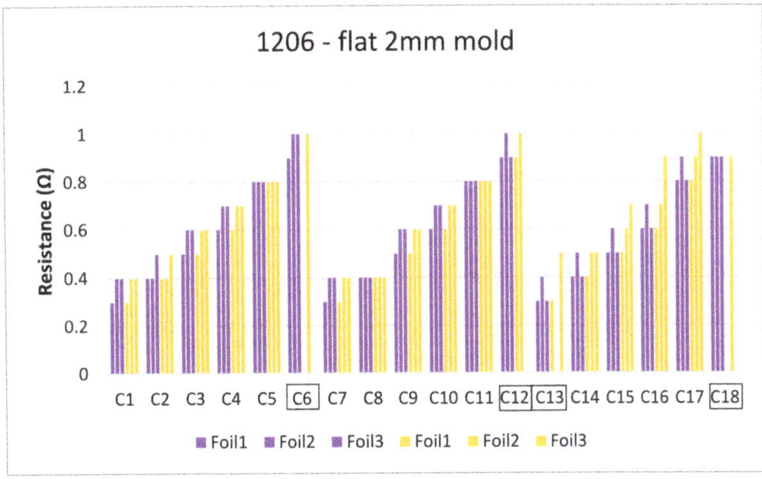

Figure 38. Resistance measurements for flat mold-2 mm.

Some components C6, C12, C13, and C18 were not included in our measurements since they were out of place after over-molding as depicted in Figure 39.

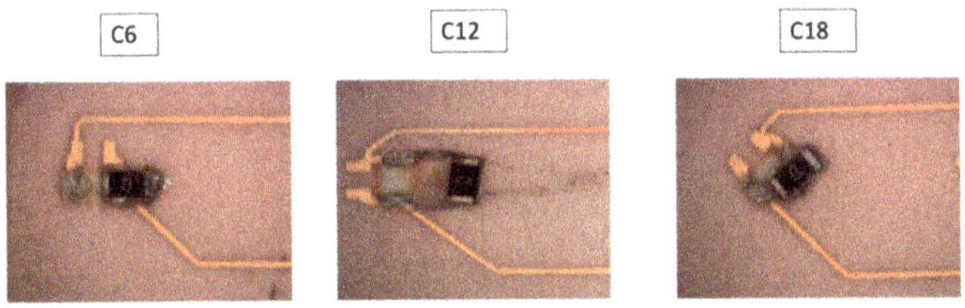

Figure 39. Out-of-place components.

Moreover, some components remained on the foil after over-molding but had high resistance measurements as was the case for C13. In this case, cross-sectional images were taken to assess the connection reliability between copper pads and low-melt solder joints, as depicted in Figure 40. This image shows a discontinuity in the copper and the PU layer.

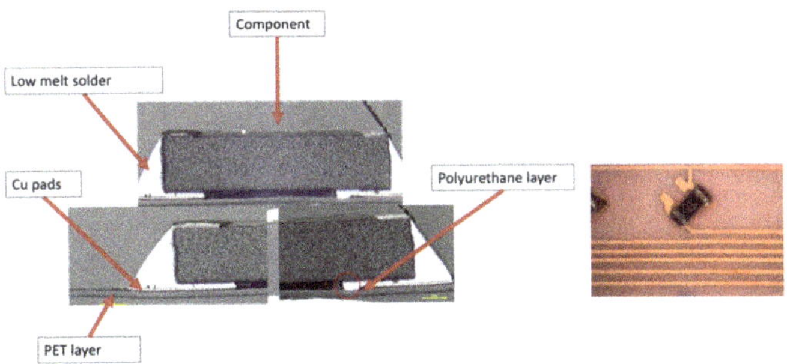

Figure 40. Cross-sectional images at 500 µm for component C13 (**Left**). Top view of the component at 2000 µm (**Right**).

As shown in Figure 41, measurements were taken in a flat mold with a thickness of 3 mm, with green bars representing testing before over-molding and orange bars representing after over-molding. Only one of each component, C13 and C14, were removed from our measurements since they led to very high resistance which caused an infinite resistance reading.

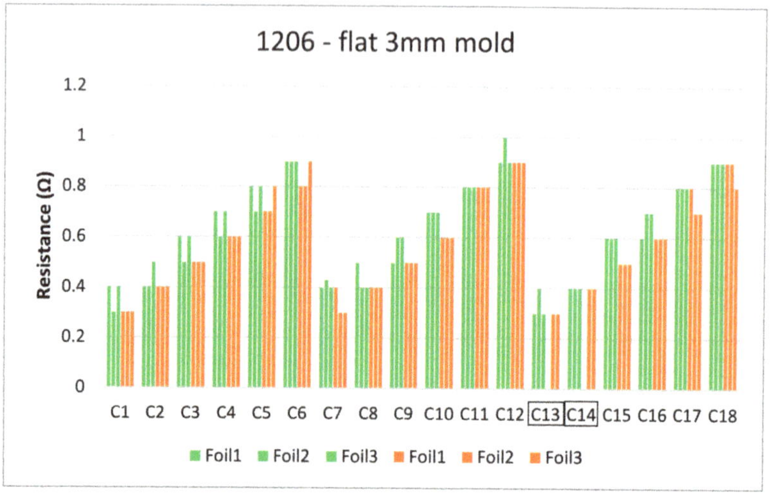

Figure 41. Resistance measurements for flat mold-3 mm for 1206 resistors.

Measurements on Samples Assembled with Conductive Adhesive

In the literature, low-temperature solder was the material used to assemble the electrical components on PET foils [33–35]. However, we considered using conductive adhesive since it also cures in low-temperature conditions at 120 °C and the degradation process of PET at such temperature is relatively slow [36]. Figure 42 depicts the resistance values in the corner mold. The change in resistance measurements was taken before (blue bars) and after (red bars) over-molding. All foils were over-molded using a high melt temperature of 300 °C. It was observed that these foils had many components that were detached after over-molding.

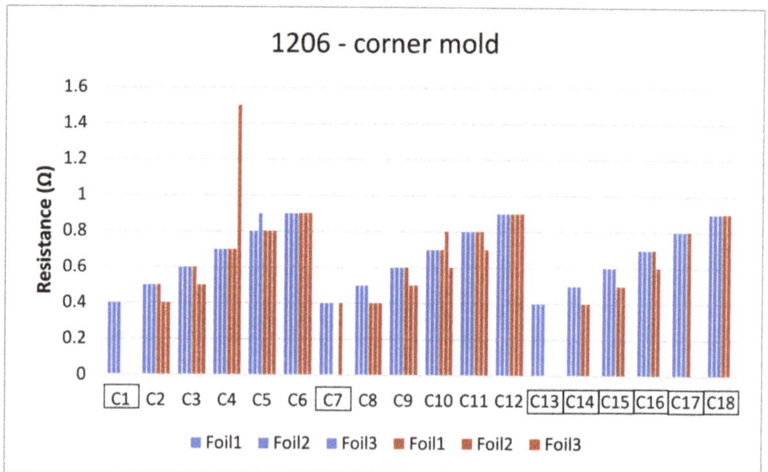

Figure 42. Resistance measurements for corner mold.

Foils 1 and 2 had some detached components (C1, C7, C13, and C17) after over-molding because these components were molded in the corner mold but also in the critical areas as well as the weak bonds of CA as depicted in Figure 43.

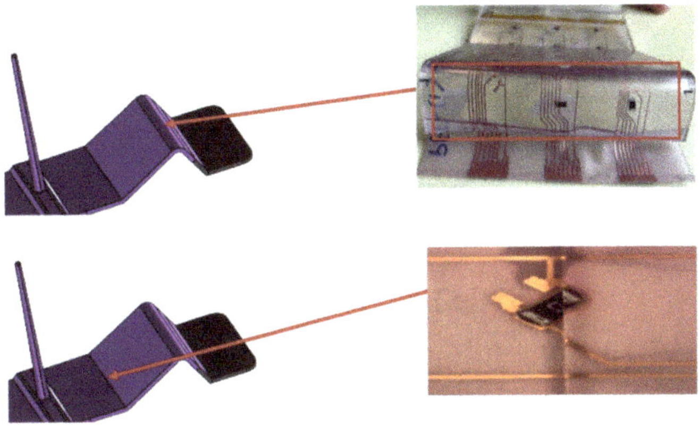

Figure 43. Corner mold critical areas.

In addition to this, the ground track of foil 3 that connects C13-C18 was broken as depicted in Figure 44.

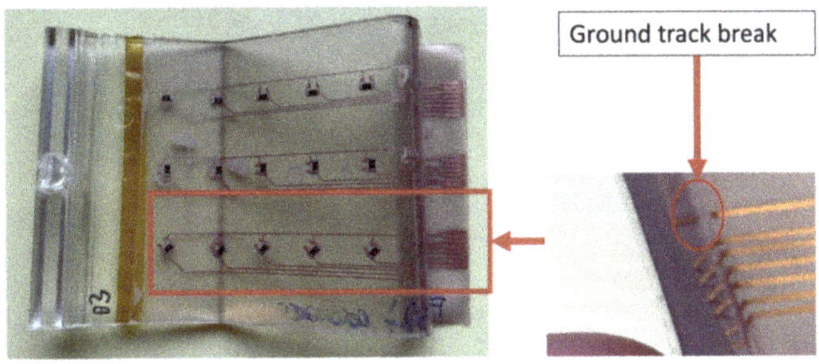

Figure 44. Ground track break in the corner mold in foil 3.

The resistance readings of the components over-molded in the 2 mm flat mold are depicted in Figure 45 before (purple bars) and after (yellow bars) and most of the components were detached after over-molding as depicted in Figure 46.

Measurements were made in a flat mold with a thickness of 3 mm, as shown in Figure 47, where green bars denote testing before over-molding and orange bars denote measurements after over-molding. The resistivity of the majority of components had changed because of the break in the ground track in foil 2, which led to non-conducting components between C7–C12. Moreover, foil 1 had a detached component (C1).

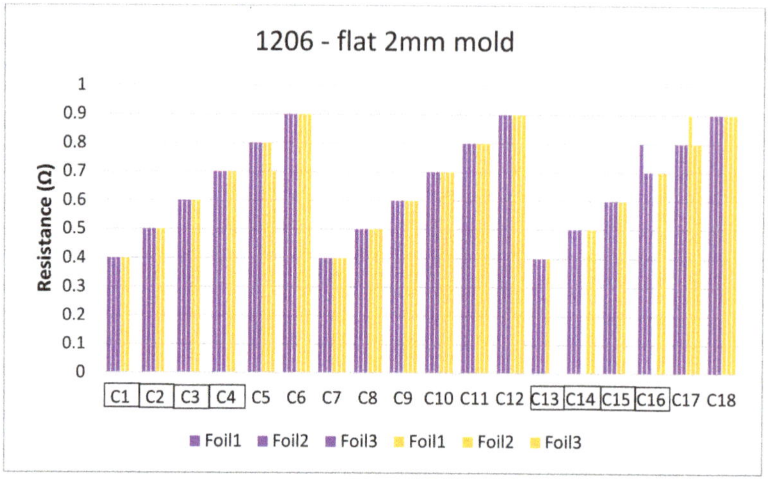

Figure 45. Resistance measurements for flat mold-2 mm for 1206 resistors.

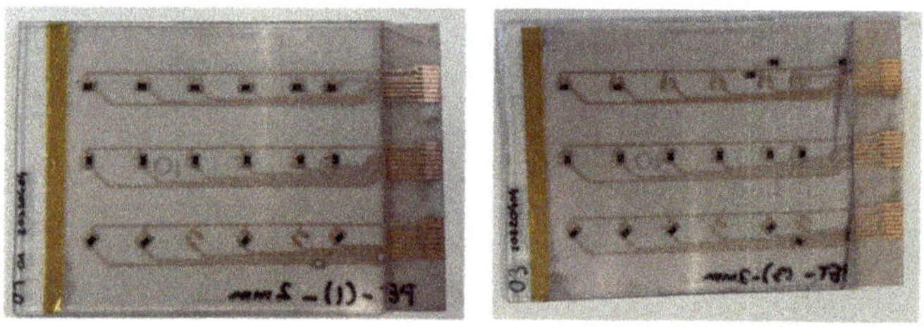

Figure 46. Example of detached components.

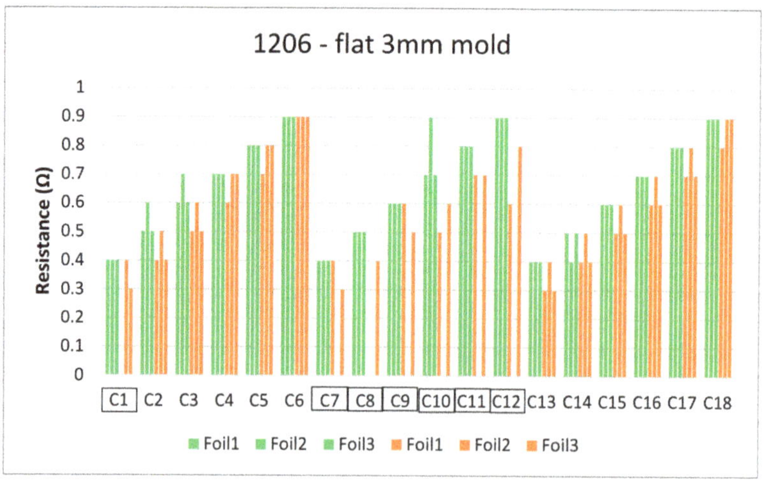

Figure 47. Resistance measurements for flat mold-3 mm for 1206 resistors.

4. Conclusion

This paper analyzed the use of different assembly materials on PI and PET copper-based foils. Regarding soldered components on PI-Cu foils, all resistors yielded 100%

success after over-molding. In contrast, the majority of the components assembled using conductive adhesive on PI-Cu and PET-Cu foils as well as low-melt soldered components on PET-Cu foils were detached after over-molding. Moreover, components should be located in areas with lower deformation and fewer stress concentrations to prevent damage, particularly when assembled using conductive adhesive and low-melt solder. However, glob-top and underfill materials could be used and may protect the components in such regions. From our experiments, failures were detected more after the over-molding of components using conductive adhesive and low-melt solder. Such failures include foil misalignment, which led to removed components and non-functional components because of the conductive adhesive and low-melt solder weak bonds on copper tracks. A relatively high mold temperature of 100 °C was found to be beneficial for good adhesion with PI (acrylic adhesive layer) and PET foils (polyurethane adhesive layer) compared to a lower mold temperature of 80 °C. Generally, it seems that a lower melt temperature, a higher injection speed, and a lower mold thickness increase the stresses on the components during over-molding which, in turn, leads to larger resistance changes after over-molding. Additionally, the electrical measurements of the over-molded foils were unaffected using varied package sizes and assembly in different orientations.

Author Contributions: Conceptualization, M.B., M.H., and F.B.; Methodology, M.B.; Software, M.H.; Validation, M.H.; Formal Analysis, M.B. and M.H.; Investigation, M.B. and M.H.; Resources, M.B., M.H., and F.B.; Data Curation, M.B. and M.H.; Writing—Original Draft Preparation, M.B. and M.H.; Writing—Review and Editing, M.B., M.H., and F.B.; Visualization, M.B. and M.H.; Supervision, F.B. and J.V. All authors have read and agreed to the published version of the manuscript.

Funding: This work was carried out in the Austrian COMET program within the project Smart@Surface No. 871495. The project received funding from BMVIT, BMDW as well as the provinces of Styria (SFG), Tyrol (Standortagentur Tirol), and Burgenland.

Institutional Review Board Statement: Not applicable.

Informed Consent Statement: Not applicable.

Data Availability Statement: Not applicable.

Acknowledgments: We especially would like to thank Covestro AG for providing the over-molding material and Polymer Competence Center Leoben GmbH (PCCL) for facilitating the shear tests.

Conflicts of Interest: The authors declare no conflict of interest. The funders had no role in the design of the study; in the collection, analyses, or interpretation of data; in the writing of the manuscript, or in the decision to publish the results.

References

1. Pötsch, G.; Michaeli, W. *Injection Molding: An Introduction*; Carl Hanser Publishers: Munich, Germany, 2008.
2. Singh, G.; Verma, A. A brief review on injection moulding manufacturing process. *Mater. Today Proc.* **2017**, *4*, 1423–1433.
3. Teh, N.J.; Prosser, S.; Conway, P.P.; Palmer, P.J.; Kioul, A. Embedding of electronics within thermoplastic polymers using injection moulding technique. In Proceedings of the Twenty Sixth IEEE/CPMT International Electronics Manufacturing Technology Symposium (Cat. No.00CH37146), Santa Clara, CA, USA, 3 October 2000; pp. 10–18. [CrossRef]
4. Bakr, M.; Bossuyt, F.; Vanfleteren, J. The integration of electronic circuits in plastics using injection technologies: A literature review. *Flex. Print. Electron.* **2022**, *7*, 023001. [CrossRef]
5. Alajoki, T.; Matti, K.; Markus, T.; Mikko, H.; Antti, K.; Kimmo, K.; Jukka-Tapani, M.; Janne, A.; Kari, R. Hybrid in-mould integration for novel electrical and optical features in 3D plastic products. In Proceedings of the 4th Electronic System-Integration Technology Conference, Amsterdam, Netherlands, 17–20 September 2012; pp. 1–6. [CrossRef]
6. Rusanen, O.; Simula, T.; Niskala, P.; Lindholm, V.; Heikkinen, M. Injection Molded Structural Electronics Brings Surfaces to Life. In Proceedings of the 2019 22nd European Microelectronics and Packaging Conference & Exhibition (EMPC), Pisa, Italy, 16–19 September 2019; pp. 1–7. [CrossRef]
7. Juntunen, E.; Ihme, S.; Huttunen, A.; Mäkinen, J. R2R process for integrating LEDs on flexible substrate. In Proceedings of the 2017 IMAPS Nordic Conference on Microelectronics Packaging (NordPac), Gothenburg, Sweden, 18–20 June 2017; pp. 12–16. [CrossRef]

8. Kololuoma, T.; Keränen, M.; Kurkela, T.; Happonen, T.; Korkalainen, M.; Kehusmaa, M.; Gomes, L.; Branco, A.; Ihme, S.; Pinheiro, C.; et al. Adopting Hybrid Integrated Flexible Electronics in Products: Case—Personal Activity Meter. *IEEE J. Electron Devices Soc.* **2019**, *7*, 761–768. [CrossRef]
9. Bakr, M.; Bauwens, P.; Bossuyt, F.; Vanfleteren, J.; Chtioui, I.; Christiaens, W. Solar cells integration in over-molded printed electronics. In Proceedings of the 2020 IEEE 8th Electronics System-Integration Technology Conference (ESTC), Tønsberg, Norway, 15–18 September 2020; pp. 1–5. [CrossRef]
10. Bakr, M.; Bossuyt, F.; Vanfleteren, J.; Su, Y. Flexible Microsystems Using Over-molding Technology. *Procedia Manuf.* **2020**, *52*, 26–31. [CrossRef]
11. Wimmer, A.; Reichel, H.; Schmidt, S. New standards for 3D-userinterfaces-manufactured by a Film Insert Molding process. In Proceedings of the 2018 13th International Congress Molded Interconnect Devices (MID), Würzburg, Germany, 25–26 September 2018; pp. 1–5. [CrossRef]
12. Gbotemi, O.; Myllymäki, S.; Jantunen, H.; Juuti, J.; Ihme, S.; Kurkinen, M.; Majava, V.; Tuhkala, M.; Kemppainen, J. Printed GNSS and Bluetooth Antennas Embedded on Flexible Low Loss Substrates for Wearable Applications. *Prog. Electromagn. Res. M* **2020**, *94*, 189–199. [CrossRef]
13. Tuomikoski, M.; Ihme, S.; Huttunen, A.; Korkalainen, M.; Yrjänä, S. Indoor air quality sensing indicators. In Proceedings of the 2016 6th Electronic System-Integration Technology Conference (ESTC), Grenoble, France, 13–15 September 2016; pp. 1–6. [CrossRef]
14. Nguyen, S.; Perez, C.J.; Desimone, M.; Pastor, J.M.; Tomba, J.P.; Carellaa, J.M. Adhesion control for injection overmolding of elastomeric propylene copolymers on polypropylene. Effects of block and random microstructures. *Int. J. Adhes. Adhes.* **2013**, *46*, 44–55. [CrossRef]
15. Stan, F.; Fetecau, C. Experimental Investigation of the Adhesion Between Thermoplastic Polyurethane and Acrylonitrile-Butadiene-Styrene Substrate. In Proceedings of the ASME 2014 International Manufacturing Science and Engineering Conference collocated with the JSME 2014 International Conference on Materials and Processing and the 42nd North American Manufacturing Research Conference. Volume 2: Processing, Detroit, MI, USA, 9–13 June 2014. [CrossRef]
16. Ott, C.; Wolf, M.; Drummer, D. Media-Tight Polymer-Polymer Assemblies By Means of Sintered Powder Layer in Assembly Injection Moulding. *Procedia Manuf.* **2020**, *47*, 362–367. [CrossRef]
17. Leong, Y.W.; Ishiaku, U.S.; Kotaki, M.; Hamada, H.; Yamaguchi, S. Interfacial characteristics of film insert molded polycarbonate film/polycarbonate-acrylonitrile-butadiene-styrene substrate, part 1: Influence of substrate molecular weight and film thickness. *Polym. Eng. Sci.* **2006**, *46*, 1674–1683. [CrossRef]
18. Leong, Y.W.; Ishiaku, U.S.; Kotaki, M.; Hamada, H.; Yamaguchi, S. Effect of crystallization and interface formation mechanism on mechanical properties of film-insert injection-molded poly(propylene) (PP) film/PP substrate. *J. Appl. Polym. Sci.* **2005**, *98*, 294–301.
19. Chen, S.C.; Li, H.M.; Huang, S.T.; Wang, Y.C. Effect of decoration film on mold surface temperature during in-mold decoration injection molding process. *Int. Commun. Heat Mass Transf.* **2010**, *37*, 501–505. [CrossRef]
20. Chen, H.L.; Chen, S.C.; Liao, W.H.; Chien, R.D.; Lin, Y.T. Effects of insert film on asymmetric mold temperature and associated part warpage during in-mold decoration injection molding of PP parts. *Int. Commun. Heat Mass Transf.* **2013**, *41*, 34–40. [CrossRef]
21. Baldan, A. Adhesion phenomena in bonded joints. *Int. J. Adhes. Adhes.* **2012**, *38*, 95–116. [CrossRef]
22. Awaja, F.; Gilbert, M.; Kelly, G.; Fox, B.; Pigram, P.J. Adhesion of polymers. *Prog. Polym. Sci.* **2009**, *34*, 948–968. [CrossRef]
23. Creton, C.; Kramer, E.J.; Brown, H.R.; Hui, C.Y. Adhesion and Fracture of Interfaces Between Immiscible Polymers: From the Molecular to the Continuum Scal. In *Molecular Simulation Fracture Gel Theory*; Springer: Berlin/Heidelberg, Germany, 2001; pp. 53–136.
24. Cole, P.J.; Cook, R.F.; Macosko, C.W. Adhesion between immiscible polymers correlated with interfacial entanglements. *Macromolecules* **2003**, *36*, 2808–2815. [CrossRef]
25. Available online: https://www.campusplastics.com/campus/de/datasheet/Makrolon%C2%AE+2805/Covestro+Deutschland+AG/22/7541f4aa (accessed on 29 September 2022).
26. LeGrand, D.G.; Bendler, J.T. Plastics engineering, Bd. 56. In *Handbook of Polycarbonate Science and Technology*; Marcel Dekker: New York, NY, USA, 2000.
27. Available online: https://www.dupont.com/content/dam/dupont/amer/us/en/products/ei-transformation/documents/EI-10113-Pyralux-FR-CCL-Data-Sheet.pdf (accessed on 29 September 2022).
28. Available online: https://www.gtsflexible.com/product-table/?tx_gtsproducts_products%5Baction%5D=list&tx_gtsproducts_products%5Bcontroller%5D=Products&cHash=b0ee19a0fdbc58db79b019bffba1c77c#productList (accessed on 29 September 2022).
29. Available online: https://www.professionalplastics.com/professionalplastics/content/AcryliteFFDataSheet.pdf (accessed on 29 September 2022).
30. Bath, J. *Lead-Free Soldering*; Springer: New York, NY, USA, 2007.
31. Available online: https://www.henkel-adhesives.com/be/en/product/electrically-conductiveadhesives/loctite_ablestikce3103wlv.html (accessed on 29 September 2022).
32. Available online: https://interflux.com/en/product/DP-5600 (accessed on 29 September 2022).

33. Wimmer, A.; Reichel, H.; Rauch, B.; Schramm, R.; Hörber, J.; Hä:ßler, B. Manufacturing of sandwich structures for the integration of electronics in in mold labelling components. In Proceedings of the 2016 12th International Congress Molded Interconnect Devices (MID), Wuerzburg, Germany, 28–29 September 2016; pp. 1–4. [CrossRef]
34. Alajoki, T.; Koponen, M.; Juntunen, E.; Petaja, J.; Heikkinen, M.; Ollila, J.; Sitomaniemi, A.; Kosonen, T.; Aikio, J.; Makinen, J.T. In-mould integration of electronics into mechanics and reliability of overmoulded electronic and optoelectronic components. In Proceedings of the 2009 European Microelectronics and Packaging Conference, Rimini, Italy, 15–18 June 2009; pp. 1–6.
35. Koponen, M.; Alajoki, T.; Kosonen, T.; Petäjä, J.; Heikkinen, M.; Vuorinen, T.; Mäkinen, J.-T. Adhesion of Flexible Printed Circuit Substrate to Overmoulded Polymer and Characterization of Overmoulded Electronic Components. In *IMAPS Nordic Annual Conference Denmark 2008*; IMAPS Nordic: Oslo, Norway, 2008; pp. 207–212.
36. Panowicz, R.; Konarzewski, M.; Durejko, T.; Szala, M.; Łazińska, M.; Czerwińska, M.; Prasuła, P. Properties of Polyethylene Terephthalate (PET) after Thermo-Oxidative Aging. *Materials* **2021**, *14*, 3833. [CrossRef] [PubMed]

micromachines

Article

Electromechanical Properties of 3D-Printed Stretchable Carbon Fiber Composites

Teemu Salo *, Donato Di Vito, Aki Halme and Jukka Vanhala

Faculty of Information Technology and Communication Sciences, Tampere University, 33720 Tampere, Finland
* Correspondence: teemu.salo@tuni.fi

Abstract: The addition of fillers has been implemented in fused filament fabrication (FFF), and robust carbon fillers have been found to improve the mechanical, electrical, and thermal properties of 3D-printed matrices. However, in stretchable matrices, the use of fillers imposes significant challenges related to quality and durability. In this work, we show that long carbon staple fibers in the form of permeable carbon fiber cloth (CFC) can be placed into a stretchable thermoplastic polyurethane (TPU) matrix to improve the system. Four CFC sample series (nominally 53–159-μm-thick CFC layers) were prepared with a permeable and compliant thin CFC layer and a highly conductive and stiff thick CFC layer. The sample series was tested with single pull-up tests and cyclic tensile tests with 10,000 cycles and was further studied with digital image correlation (DIC) analyses. The results showed that embedded CFC layers in a TPU matrix can be used for stretchable 3D-printed electronics structures. Samples with a thin 53 μm CFC layer retained electrical properties at 50% cyclic tensile deformations, whereas the samples with a thick >150-μm CFC layer exhibited the lowest resistance (5 Ω/10 mm). Between those structures, the 106-μm-thick CFC layer exhibited balanced electromechanical properties, with resistance changes of 0.5% in the cyclic tests after the orientation of the samples. Furthermore, the suitability of the structure as a sensor was estimated.

Keywords: stretchable electronics; 3D printing; carbon fibers; electromechanical testing; strain sensor

Citation: Salo, T.; Di Vito, D.; Halme, A.; Vanhala, J. Electromechanical Properties of 3D-Printed Stretchable Carbon Fiber Composites. *Micromachines* **2022**, *13*, 1732. https://doi.org/10.3390/mi13101732

Academic Editors: Libo Gao and Zhuoqing Yang

Received: 29 September 2022
Accepted: 11 October 2022
Published: 13 October 2022

1. Introduction

Additive manufacturing (AM) is widely used in several manufacturing sectors. Fused filament fabrication (FFF), especially, has advantages such as simplicity and cost-effectiveness [1]. For example, strain sensors [2], multiaxial force sensors [3], and batteries [4] have already been fabricated with this single-step FFF process. FFF can also be adopted in the textile field by printing plastics directly on textile substrates [5] or by printing the whole textile composition [6]. Moreover, it has been demonstrated that FFF with deformable plastics and substrates can be used in manufacturing stretchable electronics, which can be further laminated on textiles for wearable electronics [7].

Recently, 3D-printed stretchable and wearable electronics have gained more attention, and rigid [4] and stretchable wearables [8] have been successfully fabricated. Still, stretchable electronics that are practically integrable into clothing have not yet been manufactured via 3D printing. For integrable stretchable and wearable electronics, carbon-filled polymers are an especially promising alternative for the creation of mechanically complex and thermally and electrically conductive structures [2,9]. These polymers can be modified using carbon-based additives with different form factors and dimensions, such as carbon fibers [10], carbon nanotubes (CNTs) [2,3,9], carbon black [11], graphene [12], and others [13], with a wide variety of outcomes in terms of properties such as strength, thermal and electrical conductivity, piezoresistive behavior, and many others [14–16]. These modified materials enable the fabrication of sensors, wearables, and other end-products, e.g., by providing higher strength, fire retardancy, or electrical properties. However, FFF polymers with fillers generally require a high nozzle diameter to prevent clogging, decreasing the printing quality [2] and causing highly anisotropic printing results [17]. Fillers also rapidly increase the Young's modulus of the polymers, leading to a trade-off between stiffness and conductivity [2].

In FFF polymers, the shape and size of carbon fillers influence the formation of the fillers' conductive network, percolation threshold, and overall conductivity. For example, when the results from previous studies are converted into conductivity values, acrylonitrile butadiene styrene (ABS) filament consisting of 15 wt% nano-scale carbon black has 0.025 S/m conductivity [11], and 5.6 wt% graphene flakes with a lateral size of 3–5-µm [12] provide 0.001 S/m. CNTs were mixed into thermoplastic polyurethane (TPU) filaments (with a CNT content of 4 wt%) of in 9.5 nm diameter and 1.5 µm in length, providing 32 S/m conductivity after the 3D-printing process (with a 0.6 mm nozzle) [2]. Furthermore, Tzounis et al. blended TPU and CNTs (5 wt%, 9.5 nm diameter, 3.0 µm length), which resulted in higher conductivity, approximately 100 S/m, but this required a less accurate 0.8 mm nozzle [17]. Spoerk et al. 3D-printed polypropylene (PP) filaments filled with short carbon fibers (7 µm in diameter and 250 µm in length) at proportions of up to 10 wt% with a 0.6 mm nozzle [18] for thermally conductive structures. Even longer millimeter-scale fibers can be added during the filament manufacturing process, but these are chopped to the micrometer-scale during the process [10].

Furthermore, continuous carbon fiber filaments are used in FFF by feeding them into molten-state polymers during extrusion [19,20], increasing an object's carbon fiber content. However, this process requires a larger nozzle [19,20]. Another alternative is to impregnate carbon fiber filaments before printing, enabling more complex [21] and precise [22] printing with filaments.

Feeding carbon fillers and carbon filaments through a 3D printer's nozzle makes them compatible with readily available FFF printers. Furthermore, carbon semi-products, such as laminates and inks, can be integrated into the 3D printing process semi-automatically or automatically by pausing the process. Carbon fiber sheets can be laminated on top of objects to form durable and lightweight composite structures [23] or inside them to address porosity and layer adhesion issues [13]. Moreover, integrated carbon fiber tows can be used to monitor a matrix's structural health via the tows' resistance changes [24]. Other electrical and thermal properties can be created by spray-depositing the 3D-printed surface with CNTs, and a 19-nm layer thickness on the smoothened surface is possible [25,26]. Furthermore, the direct ink writing (DIW) method can be combined with FFF to print carbon black ink electrodes for 3D printed supercapacitors [27]. The FFF process even allows the manual or automatic integration of printed circuit board (PCB) components inside an object [8], which, along with other placement methods, enables versatile 3D-printed electronics.

In this study, stretchable and wearable 3D-printed electronics components were made by adding sparse carbon fiber cloths (CFCs) inside a TPU matrix made with FFF. The advantages of this process are that CFCs with mechanical and electrical properties were (1) integrated inside the matrix without cavities or other 3D printing design modifications, and (2) adhesives were not required for their placement—it was sufficient to change the general 3D printing settings. Furthermore, CFCs embedded in this way inside the structure (3) do not decrease the adhesion between the TPU layers, and (4) they can be cut into different shapes, which can be used as functional elements in fabricating stretchable and wearable electronics. Furthermore, integrated CFCs can create (5) more isotropic and detailed objects than those created through FFF with carbon fiber filaments.

To the authors' knowledge, permeable CFC has never been used to improve the electrical and mechanical properties of FFF objects. The measured properties of CFC compare favorably to those of carbon-filler filaments and can be used to provide stretchable and conductive composite matrices. The obtained results prove that matrices with CFC can sustain large numbers of deformation cycles with minimal changes in their resistance behavior, thanks to the combined mechanical and adhesion properties of the materials involved. We studied the characteristics and advantages of this new method by conducting quasi-static and cyclic electromechanical tests, which we further analyzed using digital image correlation (DIC) techniques to attain information about the local deformation field in the samples. Finally, the properties of the structure for sensor applications were estimated.

2. Materials and Methods

CFCs with centimeter-scale fibers are traditionally used in composite manufacturing to increase the Young's modulus of materials with a minimal increase in density, i.e., to improve the specific modulus of materials. As well as mechanical features, CFCs are thermally and electrically conductive, thus having the potential to be used in wearable electronics. The stretchability that wearable electronics require was achieved by combining sparse CFCs and a highly stretchable TPU matrix. Single pull-up tests were first used to study the mechanical properties of the CFC matrices. Then cyclic tensile tests were conducted to measure the electromechanical features. Finally, the samples' behavior was further analyzed with DIC.

2.1. Composition and Preparation of Samples

The CFCs used in this work were provided by ACP Composites [28], and the reported average single carbon fiber length was 25.4 mm. The carbon fibers were processed from polyacrylonitrile (PAN) to promote their electrochemical properties [29]. CFCs with two grades of nominal thicknesses, 0.0021″ (53 μm) and 0.006″ (153 μm), were tested. Thin CFCs were also tested as two- and three-layered plies to further improve their electrical conductivity. The single-layer CFCs and the two- and three-layered plies were laminated to make them flat and fixed together for the 3D printing process, which can decrease the nominal thicknesses of CFCs. Furthermore, plain zero samples without CFCs were tested for comparison. The composition of the carbon fiber and its nominal thickness in the electromechanical samples are presented in Table 1.

Table 1. Studied CFC compositions based on the manufacturer's data.

Series	Thickness (μm)	Weight (g/m^2)	N of Layers	Grade
1	-	-	0	-
2	53	6.8	1	800015i
3	106	13.6	2	800015i
4	159	20.3	3	800015i
5	153	17	1	800020i

A TPU filament was used as a 3D-printed backbone for the fabricated structure. TPU is a widely used material in 3D printing and stretchable electronics because of its high deformability and stability [2,3,7]. For example, TPU-based stretchable films have also been used in wearable and printed electronics in combination with screen-printable conductive silver inks. Using TPU filaments is a convenient choice for developing 3D-printed stretchable and wearable electronics because of the ease of integration in these types of systems. Blue Ultimaker TPU 95A filament (Ultimaker B.V., Utrecht, The Netherlands) by Ultimaker B.V. (nominal diameter: 2.85 mm) was used in the tests, as it is more reliable to 3D print compared to the thinner 1.75 mm diameter FFF filaments, which are prone to bend and jam during 3D printing.

A commercial Ultimaker S5 FFF printer (Ultimaker B.V., Utrecht, The Netherlands) with an official air management unit accessory, with the Cura slicer program (version 4.4.0, Ultimaker B.V., Utrecht, The Netherlands) from Ultimaker B.V., was used in the 3D printer setup. The nozzle diameter was 0.4 mm, and the layer thickness was 0.15 mm. The nozzle temperature was 240 °C and the bed temperature was 60 °C, enabling good adhesion on a clean glass building plate. The printing speed was 25 mm/s, and the cooling fan was off. The number of walls was two, the infill ratio was 100%, and the infill shape was 45° lines. Furthermore, the infill was printed before the walls so that the CFC piece was smoothly fixed on the printed surface. As well as the typical printing settings, a script was added in the middle of the printing program to pause the printing automatically to manually insert the CFC piece. Furthermore, a prime tower feature was used before printing on top of the applied CFC piece to avoid uncontrolled leaking of the molten TPU from the nozzle onto the sample.

FFF was used to fabricate samples, of which the target dimensions were 10 mm wide, 200 mm long, and 1 mm thick. In the middle of the samples, an 8-mm-wide and 230-mm-long piece of CFC was placed longitudinally. CFCs were cut in the machine direction orientation (MDO), their loose ends serving as electrical contacts. Then, the CFC was fixed with two strips of Kapton tape to avoid using an adhesive and to reinforce the electrical contacts. The placed CFCs affected the samples' dimensions, measured with a digital Vernier caliper with ± 0.01 mm measurement resolution. The sample preparation steps, drawn in Solidworks 3D design software (version 2021, Dassault Systèmes SolidWorks Corporation, Waltham, MA, USA), are presented in Figure 1.

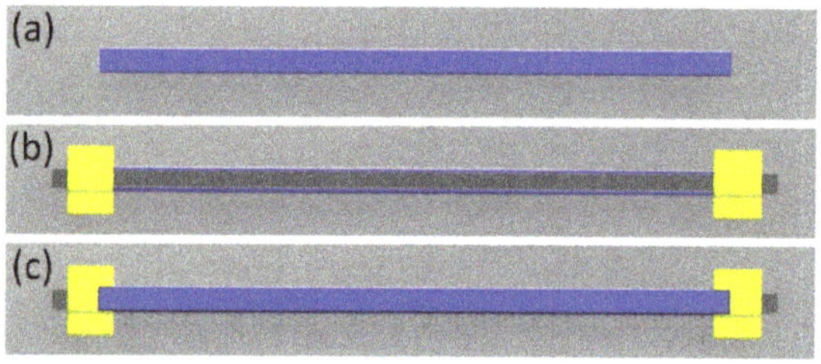

Figure 1. Preparation steps of the samples as modeled in Solidworks: (**a**) 3D printing of the sample's bottom half-layer, (**b**) placement of the CFC piece with the use of Kapton tape, and (**c**) continuing the 3D printing until the sample was ready.

2.2. Electrical Measurement of the Samples

The CFC pieces were electrically conductive. Based on the amount of carbon fibers they contained, they provided different levels of electrical conductivity in the 3D-printed matrix. In the sample preparation, the resistance of the CFC pieces was measured twice: before their placement inside the sample and after the 3D printing process. The resistance was measured using a Fluke 183 multimeter (Elfa Distrelec Oy, Helsinki, Finland) by firmly pressing the multimeter probes on the CFC pieces to achieve a stable reading (<5 Ω variation). The distance between the probes was 220 mm. The comparability of the results to those of previous studies was enabled by converting resistance to conductivity. The conductivity was calculated with the equation

$$\sigma = 1/((RA)/L), \tag{1}$$

where σ is the conductivity in S/m, R is the resistance in Ω, A is the cross-sectional area of CFC in m^2, and L is the distance between probes in m. The cross-sectional area was calculated based on the nominal width and thickness of CFCs, as reported in Table 1. Note that the conductivity here does not refer to the carbon fiber's conductivity but to the average conductivity of the macroscopic sample's CFC material.

2.3. Mechanical Tests

The mechanical behavior of the 3D-printed samples and the effect of the quantity of the integrated CFCs on the failure behavior was evaluated with an ESM303 tensile tester (Mark-10 Corporation, Copiague, NY, USA), equipped with a 500 N load cell. The distance of the clamps was 50 mm; the movement speed of the upper clamp was 25 mm/min. From each sample series, two samples were elongated 50% (25 mm) with a single pull-up test.

Because clothing-integrated stretchable and wearable components endure thousands of stretching cycles during their lifetime [30], cyclic tensile loading was chosen as a more realistic testing method. The cyclic electromechanical behavior of the samples was tested with the simultaneous use of the tensile tester and a custom-built resistance measurement system (Tampere University, Tampere, Finland). In the cyclic electromechanical tests, the movement speed of the upper clamp was 240 mm/min.

A custom-built resistance measurement system was constructed using an Arduino Uno board. The system used Arduino Uno's 10-bit AD converter for two-wire measurements to calculate real-time voltage values over the samples, which were converted to resistance values and recorded. The system had three measurement channels that used 3470 Ω resistors as a reference. Based on the reference resistor values, the system's accuracy was ±3 Ω. A threshold of 3000 Ω was used to indicate the total sample failure. The probes of the resistance measurement system were fixed to the samples' contacts with anisotropic conductive adhesive film (ACF), and were further clamped to ensure stable electrical connections.

Since the conducting material's structure is not homogeneous but is rather an interconnected network of fibers with small contact points between the fibers, the current density may influence the sample's resistance. Nevertheless, no such effect was observed with the low (less than 100 mA) measured DC currents. With higher frequencies (in MHz range) or high currents (several A), the influence of current density on the resistance would probably be observed, in alignment with previously reported results [31].

In the cyclic tests, three samples from the same category were fixed together in the 50-mm-wide clamps and simultaneously stretched 10,000 times. During testing, the tensile tester measured the average force and displacement of three samples, and the resistance measurement system measured each sample's resistance. For every sample series in Table 1, five degrees of tensile deformations were tested (10%, 20%, 30%, 40%, and 50%), and 25 cyclic tests were conducted. Figure 2 presents the clamped cyclic test samples with background light. After the cyclic tests, a strain sensor test was carried out with the cyclic test setup and a previously tested (50% elongated) cyclic test sample with a 153 μm nominally thick carbon fiber layer. The sample was elongated repeatedly up to 10% with a speed of 0.5 mm/min.

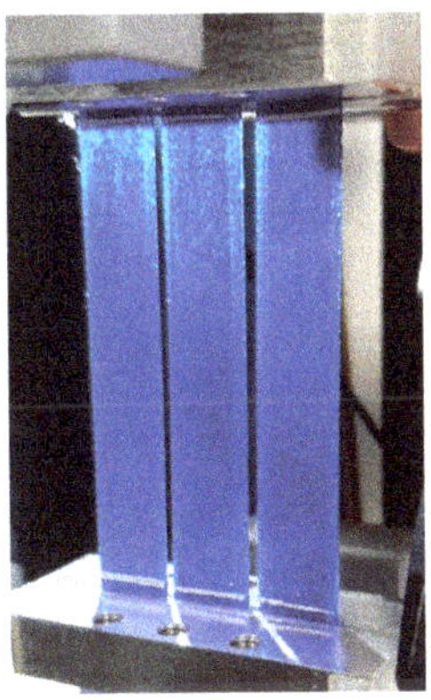

Figure 2. Clamped cyclic test samples in the ESM303 tensile tester. The samples were inspected with a background light, which revealed the integrated CFCs inside the 3D-printed TPU matrix.

The cyclic test analysis was performed using Matlab (version R2021b, MathWorks, Inc., Natick, MA, USA). Maximum and minimum resistance values during cyclic loading were extracted from the test data by averaging the steady-state results obtained after the first thousand cycles, and a high variation in the sample resistance was observed, presumably due to fiber reorientation in the CFC layer. The change in the samples' resistance was then calculated based on the two parameters extracted above. Furthermore, the average change in the resistance of the cyclically loaded samples was calculated by dividing the testing time into ten cycle periods and calculating the average resistances of each period. After that, the average resistance data from the last 5000 cycles, with which the increase in resistance versus the cycle could be linearized, were fitted using a linear relationship through the polyfit Matlab function ($n = 1$) to obtain the average resistance change per cycle.

2.4. DIC Analyzes

To further analyze the electromechanical properties of the samples, DIC was used for separate cyclic tests to inspect surface deformations of the samples in 100 early cycles, in which the largest resistance changes typically occur. The surfaces of the samples were studied and compared in the first, 50th, and 100th cycles. For this purpose, the deformation of the samples during loading was recorded using the stereo 3D DIC imaging system 3D StrainMaster Compact 5M (LaVision GmbH, Göttingen, Germany) and analyzed using DaVis software (version 10.2.1, LaVision GmbH, Göttingen, Germany). The strain measure used through the DIC analyses—and thus the one shown in the figures in this work—was the logarithmic (Hencky) strain. For DIC, the speed of the upper clamp was 50 mm/min, and the maximum elongation of the samples was 50% (25 mm).

The CFC pieces were studied with DIC without the 3D-printed TPU matrix to observe the mechanical limits of the fabricated CFC pieces more closely. In these samples, each CFC piece was laminated between two clear TPU films (Platilon U 4201 AU by Covestro). The thickness of the films was 100 μm, width 10 mm, and length 200 mm. The film was transparent and notably more deformable under the same load levels compared to the 1-mm-thick 3D-printed TPU matrix, enabling the evaluation of the deformation of the CFC pieces.

3. Results

3.1. Preparation of the Samples

The dimensions of the designed samples were 10 mm in width, 1 mm in thickness, and with a cross-sectional area of 10 mm^2. Table 2 shows the measured average dimensions of the sample series that underwent cyclic electromechanical testing, classified by amount and type of reinforcement. The sample series contained 15 parallel samples.

Table 2. Studied CFC compositions based on the manufacturer's data, and the target and measured average dimensions of the sample series.

Series	Average Width (mm)	Average Thickness (mm)	Average Area (mm^2)
Target dimensions	10	1	10
Without CFC	9.97	1.02	10.15
1 thin CFC (53 μm)	10.09	1.03	10.41
2 thin CFCs (106 μm)	10.12	1.04	10.56
3 thin CFCs (159 μm)	10.21	1.06	10.83
1 thick CFC (153 μm)	10.11	1.04	10.49

The area of the unreinforced sample series was 1.5% larger than the target dimensions. With one layer of thick CFC, the average area increased by 4.9%. With 1–3 layers of thin CFC, the average area increased by 4.1%, 5.6%, and 8.3%, correspondingly.

Figure 3 shows the resistance measurements of the CFC pieces of the samples before 3D printing, compared with the same measurements after the process. From these data, it is evident that CFCs' electrical performance improved after their integration into the 3D-printed matrix, since their resistance decreased. This differs, for example, from the currently used carbon filler filaments, which have better electrical properties before 3D printing. The resistance of the samples with thick CFC decreased by 27% after 3D printing, making the average conductivity of the sample series approximately 1000 S/m. In the sample series with one thin CFC, the resistance decreased by 33%, whereas the sample series with two thin CFCs and three thin CFCs exhibited a 38% decrease in resistance. The calculated conductivity of all thin CFC sample series was at the level of 1500 S/m, which is not a drastic variation, despite their differing thicknesses and several interfaces.

In addition to the decrease in the resistance, Figure 3 shows the scattering of the resistance values in the CFC pieces. The series with one thin CFC showed the highest dispersion of results, which decreased when more CFC plies were laminated on each other. Furthermore, although the thickness of the sample with one thick CFC and that of the sample with three thin CFCs samples were nominally similar, the resistance values and scattered results of the series with one thick CFC were at the level of the series with two thin CFCs.

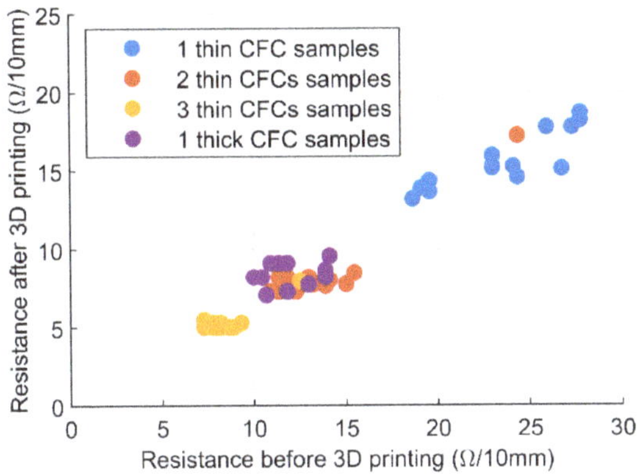

Figure 3. CFC layers' resistance per 10 mm before and after the 3D printing process. The sample series contained 15 parallel samples.

3.2. Mechanical Tests

First, the samples were elongated by 50% in the single pull-up tests; the results are shown in Figure 4. In Figure 4a, it is possible to note that in the initial elastic phase, before any crack formation in the samples, the increased amounts of CFCs added to the samples resulted in larger slopes in the elastic region of the force-displacement curve. The high amount of CFCs affected the elastic phase end, with the deformation developing into a permanent plastic phase, the transition area becoming irregular, and random force drops appearing. After the elastic phase, the decline in the force of the samples with one thick CFC was about 20 N (20%) and it was approximately 5 N (6%) in the samples with three thin CFCs. Furthermore, Figure 4b shows that the force results of the samples approached certain levels, with the force of the plain samples and the samples with one thin CFC approaching 70 N. In the samples with two thin CFCs and with three thin CFCs, the force levels were around 90 N, whereas the force in the samples with one thick CFC varied between 90 and 100 N.

After the single pull-up tests, the samples' long-term durability was examined through cyclic tensile tests. The raw data concerning resistance over time showed that three phases could be identified for each cyclic test. Data from one sample from the sample series with two thin CFCs (50% elongation) are shown in Figure 5. Initially, there was (1) a cycle in which a high increase in the sample's resistance was observed, followed by (2) a few more cycles with decreasing resistance values, which finally (3) stabilized to a specific level. In phase 3, the resistance behavior over the cycle was predictable and stable, or it varied unpredictably and accumulated damage, as shown in Figure 6.

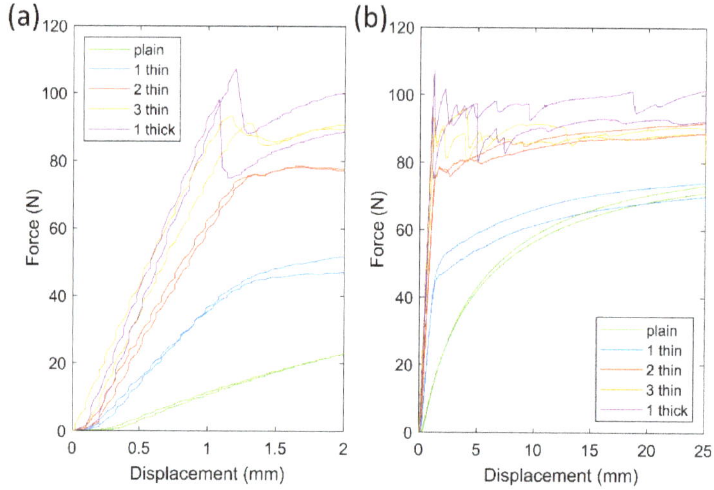

Figure 4. Force–displacement curves of the 3D-printed samples. (**a**) Results focusing on the initial 2 mm displacement (4% elongation), presenting the increase in stiffness caused by CFCs, (**b**) the samples' total displacement was 25 mm (50% elongation).

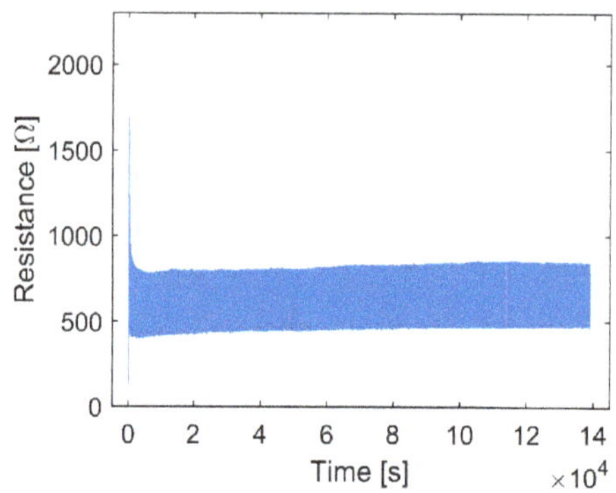

Figure 5. Raw data of a sample from the sample series with two thin CFCs, elongated 50% 10,000 times. The data show the high increase in resistance in the first cycle and the decrease in resistance and stabilization that occurred within a few cycles.

The samples' resistance varied differently depending on the CFC layer's thickness, structure, and tensile deformation level. The samples with a thin CFC layer and a low degree of elongation were more likely to have a more stable resistance variation in phase 3 (Figure 6a) than the samples with a thick layer of CFCs and high elongation (Figure 6b). In some cases, the resistance values in phase 3 exceeded the measurement limits of the test setup (3000 Ω), indicating the electrical failure of the sample.

In addition to the resistance behavior, the samples with different CFC layers exhibited different elongation and failure behavior (Figure 4). The samples with 1–2 layers of thin CFCs elongated uniformly, but those with a stiffer CFC layer elongated more locally. Thinner CFC layers tended to show delocalized damage and deformation along the whole sample through the TPU matrix; in the thicker samples, the deformation was localized where the CFC layers started forming cracks, leading to high local strains in the matrix, greatly affecting the samples' electromechanical properties.

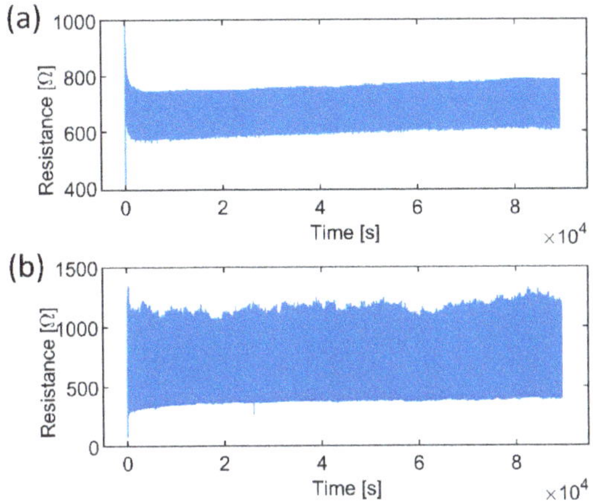

Figure 6. Cyclic behavior comparison of the samples elongated 30%. (**a**) The sample with 1 thin CFC exhibited repetitive electrical behavior in the 600 Ω–800 Ω range, whereas (**b**) the sample with 3 thin CFCs exhibited varying properties in the 350 Ω–1250 Ω range.

Figure 7 shows the resistance ranges of the cyclic sample series, with the electromechanical behavior of each measured sample regarding its resistance properties is displayed as a vertical line. The results of the three parallel samples are grouped, and the endpoints of the lines show the maximum and minimum resistances, whereas the line's length displays the resistance change during stable cyclic loading. As depicted in Figure 7a, the sample series with one thin CFC showed a moderate, gradual increase in its resistance range before the final 50% deformation level, which increased the samples' resistance range and variation. Figure 7b shows the modest resistance behavior of the series with two thin CFCs, with similar resistance values at the 30% and 40% deformation levels; at the 50% deformation level, the resistance range was less than 500 Ω.

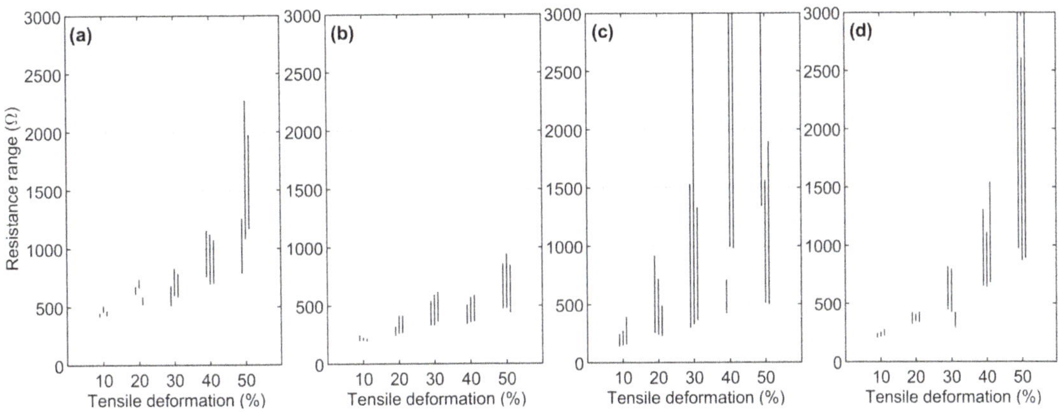

Figure 7. Resistance range lines of cyclic sample series, where three parallel samples are grouped together. The bottom and top endpoints of the lines show the minimum and maximum resistance values of the samples: (**a**) series with 1 thin CFC, (**b**) series with 2 thin CFCs, (**c**) series with 3 thin CFCs, and (**d**) series with 1 thick CFC.

As illustrated in Figure 7c, the stiff series with three thin CFCs showed more scattered results regarding the maximum resistance and resistance range values, whereas the minimum resistance level still exhibited a recognizable gradual increase. Finally, as shown in Figure 7d, the series with one thick CFC behaved similarly to the series with one thin CFC. However, the resistance was lower at the 10–20% deformation levels because of the greater thickness of the CFC layer. Conversely, this sample series showed higher resistance changes at the deformation levels of 40% and 50% because of its stiffness.

Like Figure 7, Figure 8 shows the changes in the average resistance in the samples per 5000 cycles, which indicates how much the resistance changed in the stable cyclic deformation phase of the cyclic samples (previously defined as phase 3). The failed samples that reached the 3000 Ω failure limit in Figure 7 are absent from Figure 8.

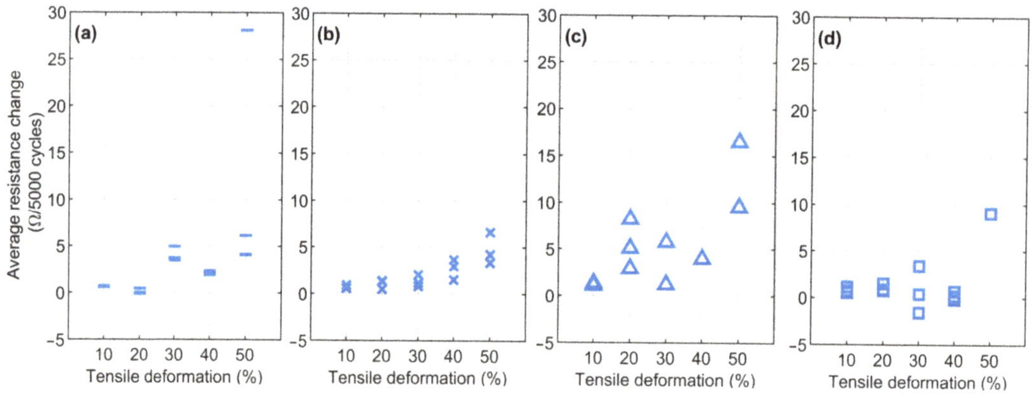

Figure 8. Average resistance changes between cycles in the last 5000 test cycles: (**a**) series with 1 thin CFC, (**b**) series with 2 thin CFCs, (**c**) series with 3 thin CFCs, and (**d**) series with 1 thick CFC.

In Figure 8a, the series with 1 thin CFC exhibited a change of approximately 5 Ω (1%) or smaller throughout 5000 cycles, except for a 50% tensile deformation level. The more random 50% tensile deformation level indicates the possible electrical limits of the sample series. In Figure 8b, the change in resistance increased as the tensile deformation increased; in the 10–30% tensile deformations, the average resistance change was always lower than 3 Ω (0.5%). At the 40–50% level, the resistance change was still small, but the low resistance range in Figure 7b increased the percentual resistance change and the results were approximately 5 Ω (lower than 1%).

In Figure 8c, the scattered resistance values show that the samples with three thin CFCs were not deformed evenly. Moreover, several samples (one sample from the 30% deformation level, two from the 40%, and one from the 50% level) were removed as failures. Despite this, the samples' percentual resistance change was lower than 1%; even with two unstable samples at 20% and 50% deformation levels, the change was approximately 2% and 1.5%, correspondingly.

Figure 8d depicts the results of the series with 1 thick CFC, which differed from the rest of the sample series in Figure 8 and which exhibited small negative resistance changes (lower than −0.3%) at the 30% and 40% tensile deformation levels. In the samples, the resistance decreased throughout the cycles, which can be explained, e.g., by the alignment of the carbon fibers. Furthermore, the average resistance changes were small at the 10–40% levels and were always less than 4 Ω (1%) due to the high thickness of the CFC layer. This, however, changed at the 50% deformation level, where the CFC layer's stiffness was high enough (compared to one of the TPU matrices) that the sample continued accumulating damage, resulting in only one intact sample at the end of the cyclic test that showed an average resistance change equal to 10 Ω per 1000 cycles.

3.3. A Strain Sensor Demonstration

Finally, the usage of 3D-printed samples with an integrated carbon fiber layer as a strain sensor was tested, with results shown in Figure 9. As an interconnection, the samples lasted well up to 50% elongation, but for sensor applications, a smaller 10% dynamic range was used after the initial 10,000 cycles, with a maximum elongation of 50%.

In the tests, rapid changes caused a dynamic error, and the response to a step change stabilized in about 200 s. The test in Figure 9a was carried out with a stretching speed of 0.5 mm/min and a maximum elongation of 10%. Figure 9a shows a linear increase in resistance, which is optimal for sensing applications. However, there was some drift in the output signal, similarly to what can be seen in the cyclic test in Figure 5. The drift stabilized after about 100 cycles.

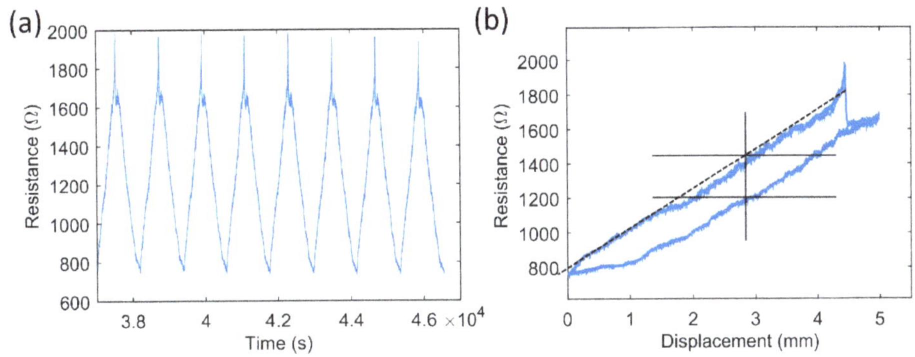

Figure 9. Sensor properties of tested cyclic test samples (153-μm-thick CFC, 10,000 cycles, 50% elongation, 240 mm/min speed). (**a**) Raw data with 10% elongation and 0.5 mm/min speed, and (**b**) a resistance-displacement curve of the first 4 cycles from (**a**).

As shown in Figure 9b, the practical dynamic range of the sensor was 750 Ω–1800 Ω. The system shown in Figure 9b was not strictly linear. The nonlinearity error and noise were <5% of the full scale. The sensor's output after the structure stabilized was repeatable, and the nonlinearity could be compensated for using software.

Figure 9b also shows hysteresis and sensitivity. The maximum difference during the hysteresis cycle was 220 Ω, which corresponds to ~10% of the full scale. The sensitivity of the sensor structure was ~200 Ω/mm.

3.4. DIC Analyses

The 3D-printed samples are also studied with DIC to identify local deformations to supplement the mechanical test results. The results of DIC analyses for the different sample series are presented in Figure 10.

Figure 10 shows that samples with different embedded CFC layers also exhibited different levels of local deformations. The non-reinforced plain sample (b1) and those with one thin CFC (a3 and b3) elongated uniformly without visual signs of deformation in the vertical direction. The stiffening effect of one thin CFC did not affect the overall surface deformation of the TPU matrix, nor did it create stress concentration areas. Instead, the sample with two thin CFCs (a1) showed multiple oblique failure bands that followed the shape of a 45° infill pattern over the entire length of the sample. Moreover, the high strain peaks observed here differed from the bulk level of deformations by only 20%.

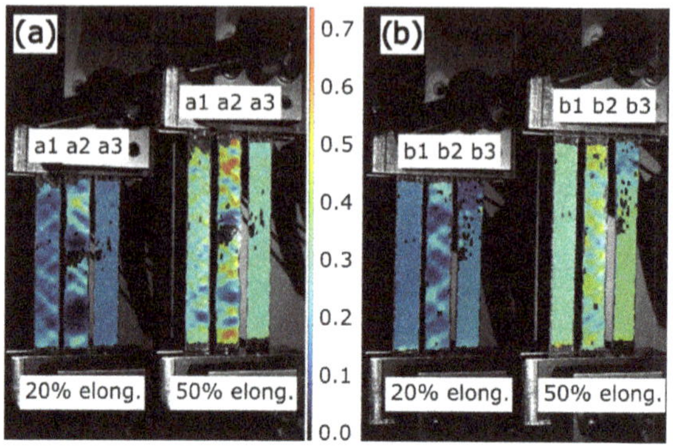

Figure 10. DIC y-axial Hencky strain results of 3D-printed samples in 20% and 50% elongations: (**a**) the a1 sample with 2 thin CFCs, the a2 sample with 3 thin CFCs, and the a3 sample with 1 thin CFC; (**b**) the b1 sample without CFCs, the b2 sample with 1 thick CFC, and the b3 sample with 1 thin CFC. The scale bar shows the level of local deformations compared to the initial undeformed phase of the samples.

The stiffest samples (a2 with three thin CFCs and b2 with one thick CFC) showed prominent failure bands, where failure was concentrated on the CFC layer. At 50% elongation, it is possible to note that some areas in the sample with three thin CFCs were almost undeformed (5%), whereas other areas exhibited axial deformation reaching over 70% due to the complete local failure of the CFC layer. The sample with one thick CFC also displayed a visible failure with a 40% deformation difference between the undeformed and failure areas.

As well as the 3D-printed samples, the deformation of CFC pieces embedded between thin and highly elastic TPU films is presented in Figure 11. In these cases, the stiffness of the CFCs and the TPU films differed considerably, making the CFC piece the load-carrying component and the TPU films only the binders of the CFC piece.

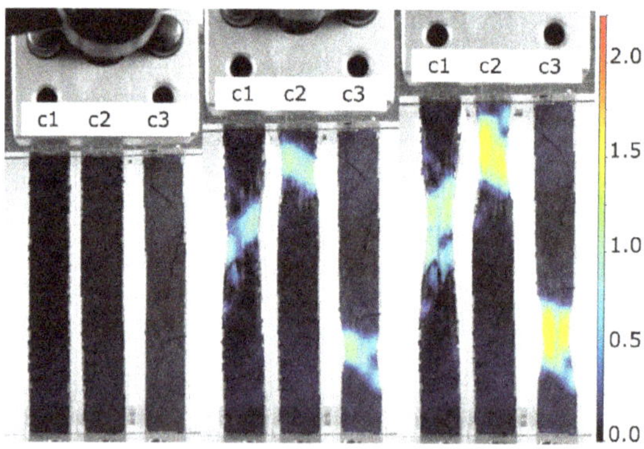

Figure 11. DIC y-axial Hencky strain analyses of the elongation of laminated CFC samples, where the c1 sample had 1 thin CFC, the c2 sample had 2 thin CFCs, and the c3 sample had 3 thin CFCs between the transparent TPU films. The deformation was mostly localized in the failure locations, whereas the rest of the sample exhibited almost no deformation under the load. The scale bar shows the level of local deformations compared to the initial undeformed phase of the samples.

The DIC examination, shown in Figure 11, indicated that the failure of the CFC piece occurred at the same deformation level for samples with different thicknesses and structures. Generally, when a failure was localized at one point on the CFC piece, the elongation was focused on that area. For instance, in the highly elongated samples shown in Figure 11, the dark-colored CFC areas were elongated by less than 10%, whereas the elongation in the cracked areas could reach over 150%. The results showed that the highly deformable TPU films around the CFC piece did not distribute loads across the CFC piece and required a stiffer TPU matrix, which can be made through FFF.

4. Discussion

4.1. Effect of CFCs on the 3D Printing Process

Adding CFCs into FFF had both negative and positive effects on the process and the prepared object. The object's dimensions changed when CFCs were added inside an object without a cavity (Table 2). The samples' dimensions increased when the amount of CFCs increased since the carbon fibers replaced molten polymer on the layer, forcing it to flow elsewhere. However, when considering FFF's general accuracy, the calculated volume increases were small, and these changes can be considered in the design phase.

In the preparation of the samples, integrating CFC pieces into the TPU matrix changed the CFC layer's electrical properties. The 3D-printed molten plastic on the CFC piece compressed the fibers on the previously printed layer in the z-direction, improving the CFCs' electrical properties by decreasing the resistance by about 30%. Furthermore, the shrinkage of the molten plastic during cooling may have compressed the CFC layers even more in the x- and y-directions, improving their electrical properties more. The measured sample series had different amounts of integrated CFC layers, and the series with one thin CFC (a nominally 53-µm-thick CFC layer) showed scattered resistance results, as shown in Figure 3. When the CFC layer's nominal thickness was at least 100 µm, the sample's final resistance value was more consistent and predictable. Moreover, the series with two thin CFCs series and that with one thick CFC had approximately the same measured electrical behavior despite their 50 µm difference in thickness, indicating that lamination and the inclusion of an interface between the two CFC plies even out the electrical properties of randomly oriented CFCs efficiently.

The conductivities of the thin and thick CFC pieces were about 1500 S/m and 1000 S/m, correspondingly. The lamination of two or three thin CFCs plies together did not change the conductivity despite its effect on the resistance values. Notably, the nominal thicknesses of the CFC plies were used in the conductivity calculations. In case of the possible compression of the plies, a lower thickness would mean even higher conductivity. Furthermore, in measuring the initial resistances (used for the conductivity calculations), the accuracy of the resistance measurements was affected by the contact resistance and surface topography of the permeable fiber matrices. The inaccuracy was the same in the case of the CFC pieces (<5 Ω). Nevertheless, sample conductivity obtained using this method was two orders of magnitude higher than that of the methods currently used to incorporate carbon particles into FFF-generated objects, which justifies the use of CFCs in FFF matrices.

Also notable is that adhesion of the TPU matrix and the CFC piece was solid, showing no traces of delamination. The CFCs were placed on the additive manufactured surface without adhesives, and the flowing molten TPU fixed the CFC pieces inside the matrix's TPU infill structure. It is also likely that the molten TPU did not fully penetrate the thick CFC piece but still encapsulated the CFC layer inside the sample.

4.2. Electromechanical Behavior of the Samples

All the integrated CFC sample series were produced using the same 3D printing process, but their electromechanical properties varied. Independently, the CFC plies had different electrical properties because of their thicknesses and fabrication procedures, which, however, did not affect properties such as their elongation at break (Figure 11). Furthermore, considering the results of the plain TPU samples, one can claim that the samples' electromechanical properties came from the unique combination of the 3D printed TPU and CFCs.

The cyclic tensile test samples exhibited three phases in their tensile deformations (Figure 4), in which the first cycles might be explained through the rearrangement of the carbon fibers in the material. During the first loading cycles, carbon fibers embedded in the 3D-printed material may experience high enough local deformation, which may break part of the conduction network; as this induces an increase in resistance, some of the fibers that are now free to move to rearrange themselves due to the highly directional deformation applied, thus creating new (and more stable) paths that lead to a substantial decrease in the overall resistance. However, this phenomenon is not fully explained and requires more analysis.

The third phase is related to the electromechanically stabilized area that is observed after the orientation of the fibers. The regularity of phase 3 is affected by the amount of tensile deformation, thickness, and the number of interfaces in the CFC piece. Moreover, phase 3′s erratic behavior commonly affects only the maximum resistance values per cycle, whereas the minimum value stays at the same level. In the series with one thin CFC, phase 3 was even, but the resistance range in the cycles was extensive. In the series with three thin CFCs series and that with one thick CFC, phase 3 was unstable because of increased damage accumulation in the CFC layer. The most promising was the series two thin CFCs, where phase 3 was stable, and the resistance varied only 500 Ω between the maximum and the minimum during cycling at 50% tensile deformation.

4.2.1. Plain Sample Series and Series with One Thin CFC

In the plain sample series and that with one thin CFC, the deformations of the samples were determined by the properties of the TPU matrix. The plain sample series exhibited smooth force-displacement curves, as shown in Figure 4, which are typical for rubber-like materials. Adding one thin CFC into the matrix changed the curves, so that there were (1) linear elastic and (2) settling plastic deformation areas. The samples with one thin CFC elongated evenly, and their force approached the same level as that of the plain samples, meaning that the thin CFC influenced the samples' initial stiffness and the TPU matrix carried a load further at higher elongations. The 53-µm-thin CFC's minimal effect on the long-term deformation of the TPU matrix can also be seen in the DIC results in Figure 10, where the plain sample series and that with one thin CFC exhibited well-distributed deformations.

The low density of fibers in the samples with one thin CFC allowed molten TPU to penetrate the fiber layer, encapsulating the fibers inside the matrix and enhancing the samples' mechanical properties, also affecting their electrical properties. The electrical properties of the series with one thin CFC were governed by the TPU's mechanical deformations, resulting in a nonlinear relationship with the average resistance, as shown in Figure 7a. At 10–40%, tensile deformations—the one thin CFC encapsulated inside the TPU matrix—the electrical connection between carbon fibers was maintained, and this was improved by the fibers' length (25.4 mm). At 50% tensile deformation, the fibers partially detached from each other, increasing the resistance range values, as shown in Figure 7a, and causing scattering, as shown in Figure 8a. Furthermore, the great increase in resistance could be caused by the random orientation of the fibers and the local thickness variations of the thin CFC, which can make parts of CFC more electrically sensitive to mechanical deformations.

4.2.2. Sample Series with Two Thin CFCs

In the sample series with two thin CFCs, the combination of the TPU matrix and CFCs was mechanically balanced, which can also be seen in the series' good electrical properties. Compared with the samples with one thin CFC, the stiffness of the samples with two thin CFCs doubled (Figure 4a), and the samples showed a small force fluctuation at the start of the plastic deformation phase (Figure 4b), indicating minor reorientation or tearing of the CFC layer. The series' closely balanced mechanical properties are also shown in Figure 10a, where the CFC layer was stiff enough to form 45° failure bands in the sample, which, however, were distributed along the samples' length, and their deformation level was only 20% higher than in the undeformed areas. The stretchability of TPU with the long length of the carbon fibers meant that the 20% deformation level of the failure bands did not negatively affect the samples' electromechanical properties.

The electrical properties of the sample series were stable. In Figure 7b, the resistance range stayed at the same level with 30–40% tensile deformations and with 50% tensile deformations of 500–1000 Ω. This stable behavior can be explained by the local failure bands, where tensile deformations were "absorbed". Furthermore, the two laminated CFC plies and the interface between them could form an impermeable layer; the molten TPU did not completely bind with the lower CFC ply. In that case, the lower CFC ply can move and be further subjected to higher tensile deformations. However, only a specific degree of movement can improve the electrical properties by preserving the conductive fiber network under the movement of the TPU matrix. In contrast, too great a movement can detach and reconnect the fibers and lead to unstable electrical properties, which likely occurred in the series with three thin CFCs. With these features, the series with two thin CFCs showed the most balanced and stable electrical properties, which can be considered for the development of stretchable interconnects or sensors for wearable applications.

4.2.3. Sample Series with Three Thin CFCs and with One Thick CFC

In the samples with three thin CFCs and those with one thick CFC, the CFC layer was rigid, surpassing the stiffness of the TPU matrix governing the deformations in the samples. The tensile deformations oriented and tore the CFC layer, leading to uneven mechanical results and further varying the samples' electrical properties at high levels of tensile deformation.

As shown in Figure 4, the failure of the sample series with three thin CFCs and that of the samples with one thick CFC resembled each other, although the samples consisting of three thin CFCs had about 10 N lower maximum tensile strength than the samples with one thick CFC. Moreover, as shown in Figure 10, the samples with three thin CFCs showed a 70% difference between the least and most elongated areas, whereas the samples with one thick CFC showed a deformation difference of 40%. These results indicate that the two interfaces of the thin CFCs negatively affected the matrix's mechanical durability and restricted the transmission of forces in the CFC layer.

During 3D printing, molten TPU likely cannot penetrate deeply into the CFC layer, although decreasing resistance due to the compression effect of TPU can be observed. The CFC and the infill area were encapsulated inside the sample, but CFC adhered only on one side, affecting the samples' mechanical and electrical properties. This partial adherence promoted the effect of the interface areas in the CFC layers, allowing them to act as the weaker points for failures and resulting in the lower tensile properties of the samples with three thin CFCs compared to those with one thick CFC, as shown in Figure 4.

Among the entire sample series, the electrical properties of the series with three thin CFCs were the lowest in regard to resistance. The samples showed low and predictable resistance values, as shown in Figure 3, with the two interface areas between the three thin CFCs making the CFC layer conductivity very even. Still, this sample series could only sustain 10–20% tensile deformations before random failures occurred, as shown in Figures 7c and 8c. Depending on the application, however, this can still be sufficient, since each sample must be subjected to an external load of approximately 100 N before reaching higher deformation levels, enabling one to use structures with three thin CFCs in low-level stretchable electronics for wearable sensor pads.

The electrical behavior of the sample series with one thick CFC and that with one thin CFC showed similar features. The difference was that the thick CFC was thicker and stiffer, making it more conductive at lower tensile deformations. However, after the breaking point, the samples exhibited higher resistance ranges because of the 45° failure bands. In fact, distributed local elongation points can orient the fibers in the tensile direction, even after thousands of cycles—whenever a failure band is formed—which can cause the resistance range to decrease, as in Figure 8d.

4.3. Sensor Properties of the Samples

Figure 9 shows the use of the already cyclically tested sample as a sensor. Despite the fact it had been elongated up to 50% 10,000 times, the sample exhibited a repeatable resistance curve at 10% elongation. The practical dynamic range was at least 10%, and the sensitivity was good, at approximately 10 Ω per 1% change of the full scale (~200 Ω/mm). Furthermore, the low amount of noise, drift, and the small nonlinearity error enables the application of the matrix in sensors. However, its high hysteresis decreases its accuracy and could limit its use.

In the sensor testing, the properties of the TPU matrix and sample dimensions also likely affected the resistance values. At the start of the testing process, the sample exhibited similar drift values to those of the cyclic test samples (Figure 5), which stabilized when the number of cycles increased. Fast mechanical deformations can cause dynamic errors; therefore, the 3D-printed matrix with the current dimensions exhibits the best performance when measuring slow changes.

5. Conclusions

Adding CFCs into the 3D printing process of a deformable TPU matrix represents a new way to make electrically conductive, stretchable, and wearable 3D-printed electronic devices. This method differs from current methods due to the possibility of creating structures with stretchability, excellent conductivity, printing accuracy, and isotropic layers. After studying the sample series during their fabrication phase, with single pull-up tests, electromechanical cyclic tests, and finally by analyzing them with DIC, the benefits of the addition of CFCs into the 3D-printed deformable matrix are obvious.

The results showed that the TPU matrix with just one nominally 53-μm-thin CFC layer provided 1500 S/m of conductivity—a result which has not been achieved with other carbon additives in the FFF process. When the thickness and number of interface areas were increased, the resistance of the CFC layers decreased to 5 Ω/10 mm, which has been conventionally achieved in the stretchable electronics field only with the use of metal-based materials.

The electromechanical properties of the CFC samples varied according to the thickness and structure of the CFC pieces, enabling versatile applications. The use of a thin and permeable CFC piece could support more stable and durable mechanical features, enabling the structure to deform based on the properties of the 3D-printed TPU matrix. The thick and multilayered CFC piece exhibited better electrical properties with low resistance and good mechanical stability, and the CFC's stiffness governed the structure's deformation. Between these extremes, the sample series with two thin CFCs and one interface area between the CFC plies demonstrated more balanced electromechanical properties. Small 45° failure bands with long carbon fibers absorbed tensile deformations while maintaining stable electrical properties.

These results show the potential of this novel production method to create 3D-printed stretchable electronics that can withstand high levels of deformation in a single pulling experiment and in numerous cycles. The use of one thin CFC increased the stiffness of the elastic phase but did not affect the plastic phase of 3D printed samples. The use of a high amount of CFCs increased the stiffness considerably and resulted in an irregular plastic phase, in which a 20% force decline was possible. In the cyclic testing, the samples with one thin CFC or two thin CFCs layers could reach even higher elongation than that measured at 50% for 10,000 cycles. This fabrication approach is easy to incorporate into current 3D printing practices, enabling the preparation of 3D-printed stretchable and wearable electronics with commercially available materials and methods. Varying the thickness and number of CFCs added into the 3D-printed matrix allows one to optimize the structures' properties for different purposes. Additive-manufactured deformable CFC electronics can be used, for example, in sensors, interconnects, and circuit boards.

The use of CFCs inside 3D-printed structures has advantages, for example, using one thin CFC in several layers in FFF so that every printing layer adheres to only one layer of a thin CFC can be implemented to make highly reinforced structures that elongate steadily and which have intermediate electrical conductivity. Moreover, the CFCs on multiple layers can be shaped differently, enabling simple 3D PCBs that are stretchable and wearable. However, the manufacturing parameters of 3D-printed TPU matrices and CFCs of differing thicknesses require further optimization for sensor applications. Furthermore, the placement method of CFCs needs to be automated, for instance, with the use of a modified pick-and-place machine.

Author Contributions: Conceptualization, A.H., T.S. and J.V.; methodology, A.H. and T.S.; software, D.D.V.; validation, A.H. and T.S.; formal analysis, D.D.V. and T.S. investigation, T.S.; resources, A.H. and D.D.V.; data curation, D.D.V. and J.V.; writing—original draft preparation, T.S.; writing—review and editing, D.D.V., T.S. and J.V.; visualization, D.D.V. and T.S.; supervision, J.V.; project administration, T.S. and J.V.; funding acquisition, J.V. All authors have read and agreed to the published version of the manuscript.

Funding: This research was partially funded by the European Union Regional Development Fund (ERDF) and city of Kankaanpää under the project SOFT3L (A73741), of which the official authority is Satakunta Regional Council under Grant 2947/31/2018. Furthermore, this research was partly funded by the Academy of Finland under the project REEL, decision number 334175.

Institutional Review Board Statement: Not applicable.

Informed Consent Statement: Not applicable.

Data Availability Statement: Not applicable.

Conflicts of Interest: The authors declare no conflict of interest.

References

1. Singh, S.; Singh, G.; Prakash, C.; Ramakrishna, S. Current status and future directions of fused filament fabrication. *J. Manuf. Process.* **2020**, *55*, 288–306. [CrossRef]
2. Hohimer, C.J.; Petrossian, G.; Ameli, A.; Mo, C.; Pötschke, P. 3D printed conductive thermoplastic polyurethane/carbon nanotube composites for capacitive and piezoresistive sensing in soft pneumatic actuators. *Addit. Manuf.* **2020**, *34*, 101281. [CrossRef]
3. Kim, K.; Park, J.; Suh, J.; Kim, M.; Jeong, Y.; Park, I. 3D printing of multiaxial force sensors using carbon nanotube (CNT)/thermoplastic polyurethane (TPU) filaments. *Sens. Actuators A Phys.* **2017**, *263*, 493–500. [CrossRef]
4. Reyes, C.; Somogyi, R.; Niu, S.; Cruz, M.A.; Yang, F.; Catenacci, M.J.; Rhodes, C.P.; Wiley, B.J. Three-Dimensional Printing of a Complete Lithium Ion Battery with Fused Filament Fabrication. *ACS Appl. Energy Mater.* **2018**, *1*, 5268–5279. [CrossRef]
5. Ahrendt, D.; Romero Karam, A. Development of a computer-aided engineering–supported process for the manufacturing of customized orthopaedic devices by three-dimensional printing onto textile surfaces. *J. Eng. Fibers Fabr.* **2020**, *15*, 1558925020917627. [CrossRef]
6. Uysal, R.; Stubbs, J.B. A New Method of Printing Multi-Material Textiles by Fused Deposition Modelling (FDM). *TEKSTILEC* **2019**, *62*, 248–257. [CrossRef]
7. Salo, T.; Halme, A.; Lahtinen, J.; Vanhala, J. Enhanced stretchable electronics made by fused-filament fabrication. *Flex. Print. Electron.* **2020**, *5*, 045001. [CrossRef]
8. Stuart, T.; Kasper, K.A.; Iwerunmor, I.C.; McGuire, D.T.; Peralta, R.; Hanna, J.; Johnson, M.; Farley, M.; LaMantia, T.; Udorvich, P.; et al. Biosymbiotic, personalized, and digitally manufactured wireless devices for indefinite collection of high-fidelity biosignals. *Sci. Adv.* **2021**, *7*, 1–18. [CrossRef]
9. Ly, S.T.; Kim, J.Y. 4D printing–fused deposition modeling printing with thermal-responsive shape memory polymers. *Int. J. Precis. Eng. Manuf. Technol.* **2017**, *4*, 267–272. [CrossRef]
10. Tekinalp, H.L.; Kunc, V.; Velez-Garcia, G.M.; Duty, C.E.; Love, L.J.; Naskar, A.K.; Blue, C.A.; Ozcan, S. Highly oriented carbon fiber-polymer composites via additive manufacturing. *Compos. Sci. Technol.* **2014**, *105*, 144–150. [CrossRef]
11. Zhang, J.; Yang, B.; Fu, F.; You, F.; Dong, X.; Dai, M. Resistivity and Its Anisotropy Characterization of 3D-Printed Acrylonitrile Butadiene Styrene Copolymer (ABS)/Carbon Black (CB) Composites. *Appl. Sci.* **2017**, *7*, 20. [CrossRef]
12. Wei, X.; Li, D.; Jiang, W.; Gu, Z.; Wang, X.; Zhang, Z.; Sun, Z. 3D Printable Graphene Composite. *Sci. Rep.* **2015**, *5*, 11181. [CrossRef]
13. Jahangir, M.N.; Billah, K.M.M.; Lin, Y.; Roberson, D.A.; Wicker, R.B.; Espalin, D. Reinforcement of material extrusion 3D printed polycarbonate using continuous carbon fiber. *Addit. Manuf.* **2019**, *28*, 354–364. [CrossRef]
14. Al-Saleh, M.H.; Sundararaj, U. Review of the mechanical properties of carbon nanofiber/polymer composites. *Compos. Part A Appl. Sci. Manuf.* **2011**, *42*, 2126–2142. [CrossRef]
15. Forintos, N.; Czigany, T. Multifunctional application of carbon fiber reinforced polymer composites: Electrical properties of the reinforcing carbon fibers—A short review. *Compos. Part B Eng.* **2019**, *162*, 331–343. [CrossRef]

16. Li, Y.; Huang, X.; Zeng, L.; Li, R.; Tian, H.; Fu, X.; Wang, Y.; Zhong, W.H. A review of the electrical and mechanical properties of carbon nanofiller-reinforced polymer composites. *J. Mater. Sci.* **2019**, *54*, 1036–1076. [CrossRef]
17. Tzounis, L.; Petousis, M.; Grammatikos, S.; Vidakis, N. 3D printed thermoelectric polyurethane/multiwalled carbon nanotube nanocomposites: A novel approach towards the fabrication of flexible and stretchable organic thermoelectrics. *Materials* **2020**, *13*, 2879. [CrossRef]
18. Spoerk, M.; Savandaiah, C.; Arbeiter, F.; Traxler, G.; Cardon, L.; Holzer, C.; Sapkota, J. Anisotropic properties of oriented short carbon fibre filled polypropylene parts fabricated by extrusion-based additive manufacturing. *Compos. Part A Appl. Sci. Manuf.* **2018**, *113*, 95–104. [CrossRef]
19. Yin, L.; Tian, X.; Shang, Z.; Wang, X.; Hou, Z. Characterizations of continuous carbon fiber-reinforced composites for electromagnetic interference shielding fabricated by 3D printing. *Appl. Phys. A.* **2019**, *125*, 266. [CrossRef]
20. Wang, Z.; Luan, C.; Liao, G.; Yao, X.; Fu, J. Mechanical and self-monitoring behaviors of 3D printing smart continuous carbon fiber-thermoplastic lattice truss sandwich structure. *Compos. Part B Eng.* **2019**, *176*, 107215. [CrossRef]
21. Hu, Q.; Duan, Y.; Zhang, H.; Liu, D.; Yan, B.; Peng, F. Manufacturing and 3D printing of continuous carbon fiber prepreg filament. *J. Mater. Sci.* **2018**, *53*, 1887–1898. [CrossRef]
22. Yu, T.; Zhang, Z.; Song, S.; Bai, Y.; Wu, D. Tensile and flexural behaviors of additively manufactured continuous carbon fiber-reinforced polymer composites. *Compos. Struct.* **2019**, *225*, 111147. [CrossRef]
23. Türk, D.-A.; Ebnöther, A.; Zogg, M.; Meboldt, M. Additive Manufacturing of Structural Cores and Washout Tooling for Autoclave Curing of Hybrid Composite Structures. *J. Manuf. Sci. Eng.* **2018**, *140*, 105001. [CrossRef]
24. Yao, X.; Luan, C.; Zhang, D.; Lan, L.; Fu, J. Evaluation of carbon fiber-embedded 3D printed structures for strengthening and structural-health monitoring. *Mater. Des.* **2017**, *114*, 424–432. [CrossRef]
25. Falco, A.; Petrelli, M.; Bezzeccheri, E.; Abdelhalim, A.; Lugli, P. Towards 3D-printed organic electronics: Planarization and spray-deposition of functional layers onto 3D-printed objects. *Org. Electron.* **2016**, *39*, 340–347. [CrossRef]
26. Abdelhalim, A.; Abdellah, A.; Scarpa, G.; Lugli, P. Fabrication of carbon nanotube thin films on flexible substrates by spray deposition and transfer printing. *Carbon N. Y.* **2013**, *61*, 72–79. [CrossRef]
27. Fieber, L.; Evans, J.D.; Huang, C.; Grant, P.S. Single-operation, multi-phase additive manufacture of electro-chemical double layer capacitor devices. *Addit. Manuf.* **2019**, *28*, 344–353. [CrossRef]
28. ACP Composites. Carbon Fiber Tissue. Product Data Sheet. 2014. Available online: https://store.acpcomposites.com/carbon-fiber-veil?quantity=1&weight=1 (accessed on 20 September 2022).
29. Hagberg, J.; Leijonmarck, S.; Lindbergh, G. High Precision Coulometry of Commercial PAN-Based Carbon Fibers as Electrodes in Structural Batteries. *J. Electrochem. Soc.* **2016**, *163*, A1790–A1797. [CrossRef]
30. Persons, A.K.; Ball, J.E.; Freeman, C.; Macias, D.M.; Simpson, C.L.; Smith, B.K.; Burch, V.R.F. Fatigue testing of wearable sensing technologies: Issues and opportunities. *Materials* **2021**, *14*, 4070. [CrossRef]
31. Guadagno, L.; Vertuccio, L.; Naddeo, C.; Raimondo, M.; Barra, G.; De Nicola, F.; Volponi, R.; Lamberti, P.; Spinelli, G.; Tucci, V. Electrical current map and bulk conductivity of carbon fiber-reinforced nanocomposites. *Polymers* **2019**, *11*, 1865. [CrossRef]

micromachines

Article

Refractive Index Sensor Based on the Fano Resonance in Metal–Insulator–Metal Waveguides Coupled with a Whistle-Shaped Cavity

Bo Li [1,2], Huarong Sun [2], Huinan Zhang [2], Yuetang Li [2], Junbin Zang [2], Xiyuan Cao [2], Xupeng Zhu [3], Xiaolong Zhao [4,*] and Zhidong Zhang [2,*]

1 School of Software, North University of China, Taiyuan 030051, China
2 Key Laboratory of Instrumentation Science & Dynamic Measurement of Ministry of Education, North University of China, Taiyuan 030051, China
3 School of Physical Science and Technology, Lingnan Normal University, Zhanjiang 524048, China
4 School of Electrical and Control Engineering, North University of China, Taiyuan 030051, China
* Correspondence: zhaoxiaolong@nuc.edu.cn (X.Z.); zdzhang@nuc.edu.cn (Z.Z.)

Abstract: A plasmonic refractive index sensor based on surface plasmon polaritons (SPPs) that consist of metal–insulator–metal (MIM) waveguides and a whistle-shaped cavity is proposed. The transmission properties were simulated numerically by using the finite element method. The Fano resonance phenomenon can be observed in their transmission spectra, which is due to the coupling of SPPs between the transmission along the clockwise and anticlockwise directions. The refractive index-sensing properties based on the Fano resonance were investigated by changing the refractive index of the insulator of the MIM waveguide. Modulation of the structural parameters on the Fano resonance and the optics transmission properties of the coupled structure of two MIM waveguides with a whistle-shaped cavity were designed and evaluated. The results of this study will help in the design of new photonic devices and micro-sensors with high sensitivity, and can serve as a guide for future application of this structure.

Keywords: plasmonic; refractive index sensor; finite element method; Fano resonance; metal–insulator–mental

Citation: Li, B.; Sun, H.; Zhang, H.; Li, Y.; Zang, J.; Cao, X.; Zhu, X.; Zhao, X.; Zhang, Z. Refractive Index Sensor Based on the Fano Resonance in Metal–Insulator–Metal Waveguides Coupled with a Whistle-Shaped Cavity. *Micromachines* **2022**, *13*, 1592. https://doi.org/10.3390/mi13101592

Academic Editors: Libo Gao and Zhuoqing Yang

Received: 2 September 2022
Accepted: 21 September 2022
Published: 25 September 2022

1. Introduction

Surface plasmon polaritons (SPPs) have the capability of overcoming the diffraction limit due to their energy evanescently confining in the perpendicular direction of the metal–insulator interference [1–3]. SPPs have been widely used in surface-enhanced Raman scattering, imaging, solar cells, sensors, optical filters [4–10], and so on. Recently, another new type of surface plasmon, Tamm plasmon modes in the metal-PhC cavity, has shown excellent performance in gas sensors and refractive index sensors due to its sensitivity to the environment and strong localization [11–13]. For the application of manipulating SPPs in chip-scale integration, metal–insulator–metal (MIM) waveguides are an excellent subwavelength photonic device [14,15]. A plasmonic interferometric biosensor based on MIM waveguides for phase-sensitive biomolecular analysis has been proposed [16]. SPP photonic devices based on MIM waveguides have received increasing attention, such as all-optical switches, sensors, and slow light devices [17–19].

Recently, the Fano resonance phenomenon was observed in the MIM waveguide-coupled resonator system [20,21]. It was first discovered by Fano Ugo, and has an asymmetric line profile due to the interference between a narrow discrete resonance and a broad spectral line or continuum [22–24]. Fano resonance, as a weak coupling and interference phenomenon, has a unique line shape, which provides a promising pathway to achieve ultrahigh sensitivity sensors, lasing, all-optical switching, and nonlinear and slow light [25–31]. Biosensors and chemical sensors based on Fano resonance have attracted much attention from researchers due to their extreme sensitivity to changes in structural parameters and the surrounding dielectrics [32,33]. These sensors exhibit good performance in terms of sensitivity and figure of merit (FOM) [34,35]. To achieve ultrahigh sensitivity, these structures and parameters need to be optimized. Therefore, the way in which to optimize the MIM waveguide-coupled resonator system to obtain a plasmonic coupled system with a Fano line shape is a key issue for designing high-sensitivity plasmonic sensors.

In this study, a whistle-shaped plasmonic structure composed of one whistle-shaped cavity and an MIM waveguide was designed to obtain high-sensitivity sensors based on Fano resonance. The transmission properties and magnetic field distributions of the whistle-shaped structures were simulated using the finite element method (FEM). The effects of the structural parameters of the whistle-shaped structure on the transmission spectrum were investigated. The refractive index sensitivity and the FOM of the whistle-shaped structure were explored. A derived plasmonic structure composed of a double whistle-shaped coupled structure was designed and evaluated. The sensitivity of the derived structure was examined.

2. Structure Model and Analytical Method

As shown in Figure 1a, the 2D schematic of the proposed plasmonic structure is composed of a whistle-shaped cavity and an MIM waveguide. The whistle-shaped cavity was composed of the input MIM waveguide on-interval and a ring cavity. As shown in Figure 1, the blue and white parts denote Ag (ε_m) and air (ε_s), respectively. The widths w of these MIM waveguides were fixed at 50 nm to support the only fundamental transverse magnetic (TM_0) in the MIM waveguides. The length of the output MIM waveguide is L, and the gap between the output MIM waveguide and the ring cavity is d_1. The inner and outer radii of the ring cavity in the whistle-shaped structure are represented by r_1 and r_3, and the center radius of the ring cavity is $r_2 = (r_1 + r_3)/2$. The red vertical dashed line passes through the center of the ring cavity and is defined as the reference line.

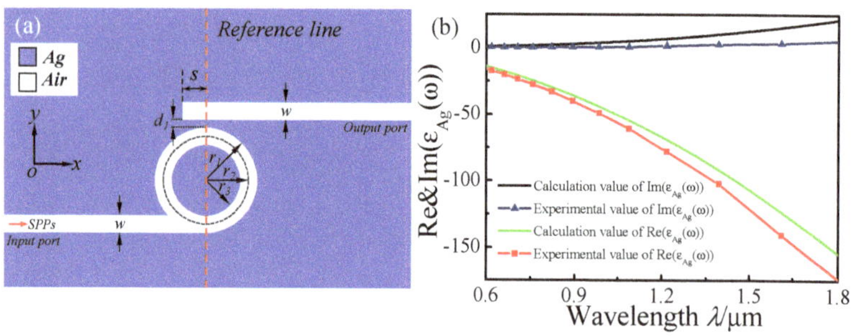

Figure 1. (**a**) Two-dimensional schematic for the MIM waveguides coupled with a whistle-shaped cavity. (**b**) The calculated and experimental values of the real and imaginary parts of the permittivity of silver.

The frequency-dependent complex relative permittivity $\varepsilon(\varpi)$ of silver is characterized by the modified Debye–Drude dispersion mode [36,37] as follows:

$$\varepsilon(\varpi) = \frac{\varepsilon_\infty + (\varepsilon_s - \varepsilon_\infty)}{1 + i\varpi\tau} + \frac{\sigma}{i\varpi\varepsilon_0} \quad (1)$$

where $\varepsilon_\infty = 3.8344$ is the infinite frequency permittivity, $\varepsilon_s = -9530.5$ represents the static permittivity, $\sigma = 1.1486 \times 10^7$ S/m is the conductivity, and $\tau = 7.35 \times 10^{-15}$ s is the relaxation time. Figure 1b shows the real part and the image part of the experiment value

and modified value using the Debye–Drude model for the silver relative permittivity. The theory mode is in good agreement with the experimental data.

The transmission properties of the single whistle-shaped structure and the double whistle-shaped coupled derived structure were investigated using FEM with perfectly matched layer-absorbing boundary conditions. The transmittance is defined as $T = |S_{21}|^2$, where S_{21} is the transmittance of the MIM waveguide [38].

For MIM waveguides, only the fundamental transverse magnetic (TM_0) mode can be supported, and its dispersion relation is expressed as [39]:

$$\tanh(\kappa w) = -2\kappa p\alpha/(\kappa^2 + p^2\alpha^2) \quad (2)$$

where κ and w are the perpendicular core wave vector and the insulator width of the MIM waveguide, respectively. The symbols in Equation (1) are defined as $p = \varepsilon_{in}/\varepsilon_m$ and $\alpha = [k_0{}^2(\varepsilon_{in} - \varepsilon_m) + \kappa]^{1/2}$, where ε_{in} and ε_m are the dielectric constants of the insulator and the metal, respectively; $k_0 = 2\pi/\lambda_0$ is the free space wave vector; κ can be solved from Equation (1) using the iterative method. Thus, the effective index n_{eff} of the MIM waveguide can be defined as $n_{eff} = (\varepsilon_m + (\kappa/k_0)^2)^{1/2}$. The wavelength of SPPs, λ_{spp}, can be expressed as $\lambda_{spp} = \lambda_0/\mathrm{Re}\ (n_{eff})$, where Re ($n_{eff}$) is the real part of n_{eff}.

3. Results and Discussions

Figure 2a shows the transmission spectrum of the whistle-shaped MIM waveguide system with $s = 30$ nm, $w = 50$ nm, $d = 10$ nm, and $r_2 = 145$ nm. As shown in Figure 2a, three asymmetrical profile peaks (0.645, 1.15, and 1.24 μm) can be observed in the transmission spectrum, which are regarded as the Fano resonance. The transmittance shows an unusual variation with the increase in wavelength from 1.15 to 1.24 μm; in particular, a steep slope curve can be observed in the transmission spectrum at the range of 1.225–1.24 μm. The H_z field distributions at $\lambda = 0.645$, 1.15, 1.225, and 1.24 mm were simulated and are displayed in Figure 2b–e to understand the physical mechanism of the asymmetrical profile peaks of the proposed structure. Steady stand wave modes can be observed, and the black arrows represent the time-average power flow distributions in the H_z field distribution graphs. For $\lambda = 0.645$ μm, $\mathrm{Re}(n_{eff}) = 1.4435$, and $\lambda_{spp} = 0.4468$ μm, so the number of wave nodes can be calculated by $2\pi r_2/\lambda_{spp} = 2$. As shown in Figure 2b, two wave nodes were formed in the H_z field distribution, and the time-average power flow distributions show an anticlockwise mode in the ring cavity. Most of the SPPs' energy was limited in the whistle-shaped cavity, and part of the SPPs' energy was coupled into the output waveguide, so a peak was formed in the transmission spectrum, as shown in Figure 2a. For $\lambda = 1.24$ μm, the Fano resonance was caused by the interference between the anticlockwise mode and the clockwise mode. The H_z field distribution (Figure 2e) is similar to Figure 2d ($\lambda_2 = 1.225$ μm), but the SPPs' energy could be coupled into the output waveguide and a resonance peak is formed in the transmission spectrum for $\lambda_1 = 1.24$ μm. For $\lambda_2 = 1.225$ μm, the SPPs' energy was not coupled into the output waveguide. The time-average power flows show that an anticlockwise mode occurred. For $\lambda_3 = 1.15$ μm, the H_z field distribution was mainly distributed in the whistle-shaped cavity, and part of SPPs was coupled into the output waveguide. The time-average power flows show a clockwise mode. Thus, the Fano resonance was due to the coupling between the anticlockwise and clockwise modes. For $\lambda_4 = 0.645$ μm, the H_z field distribution reveals that a second-order vibration mode was formed in the whistle-shaped cavity, which was a new vibration mode. The time-average power flows show an anticlockwise mode.

The transmission spectra were simulated by replacing the air of the MIM waveguide system with different refractive index media ($n = 1$, 1.33, 1.34, and 1.35), which are shown in Figure 3a, to investigate the effect of refractive index n on the transmission properties of the whistle-shaped MIM waveguide system. As shown in Figure 3a, the Fano resonance peak red shifted with the increase in n, and another new resonance peak can be observed at the short wavelength range when n was larger than 1.33. The radius r_2 of the whistle-shaped cavity was fixed because the n_{eff} value in the MIM waveguide-coupled whistle-shaped cavity decreased with the increase in n, and the number of modes increased. Therefore, a peak can be observed at the near short wave (0.60–0.7 μm) in the transmission spectra. We calculated the shift of the Fano resonance peaks (I and II) with the refractive index change. The sensitivity fitting curves of peaks I and II are shown in Figure 3b. The sensitivity of peak I is $S_1 = \delta\lambda/\delta n = 600$ nm/RIU, and the sensitivity of peak II is $S_2 = \delta\lambda/\delta n = 1229$ nm/RIU.

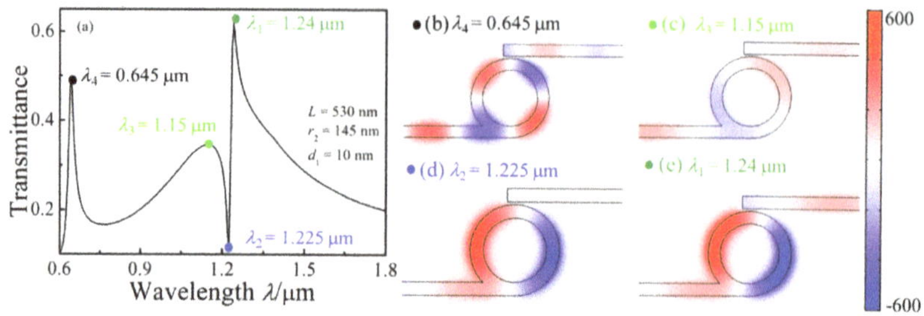

Figure 2. (**a**) Transmission spectra of the MIM waveguide-coupled whistle-shaped cavity; (**b**) contour profiles of the normalized H_z field distributions of the MIM waveguide whistle-shaped cavity: (**b**) $\lambda = 0.645$ mm, (**c**) $\lambda = 1.15$ mm, (**d**) $\lambda = 1.225$ mm, and (**e**) $\lambda = 1.24$ mm.

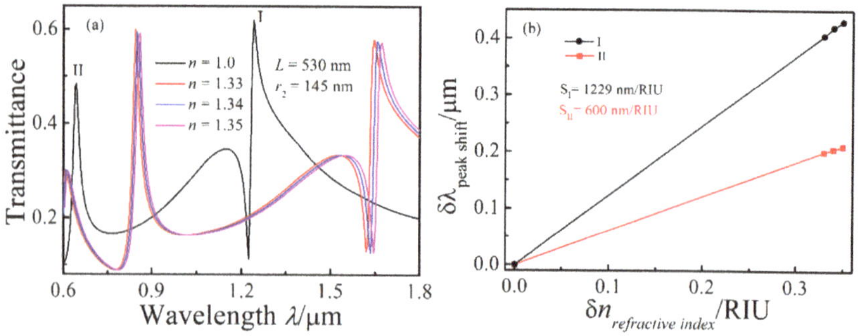

Figure 3. (**a**) Transmission spectra of the MIM waveguide-coupled whistle-shaped cavity with changing n; (**b**) shift of the Fano resonance peak as a function of the refractive index change δn.

r_1 was varied from 130 to 170 nm at intervals of 10 nm with $n = 1$, $w = 50$ nm, and $L = 530$ nm to study the effect of the different radii of the ring cavity on the Fano resonance of the MIM waveguide. With the increasing radius r_2 of the ring cavity, red shifts of the transmission spectrum and decreases in the transmittances of the Fano resonance peak can be observed in Figure 4a. Figure 4b shows the transmission spectrum with different coupled distances d between the top waveguide and the whistle-shaped ring cavity, and the other parameters were fixed as $n = 1$, $r_1 = 150$ nm, $r_3 = 100$ nm, and $d_1 = 10$ nm. From the transmission spectrum, the Fano resonance peak appears to be blue-shifted, but this was actually caused by the Fano resonance peak widening as d_1 increased, and the transmittances decreased with the increase in d_1. In contrast, the Fano resonance valley hardly shifted with as d_1 increased, which was dependent on the length of the whistle-shaped ring cavity.

In this section, we investigate the derivative structure, MIM waveguide-coupled whistle-shaped cavity, and the proposed structure, as shown in Figure 5a. Figure 5b shows the transmission spectrum of the proposed MIM waveguide-coupled double whistle-shaped cavity, and the structural parameters were fixed at $n = 1$, $R_2 = r_2 = 125$ nm, and $d_2 = 10$ nm. Although the transmission spectrum is similar to the MIM waveguide-coupled single-whisper cavity, another new peak emerged in the transmission spectrum. FR_2 was unremarkable and FR_1 and FR_2 were extremely close when $R_2 = r_2$. The H_z field distri-

butions at λ_1 = 1.09 mm, λ_2 = 1.06 mm, λ_3 = 1.05 mm, λ_4 = 1.045 mm, and λ_5 = 0.995 mm are displayed in Figure 5c. For λ_1 = 1.09 mm, the phase of the H_z field distribution in the bottom and top whistle-shaped cavities was opposite, and the output waveguide was located at the position of the strong negative time-averaged power flow, so the SPPs' energy could be passed from the output waveguide. For λ_2 = 1.06 mm, the Hz field distribution in the bottom and top whistle-shaped cavities was symmetric on the vertical axis. The output waveguide was located at the position joint between the positive and negative time-averaged power flows, so few SPPs' energy could be passed. For λ_3 = 1.05 mm, the strong H_z field distribution concentrated in the bottom whistle-shaped cavity, and few SPPs' energy were coupled into the top whistle-shaped cavity. The output waveguide was located in the abdominal of the negative time-averaged power flow, so the SPPs' energy could be passed to the output waveguide. For λ_4 = 1.045 mm, the H_z field distribution in the bottom and top whistle-shaped cavities was located on the horizontal axis. For λ_5 = 0.995 mm, the strong H_z field was mainly distributed in the bottom whistle-shaped cavity, and few SPPs' energy were coupled into the top whistle-shaped cavity.

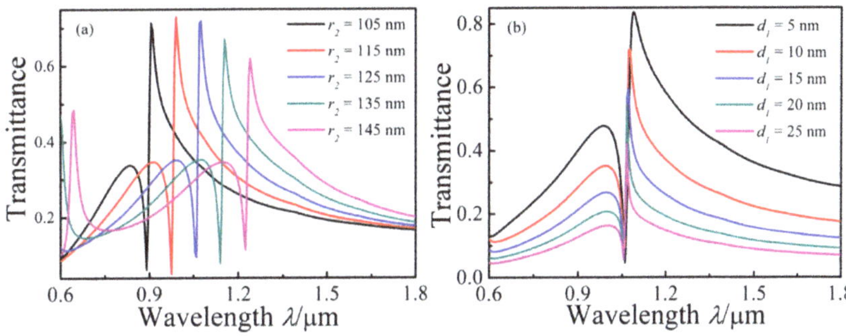

Figure 4. Transmission spectra of the MIM waveguide-coupled whistle-shaped cavity: (**a**) With changing r_2; (**b**) with changing d_1.

The value of R_2 changed and was varied from 130 to 170 nm at intervals of 10 nm, with r_1 = 150 nm and d_2 = 10 nm, to further investigate the effect of the top whisper cavity on the Fano resonance. A remarkable double Fano resonance was found in the transmission spectra, as shown in Figure 6a. With the increase in R_2, an obvious red shift can be observed in the transmission spectra. The locations of the FR_2 peak's red shift with the increase in R_2 and the transmittance of FR_2 first showed decreases and then increases. When $R_2 = r_2$, the transmittance was minimal, and with an increasing R_2, new FR_3 peaks can be observed in the transmission spectra for R_2 = 135 nm. From Figure 6b, FR_1 seems to be blue-shifted, but this was actually due to a decrease in the peak width of FR_1 with the increase in the distance (d_2) between the ring cavity from 5 to 25 nm when the values of R_2 and r_2 were fixed as 125 nm. The transmittance of FR_2 decreased, but FR_2 experienced no shift with as d_2 increased. When d_2 = 25 nm, the peak of FR_2 disappeared in the transmission spectra. With an increasing d_2, FR_3 was blue-shifted and the transmittance decreased. Figure 6c shows the transmittance spectrum when changing its refractive index n (n = 1, 1.33, 1.34, and 1.35) of the proposed double whisper cavities, and the other parameters were fixed as d_2 = 10 nm, r_2 = 125 nm, and R_2 = 115 nm. The Fano resonance exhibited a red shift with the increase in the refractive index n, and the sensitivities of 1.057 μm/RIU and 0.969 μm/RIU were obtained at $\lambda_1(FR_1)$ and $\lambda_2(FR_2)$.

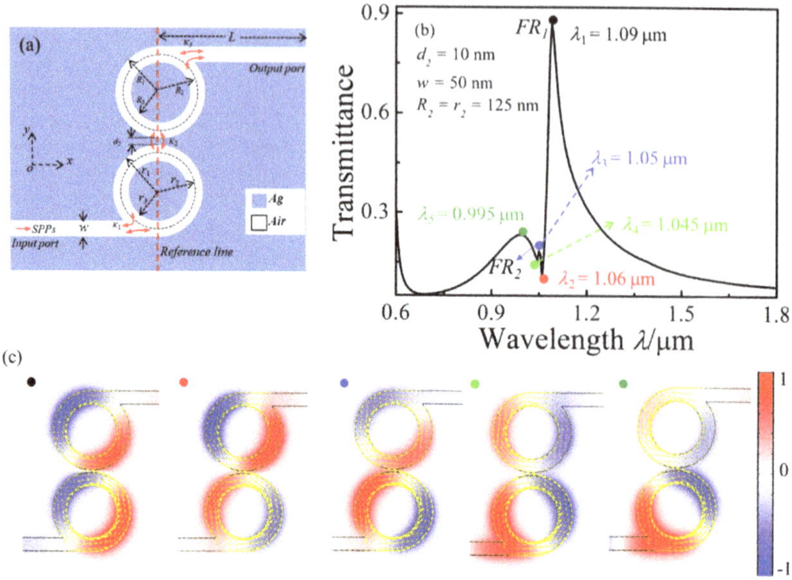

Figure 5. (**a**) Schematic of the MIM waveguide-coupled double whisper cavities; (**b**) transmission spectrum of the double whisper cavity structure; (**c**) H_z field distribution at different resonance wavelengths.

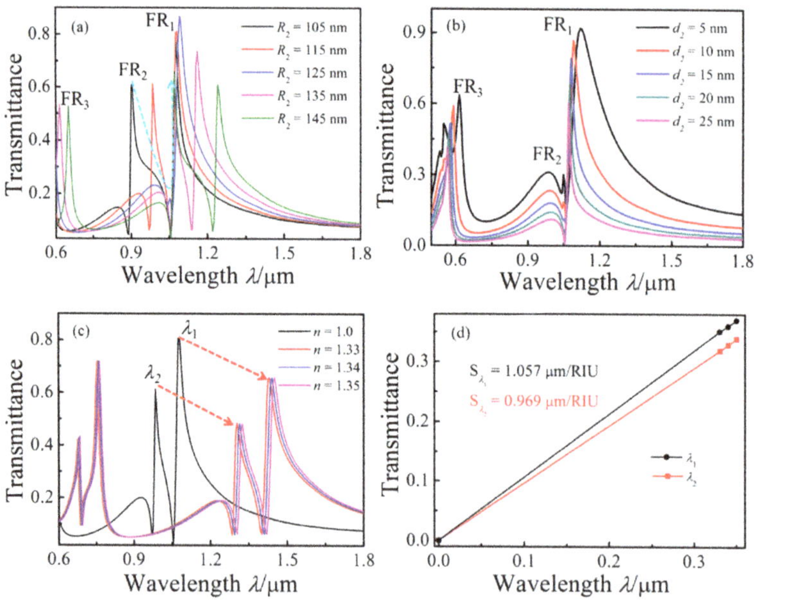

Figure 6. Transmission spectra of the different structural parameters at the MIM waveguide-coupled double whistle-shaped cavities: (**a**) R_2 changing; (**b**) d_2 changing; (**c**) n changing; (**d**) sensitivity.

4. Conclusions

In this work, a plasmonic refractive index sensor based on Fano resonance was proposed. The transmission properties were simulated numerically using FEM. Fano resonance was realized in the MIM waveguide-coupled whistle-shaped cavity. This phenomenon is due to the coupling of SPPs in the whistle-shaped cavity between the transmission along the clockwise and anticlockwise directions. Refractive index-sensing based on the Fano resonance was investigated by changing the refractive index of the insulator of the MIM waveguide. The results showed that a maximum sensitivity of 1229 RIU/nm was obtained. Compared to the single MIM waveguide-coupled whistle-shaped cavity, the double MIM with whistle-shaped cavity structure exhibited multi-Fano resonance. The refractive index sensor's sensitivity was smaller than that of the single MIM waveguide-coupled whistle-shaped cavity. The results of this study will help in the design of new photonic devices and microsensors with high sensitivity.

Author Contributions: Conceptualization, Z.Z. and B.L.; methodology, X.Z. (Xiaolong Zhao); software, H.S.; validation, Y.L. investigation, J.Z.; data curation, B.L. and H.Z.; writing—original draft preparation, B.L. and H.S.; writing—review and editing, Z.Z., X.Z. (Xupeng Zhu) and X.C. All authors have read and agreed to the published version of the manuscript.

Funding: This work was funded by the Key Research and Development Program of Shanxi Province (No. 202102030201008), the National Natural Science Foundation of China (No. 12004150), the Fundamental Research Program of Shanxi Province (No. 201901D111148), and the Scientific and Technological Innovation Programs of Higher Education Institutions in Shanxi (No. 2020L0275).

Data Availability Statement: The data presented in this study are available upon request from the corresponding author. The data are not publicly available due to being supplied by Key Laboratory of Instrumentation Science & Dynamic Measurement (North University of China), Ministry of Education, and so cannot be made freely available.

Conflicts of Interest: The authors declare no conflict of interest.

References

1. Barnes, W.L.; Dereux, A.; Ebbesen, T.W. Surface plasmon subwavelength optics. *Nature* **2003**, *424*, 824–830. [CrossRef] [PubMed]
2. Ozbay, E. Plasmonics: Merging Photonics and Electronics at Nanoscale Dimensions. *Science* **2006**, *311*, 189–193. [CrossRef] [PubMed]
3. Fang, Y.; Sun, M. Nanoplasmonic waveguides: Towards applications in integrated nanophotonic circuits. *Light-Sci. Appl.* **2015**, *4*, e294. [CrossRef]
4. Chen, J.; Sun, C.; Rong, K.; Li, H.; Gong, Q. Polarization-free directional coupling of surface plasmon polaritons. *Light-Sci. Appl.* **2015**, *9*, 419–426. [CrossRef]
5. Zhang, X.; Liu, Z. Superlenses to overcome the diffraction limit. *Nat. Mater.* **2008**, *7*, 435–441. [CrossRef] [PubMed]
6. Green, M.A.; Pillai, S. Harnessing plasmonics for solar cells. *Nat. Photonics* **2012**, *6*, 130–132. [CrossRef]
7. Zhou, F.; Qin, F.; Yi, Z.; Yao, W.T.; Liu, Z.; Wu, X.; Wu, P. Ultra-wideband and wide-angle perfect solar energy absorber based on Ti nanorings surface plasmon resonance. *Phys. Chem. Chem. Phys.* **2021**, *23*, 17041–17048. [CrossRef]
8. Zhao, F.; Lin, J.; Lei, Z.; Yi, Z.; Qin, F.; Zhang, J.; Liu, L.; Wu, X.; Yang, W.; Wu, P. Realization of 18.97% theoretical efficiency of 0.9 μm Thick c-Si/ZnO Heterojunction Ultrathin-film Solar Cells via Surface Plasmon Resonance Enhancement. *Phys. Chem. Chem. Phys.* **2022**, *24*, 4871–4880. [CrossRef]
9. Joo, Y.H.; Song, S.H.; Magnusson, R. Demonstration of long-range surface plasmon-polariton waveguide sensors with asymmetric double-electrode structures. *Appl. Phys. Lett.* **2010**, *97*, 201105. [CrossRef]
10. Zhao, L.; Zhang, X.; Wang, J.; Yu, W.H.; Li, J.D.; Su, H.; Shen, X.P. A Novel Broadband Band-pass Filter Based on Spoof Surface Plasmon Polaritons. *Sci. Rep.* **2016**, *6*, 36069. [CrossRef]
11. Zaky, Z.A.; Singh, M.R.; Aly, A.H. Tamm resonance excited by different metals/graphene. *Photonic. Nanostruct.* **2022**, *49*, 100995. [CrossRef]
12. Zaky, Z.A.; Singh, M.R.; Aly, A.H. Tamm Plasmon Polariton as Refractive Index Sensor Excited by Gyroid Metals/Porous Ta_2O_5 Photonic Crystal. *Plasmonics* **2022**, *17*, 681–691. [CrossRef]
13. Li, L.; Hao, H.Y. Evolution of high-order Tamm plasmon modes with a metal-PhC cavity. *Sci. Rep.* **2022**, *12*, 14921. [CrossRef] [PubMed]
14. Han, Z. Ultracompact plasmonic racetrack resonators in metal-insulator-metal waveguides. *Photonic. Nanostruct.* **2010**, *8*, 172–176. [CrossRef]
15. Makarenko, K.S.; Hoang, T.X.; Duffin, T.J.; Radulescu, A.; Kalathingal, V.; Lezec, H.J.; Chu, H.S.; Nijhuis, C.A. Efficient Surface Plasmon Polariton Excitation and Control over Outcoupling Mechanisms in Metal–Insulator–Metal Tunneling Junctions. *Adv. Sci.* **2020**, *7*, 1900291. [CrossRef]
16. Kozma, P.; Kehl, F.; Ehrentreich-Forster, E.; Stamm, C.; Bier, F.F. Integrated planar optical waveguide interferometer biosensors: A comparative review. *Biosens. Bioelectron.* **2014**, *58*, 287–307. [CrossRef]

17. Tian, M.; Lu, P.; Chen, L.; Liu, D.M.; Lv, C. All-optical switching in MIM waveguide resonator with an outer portion smooth bend structure containing nonlinear optical materials. *Opt. Commun.* **2012**, *285*, 4562–4566. [CrossRef]
18. Pang, S.F.; Huo, Y.P.; Xie, Y.; Hao, L.M. Fano resonance in MIM waveguide structure with oblique rectangular cavity and its application in sensor. *Opt. Commun.* **2016**, *381*, 409–413. [CrossRef]
19. Wang, G.X.; Lu, H.; Liu, X.M. Dispersionless slow light in MIM waveguide based on a plasmonic analogue of electromagnetically induced transparency. *Opt Express* **2012**, *20*, 20902–20907. [CrossRef]
20. Wen, K.H.; Hu, Y.H.; Chen, L.; Zhou, J.Y.; He, M.; Lei, L.; Meng, Z.M.; Wu, Y.J.; Li, J.F. Fano Resonance Based on End-Coupled Cascaded-Ring MIM Waveguides Structure. *Plasmonics* **2017**, *12*, 1875–1880. [CrossRef]
21. Yun, B.F.; Zhang, R.H.; Hu, G.H.; Cui, Y.P. Ultra Sharp Fano Resonances Induced by Coupling between Plasmonic Stub and Circular Cavity Resonators. *Plasmonics* **2016**, *11*, 1157–1162.
22. Fano, U. Effects of configuration interaction on intensities and phase shifts. *Phys. Rev.* **1961**, *124*, 1866–1878. [CrossRef]
23. Khanikaev, A.B.; Wu, C.H.; Shvets, G. Fano-resonant metamaterials and their applications. *Nanophotonics* **2013**, *2*, 247–264. [CrossRef]
24. Limonov, M.F.; Rybin, M.V.; Poddubny, A.N.; Kivshar, Y.S. Fano resonances in photonics. *Nat. Photonics* **2017**, *11*, 543–554. [CrossRef]
25. Stern, L.; Grajower, M.; Levy, U. Fano resonances and all-optical switching in a resonantly coupled plasmonic–atomic system. *Nat. Commun.* **2014**, *5*, 4865. [CrossRef]
26. Deng, Z.L.; Dong, J.W. Lasing in plasmon-induced transparency nanocavity. *Opt. Express* **2013**, *21*, 20291–20302. [CrossRef]
27. Wu, C.H.; Khanikaev, A.B.; Adato, R.; Arju, N.; Yanik, A.A.; Altug, H.; Shvets, G. Fano-resonant asymmetric metamaterials for ultrasensitive spectroscopy and identification of molecular monolayers. *Nat. Mater.* **2012**, *11*, 69–75. [CrossRef]
28. Wu, C.H.; Khanikaev, A.B.; Shvets, G. Broadband Slow Light Metamaterial Based on a Double-Continuum Fano Resonance. *Phys. Rev. Lett.* **2011**, *106*, 107403. [CrossRef]
29. Zhang, F.; Hu, X.Y.; Zhu, Y.; Yang, H.; Gong, Q.H. Ultralow-power all-optical tunable dual Fano resonances in nonlinear metamaterials. *Appl. Phys. Lett.* **2013**, *103*, 191116. [CrossRef]
30. Deng, Y.; Cao, G.T.; Yang, H. Tunable Fano resonance and high-sensitivity sensor with high figure of merit in plasmonic coupled cavities. *Photonic. Nanostruct.* **2018**, *28*, 45–51. [CrossRef]
31. Zhang, Y.J.; Kuang, Y.Q.; Zhang, Z.D.; Tang, Y.; Han, J.Q.; Wang, R.B.; Cui, J.G.; Hou, Y.L.; Liu, W.Y. High-sensitivity refractive index sensors based on Fano resonance in the plasmonic system of splitting ring cavity-coupled MIM waveguide with tooth cavity. *App. Phys. A* **2019**, *125*, 13. [CrossRef]
32. Lotfiani, A.; Mohseni, S.M.; Ghanaatshoar, M. High-sensitive optoelectronic SPR biosensor based on Fano resonance in the integrated MIM junction and optical layers. *Opt. Commun.* **2020**, *477*, 126323. [CrossRef]
33. Deng, Y.; Cao, G.T.; Yang, H.; Li, G.H.; Chen, X.S.; Lu, W. Tunable and high-sensitivity sensing based on Fano resonance with coupled plasmonic cavities. *Sci. Rep.* **2017**, *7*, 10639. [CrossRef] [PubMed]
34. Lan, G.Q.; Liu, S.G.; Ma, Y.; Zhang, X.R.; Wang, Y.X.; Song, Y.L. Sensitivity and figure-of-merit enhancements of liquid-prism SPR sensor in the angular interrogation. *Opt. Commun.* **2015**, *352*, 49–54. [CrossRef]
35. Chen, H.; Chen, Z.; Yang, H.; Wen, L.; Yi, Z.; Zhou, Z.; Dai, B.; Zhang, J.; Wu, X.; Wu, P. Multi-mode surface plasmon resonance absorber based on dart-type single-layer graphene. *RSC Adv.* **2022**, *12*, 7821–7829. [CrossRef] [PubMed]
36. Zayats, A.V.; Smolyaninov, I.I.; Maradudin, A.A. Nano-optics of surface plasmon polaritons. *Phys. Rep.* **2005**, *408*, 131–314. [CrossRef]
37. Gai, H.; Wang, J.; Tian, Q. Modified Debye model parameters of metals applicable for broadband calculations. *Appl. Opt.* **2007**, *46*, 2229–2233. [CrossRef]
38. Zhang, Z.D.; Luo, L.; Xue, C.Y.; Zhang, W.D.; Yan, S.B. Fano Resonance Based on Metal-Insulator-Metal Waveguide-Coupled Double Rectangular Cavities for Plasmonic Nanosensors. *Sensors* **2016**, *16*, 642. [CrossRef]
39. Zhang, Z.Y.; Wang, J.D.; Zhao, Y.A.; Lu, D.; Xiong, Z.H. Numerical investigation of a branch-shaped filter based on metal-insulator-metal waveguide. *Plasmonics* **2011**, *6*, 773. [CrossRef]

micromachines

MDPI

Article

Additive-Manufactured Platinum Thin-Film Strain Gauges for Structural Microstrain Testing at Elevated Temperatures

Xiaochuan Pan †, Fan Lin †, Chao Wu, Yingjun Zeng, Guochun Chen, Qinnan Chen *, Daoheng Sun * and Zhenyin Hai *

Department of Mechanical and Electrical Engineering, School of Aerospace Engineering, Xiamen University, Xiamen 361005, China
* Correspondence: chenqinnan@xmu.edu.cn (Q.C.); sundh@xmu.edu.cn (D.S.); haizhenyin@xmu.edu.cn (Z.H.)
† These authors contributed equally to this work.

Abstract: This paper investigates the feasibility and performance of the fabrication of platinum high-temperature thin-film strain sensors on nickel-based alloy substrates by additive manufacturing. The insulating layer was made of a dielectric paste by screen printing process. A 1.8-micron-thick platinum film was deposited directly on the insulating layer. The four-wire resistance measurement method was used to eliminate the contact resistance of the solder joints. Comprehensive morphological and electrical characterization of the platinum thin-film strain gauge was carried out, and good static and dynamic strain responses were obtained, which confirmed that the strain gauge was suitable for in situ strain monitoring of high-temperature complex components.

Keywords: thin film strain gauge; additive manufacturing; thin film; high temperature

Citation: Pan, X.; Lin, F.; Wu, C.; Zeng, Y.; Chen, G.; Chen, Q.; Sun, D.; Hai, Z. Additive-Manufactured Platinum Thin-Film Strain Gauges for Structural Microstrain Testing at Elevated Temperatures. *Micromachines* **2022**, *13*, 1472. https://doi.org/10.3390/mi13091472

Academic Editors: Libo Gao and Zhuoqing Yang

Received: 9 August 2022
Accepted: 28 August 2022
Published: 5 September 2022

1. Introduction

As one of the most widely used sensors, strain gauges can be used to detect load, pressure, strain, and torque [1–3] and have a wide range of applications in structural health monitoring, robotics, electronic skin, and other fields. Strain gauges also have a wide range of applications in flexible electronics, where they can be used to monitor human respiration, movement, and other factors [4,5]. High-temperature thin-film strain gauges (TFSGs) have been widely used to monitor the in situ strain of hot components [6–11]. TFSGs are devices capable of measuring strain on the surface of structural components by converting mechanical deformation into electrical signals [12]. They are typically only a few microns thick and can be deposited directly on the components, which work in high temperatures. They do not damage the structure under test, hardly interfere with the flow field characteristics of the surface under test, and due to their negligible mass, they have minimal influence on the inertial vibration of the structure [13,14]. The thin-film sensor has in situ sensing properties, so no bonding process is required, which greatly reduces creep error. In addition, the strain transferability and response speed of thin-film sensors are particularly good, so they are more suitable for dynamic strain testing [9,15,16]. As a high-temperature strain-sensitive material with great research value, metal/alloy materials, such as PdCr and NiCr, show desired linear resistance characteristics versus temperature, which is very important for TFSGs temperature compensation and improving their test accuracy [17]. Metal foil TFSGs are mainly fabricated by sputtering, vapor deposition, and other conditions [17,18]. These coating techniques are difficult to apply to large structures due to the limited size of the processing chamber. In addition, these processes deposit thin films on curved surfaces, resulting in uneven film thickness [6]. Pin-coating and screen-printing, among others, are low-cost and easy-to-implement thin-film processes, but different screens need to be prepared for different patterns. It means their printing flexibility are limited, and less economical for small batch production and multi-step processes [19–22]. Lithography-based techniques are another traditional method of thin-film sensor fabrication. Although lithographic patterns provide high structural resolution quality, their conformality is insufficient for curved surfaces [12]. Inkjet printing, aerosol jet printing, and direct ink writing (DIW) enable patterning with line widths of hundreds of microns and thicknesses of several microns [1,6,23–26]. Based on the diversity of printing structures and material choices, DIW provides an alternative fabrication method for the rapid fabrication of thin-film sensors. This method does not require expensive cleanroom facilities and high-end equipment and is proven to be more robust and flexible micro fabrication process [27]. Additionally, this technique enables patterning on three-dimensional surfaces with high structural resolution [28–31]. Platinum (Pt) is widely known for its high thermal and chemical stability, so it is commonly used in micromechanical devices that operate in harsh environments, such as thermal flow sensors [32] and thermistor temperature sensors [7]. In this study, Pt TFSG was fabricated via DIW based on the Weisenberg's effect. The DIW technique based on the Weisenberg's effect is considered one of the most promising patterning methods for solutions with rheological properties [23]. Under the shear thinning effect of the high-speed rotating microneedles, the solution is rapidly transported to the printing needle. Pt TFSG with relatively stable electrical properties was prepared by DIW on GH3536 nickel-based superalloy. The static and dynamic strain responses of Pt TFSG were investigated and tested from 25 °C to 800 °C.

2. Materials and Methods

2.1. Materials

A high-temperature nickel-based alloy (GH3536, Shenzhen Baishun Metal Materials Co., Ltd., Shenzhen, China), which is commonly used in engineering, was selected as the substrate. A dielectric paste (07HGT90, Shenzhen Sryeo Electronic Paste Co., Ltd., Shenzhen, China) (silicon aluminum calcium oxide) was used as the insulating layer. Pt paste was used for the DIW and silver palladium paste (55H-1805, Shenzhen Sryeo Electronic Paste Co., Ltd., Shenzhen, China) was used to connect the strain grid and Pt leads.

2.2. Strain Gauge Fabrication

A schematic diagram of the Pt TFSG fabrication process is shown in Figure 1. First, the GH3536 nickel-based alloy substrate was ultrasonically cleaned sequentially with acetone, deionized water, and alcohol. The dielectric paste was then screen printed on a nickel-based alloy and annealed at 850 °C for 0.5 h. The Pt paste was diluted with butyldiglycol (9:1 mass ratio), followed by patterning using a DIW platform based on the Weissenberg effect. To eliminate the influence of solder joints and leads during the test, a four-wire structure was used. Silver palladium paste was used to connect the Pt strain grid and Pt leads. Finally, the Pt TFSG was sintered at 850 °C for 1 h in an air atmosphere.

(a)

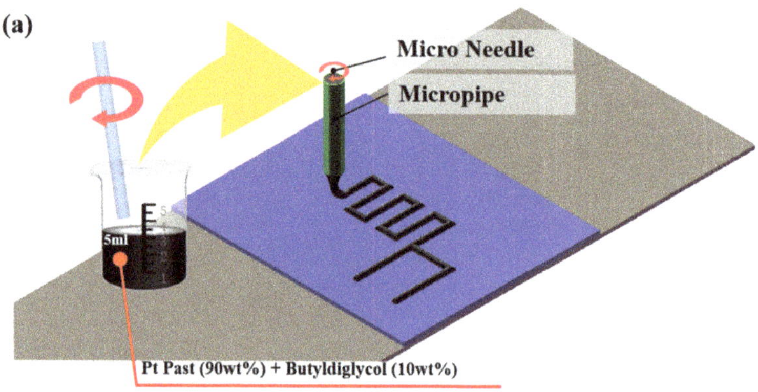

Figure 1. *Cont.*

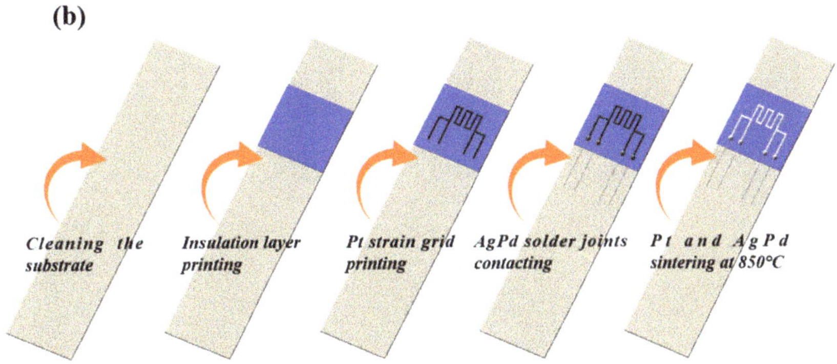

Figure 1. (**a**) Schematic diagram of direct ink writing (DIW) process of Pt paste. (**b**) Process flow diagram of the Pt thin film strain gauge.

2.3. Experiment Setup

The strain response of the Pt TFSG was investigated using the cantilever beam method [6], as shown in Figure 2b.The strain of the cantilever is controlled by applying displacement at the free end by a stepper motor. The value of strain ε was calculated using the following equation [33]:

$$\varepsilon = \frac{3xyh}{2l^3} \tag{1}$$

where ε is the strain developed at the location of the TFSGs, y is the deflection of the free end of the substrate, h is the thickness of the substrate, x is the distance from the center of the strain gauge to the point where the load is applied, and l is the length of the substrate.

In this work, one end of the substrate of the strain gauge was clamped and fixed, and the other end was controlled by a stepper motor with an accuracy of 1 micron. By adjusting the stepping distance (y) of the stepper motor, because other parameters (x, h, and l) can be measured with a ruler, the strain on the strain gauge can be calculated.

The indicator of strain gauge sensitivity, also known as the gauge factor (GF), and can be expressed by the following equation:

$$\mathrm{GF} = \frac{\Delta R/R}{\varepsilon} \tag{2}$$

where R is the initial resistance of the TFSG and ΔR is the change in the resistance due to the strain. The high temperature resistance stability of Pt TFSG was tested using a temperature resistance test system, which consists of a tube furnace, a standard thermocouple, data acquisition instruments and a computer (Figure 2a).

The high temperature stability experiment was realized by the temperature resistance test system. The Pt TFSG was placed in the tube furnace, and the tube furnace was heated from room temperature to a specified temperature (400 °C, 500 °C, 600 °C, 700 °C, and 800 °C) at a heating rate of 5 °C/min and then kept for 1 hour, and its high temperature stability was judged by calculating the resistance change rate of the strain gauge.

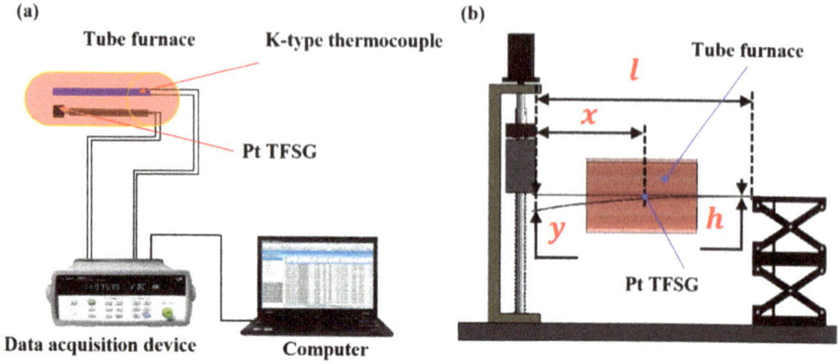

Figure 2. Experiment setups for (**a**) temperature-resistance detection and (**b**) micro-strain detection.

Compared with the normal temperature strain test system, the high temperature strain test system added a tube furnace. The tube furnace was heated to the specified temperature (400 °C, 500 °C, 600 °C, and 700 °C) at a heating rate of 5 °C/min and maintained for more than half an hour to establish thermal equilibrium. A stepper motor was used to displace the free end of the substrate (applied y) during thermal equilibration, enabling high temperature strain testing.

2.4. Characterisation Instruments

The thicknesses of Pt TFSG were measured by a profilometer (Dektak XT). The surface morphology of platinum film was characterized by scanning electron microscopy (SEM, JSM-IT500A). The elements of Pt TFSG were analyzed with energy dispersive spectroscopy (EDS).

3. Results and Discussion

3.1. Characterizations of the Pt TFSGs

Figure 3a shows the optical image of the printed Pt TFSG. The direct writing process was affected by parameters such as ink viscosity, substrate surface roughness, substrate moving speed, and microneedle rotation speed, so the line width and film thickness of the fabricated samples were different, resulting in different resistances between samples. The resistance of 10 printed platinum sensors was measured at room temperature of 25 °C, and the average resistance was calculated to be 70.5 Ω, the maximum resistance was 75.7 Ω, and the minimum resistance was 65.3 Ω. The Pt TFSG selected for electrical characterization had a resistance of 71.8 Ω at room temperature. The thickness of the strain grid was 1.8 μm and the line width was 620 μm (Figure 3b). Figure 3c shows the micromorphology of the Pt TFSGs. The continuous, uniforms and dense structure ensured that the sensitive film had excellent strain response and electrical conductivity. The EDS analysis (Figure 3d) proved that the main component of the film is Pt.

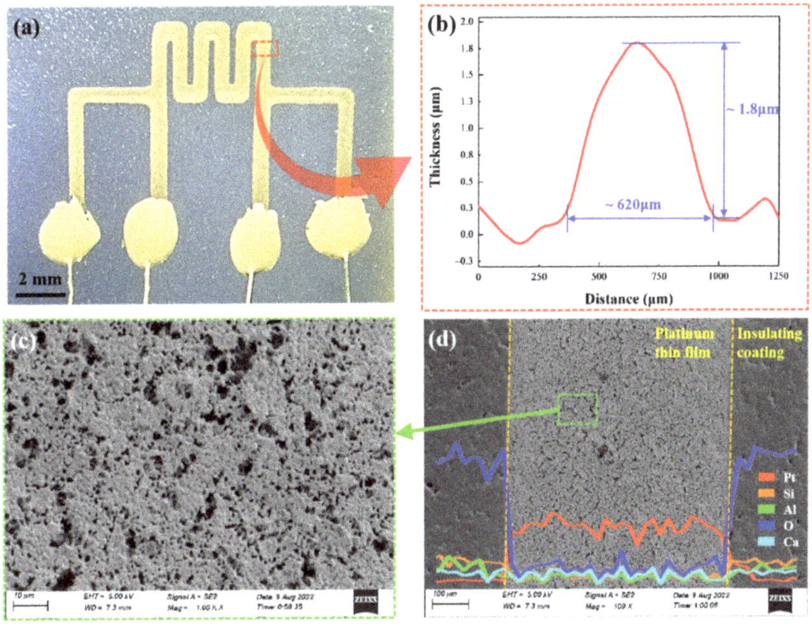

Figure 3. (**a**) Optical image of a printed Pt thin-film strain gauge. (**b**)Thickness of the Pt TFSG. (**c**) SEM images of the Pt sensitive thin film. (**d**) EDS analysis of the Pt TFSG trace.

3.2. Strain Testing at Room Temperature

To assess the applicability of Pt as a resistive sensing material, the piezoresistive response characteristics of Pt TFSG were first evaluated by applying different strains at room temperature. Figure 4a shows a plot of the resistance change of Pt TFSG in the strain range of 0 με to 1000 με; 200 με, 400 με, 600 με, 800 με, and 1000 με were sequentially applied to TFSG and kept for 10 s and then unloaded in sequence with a strain amount of 200 με. A recoverable resistance change was clearly observed throughout the loading and unloading process. The distinguishable and stable step signal indicated the excellent static strain response capability of Pt TFSG. The Pt TFSG exhibits a positive GF with an increase in resistance with applied positive strain. Figure 4b depicts the pulsed signals under different strain amounts at a constant strain rate (200 με/s). Resolvable resistance changes indicated the good response of Pt TFSG to strain. Figure 4c shows the dynamic strain response of Pt TFSG at 1000 με. The maximum and minimum values of the pulse signal were almost kept at the same level, indicating the good consistency in dynamic strain response of Pt TFSG. Figure 4d shows the independence of Pt TFSG on strain velocity (strain frequency). Then, 1000 με were applied at strain rates of 20 με/s, 50 με/s, 100 με/s, 200 με/s, and 500 με/s. Despite the different strain velocities, the resistances of Pt TFSG were almost uniform at 1000 με, indicating that Pt TFSG was strain-rate independent.

Figure 5a–d shows the strain response of Pt TFSG under negative strain (compressive strain). Similarly, Pt TFSG exhibited good stability under negative static strain. Under negative dynamic strain, the Pt TFSG exhibited a consistent strain pulse signal. Under negative variable-speed strain signals, Pt TFSG exhibited strain-rate independence. This indicated that Pt TFSG prepared by DIW can be used for dynamic/static strain monitoring and sensing of structural components.

The gauge factor of Pt TFSG could be calculated from the ratio of resistance change rate to strain. In the Figure 6, the slope of the straight line was the GF of Pt TFSG. By calculation, its gauge factor was 1.94. This value was a comprehensive representation of the strain at the Pt TFSG, including the dimensional change and the resistivity change of the strain gauge when it was strained, and the ability of the strain to transfer from the nickel-based alloy substrate to the metal layer was also included [12]. The GF is reported to be in the range of 1.95–2.5 for most metal/alloy strain sensors [12,34,35].

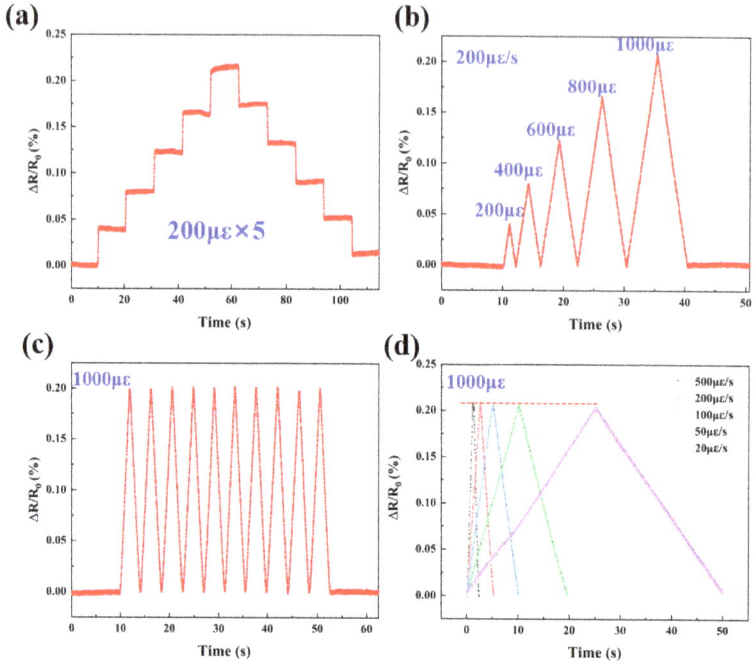

Figure 4. (**a**–**d**) are tensile strain response of Pt TFSGs.

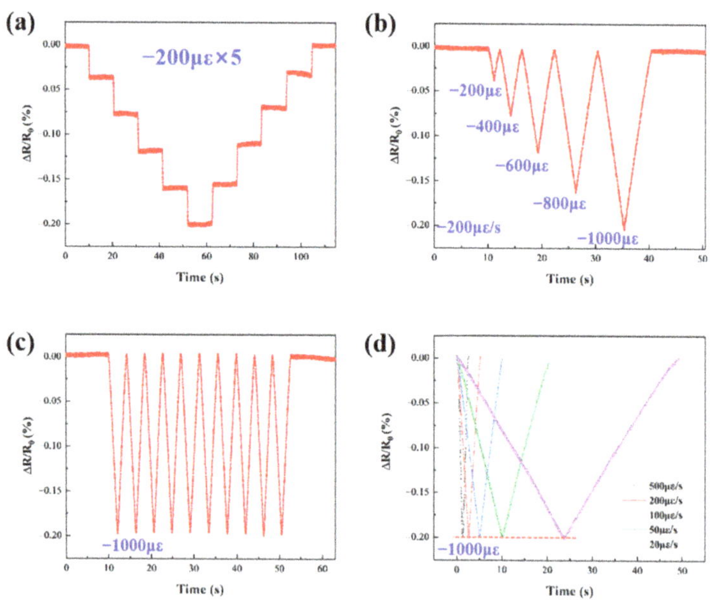

Figure 5. (**a**–**d**) are compressive strain response of Pt TFSGs.

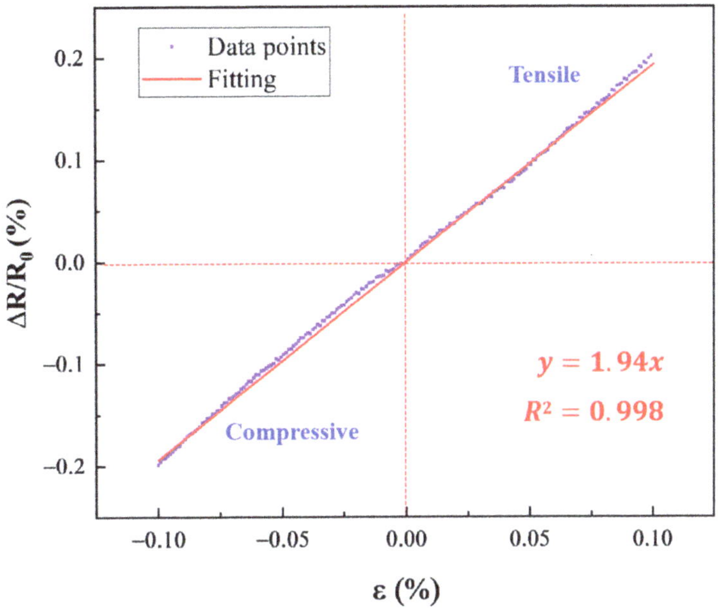

Figure 6. Linear fitting of tensile and compressive strain.

3.3. Strain Testing at High Temperature

High temperature resistance stability is also an important capability of TFSG. The resistance as a function of heating and holding time is plotted in Figure 7. For the heating stage, the resistance of Pt TFSG increased with the increase in temperature, showing a positive temperature coefficient of resistance. For the holding stage below 700 °C, there was little change in resistance. For example, at 700 °C for 1 h, the resistance only increased by 0.017%. Since the GF of Pt TFSG was 1.94, the strain error caused by resistance drift at 700 °C for 1 h was only 87 με, indicating that Pt TFSG has good thermal stability. After holding at 800 °C for 1h, the resistance of Pt TFSG increased by 8.2%, and the increase of resistance of Pt film was related to high temperature oxidation [7]. All the above test results demonstrate that the Pt TFSG fabricated by DIW had good thermal stability in the temperature range from room temperature to 700 °C, and the thermal stability at high temperature was the first design principle of high temperature TFSG [6].

The strain responses of Pt TFSGs were performed at 400 °C, 500 °C, 600 °C, and 700 °C using a high-temperature strain testing system. As shown in Figure 8, a strain of 1000 με was applied at a frequency of 0.25 Hz at high temperature and the change in resistance was recorded. It can be observed from the figure that the GF of Pt TFSG also did not decrease significantly up to 600 °C, which was about 1.9. When the temperature was increased to 700 °C, the GF of the TFSG dropped to 1.7. The synchronized changes in relative resistance at the four temperatures indicated the fast response of the Pt TFSG. The relative resistance change is positively linearly related to the strain. Although its high temperature dynamic strain response is not as good as room temperature, it can be seen from the temperature curve in the figure that the resistance fluctuation was mainly caused by temperature fluctuation. This confirms that the DIW-prepared Pt TFSG can be used for dynamic strain monitoring of high temperature up to 700 °C. As shown in Table 1, its high-temperature withstanding capability was better than most TFSGs that have been reported [16,17,36–42].

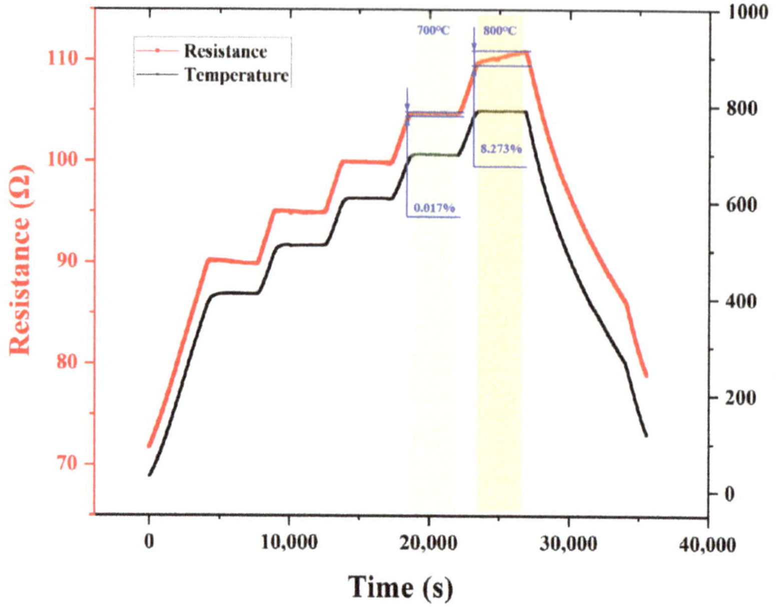

Figure 7. Resistance changes curves of Pt TFSGs at different temperatures.

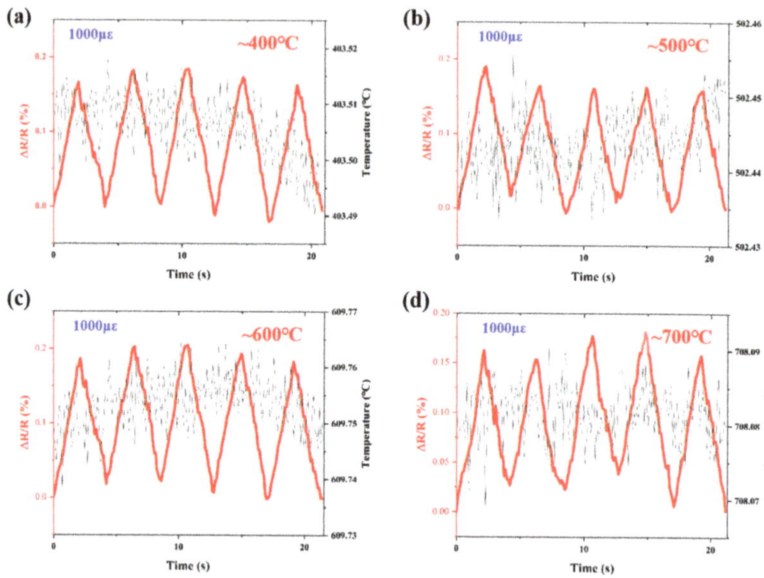

Figure 8. *Cont.*

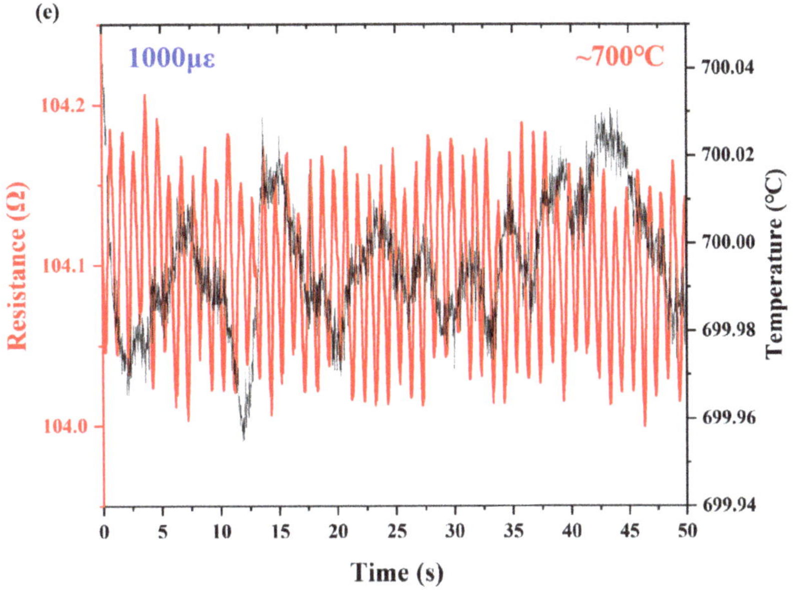

Figure 8. Strain response of Pt TFSGs: (**a**) 400 °C, 5 cycles; (**b**) 500 °C, 5 cycles; (**c**) 600 °C, 5 cycles; (**d**) 700 °C, 5 cycles; (**e**) 700 °C, 50 cycles.

Table 1. Summary of recently released strain gauges.

Material	Maximum Temperature (°C)	Manufacturing Method	GF	Substrate	Reference
Pt	700	DIW	1.7–1.9	Ni-base superalloy	This work
$TiAlN_xO$	500	Puttering	2.2–2.5	Sapphire	[37]
AlN/Pt	500	Sputtering	≤4.7	Al_2O_3	[16]
Pt	500	Sputtering	1.9–2.5	Sapphire	[38]
Pt/SiO_2	250	Sputtering	18	Si-wafers	[39]
Pt	440	Sputtering	-	Sapphire	[42]
TiAlN	350	Sputtering	2.5	Sapphire	[36]
Invar36	150	Sputtering	2.5–4.5	Microslides	[40]
NiCr	700 (with a protective layer)	Sputtering	2.5	Ni-base superalloy	[41]

4. Conclusions

In this paper,

- By DIW technology based on Weisenberg's effect, Pt TFSG with thickness less than 2 μm were fabricated.
- The strain response of the Pt TFSG at room temperature was well verified with a GF of 1.9 at room temperature.
- The high temperature stability experiment of the strain gauge showed that its maximum working temperature could reach 700 °C, and the resistance change rate for one hour at this temperature was only 0.017%.
- The strain test at high temperature showed that the GF of the Pt TFSG was almost unchanged when it was less than 600 °C and decreased to about 1.7 at 700 °C.

Therefore, it can be considered that this strain gauge can be applied to high temperature strain tests below 700 °C. If the printing device for direct writing is integrated with the five-axis processing equipment, it is possible to manufacture thin films on complex curved surfaces, and future work will also focus on this.

Author Contributions: Conceptualization, X.P.; methodology, X.P. and F.L.; software, X.P.; validation, C.W. and F.L.; formal analysis, F.L.; investigation, F.L.; resources, C.W.; data curation, Y.Z.; writing—original draft preparation, X.P.; writing—review and editing, Z.H.; visualization, G.C.; supervision, Z.H.; project administration, Q.C.; funding acquisition, D.S. All authors have read and agreed to the published version of the manuscript.

Funding: This research was funded by China Aviation Research Institute, Shenyang Engine Design and Research Institute, grant number JC3602007026.

Institutional Review Board Statement: Not applicable.

Informed Consent Statement: Not applicable.

Data Availability Statement: Not applicable.

Acknowledgments: The authors would like to thank the China Aviation Research Institute, Shenyang Engine Design and Research Institute (Grant number JC3602007026).

Conflicts of Interest: The authors declare no conflict of interest.

References

1. Zhang, Y.Z.; Anderson, N.; Bland, S.; Nutt, S.; Jursich, G. All-printed strain sensors: Building blocks of the aircraft structural health monitoring system. *Sens. Actuators A Phys.* **2017**, *253*, 165–172. [CrossRef]
2. Liao, X.Q.; Liao, Q.L.; Yan, X.Q.; Liang, Q.J.; Si, H.N.; Li, M.H.; Wu, H.L.; Cao, S.Y.; Zhang, Y. Flexible and Highly Sensitive Strain Sensors Fabricated by Pencil Drawn for Wearable Monitor. *Adv. Funct. Mater.* **2015**, *25*, 2395–2401. [CrossRef]
3. Siddique, S.; Park, J.G.; Andrei, P.; Liang, R. M3D aerosol jet printed buckypaper multifunctional sensors for composite structural health monitoring. *Results Phys.* **2019**, *13*, 102094. [CrossRef]
4. Jin, H.; Abu-Raya, Y.S.; Haick, H. Advanced Materials for Health Monitoring with Skin-Based Wearable Devices. *Adv. Healthc. Mater.* **2017**, *6*, 1700024. [CrossRef]
5. Augustin, T.; Karsten, J.; Kotter, B.; Fiedler, B. Health monitoring of scarfed CFRP joints under cyclic loading via electrical resistance measurements using carbon nanotube modified adhesive films. *Compos. Part A Appl. Sci. Manuf.* **2018**, *105*, 150–155. [CrossRef]
6. Wu, C.; Pan, X.C.; Lin, F.; Cui, Z.F.; He, Y.P.; Chen, G.C.; Zeng, Y.J.; Liu, X.L.; Chen, Q.N.; Sun, D.H.; et al. TiB2/SiCN Thin-Film Strain Gauges Fabricated by Direct Writing for High-Temperature Application. *IEEE Sens. J.* **2022**, *22*, 11517–11525. [CrossRef]
7. Shen, A.; Kim, S.B.; Bailey, C.; Ma, A.W.K.; Dardona, S. Direct Write Fabrication of Platinum-Based Thick-Film Resistive Temperature Detectors. *IEEE Sens. J.* **2018**, *18*, 9105–9111. [CrossRef]
8. Wu, C.; Lin, F.; Pan, X.C.; Cui, Z.F.; He, Y.P.; Chen, G.C.; Liu, X.L.; He, G.H.; Chen, Q.N.; Sun, D.H.; et al. TiB2-Modified Polymer-Derived Ceramic SiCN Double-Layer Thin Films Fabricated by Direct Writing for High-Temperature Application. *Adv. Eng. Mater.* **2022**, 2200228. [CrossRef]
9. Wang, H.Y.; Liu, K.; An, Z.H.; Wu, X.; Huang, X. Ion-Beam Sputtered Thin-Film Strain-Gauge Pressure Transducers. *Sens. Actuators A Phys.* **1993**, *35*, 265–268.
10. Chung, G.S. Characteristics of tantalum nitride thin film strain gauges for harsh environments. *Sens. Actuators A Phys.* **2007**, *135*, 355–359. [CrossRef]
11. Gregory, O.J.; You, T. Piezoresistive properties of ITO strain sensors prepared with controlled nanoporosity. *J. Electrochem. Soc.* **2004**, *151*, H198–H203. [CrossRef]
12. Mao, N.Y.; Enrique, P.D.; Chen, A.I.H.; Zhou, N.Y.; Peng, P. Dynamic response and failure mechanisms of a laser-fabricated flexible thin film strain gauge. *Sens. Actuators A Phys.* **2022**, *342*, 113655. [CrossRef]
13. Koppelhuber, A.; Bimber, O. LumiConSense A Transparent, Flexible, and Scalable Thin-Film Sensor. *IEEE Comput. Graph.* **2014**, *34*, 98–102. [CrossRef]
14. Yan, L.T.; Si, W.J.; Liu, L.Y.; Zheng, Y.P.; Dai, C.N. Development of nanostructured thin film sensors. *Rare Metal. Mat. Eng.* **2006**, *35*, 21–24.
15. Nayak, M.M.; Gunasekaran, N.; Muthunayagam, A.E.; Rajanna, K.; Mohan, S. Diaphragm-Type Sputtered Platinum Thin-Film Strain-Gauge Pressure Transducer. *Meas. Sci. Technol.* **1993**, *4*, 1319–1322. [CrossRef]
16. Schmid, P.; Triendl, F.; Zarfl, C.; Schwarz, S.; Artner, W.; Schneider, M.; Schmid, U. Influence of the AlN/Pt-ratio on the electromechanical properties of multilayered AlN/Pt thin film strain gauges at high temperatures. *Sens. Actuators A Phys.* **2020**, *302*, 111805. [CrossRef]
17. Liu, H.; Mao, X.L.; Yang, Z.B.; Cui, J.T.; Jiang, S.W.; Zhang, W.L. High temperature static and dynamic strain response of PdCr thin film strain gauge prepared on Ni-based superalloy. *Sens. Actuators A Phys.* **2019**, *298*, 111571. [CrossRef]
18. Hans, M.; Baben, M.T.; Music, D.; Ebenhoch, J.; Primetzhofer, D.; Kurapov, D.; Arndt, M.; Rudigier, H.; Schneider, J.M. Effect of oxygen incorporation on the structure and elasticity of Ti-Al-O-N coatings synthesized by cathodic arc and high power pulsed magnetron sputtering. *J. Appl. Phys.* **2014**, *116*, 093515. [CrossRef]

19. Bose, A.K.; Zhang, X.Z.; Maddipatla, D.; Masihi, S.; Panahi, M.; Narakathu, B.B.; Bazuin, B.J.; Williams, J.D.; Mitchell, M.F.; Atashbar, M.Z. Screen-Printed Strain Gauge for Micro-Strain Detection Applications. *IEEE Sens. J.* **2020**, *20*, 12652–12660. [CrossRef]
20. Fasolt, B.; Hodgins, M.; Rizzello, G.; Seelecke, S. Effect of screen printing parameters on sensor and actuator performance of dielectric elastomer (DE) membranes. *Sens. Actuators A Phys.* **2017**, *265*, 10–19. [CrossRef]
21. Qi, X.; Lim, S. A Screen-Printed Metal Hybrid Composite for Wireless Wind Sensing. *Nanomaterials* **2022**, *12*, 972. [CrossRef] [PubMed]
22. Han, T.; Kundu, S.; Nag, A.; Xu, Y.Z. 3D Printed Sensors for Biomedical Applications: A Review. *Sensors* **2019**, *19*, 1076. [CrossRef] [PubMed]
23. Wu, C.; Pan, X.C.; Lin, F.; Cui, Z.F.; Li, X.; Chen, G.C.; Liu, X.L.; He, Y.P.; He, G.H.; Hai, Z.Y.; et al. High-temperature electrical properties of polymer-derived ceramic SiBCN thin films fabricated by direct writing. *Ceram. Int.* **2022**, *48*, 15293–15302. [CrossRef]
24. Karas, B.; Beedasy, V.; Leong, Z.; Morley, N.A.; Mumtaz, K.; Smith, P.J. Integrated Fabrication of Novel Inkjet-Printed Silver Nanoparticle Sensors on Carbon Fiber Reinforced Nylon Composites. *Micromachines* **2021**, *12*, 1185. [CrossRef] [PubMed]
25. Zhou, P.Y.; Liao, Y.Z.; Li, Y.H.; Pan, D.Y.; Cao, W.X.; Yang, X.B.; Zou, F.X.; Zhou, L.M.; Zhang, Z.; Su, Z.Q. An inkjet-printed, flexible, ultra-broadband nanocomposite film sensor for in-situ acquisition of high-frequency dynamic strains. *Compos. Part A Appl. Sci. Manuf.* **2019**, *125*, 105554. [CrossRef]
26. Correia, V.; Caparros, C.; Casellas, C.; Francesch, L.; Rocha, J.G.; Lanceros-Mendez, S. Development of inkjet printed strain sensors. *Smart Mater. Struct.* **2013**, *22*, 105028. [CrossRef]
27. Guo, Z.Y.; Yu, P.S.; Liu, Y.; Zhao, J.H. High-precision resistance strain sensors of multilayer composite structure via direct ink writing: Optimized layer flatness and interfacial strength. *Compos. Sci. Technol.* **2021**, *201*, 108530. [CrossRef]
28. Sullivan, K.T.; Zhu, C.; Tanaka, D.J.; Kuntz, J.D.; Duoss, E.B.; Gash, A.E. Electrophoretic Deposition of Thermites onto Micro-Engineered Electrodes Prepared by Direct-Ink Writing. *J. Phys. Chem. B* **2013**, *117*, 1686–1693. [CrossRef] [PubMed]
29. Tian, X.C. Direct ink writing of 2D material-based supercapacitors. *2D Mater.* **2022**, *9*, 012001. [CrossRef]
30. Hao, L.; Tang, D.N.; Sun, T.; Xiong, W.; Feng, Z.Y.; Evans, K.E.; Li, Y. Direct Ink Writing of Mineral Materials: A review. *Int. J. Pr. Eng. Man.-Gt.* **2021**, *8*, 665–685. [CrossRef]
31. Lewis, J.A.; Smay, J.E.; Stuecker, J.; Cesarano, J. Direct ink writing of three-dimensional ceramic structures. *J. Am. Ceram. Soc.* **2006**, *89*, 3599–3609. [CrossRef]
32. Kuo, J.T.W.; Yu, L.; Meng, E. Micromachined Thermal Flow Sensors-A Review. *Micromachines* **2012**, *3*, 550–573. [CrossRef]
33. Ren, S.; Jiang, S.W.; Liu, H.; Zhang, W.L.; Li, Y.R. Investigation of strain gauges based on interdigitated Ba0.5Sr0.5TiO3 thin film capacitors. *Sens. Actuators A Phys.* **2015**, *236*, 159–163. [CrossRef]
34. Kazi, I.H.; Wild, P.M.; Moore, T.N.; Sayer, M. The electromechanical behavior of nichrome (80/20 wt.%) film. *Thin Solid Film.* **2003**, *433*, 337–343. [CrossRef]
35. Liu, H.; Mao, X.L.; Cui, J.T.; Jiang, S.W.; Zhao, X.H.; Jiang, H.C.; Zhang, W.L. Effect of thickness on the electrical properties of PdCr strain sensitive thin film. *J. Mater. Sci.-Mater. El.* **2019**, *30*, 10475–10482. [CrossRef]
36. Zarfl, C.; Schwab, S.; Schmid, P.; Hutter, H.; Schmid, U. Influence of the sputter gas composition on the electromechanical properties and on the stability of TiAlNxO1-x thin films. *Sens. Actuators A Phys.* **2017**, *267*, 552–559. [CrossRef]
37. Zarfl, C.; Schmid, P.; Balogh, G.; Schmid, U. Electro-mechanical properties and oxidation behaviour of TiAlNxOy thin films at high temperatures. *Sens. Actuators A Phys.* **2015**, *226*, 143–148. [CrossRef]
38. Schmid, P.; Triendl, F.; Zarfl, C.; Schwarz, S.; Artner, W.; Schneider, M.; Schmid, U. Electro-mechanical properties of multilayered aluminum nitride and platinum thin films at high temperatures. *Sens. Actuators A Phys.* **2019**, *293*, 128–135. [CrossRef]
39. Schmid-Engel, H.; Uhlig, S.; Werner, U.; Schultes, G. Strain sensitive Pt-SiO2 nano-cermet thin films for high temperature pressure and force sensors. *Sens. Actuators A Phys.* **2014**, *206*, 17–21. [CrossRef]
40. Kalpana, H.M.; Prasad, V.S. Development of the invar36 thin film strain gauge sensor for strain measurement. *Meas. Sci. Technol.* **2014**, *25*, 065102. [CrossRef]
41. Kayser, P.; Godefroy, J.C.; Leca, L. High-Temperature Thin-Film Strain-Gauges. *Sens. Actuators A Phys.* **1993**, *37*, 328–332. [CrossRef]
42. Fricke, S.; Friedberger, A.; Seidel, H.; Schmid, U. A robust pressure sensor for harsh environmental applications. *Sens. Actuators A Phys.* **2012**, *184*, 16–21. [CrossRef]

micromachines

Article

ZrB_2/SiCN Thin-Film Strain Gauges for In-Situ Strain Detection of Hot Components

Fan Lin †, Xiaochuan Pan †, Chao Wu, Yingjun Zeng, Guochun Chen, Qinnan Chen *, Daoheng Sun * and Zhenyin Hai *

Department of Mechanical and Electrical Engineering, School of Aerospace Engineering, Xiamen University, Xiamen 361005, China
* Correspondence: chenqinnan@xmu.edu.cn (Q.C.); sundh@xmu.edu.cn (D.S.); haizhenyin@xmu.edu.cn (Z.H.)
† These authors contributed equally to this work.

Abstract: The in-situ strain/stress detection of hot components in harsh environments remains a challenging task. In this study, ZrB_2/SiCN thin-film strain gauges were fabricated on alumina substrates by direct writing. The effects of ZrB_2 content on the electrical conductivity and strain sensitivity of ZrB_2/SiCN composites were investigated, and based on these, thin film strain gauges with high electrical conductivity (1.71 S/cm) and a gauge factor of 4.8 were prepared. ZrB_2/SiCN thin-film strain gauges exhibit excellent static, cyclic strain responses and resistance stability at room temperature. In order to verify the high temperature performance of the ZrB_2/SiCN thin-film strain gauges, the temperature-resistance characteristic curves test, high temperature resistance stability test and cyclic strain test were conducted from 25 °C to 600 °C. ZrB_2/SiCN thin-film strain gauges exhibit good resistance repeatability and stability, and highly sensitive strain response, from 25 °C to 600 °C. Therefore, ZrB_2/SiCN thin-film strain gauges provide an effective approach for the measurement of in-situ strain of hot components in harsh environments.

Keywords: thin film strain gauge; direct ink writing; polymer-derived ceramics; conductive composites; high temperature

Citation: Lin, F.; Pan, X.; Wu, C.; Zeng, Y.; Chen, G.; Chen, Q.; Sun, D.; Hai, Z. ZrB_2/SiCN Thin-Film Strain Gauges for In-Situ Strain Detection of Hot Components. *Micromachines* **2022**, *13*, 1467. https://doi.org/10.3390/mi13091467

Academic Editors: Libo Gao and Zhuoqing Yang

Received: 10 August 2022
Accepted: 27 August 2022
Published: 4 September 2022

Publisher's Note: MDPI stays neutral with regard to jurisdictional claims in published maps and institutional affiliations.

1. Introduction

Thin film strain gauges (TFSGs) are widely used for in-situ strain detection of various components and structures in the aerospace, transportation and automobile industries, civil engineering and even the medical field due to their advantages of non-interference, small size, fast response and in-situ integration [1–7]. TFSGs are mainly fabricated by depositing alloy/metal films such as NiCr, PdCr and TaN-Cu on the surface of components [1,8,9]. An effective TFSG must exhibit a number of appropriate properties (e.g., adequate operating range, reasonable conductivity, lack of frequency dependence), probably the most important property being the sensitivity or gauge factor (GF) [10]. However, the gauge factor (GF) of traditional metal/alloy iso piezoresistive materials is about two, resulting in low sensitivity and difficulty in detecting tiny strains.

To solve this issue, many researchers have turned to conductive composites. Conductive composites contain an insulating matrix and conductive nanoparticles dispersed therein, and the conduction mechanism and the strain sensitivity are primarily dominated by the tunneling effect [10,11]. Flexible sensors based on conductive polymer composites have stretchability and strain factors that far surpass foil strain gauges, and are widely used in wearable devices, electronic skins and human motion detection, etc. [12–17]. However, most TFSGs based on conductive composites are limited to room temperature. They are not thermally stable at high temperatures.

To construct thermally stable conductive composites with a highly sensitive piezoresistive response, the high-temperature thermal stability of the insulating matrix and the conductive phase is the first design principle. Compared to metallic and polymeric materials, most ceramics, such as polymer-derived ceramics (PDCs) pyrolyzed at high temperature are electrically insulating and thermally stable at high temperatures [18]. Existing PDCs high temperature sensors such as temperature sensors and pressure sensors are still dominated by discrete bulk devices [19–21]. Compared with larger discrete devices, in-situ integrated TFSGs with a thickness of micrometers exhibit reproducible faster response [22]. However, the huge volume shrinkage during the pyrolysis of PDCs will cause stress mismatch at the interface of the film, leading to cracking or peeling off [23]. The use of particle fillers such as SiC, ZrO_2 and TiB_2 can not only reduce the shrinkage of the PDCs during the pyrolysis process, but also modify the properties of PDCs, such as electrical properties and mechanical properties [24]. Boride ceramics are excellent conductors of electricity with good mechanical properties, and oxidation resistance [25,26]. This makes boride ceramics promising for high temperature sensors and electrical functional devices [2,27].

In this study, PDC-SiCN was used as the insulating matrix, and the dispersed ZrB_2 conductive particles acted as the conductive phase. TFSG based on ceramic conductive composites was fabricated by the direct ink writing (DIW) technique based on the Weissenberg effect. Herein, the morphologies and microstructure of the ZrB_2/SiCN TFSGs were characterized. The effects of ZrB_2 content amount on the electrical conductivity and GF of ZrB_2/SiCN films were investigated. The piezoresistive response of ZrB_2/SiCN TFSGs at room temperature was tested. Ultimately, high temperature performance of the ZrB_2/SiCN TFSG was investigated from 25 °C to 600 °C.

2. Materials and Methods

2.1. Materials and Preparation Process

As shown in Figure 1a, commercially available PSN2 (Chinese Academy of Sciences, China) filled with ZrB_2 nanopowder (average diameter: 50 nm, Shanghai Chaowei Nano Technology Co., Ltd., Shanghai, China) was utilized as printing ink. The filling weight percent of ZrB_2 nanopowder is 40~60 wt%. The ZrB_2 nanopowder were uniformly dispersed in PSN2 by magnetic stirring for more than 2 h. Briefly, as shown in Figure 1b, the ink was printed by a Weissenberg-based DIW platform, which consisted of three key components: an x–y high-precision moving platform, a homemade printing setup including a printing head and a charged–coupled device camera. The printing head consists of micron tube and microneedle. The solution is quickly transported to the printing needle through the micron tube under the high-speed rotation of the microneedles. Then, the prepared thin-film strain grids were pyrolyzed in a tube furnace under nitrogen atmosphere (−0.1 MPa is evacuated before introducing nitrogen) at 800 °C for 4 h (heating rate 2 °C/min and cooling rate 3 °C/min). Finally, Ag paste was used to prepare solder joints to connect the thin-film strain grids and the Pt leads.

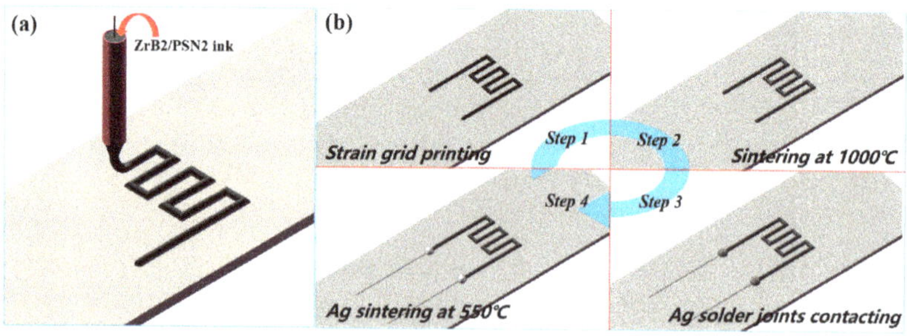

Figure 1. (**a**) Direct ink writing patterning of the strain gauge; (**b**) Schematic of the ZrB_2/SiCN thin-film strain gauge fabrication process.

2.2. Experiment Setup

Strain grid thicknesses were determined by a profilometer (Dektak XT, BRUKER, Billerica, MA, USA). SEM (SUPRA55 SAPPHIRE, CARL ZEISS, Oberkochen, Battenburg,

Germany) coupled with EDS was used to characterize the morphology of the obtained samples. High temperature furnace (GSL-1700X, HF. Kejing, Hefei, China) was used to pyrolysis and high temperature furnace (OTF-1200X, HF. Kejing, Hefei, China) was used to test in high temperature.

A cantilever beam arrangement was used to investigate the strain response behavior of the ZrB_2/SiCN TFSGs, as shown in Figure 2a. One end of the beam was clamped, and the sensor was subjected to strain by applying displacement at the free end of the cantilever [2]. The corresponding resistance changes of the TFSG were recorded using data acquisition equipment. Calculate the strain at the location of the strain gauge according to the following Equation (1) [28]:

$$\varepsilon = \frac{3yhx}{2l^3} \quad (1)$$

where ε is the strain at the location of the TFSGs, y is the deflection at the free end, l is the length of the cantilever beam, x is the distance from center of strain gauge to the point of application of load and h is the thickness of the beam. The indicator for strain sensitivity of the strain gauge is defined as:

$$\mathrm{GF} = \frac{\Delta R/R_0}{\varepsilon} \quad (2)$$

where ΔR is the change of TFSG resistance when strain ε is applied and R_0 is the initial resistance of TFSG. The piezoresistive response of TFSG at high temperatures was done in a high temperature furnace. The tube furnace is heated to 600 °C at 12 °C/min. In the meantime, the stepper motor is applied strain to the free end of the cantilever beam to obtain the strain response at high temperatures.

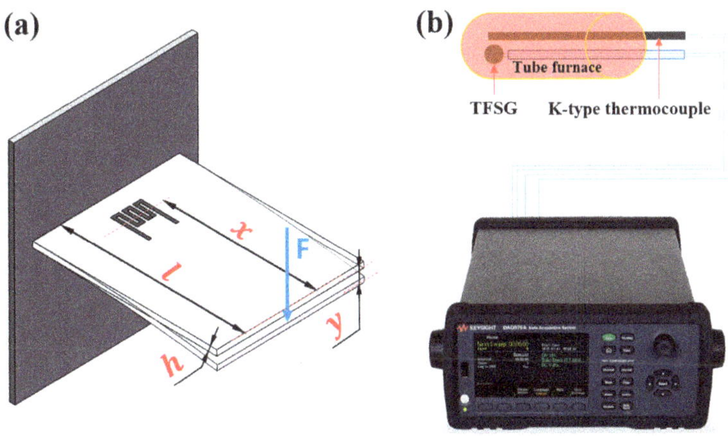

Figure 2. Schematic of experiment setups for (**a**) strain response testing and (**b**) temperature resistance characteristics testing.

The high temperature resistance test system of the strain gauge is shown in Figure 2b, which consists of a tube furnace and a standard k-type thermocouple. The unit of TCR is ppm/°C and is used to express the relationship between the resistance of the strain gauge and the temperature. TCR can be calculated by the following Equation (3) [29]:

$$\mathrm{TCR} = \frac{dR}{RdT} \times 10^6 \quad (3)$$

3. Results

3.1. Microstructural Characterisation of ZrB_2/SiCN TFSG

The fabricated ZrB_2/SiCN TFSGs on Al_2O_3 substrate are shown in Figure 3a. The length and width of ZrB_2/SiCN TFSGs are 7 mm and 5 mm, respectively. Its line width and thickness were determined by the profilometer, and were 600 μm and 15 μm, respectively. Porosity, cracks and inhomogeneity are the main factors affecting the electrical conductivity and thermal stability of TFSG. The low-magnification SEM image of ZrB_2/SiCN TFSGs is shown in Figure 3c. There are no obvious cracks on the surface of ZrB_2/SiCN TFSGs. The high-magnification SEM image of ZrB_2/SiCN TFSGs in Figure 3d shows a dense and crack-free surface. The SEM cross-sectional image presented in Figure 3e shows that the interface is clearly visible, and the sensitive grid is tightly bonded to the substrate without an obvious gap.

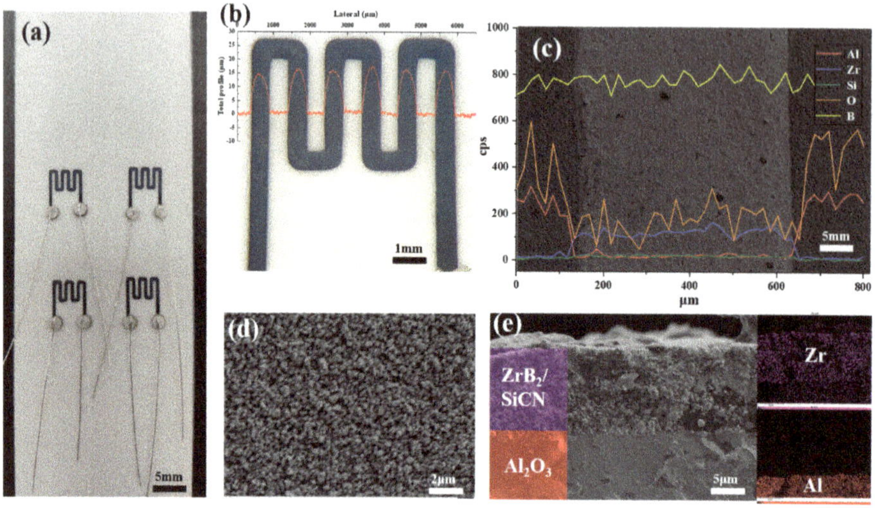

Figure 3. (**a**) Samples of the thin film strain gauge; (**b**) Thickness and width of strain grids; (**c**) Low-magnification SEM image and EDS analysis of TFSGs; (**d**) High-magnification SEM image of TFSGs; (**e**) The cross-sectional SEM images and EDS analysis of TFSGs.

3.2. Piezoresistive Response of ZrB_2/SiCN TFSG

The electrical conductivities of the printed ZrB_2(40 wt%)/SiCN, ZrB_2(50 wt%)/SiCN, ZrB_2(60 wt%)/SiCN strained grids are 0.036 S/cm, 0.077 S/cm and 1.71 S/cm, respectively. With the increase of ZrB_2 nanopowder filling, the electrical conductivity of the strain grids increases significantly, which is related to the conductive network composed of ZrB_2 in ZrB_2/SiCN composites. To investigate the piezoresistive behavior of ZrB_2/SiCN TFSGs in detail, their strain responses were tested using the deflection method at room temperature. The strain responses of the ZrB_2(40 wt%)/SiCN TFSG at room temperature are shown in Figure 4a–d. Figure 4a shows the static strain response of the ZrB_2(40 wt%)/SiCN TFSG. Strain was applied sequentially in 100 με increments, and the change in resistance was consistent with the strain applied to the sensor as time progresses. The ZrB_2(40 wt%)/SiCN TFSG exhibits a good response that stepwise applied strain leads to a distinguishable, recoverable step change in the resistance of the TFSG. The ZrB_2(40 wt%)/SiCN TFSG exhibits a positive GF, that is, the resistance increases with increasing positive strain and decreases with increasing negative strain. Figure 4b shows the strain responses of ZrB_2(40 wt%)/SiCN TFSG under different strain amounts, where strains of 100 με, 200 με, 300 με, 400 με, and 500 με were sequentially applied to the TFSG at a constant strain rate of 100 με/s. Figure 4c,d show the strain response of 500 με at different strain rates (20 με/s, 50 με/s, 100 με/s, 200 με/s, and 400 με/s) and the cyclic strain response with a period of 4 s, respectively. The applied strains were all 500 με. Consistent changes in relative resistance indicate that the ZrB_2(40 wt%)/SiCN TFSG has a stable and strain-rate independent strain response.

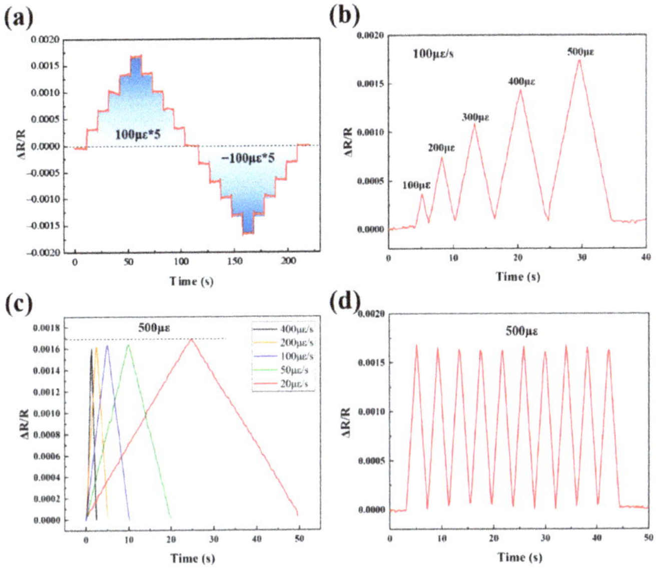

Figure 4. Piezoresistive response of the ZrB_2(40 wt%)/SiCN TFSG: (**a**) The static response; (**b**) The response of different strain quantities at a constant strain rate (100 με/s); (**c**) The response at different strain rates; (**d**) The cyclic response.

The strain responses of ZrB_2(50 wt%)/SiCN TFSG at room temperature are shown in Figure 5a–d. Similar to ZrB_2(40 wt%)/SiCN TFSG, ZrB_2(50 wt%)/SiCN TFSG exhibits stable, distinguishable strain responses.

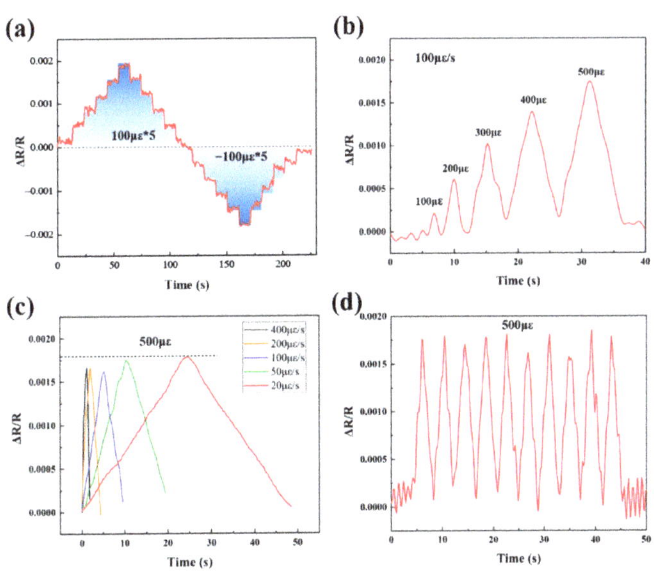

Figure 5. Piezoresistive response of the ZrB_2(50 wt%)/SiCN TFSG: (**a**) The static response; (**b**) The response of different strain quantities at a constant strain rate (100 με/s); (**c**) The response at different strain rates; (**d**) The cyclic response.

The strain responses of ZrB_2(60 wt%)/SiCN TFSG at room temperature are shown in Figure 6a–d. Compared with ZrB_2(40 wt%)/SiCN and ZrB_2(50 wt%)/SiCN TFSGs, the strain signal of ZrB_2(60 wt%)/SiCN TFSG is more obvious.

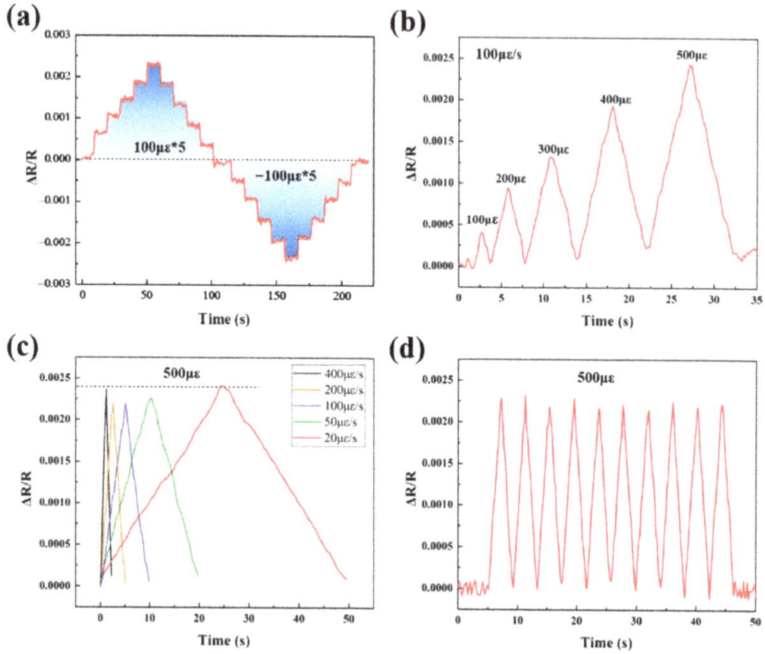

Figure 6. Piezoresistive response of the ZrB_2(60 wt%)/SiCN TFSG: (**a**) The static response; (**b**) The response of different strain quantities at a constant strain rate (100 με/s); (**c**) The response at different strain rates; (**d**) The cyclic response.

The GFs of the ZrB_2/SiCN TFSGs were calculated according to Equation (2). The GFs of ZrB_2(40 wt%)/SiCN, ZrB_2(50 wt%)/SiCN and ZrB_2(60 wt%)/SiCN at room temperature are 3.4, 3.3 and 4.8, respectively (see Figure 7). Comparing the GF of TFSG with different ZrB_2 filling amount, the GF of ZrB_2(60 wt%)/SiCN is the highest, which is mainly owing to the change in resistivity by the concentration of ZrB_2 conductive phase, which leads to the increase of the effect of piezoresistive response. Guenter Schultes et al. fabricated boride TFSG on a Al_2O_3 substrate by DC magnetron sputtering and obtained 0.7 GF [26]. In contrast, the GF of TFSG based on ZrB_2/SiCN conductive ceramic composite is several times higher than that of single boride TFSG.

3.3. *High Temperature Performance of the ZrB_2/SiCN TFSG*

Since ZrB_2(60 wt%)/SiCN TFSG exhibited higher conductivity and strain sensitivity, its high temperature performance was tested. In practical applications, the consistency of the temperature-resistance characteristics during the cycle temperature and high temperature resistance stability of the TFSG are very important, and it reflects the stability and oxidation resistance of the TFSG at high temperatures. In order to study the repeatability of the temperature resistance of ZrB_2(60 wt%)/SiCN TFSG, we tested the temperature-resistance characteristic curves of two times of heating and cooling (Figure 8a). ZrB_2(60 wt%)/SiCN TFSG exhibited a negative temperature coefficient of resistance of −428 ppm/°C and good repeatability in the range of 25–600 °C. Figure 8b shows the resistance change curves of ZrB_2(60 wt%)/SiCN TFSG at 600 °C for 2 h. ZrB_2(60 wt%)/SiCN TFSG exhibited excellent resistance stability and antioxidant qualities at 600 °C, and resistance increased by 1.3% after two hours of oxidation. The good repeatability, stability and consistency of the resistance of ZrB_2(60 wt%)/SiCN TFSG are attributed to the oxide layer formed on the surface of the film, which prevents the further diffusion of oxygen [30].

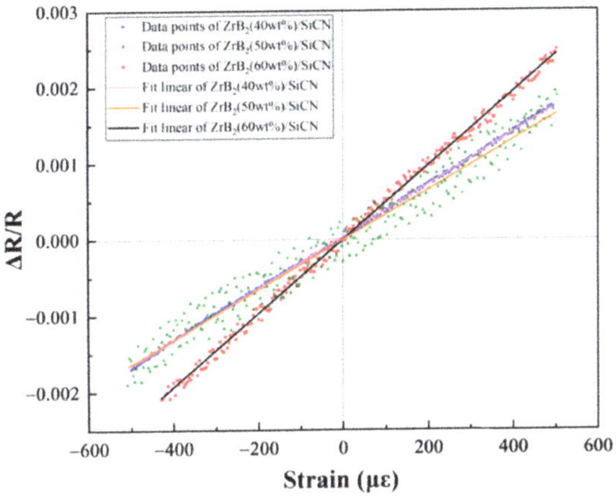

Figure 7. The GFs of the ZrB_2/SiCN TFSGs.

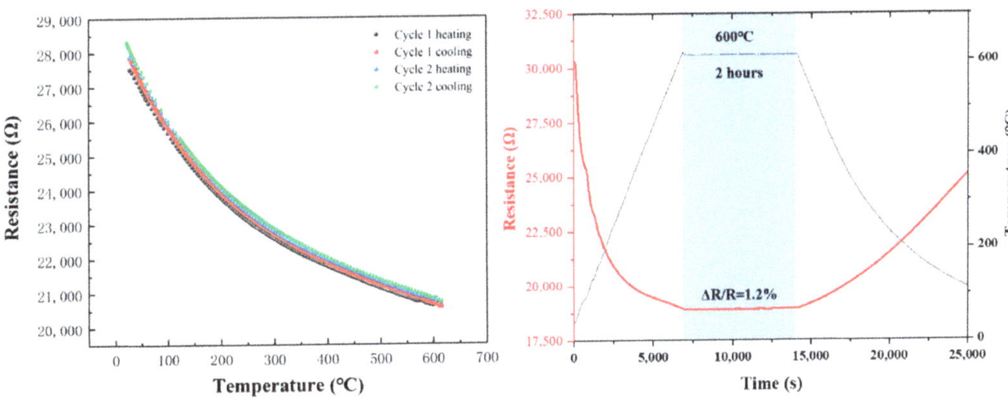

(a) (b)

Figure 8. (**a**) R–T curves of ZrB_2(60 wt%)/SiCN TFSG from room temperature to 600 °C; (**b**) Resistance change curves of ZrB_2(60 wt%)/SiCN TFSG at 600 °C for 2 h.

To evaluate the strain response of ZrB_2(60 wt%)/SiCN TFSG at high temperatures, cyclic strain tests were carried out from room temperature to 600 °C, as shown in Figure 9a. The curves of the cyclic strain response were intercepted at 400 °C, 500 °C and 600 °C, respectively, as shown in the Figure 9b–d. ZrB_2(60 wt%)/SiCN TFSG exhibits good repeatability and stability resistance at high temperatures. Although the overall resistance decreases with increasing temperature due to the temperature-resistance effect, the pulse signal caused by the cyclic strain is clearly visible. The above high-temperature test results show that ZrB_2(60 wt%)/SiCN TFSG has good resistance stability and highly sensitive strain response in the temperature range from room temperature to 600 °C, and has potential application in the field of hot component strain monitoring/sensing.

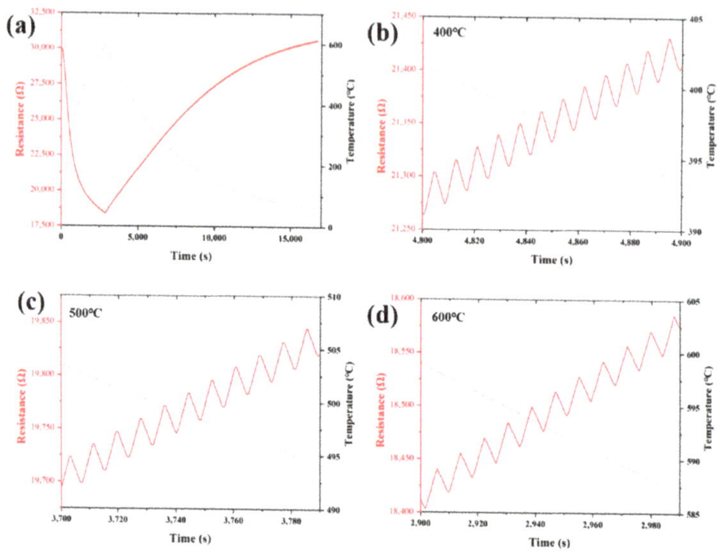

Figure 9. (**a**)The cyclic response from 25 °C to 600 °C; (**b**) The cyclic response at 400 °C; (**c**) The cyclic response at 500 °C; (**d**) The cyclic response at 600 °C.

4. Discussion

ZrB_2/SiCN TFSGs were fabricated on alumina substrates by DIW of the Weissenberg effect. The used DIW process enabled thin-film patterning and in situ strain/stress sensing of high-temperature components. The ZrB_2/SiCN TFSGs were characterized to determine the structural dimensions, surface topography and cross-sectional structure by SEM. The piezoresistive behavior of ZrB_2/SiCN TFSGs at room temperature was investigated by the deflection method. ZrB_2/SiCN TDSGs exhibited excellent strain responses and resistance stability at room temperature. The effects of ZrB_2 content on the electrical conductivity and strain sensitivity of ZrB_2/SiCN composites were investigated. Finally, ZrB_2(60 wt%)/SiCN film with high conductivity (1.71 S/cm) and GF of 4.8 was used as the sensitive material for high-temperature thin-film strain gauges. The temperature-resistance characteristic curves of ZrB_2(60 wt%)/SiCN TFSGs were tested, and the TFSGs exhibited a negative temperature coefficient of resistance of −428 ppm/°C and good repeatability in the range of 25–600 °C. The resistance change curves of ZrB_2(60 wt%)/SiCN TFSGs were tested at 600 °C. The TFSGs have good resistance stability with resistance increasing by 1.3% after two hours of oxidation at 600 °C. Finally, the strain response verification was conducted from 25 °C to 600 °C. ZrB_2(60 wt%)/SiCN TFSG has a highly sensitive strain response from 25 °C to 600 °C. Therefore, ZrB_2/SiCN TFSGs based on the Weissenberg DIW can be applied to micro-strain detection from room temperature to 600 °C. Further research is underway to improve the antioxidant nature of ZrB_2/SiCN TFSGs for applying to higher temperatures.

Author Contributions: Conceptualization, F.L. and X.P.; methodology, F.L. and X.P.; software, F.L.; validation, C.W. and F.L.; formal analysis, X.P.; investigation, F.L.; resources, C.W.; data curation, G.C.; writing—original draft preparation, F.L.; writing—review and editing, Z.H.; visualization, F.L. and Y.Z.; supervision, Z.H.; project administration, Q.C.; funding acquisition, D.S. All authors have read and agreed to the published version of the manuscript.

Funding: This research was funded by China Aviation Research Institute, Shenyang Engine Design and Research Institute (Grant number JC3602007026).

Institutional Review Board Statement: Not applicable.

Informed Consent Statement: Not applicable.

Data Availability Statement: Not applicable.

Conflicts of Interest: The authors declare no conflict of interest.

References

1. Liu, H.; Mao, X.L.; Yang, Z.B.; Cui, J.T.; Jiang, S.W.; Zhang, W.L. High temperature static and dynamic strain response of PdCr thin film strain gauge prepared on Ni-based superalloy. *Sens. Actuators A Phys.* **2019**, *298*, 111571. [CrossRef]
2. Wu, C.; Pan, X.C.; Lin, F.; Cui, Z.F.; He, Y.P.; Chen, G.C.; Zeng, Y.J.; Liu, X.L.; Chen, Q.N.; Sun, D.H.; et al. TiB2/SiCN Thin-Film Strain Gauges Fabricated by Direct Writing for High-Temperature Application. *IEEE Sens. J.* **2022**, *22*, 11517–11525. [CrossRef]
3. Ayerdi, I.; Castano, E.; Garciaalonso, A.; Gracia, F.J. Characterization of tantalum oxynitride thin-films as high-temperature strain-gauges. *Sens. Actuators A Phys.* **1995**, *46*, 218–221. [CrossRef]
4. Yang, S.Y.; Zhang, C.C.; Yang, Z.Q.; Wang, H.; Zhao, X.L.; Ding, G.F. In-situ self-compensated pt-ito thin film strain gage with a nanolaminated structure. In Proceedings of the 20th International Conference on Solid-State Sensors, Actuators and Microsystems and Eurosensors XXXIII (TRANSDUCERS and EUROSENSORS), Berlin, Germany, 23–27 June 2019; pp. 2005–2008.
5. Gregory, O.J.; Luo, Q.; Crisman, E.E. High temperature stability of indium tin oxide thin films. *Thin Solid Films* **2002**, *406*, 286–293. [CrossRef]
6. Chung, G.S. Characteristics of tantalum nitride thin film strain gauges for harsh environments. *Sens. Actuators A Phys.* **2007**, *135*, 355–359. [CrossRef]
7. Fiorillo, A.S.; Critello, C.D.; Pullano, S.A. Theory, technology and applications of piezoresistive sensors: A review. *Sens. Actuators A Phys.* **2018**, *281*, 156–175. [CrossRef]
8. Wang, H.Y.; Liu, K.; An, Z.H.; Wu, X.; Huang, X. Ion-Beam Sputtered Thin-Film Strain-Gauge Pressure Transducers. *Sens. Actuators A Phys.* **1993**, *35*, 265–268.
9. Wang, C.M.; Hsieh, J.H.; Li, C. Electrical and piezoresistive properties of TaN-Cu nanocomposite thin films. *Thin Solid Films* **2004**, *469*, 455–459. [CrossRef]
10. Garcia, J.R.; O'Suilleabhain, D.; Kaur, H.; Coleman, J.N. A Simple Model Relating Gauge Factor to Filler Loading in Nanocomposite Strain Sensors. *ACS Appl. Nano Mater.* **2021**, *4*, 2876–2886. [CrossRef]
11. Sanli, A.; Benchirouf, A.; Muller, C.; Kanoun, O. Piezoresistive performance characterization of strain sensitive multi-walled carbon nanotube-epoxy nanocomposites. *Sens. Actuators A Phys.* **2017**, *254*, 61–68. [CrossRef]
12. Boland, C.S.; Khan, U.; Ryan, G.; Barwich, S.; Charifou, R.; Harvey, A.; Backes, C.; Li, Z.; Ferreira, M.S.; Mobius, M.E.; et al. Sensitive electromechanical sensors using viscoelastic graphene-polymer nanocomposites. *Science* **2016**, *354*, 1257–1260. [CrossRef] [PubMed]
13. Wu, Y.T.; Yan, T.; Zhang, K.Q.; Pan, Z.J. A Hollow Core-Sheath Composite Fiber Based on Polyaniline/Polyurethane: Preparation, Properties, and Multi-Model Strain Sensing Performance. *Adv. Mater. Technol.* **2022**, 2200777. [CrossRef]
14. Lin, Y.K.; Yin, Q.; Wang, J.; Jia, H.B.; Yuan, G.L.; Wang, J.Y. Sensitivity enhanced, highly stretchable, and mechanically robust strain sensors based on reduced graphene oxide-aramid nanofibers hybrid fillers. *Chem. Eng. J.* **2022**, *443*, 136468. [CrossRef]
15. Rinaldi, A.; Tamburrano, A.; Fortunato, M.; Sarto, M.S. A Flexible and Highly Sensitive Pressure Sensor Based on a PDMS Foam Coated with Graphene Nanoplatelets. *Sensors* **2016**, *16*, 2148. [CrossRef] [PubMed]
16. Chen, H.T.; Miao, L.M.; Su, Z.M.; Song, Y.; Han, M.D.; Chen, X.X.; Cheng, X.L.; Chen, D.M.; Zhang, H.X. Fingertip-inspired electronic skin based on triboelectric sliding sensing and porous piezoresistive pressure detection. *Nano Energy* **2017**, *40*, 65–72. [CrossRef]
17. Yamada, T.; Hayamizu, Y.; Yamamoto, Y.; Yomogida, Y.; Izadi-Najafabadi, A.; Futaba, D.N.; Hata, K. A stretchable carbon nanotube strain sensor for human-motion detection. *Nat. Nanotechnol.* **2011**, *6*, 296–301. [CrossRef]
18. Wu, C.; Pan, X.C.; Lin, F.; Cui, Z.F.; Li, X.; Chen, G.C.; Liu, X.L.; He, Y.P.; He, G.H.; Hai, Z.Y.; et al. High-temperature electrical properties of polymer-derived ceramic SiBCN thin films fabricated by direct writing. *Ceram Int.* **2022**, *48*, 15293–15302. [CrossRef]
19. Zhang, L.G.; Wang, Y.S.; An, L.N. Piezoresistivity of polymer-derived AlSiCN ceramics. In *Applied Materials and Technologies for Modern Manufacturing, Pts 1–4*; Trans Tech Publications Ltd.: Bäch, Switzerland, 2013; Volume 423–426, pp. 89–92. [CrossRef]
20. Yu, Y.D.; Li, J.P.; Niu, J.H.; Yi, F.J.; Meng, S.H. The stability and repeatability of high temperature electrical properties of SiAlCN ceramic sensor heads. *Ceram Int.* **2019**, *45*, 7588–7593. [CrossRef]
21. Zhao, R.; Shao, G.; Cao, Y.J.; An, L.N.; Xu, C.Y. Temperature sensor made of polymer-derived ceramics for high-temperature applications. *Sens. Actuators A Phys.* **2014**, *219*, 58–64. [CrossRef]
22. Fu, X.L.; Lin, Q.Y.; Peng, Y.Q.; Liu, J.H.; Yang, X.F.; Zhu, B.P.; Ouyang, J.; Zhang, Y.; Xu, L.C.; Chen, S. High-Temperature Heat Flux Sensor Based on Tungsten-Rhenium Thin-Film Thermocouple. *IEEE Sens. J.* **2020**, *20*, 10444–10452. [CrossRef]
23. Lu, K.; Erb, D. Polymer derived silicon oxycarbide-based coatings. *Int. Mater. Rev.* **2018**, *63*, 139–161. [CrossRef]
24. Barroso, G.; Li, Q.; Bordia, R.K.; Motz, G. Polymeric and ceramic silicon-based coatings—A review. *J. Mater. Chem. A* **2019**, *7*, 1936–1963. [CrossRef]
25. Ni, D.W.; Cheng, Y.; Zhang, J.P.; Liu, J.X.; Zou, J.; Chen, B.W.; Wu, H.Y.; Li, H.J.; Dong, S.M.; Han, J.C.; et al. Advances in ultra-high temperature ceramics, composites, and coatings. *J. Adv. Ceram* **2022**, *11*, 1–56. [CrossRef]
26. Schultes, G.; Schmitt, M.; Goettel, D.; Freitag-Weber, O. Strain sensitivity of TiB2, TiSi2, TaSi2 and WSi2 thin films as possible candidates for high temperature strain gauges. *Sens. Actuators A Phys.* **2006**, *126*, 287–291. [CrossRef]

27. Wu, C.; Lin, F.; Pan, X.C.; Cui, Z.F.; He, Y.P.; Chen, G.C.; Liu, X.L.; He, G.H.; Chen, Q.N.; Sun, D.H.; et al. TiB2-Modified Polymer-Derived Ceramic SiCN Double-Layer Thin Films Fabricated by Direct Writing for High-Temperature Application. *Adv. Eng. Mater.* **2022**, 2200228. [CrossRef]
28. Ren, S.; Jiang, S.W.; Liu, H.; Zhang, W.L.; Li, Y.R. Investigation of strain gauges based on interdigitated Ba0.5Sr0.5TiO3 thin film capacitors. *Sens. Actuators A Phys.* **2015**, *236*, 159–163. [CrossRef]
29. Wen, M.; Guan, X.C.; Li, H.; Ou, J.P. Temperature characteristics of thick-film resistors and its application as a strain sensor with low temperature-sensitivity. *Sens. Actuators A Phys.* **2020**, *301*, 111779. [CrossRef]
30. Parthasarathy, T.A.; Rapp, R.A.; Opeka, M.; KeranS, R.J. A model for the oxidation of ZrB2, HfB2 and TiB2. *Acta Mater.* **2007**, *55*, 5999–6010. [CrossRef]

micromachines

Article

A SiCN Thin Film Thermistor Based on DVB Modified Polymer-Derived Ceramics

Chao Wu, Fan Lin, Xiaochuan Pan, Yingjun Zeng, Guochun Chen, Lida Xu, Yingping He, Daoheng Sun * and Zhenyin Hai *

Department of Mechanical and Electrical Engineering, School of Aerospace Engineering, Xiamen University, Xiamen 361005, China
* Correspondence: sundh@xmu.edu.cn (D.S.); haizhenyin@xmu.edu.cn (Z.H.)

Abstract: Carbon-rich SiCN ceramics were prepared by divinylbenzene (DVB)-modified polysilazane (PSN2), and a high-conductivity SiCN thin film sensor suitable for medium-low temperature sensing was fabricated. The modified liquid precursors were patterned by direct ink writing to produce SiCN resistive grids with line widths of several hundreds of micrometers and thicknesses of several micrometers. The introduction of DVB not only increases the critical thickness of SiCN ceramics several times, but also significantly improves the conductivity of SiCN, making it meet the conductivity requirements of sensing applications in the mid-low temperature range. The electrical conductivity and microstructure of DVB-modified SiCN ceramics were studied in detail. In the temperature range of 30~400 °C, the temperature resistance performance of DVB modified SiCN resistance grid was measured. The SiCN ceramics with low DVB content not only have excellent electrical conductivity, but also have good oxidation resistance.

Keywords: polymer-derived ceramics; thermistor; SiCN; conductivity

Citation: Wu, C.; Lin, F.; Pan, X.; Zeng, Y.; Chen, G.; Xu, L.; He, Y.; Sun, D.; Hai, Z. A SiCN Thin Film Thermistor Based on DVB Modified Polymer-Derived Ceramics. *Micromachines* **2022**, *13*, 1463. https://doi.org/10.3390/mi13091463

Academic Editors: Libo Gao and Zhuoqing Yang

Received: 11 August 2022
Accepted: 27 August 2022
Published: 3 September 2022

1. Introduction

Polymer-derived ceramics (PDCs), such as SiCN, SiOC and SiBCN, are considered for high-temperature sensor applications due to their semiconducting behavior and excellent thermal stability [1–5]. The unique polymer-to-ceramic (liquid-solid) transition process allows for attractive patterning options, such as direct ink writing (DIW) and soft lithography, making PDCs a promising candidate for high-temperature micro-electromechanical systems (MEMS) and thin-film sensors (TFSs) candidate material [6–8]. Temperature sensors are a promising application of PDCs as thermistors. A thermistor is a device that measures temperature by measuring its resistance [9]. Manganate, SiCN, SiOC and other thermistors with negative temperature coefficient (NTC) show exponential reduction in resistance with temperature, while positive temperature coefficient (PTC) thermistors such as Pt and NiCr show a linear increase in resistance with temperature [10–12]. The in-situ generated free carbon phase plays a key role in the electrical conductivity and semiconducting behavior of PDCs [13–16], which enables the electrical resistance of PDCs to have a pronounced temperature response.

Although the temperature-resistance behavior of some PDCs has been developed, it is not easy to fabricate TFSs using PDC materials. The poor electrical conductivity of PDCs and the low critical thickness of constrained sintered PDCs films remain huge obstacles for the wider application of PDC TFSs. For example, poor electrical conductivity makes PDCs electrically insulating at room temperature, which limits their operating temperature range to above 400 °C [17,18]. The critical thickness of constrained sintering is about 3 μm, which limits the thin-film of PDCs device [1]. Although the relatively good conductivity can be obtained by molecular design of precursor solutions and optimization of pyrolysis parameters, the pyrolysis temperature is usually higher than 1400 °C [13]. Another effective strategy is to fill the PDC with conductive particles that exceed the penetration threshold [6,7]. This strategy not only improves the electrical conductivity of PDCs film but also improves their critical thickness. However, the dispersibility and stability of solid-phase particles in the precursor solution are difficult to guarantee. Therefore, liquid-phase modified PDCs would be preferred. Divinylbenzene (DVB), as a widely used precursor of carbon sources, has been used to improve the electrical conductivity of PDCs [19–21]. For example, the conductivity of DVB-modified PDC-SiCN is improved by 5 orders of magnitude [11]. However, currently, DVB-modified PDC devices are usually millimeter-scale and difficult to integrate with structural components, which significantly decreased the sensitivity and response time of the sensor [11]. The advantages of PDCs in additive manufacturing and controlled molding have not been well utilized.

In this study, the electrical conductivity of PDC-SiCN was increased by adding a DVB carbon source precursor to a commercially available polysilazane (PSN). An in-situ integrated thin-film temperature sensor was successfully fabricated via DIW platform based on the Weissenberg effect and tested from room temperature (30 °C) to 400 °C, expanding the application of PDC sensors in the mid-low temperature range.

2. Materials and Methods

2.1. Materials and Fabrication Methods

In this study, a commercially available polysilazane (PSN, Institute of Chemistry, Chinese Academy of Sciences, Beijing, China) was utilized as the SiCN precursor, DVB (technical grade 80%, Sigma-Aldrich, St. Louis, MO, USA) was selected as the carbon source to improve the conductivity of SiCN. The fabricated process of TFSs was illustrated in Figure 1a. First, different mass fractions of DVB were added to PSN and magnetically stirred at 100 °C for about a few minutes to form a uniform printable mixture (The contents of DVB were 10 wt%, 20 wt%, 30 wt%, 40 wt%, 50 wt%, respectively). Briefly, DVB-modified PSN2 ink was printed by a Weissenberg-based DIW platform, which consisted of three key components: an x–y high-precision moving platform, a homemade printing setup including a printing head, and a charged–coupled device camera. Then, the printed thin-film resistor grids were pyrolyzed in a tube furnace under high-purity nitrogen atmosphere at 1100 °C for 4 h.

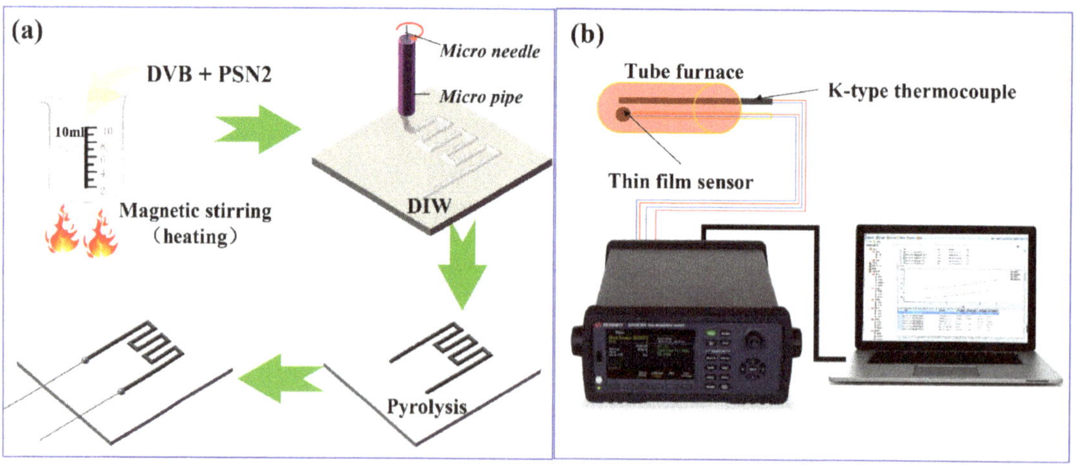

Figure 1. Schematic illustration of (**a**) process and (**b**) temperature-resistance test device.

2.2. Characterization Techniques

Resistance grid thicknesses were determined by a profilometer (Dektak XT). SEM (SUPRA55 SAPPHIRE) coupled with EDS was used to characterize the morphology and elemental content of the obtained films. XPS (Thermo Scientific ESCALAB Xi+) measurements were performed to determine chemical bonds. Free carbon in SiCN was characterized using confocal in situ Raman spectroscopy (LabRAM HR Evolution). As shown in Figure 1b, the utilised temperature–resistance.

3. Results

3.1. Film Morphology

Pores and cracks are the main factors affecting PDCs conductivity [1]. Therefore, the critical thickness of the DVB-modified SiCN film (defined as the maximum thickness of monolayer deposition) was first determined. In the DIW process, the line width of the resistance grids was fixed at about 500 μm, and the line thickness was successively increased. After pyrolysis, the thickness of the resistance grids and the shape of the surface profile were determined by a profilometer. As shown in Figure 2, the critical thickness of SiCN film gradually increased from 3.7 μm of SiCN to 5.7 μm of 10 wt% DVB-SICN, 6.0 μm of 30 wt% DVB-SiCN, 6.9 μm of 50 wt% DVB-SiCN. The addition of DVB provides more carbon sources for SiCN ceramics. The high critical thickness may be related to the carbon content of SiCN. The increased critical thickness reduces the risk of film cracking and peeling during DIW patterning and pyrolysis, which is beneficial for maintaining the structural integrity of the resistive grids.

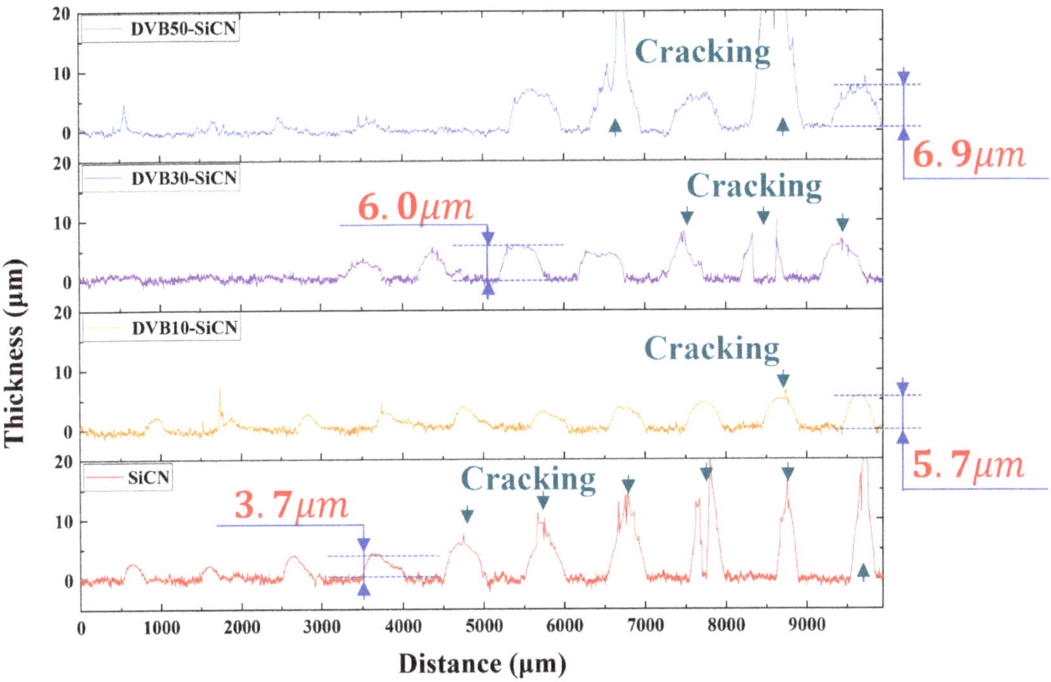

Figure 2. Surface profile of SiCN and DVB-modified SiCN lines.

Figure 3a shows an optical image of the thin film sensor. Figure 3b is a SEM image showing a SiCN resistive grid line, indicating that its line width is about 550 μm. Figure 3c shows the surface morphologies of DVB20-SiCN film. It can be seen that the film is dense without obvious defects. The SEM image of the cross-section of the SiCN film shown in Figure 3d shows that the SiCN film is well bonded to the Al_2O_3 substrate. The results of elemental analysis of all films are summarized in Table 1, and it can be clearly seen that the addition of DVB significantly increases the carbon content in the SiCN films. The higher oxygen content may be related to the oxygen contamination during the DIW process, and the oxygen adsorption on the surface of the SiCN film.

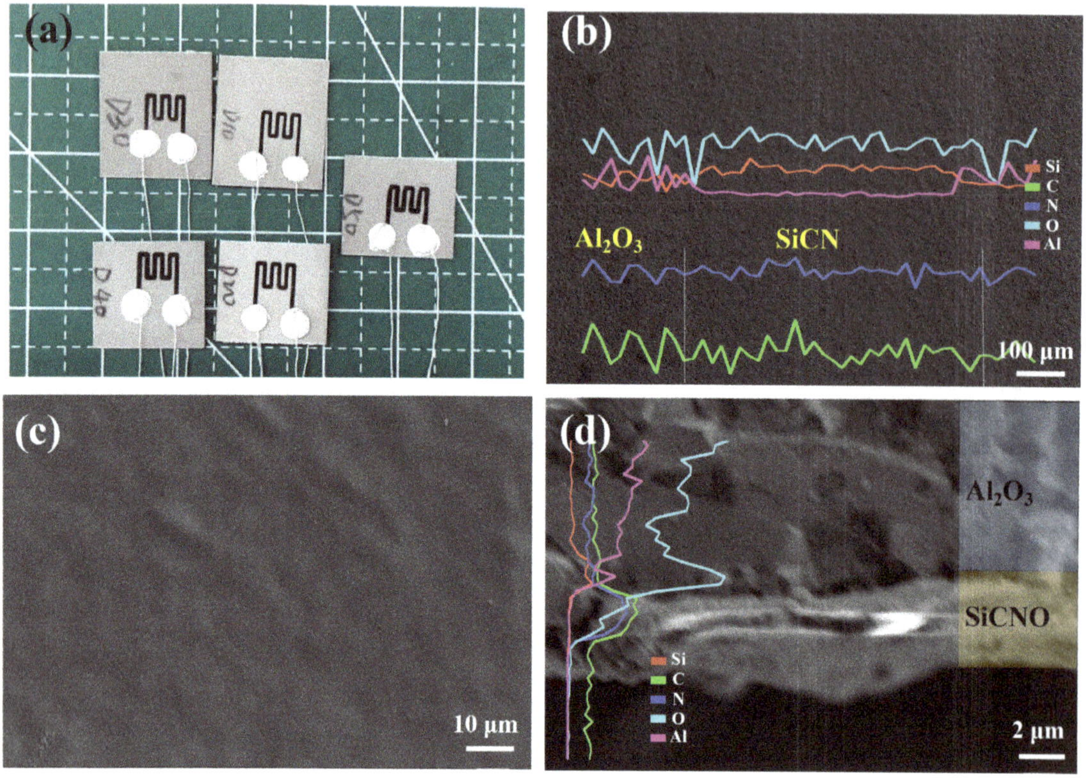

Figure 3. (**a**) Optical image showing SiCN thin film sensor. (**b**) SEM image showing a single DIW-SiCN line (inset shows EDS analysis conducted in the line scan mode). (**c**) Surface morphology of the DVB20-SiCN film. (**d**) SEM image showing the cross section of the DVB20-SiCN film (inset shows EDS analysis conducted in the line scan mode).

Table 1. Results of elemental analysis.

Sample	Si (wt%)	C (wt%)	N (wt%)	O (wt%)
SiCN	32.75	17.89	2.02	47.34
D10-SiCN	31.87	19.24	3.23	45.67
D20-SiCN	32.64	18.99	4.84	43.53
D30-SiCN	33.35	20.01	2.31	44.33
D40-SiCN	29.46	23.29	5.52	41.72
D50-SiCN	31.20	22.69	3.92	42.19

3.2. Film Composition

The XPS was employed to analyze the chemical composition and bonding characteristics within the film. The XPS spectra of DVB10-SiCN film pyrolyzed at 1100 °C are shown in Figure 4. The spectrum of Si (2p) shows peaks at 102.35 eV and 104.19 eV due to the formation of Si-N, Si-O bonds. For the C (1 s) spectrum, peaks were observed at 284.8 eV, 286.06 eV, and 288.25 eV and are attributed to C-C/C-H, C-O/C-N, and C=O, respectively. The highest intensity of the C-C peak indicates the completion of the pyrolysis process and the formation of free carbon. Similarly, for N (1 s) spectrum, the peaks at 399.35 eV, and

400.95 eV correspond to C-N and Si-N, respectively. As for the O 1 s spectrum, two peaks centered at 532.4 eV and 533.87 eV, respectively, are associated with the Si-O band and adsorbed -OH, which is related to the oxygen contamination during the preparation [1]. The above results indicate the formation of free carbon-rich SiCN ceramics.

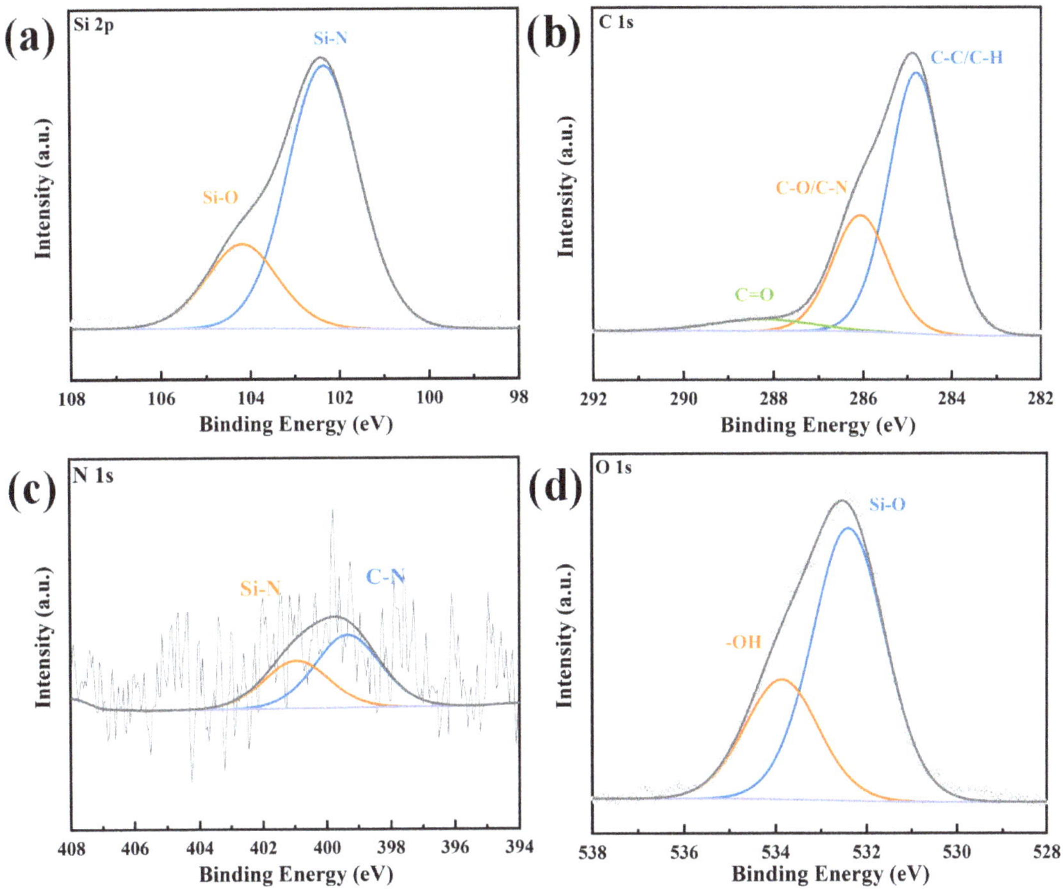

Figure 4. XPS spectra of the as-prepared DVB20-SiCN film. (**a**) Si 2P spectrum. (**b**) C 1s spectrum. (**c**) N 1s spectrum. (**d**) O 1s spectrum.

Raman spectroscopy was used to further analyze the carbon in the SiCN films. The Raman spectra of the investigated SiCN ceramic films are shown in Figure 5. The spectra of SiCN modified with different DVB concentrations exhibit similar shape, which contains a D peak at ~1333 cm^{-1}, and G peak at ~1610 cm^{-1}, indicating a strong disorder state of the amorphous carbon. The intensity, position, and width of the D and G bands may vary, depending on the structural organization of the sample under study [22]. The intensity ratio of the D and G modes, I_D/I_G, enables the evaluation of the carbon nanoparticle size by using the Ferrari-Robertson equation [22,23]:

$$\frac{I_D}{I_G} = C'(\lambda)L_a^2 \quad (1)$$

$C'(\lambda)$ is a coefficient depending on the excitation wavelength of the laser. The value of $C'(\lambda)$ for the wavelength of 532 nm is assigned to 0.6195 nm^{-2} [24]. The L_a values are listed in

Table 2. The lateral cluster size La in all samples is between 1 nm and 2 nm, indicating the nanostructural nature of the free carbon in SiCN. Previous studies have shown that carbon in PDC SiCN will undergo precipitation during the pyrolysis process, and then evolve into free carbon [24]. The value of I_D/I_G is used to measure the disorder degree of free carbon, and the higher the ratio is, the higher the defect carbon content would be [25]. When the DVB content is 20 wt%, the I_D/I_G value suddenly increases, implying that the content of disordered carbon in SiCN gradually increases. The increase in L_a may be due to the in-plane growth of nano-polycrystalline graphite [23].

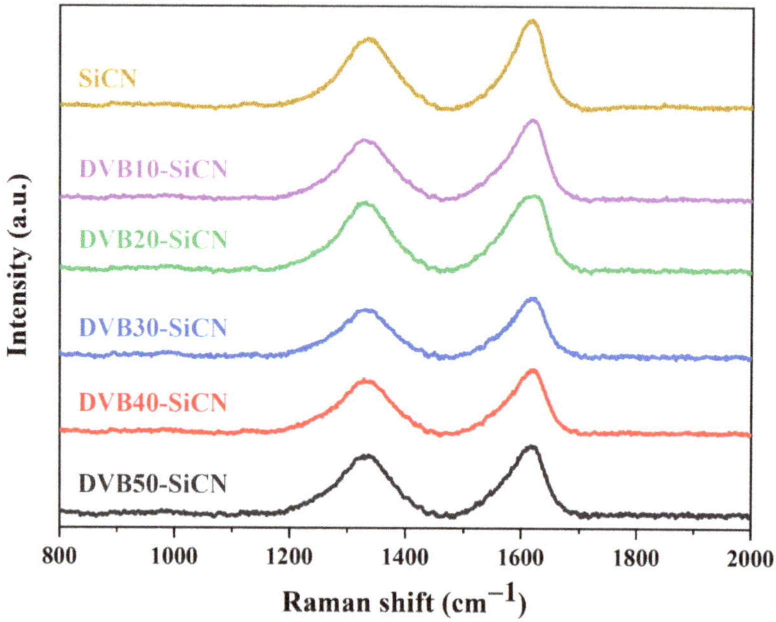

Figure 5. Raman spectra of SiCN films modified with different concentrations of DVB.

Table 2. The intensity ratio of D peak to G peak I_D/I_G and carbon cluster size L_a obtained from the curve-fitting of Raman spectra.

Sample	SiCN	DVB10-SiCN	DVB20-SiCN	DVB30-SiCN	DVB40-SiCN	DVB50-SiCN
I_D/I_G	0.77	0.76	0.89	0.80	0.84	0.87
L_a (nm)	1.118	1.106	1.198	1.133	1.630	1.183

3.3. Electrical Performance

The sensing properties of the thin film resistance grids were characterized by measuring the temperature-dependent resistance. The SiCN resistance grid pyrolyzed at 1100 °C is insulated at room temperature (conductivity: 10^{-6} S/m–10^{-4} S/m) [5,26]. The electrical conductivity of the DVB-modified SiCN films was calculated from the resistance and size of the resistive grid. The room temperature conductivities of the DVB10-SiCN, DVB20-SiCN, DVB30-SiCN, DVB40-SiCN and DVB50-SiCN resistor grids are 9.4 S/m, 39.8 S/m, 32.3 S/m, 24.8 S/m and 30 S/m, respectively (Figure 6a). Compared with SiCN, the electrical conductivity of DVB-modified SiCN is significantly improved. However, when the DVB content was increased to 20%, the conductivity did not further improve. This is related to the content of free carbon in SiCN and the percolative network formed by it. In conductive composites, when the conductive phase reaches the percolation threshold, the conductivity increases by orders of magnitude. Further increasing the concentration of the conductive phase, the conductivity increases slowly and becomes stable [27]. The percolation behavior can be attributed to the formation of free carbon network. In DVB20/SiCN the percolative network is already completely formed and it does not improve significantly in DVB50/SiCN [28]. This explains that the electrical conductivity of the two compositions is quite similar despite their DVB amounts of 20 wt% and 50 wt%, respectively.

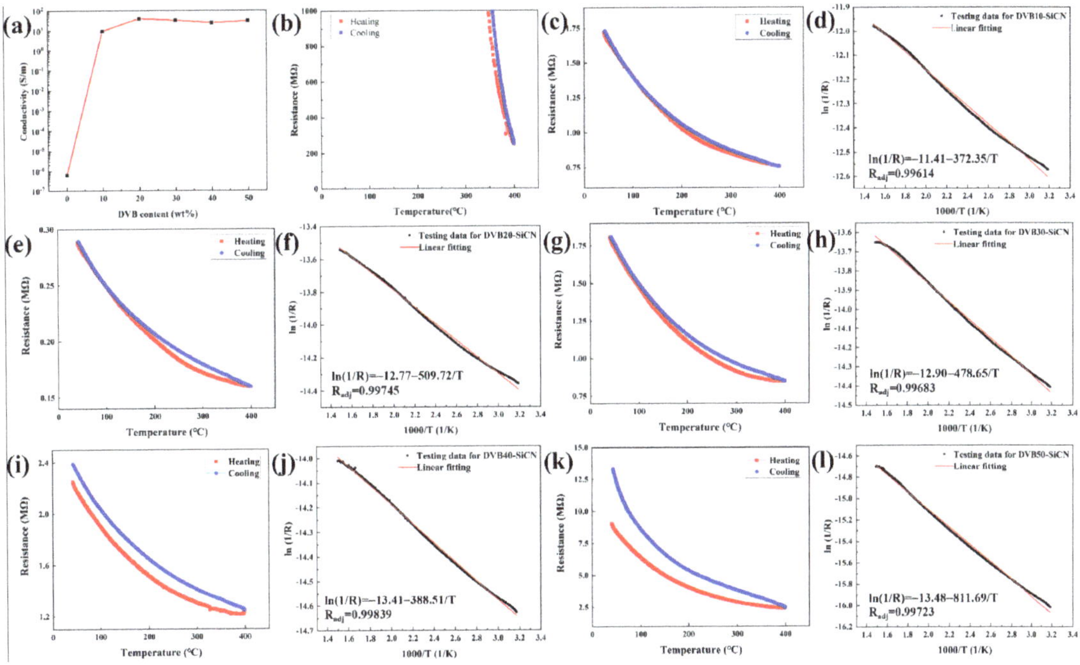

Figure 6. (**a**) Variation of SiCN conductivity at room temperature with DVB content. The resistance of film as a function of temperature: (**b**) SiCN film, (**c**) DVB10-SiCN film, (**e**) DVB20-SiCN film, (**g**) DVB30-SiCN film, (**i**) DVB40-SiCN film, (**k**) DVB50-SiCN film. A plot of resistance vs. temperature in the format ln(1/R) vs. 1000/T: (**d**) DVB10-SiCN film, (**f**) DVB20-SiCN film, (**h**) DVB30-SiCN film, (**j**) DVB40-SiCN film, (**l**) DVB50-SiCN film.

The temperature sensing properties of thin-film resistive grids were characterized by measuring temperature-dependent resistance (Figure 6b–l). Below 350 °C, the resistance of the SiCN resistive grid is greater than 1 GΩ, which limits the application of SiCN TFSs in the mid-low temperature range. The high conductivity of DVB-modified SiCN resistive grid makes it more suitable for sensing in the mid-low temperature range. In the temperature range of 30–400 °C, all DVB-modified SiCN resistive grids exhibit negative temperature coefficient of resistance, that is, the resistance decreases monotonically with increasing temperature, presenting a good sensitivity of DVB-modified SiCN resistive grids under high temperature (T) environments. The ln(1/R)–1000/T curves for the DVB-modified SiCN resistive grids were obtained according to the thermistor equation [17,18]:

$$ln\frac{1}{R} = c_1\frac{1}{T} - c_2 \tag{2}$$

where c_1 and c_2 are constants. For all SiCN resistive grids, linear behavior following the thermistor equation was obtained. The constant c_1 values of DVB10-SiCN, DVB20-SiCN, DVB30-SiCN, DVB40-SiCN, and DVB50-SiCN thermistors are −372.35, −509.72, −478.65,

−388.51 and −811.69, respectively, indicating that they all have excellent sensitivity to temperature change. However, during one cycle of heating and cooling tests, it can be found that SiCN films modified with low DVB content (DVB: 10–30 wt%) have better repeatability. The incomplete coincidence of the heating and cooling resistance curves in the temperature range of 200–400 °C is related to the response speed of the thermocouple and the thermal conduction behavior of thermal components such as Al_2O_3 substrates. In other words, the thermocouple measures the ambient temperature of the Al_2O_3 substrate, while the temperature experienced by the TFS is closer to the surface temperature of the Al_2O_3. During the heating and cooling process, their thermal conduction behaviors are different, resulting in incompletely consistent resistance curves. However, when the heating and cooling are completed, the resistance of the TFS can return to its original value. For the SiCN films modified with high DVB content (DVB: 40–50 wt%), it can be found that the resistance of TFS increases significantly after one thermal cycle. This may be related to the oxidation of carbon. The modified SiCN films with low DVB content not only significantly improved the electrical conductivity, but also exhibited better resistance repeatability.

4. Conclusions

In this study, a DVB-modified PDC thin-film resistive grid with a line width of about 550 μm and a thickness of less than 7 μm was fabricated for by a DIW process. Compared with PDC-SiCN, the introduction of more carbon sources by adding DVB not only significantly improves the electrical conductivity of SiCN, but also increases the critical thickness of SiCN film several times. This provides another method for the thinning, patterning and high conductivity of PDC devices. A thin thermistor integrated with the Al_2O_3 substrate was successfully fabricated using DVB-modified PDC-SiCN as the sensing element, which demonstrated the feasibility of the proposed method. The conductive behavior and microstructure of DVB-modified PDC-SiCN ceramics were studied in detail. The temperature resistance behavior of the DVB-modified SiCN resistive grid in the mid-low temperature range from room temperature to 400 °C was measured. The SiCN ceramics modified with low DVB content not only effectively improved the electrical conductivity but also had better oxidation resistance.

Author Contributions: Conceptualization, D.S. and Z.H.; methodology, C.W.; software, X.P.; validation, X.P., F.L. and G.C.; formal analysis, G.C.; investigation, F.L.; resources, L.X.; data curation, X.P.; writing—original draft preparation, C.W.; writing—review and editing, Z.H.; visualization, Y.Z.; supervision, Y.H.; project administration, Y.Z.; funding acquisition, D.S. and Z.H. All authors have read and agreed to the published version of the manuscript.

Funding: This research was funded by Shenyang Engine Design and Research Institute, grant number JC3602007026 and National Natural Science Foundation of China, grant number 62101469.

Data Availability Statement: The data used to support the findings of this study are included within the article.

Acknowledgments: We thank Tan Kah Kee Innovation Laboratory for providing access to characterisation instruments.

Conflicts of Interest: There are no conflict to declare.

References

1. Wu, C.; Pan, X.C.; Lin, F.; Cui, Z.F.; Li, X.; Chen, G.C.; Liu, X.L.; He, Y.P.; He, G.H.; Hai, Z.Y.; et al. High-temperature electrical properties of polymer-derived ceramic SiBCN thin films fabricated by direct writing. *Ceram. Int.* **2022**, *48*, 15293–15302. [CrossRef]
2. Wang, K.S.; Gunthner, M.; Motz, G.; Bordia, R.K. High performance environmental barrier coatings, Part II: Active filler loaded SiOC system for superalloys. *J. Eur. Ceram. Soc.* **2011**, *31*, 3011–3020. [CrossRef]
3. Zhao, W.Y.; Shao, G.; Han, S.X.; Cai, C.H.; Liu, X.C.; Sun, M.R.; Wang, H.L.; Li, X.J.; Zhang, R.; An, L.N. Facile preparation of ultralight polymer-derived SiOCN ceramic aerogels with hierarchical pore structure. *J. Am. Ceram. Soc.* **2019**, *102*, 2316–2324. [CrossRef]

4. Gao, Y.; Mera, G.; Nguyen, H.; Morita, K.; Kleebe, H.J.; Riedel, R. Processing route dramatically influencing the nanostructure of carbon-rich SiCN and SiBCN polymer-derived ceramics. Part I: Low temperature thermal transformation. *J. Eur. Ceram. Soc.* **2012**, *32*, 1857–1866. [CrossRef]
5. Ma, B.S.; Zhu, Y.; Wang, K.W.; Sun, Z.Z. PIP process greatly influencing the microstructure and electrical conductivity of polymer-derived SiCN ceramics. *J. Alloy. Compd.* **2019**, *784*, 1084–1090. [CrossRef]
6. Wu, C.; Pan, X.C.; Lin, F.; Cui, Z.F.; He, Y.P.; Chen, G.C.; Zeng, Y.J.; Liu, X.L.; Chen, Q.N.; Sun, D.H.; et al. TiB2/SiCN Thin-Film Strain Gauges Fabricated by Direct Writing for High-Temperature Application. *IEEE Sens. J.* **2022**, *22*, 11517–11525. [CrossRef]
7. Wu, C.; Lin, F.; Pan, X.C.; Cui, Z.F.; He, Y.P.; Chen, G.C.; Liu, X.L.; He, G.H.; Chen, Q.N.; Sun, D.H.; et al. TiB2-Modified Polymer-Derived Ceramic SiCN Double-Layer Thin Films Fabricated by Direct Writing for High-Temperature Application. *Adv. Eng. Mater.* **2022**, 2200228. [CrossRef]
8. Nagaiah, N.R.; Kapat, J.S.; An, L.; Chow, L. Novel Polymer Derived Ceramic-High Temperature Heat Flux Sensor for Gas Turbine Environment. *J. Phys. Conf. Ser.* **2006**, *34*, 458–463. [CrossRef]
9. Kim, T.J.; Davis, K.L.; Liu, Y.P.; Bredemann, J.R.; Ma, Z.Q.; Anderson, M.; Corradini, M.L. Development of a Stable High-Temperature Diamond Thermistor Using Enhanced Supporting Designs. *IEEE Sens. J.* **2019**, *19*, 6587–6594. [CrossRef]
10. Shen, A.; Kim, S.B.; Bailey, C.; Ma, A.W.K.; Dardona, S. Direct Write Fabrication of Platinum-Based Thick-Film Resistive Temperature Detectors. *IEEE Sens. J.* **2018**, *18*, 9105–9111. [CrossRef]
11. Ma, B.S.; Cao, Y.J.; Gao, Y.; Wang, Y.G. Fabrication of a thin double-layer thermistor based on DVB-modified polymer-derived SiCN ceramics. *J. Alloy. Compd.* **2018**, *732*, 491–497. [CrossRef]
12. Yang, B.W.; He, M.; Wen, K.H.; Xiong, D.P.; Feng, Y.F.; Ta, S.W.; Yang, Z. Comparison of morphology, electrical properties and sensitivity between bulk and thin-film Mn1.5Co1Ni0.5O4 thermistors. *Ceram. Int.* **2020**, *46*, 27134–27142. [CrossRef]
13. Wen, Q.B.; Yu, Z.J.; Riedel, R. The fate and role of in situ formed carbon in polymer-derived ceramics. *Prog. Mater. Sci.* **2020**, *109*, 63. [CrossRef]
14. Li, F.P.; Zhao, L.N.; Dang, W.; Xu, Z.L.; Zhao, K.; Xue, M.J.; Tang, Y.F. Pore structure and thermal oxidation curing behavior of porous polymer derived ceramics with superhigh porosity fabricated by freeze casting. *Ceram. Int.* **2021**, *47*, 31114–31121. [CrossRef]
15. Kleebe, H.J.; Blum, Y.D. SiOC ceramic with high excess free carbon. *J. Eur. Ceram. Soc.* **2008**, *28*, 1037–1042. [CrossRef]
16. Mera, G.; Navrotsky, A.; Sen, S.; Kleebe, H.J.; Riedel, R. Polymer-derived SiCN and SiOC ceramics—Structure and energetics at the nanoscale. *J. Mater. Chem.* **2013**, *1*, 3826–3836. [CrossRef]
17. Chen, Y.G.; Cao, Y.J.; Wang, Y.G. Electrical property of joints made of polymer-derived SiAlCN ceramic via adhesive joining. *Ceram. Int.* **2021**, *47*, 3649–3656. [CrossRef]
18. Zhao, R.; Shao, G.; Cao, Y.J.; An, L.N.; Xu, C.Y. Temperature sensor made of polymer-derived ceramics for high-temperature applications. *Sens. Actuator-Phys.* **2014**, *219*, 58–64. [CrossRef]
19. Lu, K.; Erb, D.; Liu, M.Y. Thermal stability and electrical conductivity of carbon-enriched silicon oxycarbide. *J. Mater. Chem.* **2016**, *4*, 1829–1837. [CrossRef]
20. Liu, D.B.; Shi, B.L.; Wang, C.Q.; Li, Z.S.; Wang, X.B.; Xu, B.S.; Qu, L.J. Polymer-derived SiC ceramic aerogels with in-situ growth of SiC nanowires. *Ceram. Int.* **2022**, *48*, 9157–9163. [CrossRef]
21. Greenough, M.; Zhao, Z.Y.; Jacobsohn, L.G.; Tong, J.H.; Bordia, R.K. Low/intermediate temperature pyrolyzed polysiloxane derived ceramics with increased carbon for electrical applications. *J. Eur. Ceram. Soc.* **2021**, *41*, 5882–5889. [CrossRef]
22. Reinold, L.M.; Graczyk-Zajac, M.; Gao, Y.; Mera, G.; Riedel, R. Carbon-rich SiCN ceramics as high capacity/high stability anode material for lithium-ion batteries. *J. Power Sources* **2013**, *236*, 224–229. [CrossRef]
23. Cao, Y.J.; Yang, X.P.; An, L.A. Electric conductivity and microstructure evolution of polymer-derived SiAlCO ceramics. *Ceram. Int.* **2016**, *42*, 4033–4038. [CrossRef]
24. Chen, Y.H.; Yang, X.P.; Cao, Y.J.; Gan, Z.H.; An, L.A. Quantitative study on structural evolutions and associated energetics in polysilazane-derived amorphous silicon carbonitride ceramics. *Acta Mater.* **2014**, *72*, 22–31. [CrossRef]
25. Liu, T.; Xie, X.B.; Pang, Y.; Kobayashi, S. Co/C nanoparticles with low graphitization degree: A high performance microwave-absorbing material. *J. Mater. Chem.* **2016**, *4*, 1727–1735. [CrossRef]
26. Li, N.; Cao, Y.J.; Zhao, R.; Xu, Y.J.; An, L.A. Polymer-derived SiAlOC ceramic pressure sensor with potential for high-temperature application. *Sens. Actuator-Phys.* **2017**, *263*, 174–178. [CrossRef]
27. Stankovich, S.; Dikin, D.A.; Dommett, G.H.B.; Kohlhaas, K.M.; Zimney, E.J.; Stach, E.A.; Piner, R.D.; Nguyen, S.T.; Ruoff, R.S. Graphene-based composite materials. *Nature* **2006**, *442*, 282–286. [CrossRef]
28. Dalcanale, F.; Grossenbacher, J.; Blugan, G.; Gullo, M.R.; Lauria, A.; Brugger, J.; Tevaearai, H.; Graule, T.; Niederberger, M.; Kuebler, J. Influence of carbon enrichment on electrical conductivity and processing of polycarbosilane derived ceramic for MEMS applications. *J. Eur. Ceram. Soc.* **2014**, *34*, 3559–3570. [CrossRef]

micromachines

Review

Review of Flexible Wearable Sensor Devices for Biomedical Application

Xueli Nan [1,2,*], Xin Wang [1], Tongtong Kang [1], Jiale Zhang [1], Lanxiao Dong [1], Jinfeng Dong [1], Peng Xia [3] and Donglai Wei [3]

1 School of Automation and Software Engineering, Shanxi University, Taiyuan 030006, China
2 School of Biomedical Engineering, Shanghai Jiao Tong University, Shanghai 200030, China
3 School of Mathematical Sciences, Shanxi University, Taiyuan 030006, China
* Correspondence: nanxueli@sxu.edu.cn; Tel.: +86-15835128140

Abstract: With the development of cross-fertilisation in various disciplines, flexible wearable sensing technologies have emerged, bringing together many disciplines, such as biomedicine, materials science, control science, and communication technology. Over the past few years, the development of multiple types of flexible wearable devices that are widely used for the detection of human physiological signals has proven that flexible wearable devices have strong biocompatibility and a great potential for further development. These include electronic skin patches, soft robots, bio-batteries, and personalised medical devices. In this review, we present an updated overview of emerging flexible wearable sensor devices for biomedical applications and a comprehensive summary of the research progress and potential of flexible sensors. First, we describe the selection and fabrication of flexible materials and their excellent electrochemical properties. We evaluate the mechanisms by which these sensor devices work, and then we categorise and compare the unique advantages of a variety of sensor devices from the perspective of in vitro and in vivo sensing, as well as some exciting applications in the human body. Finally, we summarise the opportunities and challenges in the field of flexible wearable devices.

Keywords: flexible wearable; biomedical; flexible materials; sensors

Citation: Nan, X.; Wang, X.; Kang, T.; Zhang, J.; Dong, L.; Dong, J.; Xia, P.; Wei, D. Review of Flexible Wearable Sensor Devices for Biomedical Application. *Micromachines* **2022**, *13*, 1395. https://doi.org/10.3390/mi13091395

Academic Editors: Libo Gao and Zhuoqing Yang

Received: 3 August 2022
Accepted: 23 August 2022
Published: 26 August 2022

1. Introduction

Flexible wearable devices, which target the flexibility of devices, have unique features and advantages, such as being light weight, having good flexibility [1,2], and potential for miniaturisation [3]. They are widely used in biomedicine [4], information acquisition [5], human–machine interaction [6], and robotics [7,8], triggering a new round of technological innovation in the biomedical device industry. The wearable devices that are applied to the human body aim to achieve the real-time monitoring of human body information, personalised diagnosis and treatment, and have great potential for further development. Convenient, at-home personalised medical devices can provide real-time information on human physiological conditions [9]; simultaneously, they can reduce the pressure on hospital treatment and can save resources. Wearable biomedical products are becoming increasingly popular and recognised by the public.

According to an industry survey report that was released by the Xinsijie Industrial Research Centre, the global biosensor market will reach USD 23.62 billion by 2021. It is expected that, from 2022 to 2026, the global biosensor market will increase at an average annual growth rate of approximately 9%. As the main downstream market, the medical field accounts for 78% of the demand and is primarily used in wearable medical devices. In the context of the new crown epidemic, various places have taken epidemic prevention measures, to a certain extent, which has stimulated the development of personalised wearable devices. Coupled with the aging population and an increasing prevalence of chronic diseases annually, biomedical wearable devices have ushered in huge development opportunities. Currently, the established wearable devices include smart watches, sports bracelets, hearing aids, and vision correction devices. These wearable devices play an important role in the real-time monitoring of physiological information, auxiliary treatment, and healthcare. We believe that, with the development of society, flexible wearable devices for biomedicine will become popular.

In the past few years, flexible sensing, which is the core of wearable devices, has enabled the quantification of conventional external stimuli [10], such as pressure [11], tensile forces [12], and temperature [13]. With the development of disciplines such as materials science, control science, and communication technology, there has been a series of breakthroughs in flexible electronics [14]. These sensors have achieved multi-angle and multi-directional information acquisition and can maintain excellent performance in complex environments [15], with varying degrees of sensitivity, minimum detection values, and detection ranges [16]. These performance enhancements provide better stability, speed, and accuracy for the detection of weak signals in the human body. Additionally, the development of implantable [17], adaptive [18], and degradable [19] materials and the emergence of self-powered bio-cells [20] have led to the emergence of a wide range of in vivo sensor devices. However, there is a lack of comprehensive summaries of flexible wearable devices in the context of in vitro- and in vivo-oriented biomedicine [21,22]; therefore, we present a systematic overview of a wide range of wearable devices from a sensor perspective, as well as some exciting applications.

The sensitivity of the sensor reflects the lowest detection limit of the device, corresponding to the signal strength that can be perceived and measured. The detection range corresponds to the sensing signal that can be applied, and the response time and relaxation time are related to the real-time and rapid response ability of the wearable device. These key metrics determine the applicability and sensing capabilities of the wearable device. Additionally, the durability [23], cycling stability, and biocompatibility [24] of the sensor are fundamental for the stable operation of wearable devices over time. These properties are widely regarded as essential characteristics of flexible sensor devices. The establishment of flexible wearable electronic systems with a high degree of flexibility [25] (stretching, bending, and folding) and excellent performance metrics (conductivity [26], sensitivity [27], detection range [28], and stability [29]) is necessary for the development of personalised medicine in society.

It is worth noting that flexible wearable sensors for biomedicine usually have excellent flexibility and compatibility, but also have low quality. The excellent flexibility can meet the requirements of a complex body surface, and the sensor deformation that is caused by different limb movements does not affect the performance of the device. Excellent compatibility can prevent adverse reactions (inflammation and allergy) between the device and human body; compatibility is a necessary property of implantable sensors. Low quality can be carried easily and increases people's sense of experience, which is an important indicator of sensor production. However, wearable sensors for biomedicine still have limitations, and their production cost is relatively high, which is not conducive for the promotion of the products. There are many disposable wearable sensors that can cause environmental pollution. Cost reduction, recycling, and post-use degradation of sensors are important directions for current wearable-device research.

This paper has reviewed flexible sensor devices for biomedical applications, has summarised some synthetic materials that are commonly used for flexible sensing, has focused on their microstructure and sensing mechanisms, and has described the different ways in which sensor devices work in vitro and in vivo. The synthesis, doping, and modification of flexible materials exhibit excellent synergistic effects, enabling the integration of advantageous properties [30]. In vitro sensing is typically performed by using electronic skin [31] and medical devices [32] for human signals, whereas in vivo capture is typically performed using implantable and degradable [33] sensors for the information acquisition [34], wireless transmission systems [35,36], and self-powered electrical signal inputs [37]. Flexible electronics offer good compatibility with biological tissue materials, environmental adaptability, and harmless working behaviours to the human body, thereby promising a higher standard of disease monitoring and treatment. The general framework of this study is shown in Figure 1.

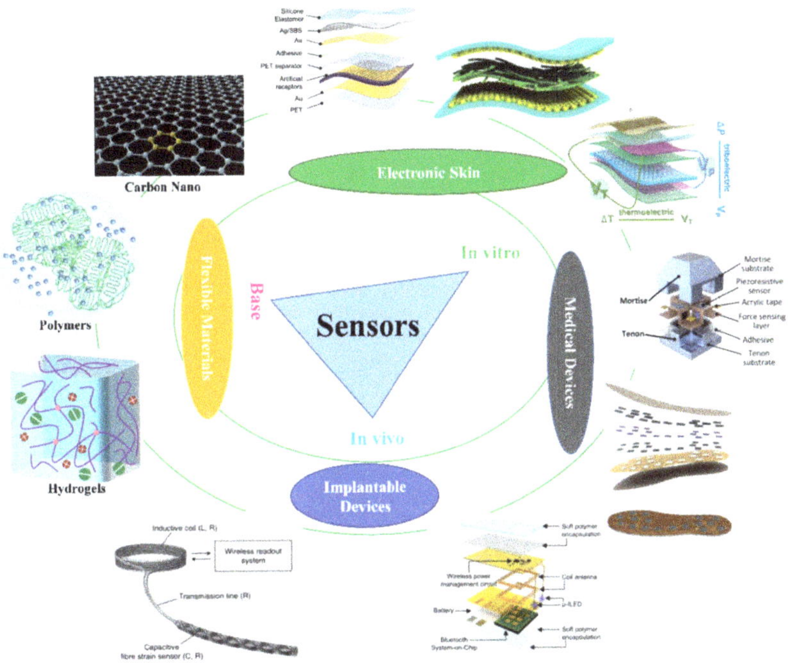

Figure 1. Overview of research in flexible wearable sensor devices for biomedical applications.

2. Selection and Production of Flexible Materials

The selection and fabrication of materials can significantly affect the performance of the sensors [38]. Biomedically oriented sensor devices place higher demands on the flexibility, conductivity, and durability of the materials. Existing flexible materials include polyethylene terephthalate (PET), polyimide (PI), polyvinyl alcohol (PVA), PDMS (Polydimethylsiloxane), and PEN (Polyethylene naphthalate).

2.1. Commonly Used Flexible Materials

PET, which is commonly used in flexible printed circuit boards, exhibits excellent mechanical strength. Simultaneously, PET has excellent electrical insulating properties and can be used as a device encapsulation layer. PET also exhibits light transmission properties and can be used as a protective layer for optical sensors.

PI has a few advantages over PET materials. Its biggest advantage is that it has excellent heat resistance. PI has excellent environmental stability and is unaffected by temperature or humidity. It can be used as a protective layer for sensors, preventing non-detectable stimuli from affecting the device performance.

PVA, which has excellent hydrophilicity, is often used in device adhesives and synthetic fibres. PVA has excellent film-forming properties, is easily degradable, and is popular in the manufacture of disposable devices.

Polydimethylsiloxane (PDMS) is the most commonly used flexible material. It has good flexibility and is an excellent force-sensitive material in both high- and low-temperature environments. Additionally, PDMS has good dielectric properties and can be used as a medium for pressure sensors. PDMS also has a certain air permeability, and is widely used in pressure sensing, flexible wearables, and device packaging.

PEN is commonly used in flexible printed circuit boards (PCBs) and capacitor films. It is more advantageous than PET in terms of heat resistance and mechanical strength and is a device material that comes in the form of a film. It exhibits good performance in blocking gas and water and can be used as a packaging material for devices.

Although PET, PI, and PEN are not flexible enough, these commonly used flexible device materials gain a certain degree of flexibility owing to the decrease in their thickness. PVA and PDMS are frequently used in patch-type wearable sensors.

However, these materials cannot meet the requirements of sensor fabrication. The synthesis, doping, and modification of a variety of flexible materials that exhibit excellent synergistic effects (1 + 1 > 2), and the integration of advantageous properties resulting from the reaction, are major current studies in the field of material synthesis [39]. This paper summarises the selection and the fabrication of the following three materials: carbon nanomaterials [40], polymeric materials [41], and hydrogel materials [42]. The typical examples of material syntheses are listed in Table 1.

Table 1. Preparation methods, advantageous properties, modes of action, and some unique applications of synthetic materials.

	Carbon Nanomaterials			Polymeric Materials		Hydrogel Materials	
Selection of Materials	**Graphene Nanoribbons in 1D, Ti3C2Tx MXene in 2D**	**Carbon Nanotubes in 1D, Graphene in 2D**	**1H,1H,1H, 2H-Pefluorodecyltrimethoxysilane, Ti3C2Tx Mxene**	**Styrene Monomer, Styrene Radical, Hydroxyl Group**	**Polyaniline, Dimethyl Formamide**	**PAAm Hydrogel, 3,4-Ethylenedioxythiophene, and Polystyrene Sulfonate**	**SiO_2 Nanoparticles and PAAm Hydrogel**
Synthesis and preparation methods	Doping-induced hybridisation, nitrogen doping	Controlling the dynamic injection of charged droplets	Superficial fluorine functionalisation, grafting modification	Solution polymerisation	Solvent-exchange strategy, wet-spinning technology	Dissolution diffusion	Adjustment of network structure and cross-linking mechanisms
Advantageous performance	High cycle stability, low hysteresis, durability	Nanoscale structure, nanoscale conductivity, lateral infinity	Surface hydrophobicity, resistant to oxidation	Powerful underwater absorbency, durability	High tensile ratio, high energy, excellent mechanical properties	Porous structure, ultra-thin thickness, high permeability, low impedance	High elasticity, no temporary entanglement, high sensitivity
Mode of action	Through the synthesis, doping, and modification of materials, the advantageous properties of multiple materials are combined, and new properties are generated through reactions, resulting in excellent composite materials, sensing media, and interactive interfaces.						
Applications	Improved low-dimensional material properties for health detection systems	Pseudo-3D macroforms and core fillers for high performance nanocomposites	Perception in a liquid environment, as a sensing medium layer	Bonding of devices in liquid environments	Weaving, energy harvesting, and charge storage	Acts as a liquid electrolyte, forming a conformable and low impedance interface	Used in repetitive motion with minimal energy dissipation
Main weaknesses	Inherently fragile	Easily tied a knot	Inherently fragile	Steric hindrance between hydroxyl groups	Easy to break	Poor mechanical properties and easy to tear	Not close contact
References	[43]	[44]	[45]	[46]	[47]	[48]	[49]

2.2. Carbon Nanomaterials

Carbon nanomaterials, which exhibit excellent electrical conductivity, such as graphene [50], carbon nanotubes [51], and MXene [52], have received considerable attention from scientists because of their excellent performance indicators and unique nanostructures. In more than 10 years since the discovery of graphene, the organic combination of different biomolecules and graphene has shown excellent functionalisation and a strong carrying ability, which stems from its ultra-high lateral space and excellent electrical conductivity. Carbon nanotubes and MXene have a natural advantage in the hybridisation of carbon-based materials because of their considerable biocompatibility, their inherent degradability of carbon-based materials, and their matching electrical conductivity. The difference is that MXene is more likely to produce different biological effects, owing to its hydrophilicity, diffusivity, and controllability caused by functional groups [53], which is particularly important in biomedicine. The excellent stability of carbon nanomaterials [54] makes them uniquely competitive in device cycle testing.

Lee et al. evaluated a new dopant-induced heterodimensional hybridisation method for 1D/2D materials (Figure 2a) by constructing hybrid structures of 2D Ti3C2Tx-type MXene and 1D graphene nanoribbons through nitrogen doping [43]. The structure has an excellent viscosity, a high resistance, and fabricates as a piezoresistive pressure sensor with hysteresis as low as 1.33% and a stable cycle count of more than 10,000 cycles under high pressure. Particularly, the authors have improved the connection properties of a low-dimensional hybrid material [55] via elemental doping in a gaseous environment, which resulted in better interfacial adhesion and substantially improved the hysteresis performance and high-pressure cycling stability of the sensor.

Ding et al. reported a self-assembled 2D carbon nanostructure network (Figure 2b) that was based on an electrospray/mesh technique [44]. Low-dimensional carbon nanomaterials exhibit excellent optoelectronic properties; however, their nano properties are significantly degraded when they are assembled into blocks. By controlling the dynamic injection of charged droplets and combining the nanoscale diameter of 1D carbon nanotubes and the lateral infinity of 2D graphene, a 2D nanostructured network was formed. It can be used as a pseudo-3D structure that displays an ordered network of nanofibres [56] and is laterally infinite, with excellent flexibility and nanoscale electrical conductivity.

Yang et al. prepared a Ti3C2Tx-type MXene film that was resistant to oxidation (Figure 2c) in a liquid environment [45], and formed a stable hydrophobic protective interface via fluorine functionalisation. Thus, by embedding bacterial nanocellulose, a highly sensitive sensing medium layer can be prepared with excellent detection performance in pressure sensing. In a liquid environment, its conductivity remains almost unchanged, and this excellent waterproofing performance is expected to realise invasive-force detection. The excellent hybridisation ability [57] and unique sensing properties [58] of MXene have attracted the attention of scientists in the field of biosensing.

These carbon nanomaterials have been successfully hybridised using a unique synthetic approach in order to obtain low-dimensional carbon materials, successfully expanding the 2D structure of carbon nanomaterials and enriching the potential mechanisms of the structure–property relationships [59]. It is worth noting that, when carbon nanomaterials are used alone, especially in bulk, their nanometre properties are severely compromised. Carbon nanotubes are prone to an uneven distribution during their application. Ti3C2Tx-type MXenes also have the problems of easy oxidation and recombination, and the hybrid interface is unstable. These limitations require the reasonable doping of carbon nanomaterials in order to optimise the device performance. However, the excellent properties of carbon nanomaterials, including adsorption, electrical conductivity, thermal conductivity, flexibility, low mass, and a high specific surface area, make them irreplaceable in biomedical wearable sensors.

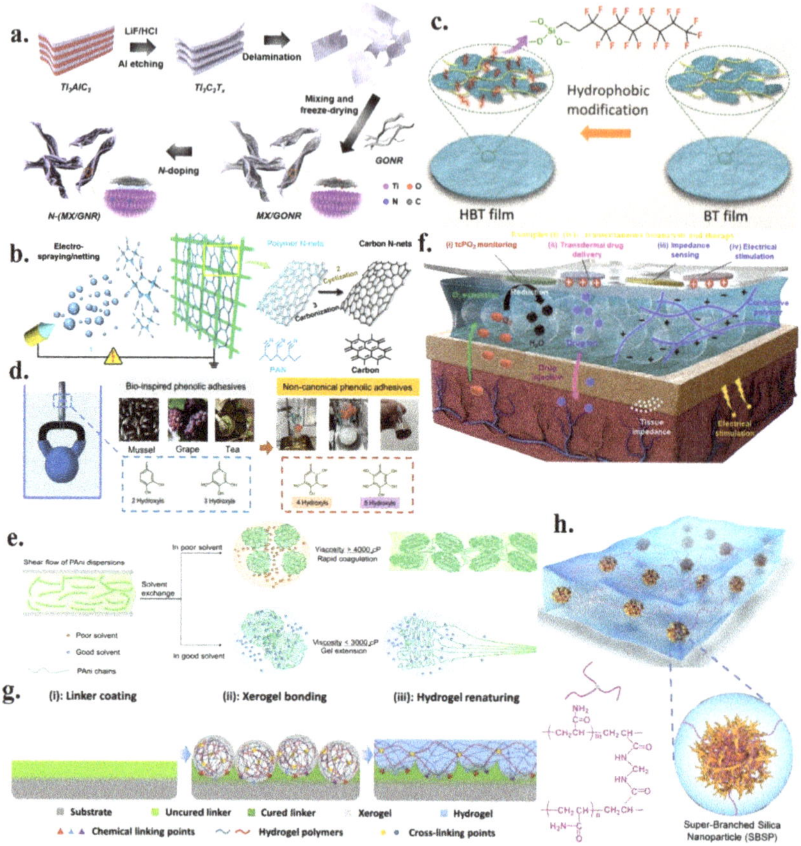

Figure 2. (**a**) New dopant-induced hybridisation method for 1D/2D materials; reproduced with permission from the American Chemical Society (2021) [43]; (**b**) Self-assembled 2D carbon nanostructure network; reproduced with permission from Springer Nature (2020) [44]; (**c**) Ti3C2Tx-type MXene film that is resistant to oxidation; reproduced with permission from the American Chemical Society (2022) [45]; (**d**) Non-standard phenolic polymer; reproduced with permission from Springer Nature (2022) [46]; (**e**) Ultra-fine polyaniline fibre; reproduced with permission from Springer Nature (2022) [47]; (**f**) Ultra-soft, highly permeable, low-impedance ultrathin hydrogel; reproduced from Ref. [48]; (**g**) Method for adhering hydrogels to various solid interfaces; reproduced from Ref. [60]; (**h**) Highly solvated, hyperbranched nanoparticle-reinforced polymer hydrogel; reproduced with permission from Wiley (2022) [49].

2.3. Polymeric Materials

Polymeric materials [61] are often used as flexible substrates and coating materials in order to improve the performance of sensors during fabrication. Polymer materials have been doped and modified in order to achieve different types of organisation and unique properties, including higher electrical conductivity [62] and ion mobility [63], as well as more stable structural and mechanical properties.

Recently, Ejima et al. reported a non-standard phenolic polymer [46] that exhibited strong underwater adsorption (Figure 2d). The addition of a hydroxyl group to the catechol moiety enhanced the adhesion capacity. The synthesis of phenolic polymers with four or five hydroxyl groups on styrene monomers results in superb underwater adsorption. These non-standard phenolic groups have excellent adsorption properties and need only be added to polystyrene in small quantities to produce strong underwater adhesion on different substrates. Particularly, this adhesion exhibits excellent stability and durability for long-term applications in bonding devices in liquid environments.

Tao et al. reported an ultrafine polyaniline fibre [47] that was wet-woven using a solvent-exchange strategy (Figure 2e). The polymer fibre reduced the viscosity of the fibre by diffusion between the solvents while enhancing the tensile capacity of the fibre. During stretching, the polymeric fibre maintained excellent mechanical properties, good electrical conductivity, and was less than 5 μm in diameter. The fibre is an excellent electrode material, which is even better than some carbon nanomaterials in energy storage [64] and current conduction [65]. The stability of its electrochemical performance is expected to be a major breakthrough in the self-supply of the system.

These polymeric materials have been developed in order to provide a more comprehensive range of performance requirements and to have significant applications in sensor integration and wearable fabrics [66]. Biomedical polymer materials are rapidly developing in terms of functionalisation, refinement, and fibrosis. The synthesised polymers have achieved remarkable results in implantable, degradable, catalysed, energy storage, and energy conversion applications, and are widely used. However, it is difficult to change the complexity of the preparation of polymer materials and the high cost; the effective improvement of this limitation will bring a powerful boost to the promotion of the research products.

2.4. Hydrogel Materials

Small and thin flexible devices should be more compatible and intelligent as alternatives to traditional medical devices. Carbon nanomaterials and polymeric materials do not fully meet the requirements of future personalised medical devices. The emergence of modified hydrogels [67] provides more options for personalised medical devices, which are widely used in areas such as electronic skin and human–machine [68] interactions. The porous structure of hydrogels is conducive for enhancing the transfer rate at the interface and has a wide range of applications in ion transfer and signal transmission [69]. The prepared hydrogels are extremely biocompatible by dissolution, reaction, and modification in order to obtain various advantageous properties [70].

For example, Kim et al. developed an ultra-soft, highly permeable, low-impedance ultra-thin hydrogel (Figure 2f) that acts as a liquid electrolyte on the skin, forming a tissue-like spatial- and quasi-solid interface [48]. This interface has a high permeability and low impedance properties and is extremely 'conformable' to different body parts. The porous structure [71] realises the diffusion and propagation of biological macromolecules, and the low impedance improves the injection efficiency of charges under external stimulation. Additionally, Wang et al. proposed a method to adhere hydrogels to various solid interfaces [60], providing a conformal contact method between the hydrogel and the torso sites (Figure 2g). The hydrogel was dehydrated, was glued on to a solid interface with a small amount of glue and was hydrated in order to form a new gel layer. This method enables the organic bonding of hydrogels to human tissues and has great potential for applications in bioassays and healthcare. Moreover, the conformal adhesion [72] of hydrogel significantly changes the physicochemical properties and electrical conductivity of the solid surface, making it possible for the same device to be reused in different complex environments.

Qiu et al. developed a highly soluble, hyperbranched nanoparticle-reinforced polymer hydrogel (Figure 2h) with no temporary entanglement, hysteresis-free material properties during cyclic loading, and fatigue-free properties that can significantly reduce energy dissipation and can exhibit excellent stretchability and flexibility [49]. Minimal material energy dissipation [73] is essential for achieving high device performance at sites of repetitive human actions (e.g., heartbeat, breathing, and movement). In hydrogel networks, hyperbranched nanoparticles act as the dominant cross-linked, highly expanded polymer chains. They are linked without temporary entanglement [74] and have excellent elastic properties, exhibiting great potential in strain and impedance sensors.

As special polymer materials, hydrogels exhibit extremely competitive properties in terms of biocompatibility. Hydrophilic polymers are often used as functional groups in hydrogels, which produce different physical and chemical properties under different environmental stimuli through physical or chemical cross-linking modification. The swellability of the hydrogel medium and the resulting network voids are greatly beneficial for the penetration and transportation of macromolecules. Additionally, hydrogels inherit the disadvantage of the high cost of polymer materials, while increasing the complexity of integration. It is undeniable that, although hydrogel is slightly insufficient in terms of cost and yield, its performance is excellent and cost-effective.

In summary, carbon nanomaterials, polymeric materials, and hydrogel materials were doped and synthesised rationally in order to obtain exciting and excellent properties. The difference is that carbon-based materials tend to acquire excellent piezoelectric properties [75], whereas hydrogel materials tend to focus more on the flexibility and permeability [76] of the material. The recently emerging metal-organic frameworks (MOFs) [77] are also highly sought after, and possess functionalities that are similar to carbon nanomaterials and porous structures that are similar to hydrogel materials. The different organic frameworks can be formed by self-assembly. Cu-based [78], Zr-based [79], porphyrin-based [80], and MXene-derived [81] MOFs have emerged and have begun to show their cutting-edge applications in biosensing, medical detection, and gas monitoring. The development of MOF materials with different groups is expected to achieve further breakthroughs and optimisation in the field of biomedical device fabrication. It is worth noting that different materials may produce nonlinear deviations under different working environments, and the choice of a reasonable working environment for synthesis is significant for the retention of different advantageous properties and the modification of inferior ones, which may result in unexpected outcomes.

3. Application of Sensors in the Human Body

Flexible wearable sensor devices are widely used in human–machine interactions, healthcare [82], and personalised diagnosis and treatment systems [83] in order to determine the physiological level of the human body. The prevalence of various high-precision preparation processes, such as 3D printing [84], photolithography [85], and printed coating techniques [86], has dramatically increased both the accuracy and the yield, facilitating the ease of device fabrication. The use of other auxiliary equipment and test systems, such as centrifuges, thermostatic heating chambers, ultrasonic devices, and LCR testers, enables more efficient device fabrication. Simultaneously, wearable devices should have good flexibility, transportability, and excellent biocompatibility [24], contributing to more sensitive feedback, stable control, and efficient transmission. From the perspective of the detection location, sensors can be classified as in vitro [87] and in vivo [88]. In vitro sensor devices are further divided into electronic skin classes and other medical devices based on the appearance of the package and the application of the device. Compared with in vitro sensing, the application environment [89] for in vivo sensing is more stringent, but many professors and scholars have made unique attempts and products. In this review, we present a comprehensive overview of in vitro and in vivo sensing techniques.

3.1. In Vitro Sensing

In recent years, wireless communication [90] and human–machine interaction technology have developed rapidly, and signal transmission is highly consistent and instantaneous, making the in vitro sensing of wearable devices possible. In vitro flexible wearable devices that are used to track the movement of various parts [91] of the human body, assist in the diagnosis of various diseases [92], and can establish a higher level of human–machine interaction interface, have become a major focus in in vitro sensing. Wireless medical devices that are based on wireless data transmission, which can be applied to flexible platforms and environments with excellent information output and control stability, have become one of the main trends in biomedical fields. In vitro sensor devices usually exist in the form of electronic skin or medical devices. As such, flexible wearable devices can adapt to human limb movements and have good bending and stretching properties [93].

3.1.1. Relying on Electronic Skin for In Vitro Sensing

Electronic skin serves as an excellent integrated platform with a thin and soft interface that can be applied to most parts of the human surface. Electronic skin can sense internal physiological information about the human body, can make sound judgments, and can prevent and monitor diseases. The inherently weak properties of the human signal dictate that the built-in sensors must be sufficiently sensitive [94], and the persistence of the human signal requires an excellent robustness [95] of the electronic skin. The sensor transmits the human body signal as an optoelectronic signal in order to enable information interactions.

Deng et al. reported a patch-type piezoresistive sensor with both high sensitivity and a wide detection range (Figure 3a) for monitoring human physiological signals [96]. Inspired by the microstructure of a rose petal, the sensor exhibits excellent mechanical flexibility and electrical properties using a layered microstructure with polyaniline/polyvinylidene fluoride nanofibre films on the top and bottom layers and interlocking electrodes with a domed structure in the middle. Structurally, this domed interlocking structure [97] can significantly increase the contact area and reduce the resistance when subjected to pressure and exhibits a low relaxation time, enabling a sensitive adjustment of the electrical signal output of the sensor (Figure 3b). The structure can significantly expand the contact area of the nanofibres under pressure and can detect small pressures. In terms of performance, the sensor has a sensitivity of 53 kPa-1, a response time of 38 ms, a recovery time of 19 ms, a very low power consumption, and an excellent robustness (over 50,000 cycles). The designed skin patch has a significant ability to perceive vibrational forces and can monitor subtle motion signals (e.g., breathing, swallowing, muscle vibration, and foot pressure).

Wen et al. designed an electronic skin that senses pressure and temperature independently (Figure 3c) and can convert pressure into a voltage difference between polar plates [98]. Similarly, the sensor uses a tapered microstructure for sensitive pressure sensing. In contrast, the sensor employs a triboelectric nanogenerator (TENG) whose tapered structure is applied to the triboelectric layer. Particularly, the electronic skin can express the temperature signal independently while also sensing pressure (Figure 3d). The temperature signal is detected using a specially designed thermocouple membrane, and the thermoelectric coupling effect [99] can be expressed. The triboelectric sensor senses the pressure change from the vertical direction, whereas the thermocouple film senses the temperature change from the horizontal direction. This vertical detection method avoids the complicated decoupling process, enabling dynamic sensing of pressure and temperature from external signals.

In the above-mentioned electronic skin research, the perception of the magnitude of the force was realised, but the perception of the direction of the force was not specific. This is because the relatively parallel hierarchical structure of the electronic skin has unique advantages in terms of the perception of force magnitude, and the perception of force direction [100] often needs to be performed through an array. Therefore, integrating sensor units into electronic skin arrays is a widely used approach in sensor research.

Notably, Yun et al. reported a stand-alone skin-like health care patch (Figure 3e) that detects changes in the heart rate through conformal contact with the wrist [101]. The body of the patch is a flexible light-volume tracing sensor array consisting of an organic photodiode (OPD) and two organic light-emitting diodes (OLED). The OLED emits a red light. After being reflected by the arm, a change in the reflected light intensity is detected by the OPD. Notably, the intensity of the reflected light varies owing to the changes in the heart rate. It also carries a soft display consisting [102] of a diode array (17×7) with a variable microcrack. Au interconnects combined with a stress relief layer in order to allow stable operation under folding, twisting, and 30% stretching. The signal is processed by the microcontroller, is filtered, counted, and displayed on the diode array in real-time. The patch can be scaled up to monitor biosignals in daily life, contributing to more comprehensive health monitoring.

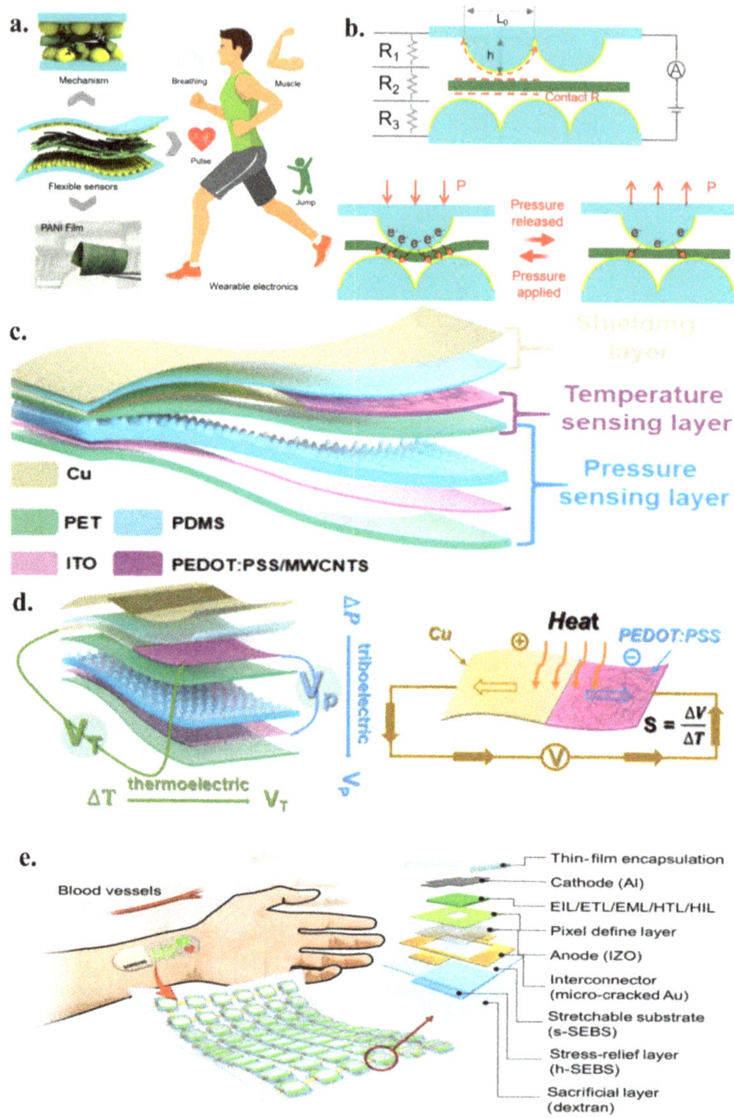

Figure 3. (**a**) Patch−type piezoresistive sensor; reproduced with permission from the American Chemical Society (2021) [96]; (**b**) Electrical signal output from the sensor; reproduced with permission from the American Chemical Society (2021) [96]; (**c**) Electronic skin that senses pressure and temperature independently; reproduced with permission from Elsevier (2022) [98]; (**d**) Temperature signal from the electronic skin; reproduced with permission from Elsevier (2022) [98]; (**e**) A self-contained skin−like health care patch; reproduced from Ref. [101].

These studies on sensing human signals have demonstrated that sensors rely on electronic skin in order to establish a sensing platform [103] for human signals, which can effectively capture human body information and is extremely important for implementing the health monitoring of the human body. However, this information does not enable a comprehensive judgment of human health, and the monitoring of some hidden signals, particularly in vivo signals [104], is one of the current research areas in in vitro diagnostics.

In addition to these traditional patch-type electronic skins, the application of electronic skin with integrated triboelectric nanogenerators (TENGs) has recently emerged, which realises the self-powering of wearable devices. The application of TENGs makes the integration of wearable devices and other sensor communication technologies more convenient, and the self-powered feature has unique advantages for device miniaturisation.

Recently, Wang et al. reported an intelligent tactile sensing system combining TENGs and deep learning technology [105] (Figure 4a), which has an important practical value in the cognitive learning of visually impaired patients and the fabrication of bionic prostheses. The system integrates three separate tactile sensors that are doped with different materials and uses a raised surface structure for better sensitivity and perception. By extracting features from the electrical signal of the tactile sensor and through coupling calculations, the typical features of an unknown material can be obtained. The relationship and normalisation of the three electrical signals were established in order to realise the tactile perception of size, position, pressure, temperature, and humidity. By integrating a convolutional neural network, the features were visualised, and a high-precision detection of 96.62% was achieved. The application of deep learning technology can achieve more subtle material feature recognition than that of the human eye.

Wang et al. designed a badge-reel-like, stretchable, wearable, self-driving sensor, and its system [106] (Figure 4b). TENGs, utilising grating structures, exhibit extremely low hysteresis and ultra-high durability (over 120,000 duty cycles). When the sensor is bent or stretched, the triboelectric layer is squeezed in order to produce an electrical signal. On this basis, the use of peak counts for the dynamic monitoring of body posture, especially in places where bending and stretching of the spine, joints, neck, and other locations often occur, is conducive for reducing the risk of disease in the spine and joints that may be caused by long-term bad posture. The system adopts mature 3D printing technology and flexible printing technology, making it possible to realise the popularisation of the products.

These studies on the perception of external stimuli have proven that flexible electronic skin can effectively interact with the external environment. On the basis of breaking the original balance by external stimulation, the triboelectric layer is contacted and squeezed in order to realise the output of the electrical signals. The TENGs enable the e-skin to achieve self-sufficient power requirements, which can quickly provide detectable responses. Additionally, avoiding the influence of interfering stimuli is still an urgent problem that needs to be solved, which determines the environment in which they can work.

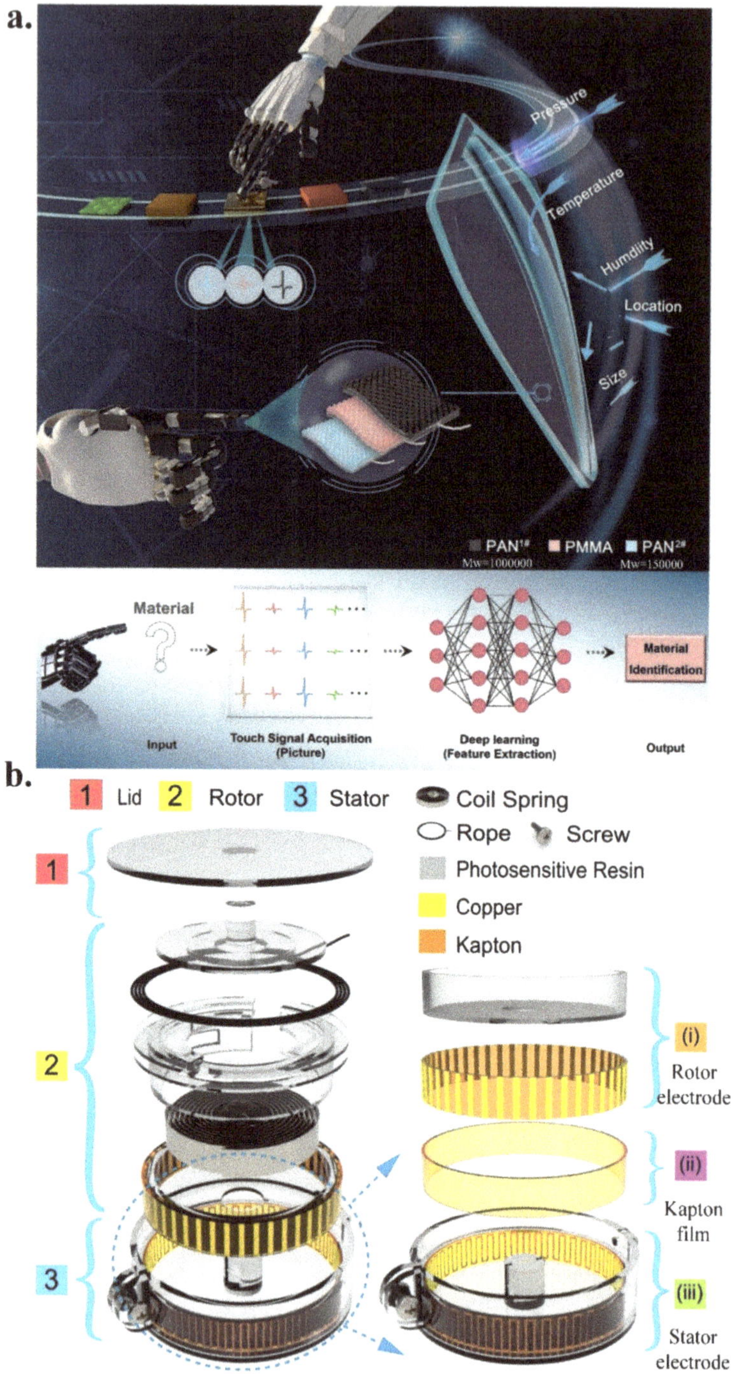

Figure 4. (**a**) Intelligent tactile sensing system combining TENGs and deep learning technology; reproduced with permission from Wiley (2022) [105]; (**b**) Badge-reel-like, stretchable, wearable, self-driving sensor, and its system; reproduced with permission from Springer Nature (2021) [106].

3.1.2. Relying on Medical Devices for In Vitro Sensing

Flexible medical devices [107] are worn on the human body and they interact with information using built-in sensors. Compared with electronic skins, medical devices have a variety of packaging forms and can carry sensors with various geometrical appearances [108], more diverse sensing forms, and more comprehensive sensing information, promising better diagnosis and treatment.

Wang et al. designed an active sensing array [109] for the diagnostic assessment of lumbar degenerative diseases by detecting plantar pressure distribution (Figure 5a). A PVDF piezoelectric sensor was used to convert force to electricity. The array performed well in pressure cycling and response time tests with a high degree of robustness, supported vector machine supervised learning algorithms [110], collected walking foot pressure distributions from patients with lumbar degenerative diseases for common human limb movements, built a database sample, classified and identified based on plantar pressure, and achieved a recognition accuracy of 99.2%. The monitoring system that was established based on the sensing array allows information interaction with mobile phones via Bluetooth in order to display the wearer's foot pressure distribution in real-time. The patient's foot pressure data can be effectively uploaded to the client. It facilitates the diagnosis and the rehabilitation assessment of patients with degenerative lumbar spine diseases. This back-end data processing method has been explored for the establishment of the medical Internet of Things.

In another study, Koh et al. reported a flexible gripper [111] that is capable of gripping millimetre-sized organisms while enabling temperature and pressure sensing (Figure 5b). The gripper enables bidirectional sensing and stimulus interaction through silver nanowires [112], while acting as a mediator for Joule heating. The strain and the vibration are detected through the addition of a cracked strain sensor, which changes the structure of the strain gauge by stretching and bending the gripper. This crack-type strain sensor is located on the back of the gripper, and any mechanical deformation that is experienced by the soft gripper can be sensed, which causes structural changes in the crack and affects the resistance of the sensor. By carrying a heater, temperature sensing and thermal motion are realised using AgNWs. Additionally, the gripper can provide temperature and pressure stimulation, which can contribute to the maintenance of human tissues, such as muscle soreness and bone misalignment.

Wu et al. reported a six-dimensional force sensor with a flexible mortise and tenon interlocking structure (Figure 5c), which was studied and decoded using a neural network [113]. The six-dimensional forces are forces F_x, F_y, and F_z and moments M_x, M_y, and M_z. From the perspective of control science, the sensor performs neural network and finite element analyses on the forces at different angles, showing that the sensor has different states under multi-angle forces. The sensor, with a minimum size of 7 mm × 7 mm × 7 mm, is used in the precise orthodontic treatment of teeth. It has 12 sensing units that can quantitatively detect external stimuli in different directions and output a voltage signal when the sensor is squeezed or twisted.

In addition to these traditional force tests, "non-force" tests are favoured by researchers. For example, temperature detection [114] of the human body surface and body and the detection of breathing gas [115]. In this era of the COVID-19 pandemic, these tests have attracted special attention in medical surveillance.

Recently, Ding et al. reported a fibre-optic temperature sensor [116] that is based on photoelectric up conversion (Figure 6a), which exhibits strong temperature dependence from the conversion of infrared light to visible light. The sensing unit is mainly composed of InGaP-based double-junction light emitting diodes and GaAs-based light emitting diodes. A large bandgap is connected in a series in the middle, exhibiting excellent sensitivity to temperature. With an increase in the temperature, the optical wavelength also increases. Simultaneously, an increase in the temperature leads to the narrowing of the bandgap, which leads to a reduction in the voltage. The optical signal exhibited excellent robustness in the cyclic heating and cooling tests. Additionally, the sensor has a strong anti-electromagnetic interference ability, and when combined with thermal imaging [117] and nuclear magnetic resonance [118], it is expected to be able to measure the local temperature in the body. The wavelength of the emitted light of the sensor is still limited to the red-light band, and the development of emitted light in the longer wavelength band is a research hotspot in the field of photoelectric up conversion in the future.

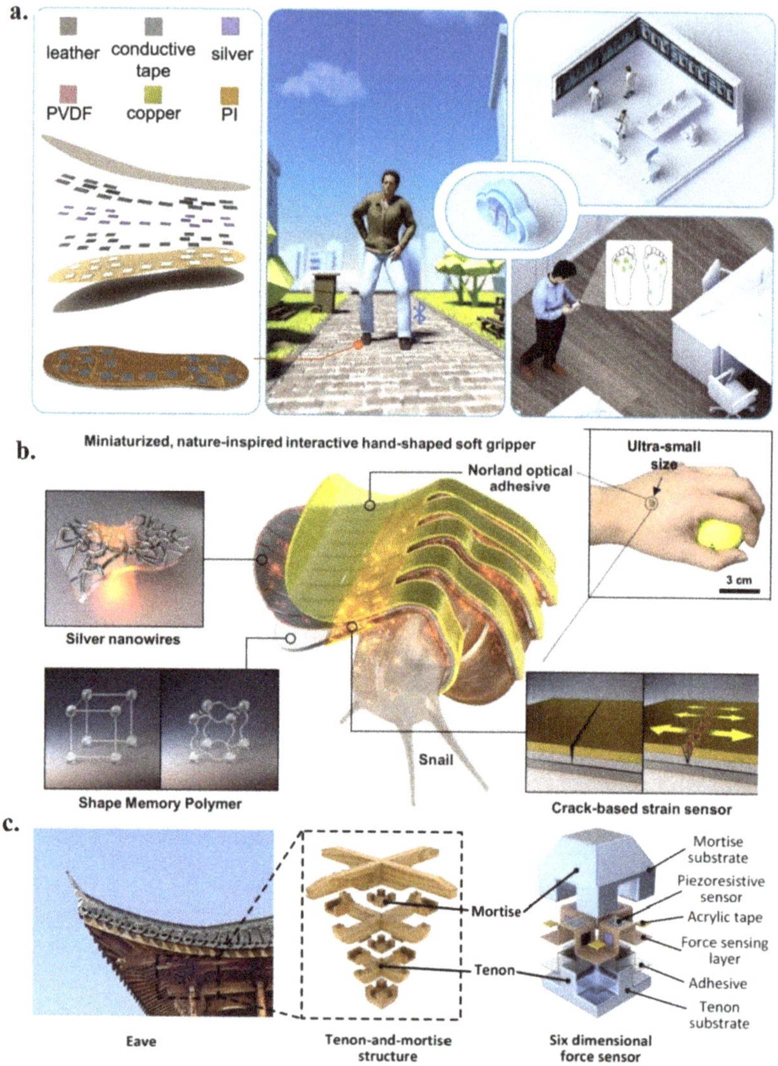

Figure 5. (**a**) Active sensing array; reproduced with permission from Wiley (2022) [109]; (**b**) Flexible gripper; reproduced with permission from the American Association for the Advancement of Science (2021) [111]; (**c**) Six-dimensional force sensor with a flexible mortise and tenon interlock structure; reproduced with permission from Elsevier (2022) [113].

Sheng et al. reported a polyimide-sputtered and polymerised [119] gas humidity sensor (Figure 6b); it broke the perception that the traditional polyimide capacitive humidity sensor has low sensitivity and a slow response, and at a large detection range, low humidity (<40%) could still achieve a humidity measurement. Polyimide-sputtered and polymerised materials have excellent hygroscopicity and hydrophilicity, and their porous spatial structure realises rapid gas transfer. When the sensor is in a humid environment, the film is gradually oxidised, which directly affects the capacitance of the sensor, thereby realising humidity detection. The sensor is applied to respiratory humidity monitoring [120], which detects the humidity of inhaled and exhaled air, and is very competitive in the mask, face shield, and helmet industries.

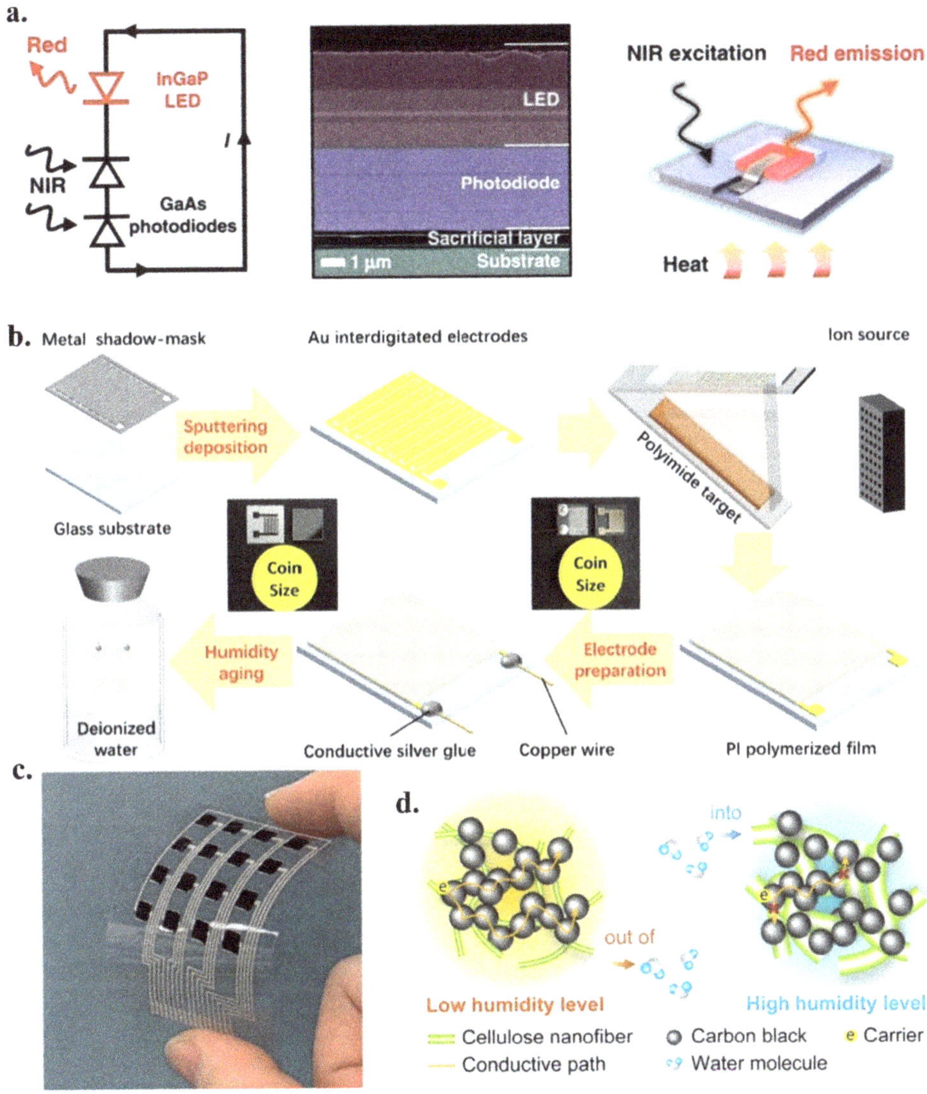

Figure 6. (**a**) Fibre-optic temperature sensor based on photoelectric up conversion; reproduced with permission from Springer Nature (2022) [116]; (**b**) Polyimide-sputtered and polymerised gas humidity sensor; reproduced with permission from the American Chemical Society (2022) [119]; (**c**) Ink-based printing flexible humidity sensor; reproduced with permission from the American Chemical Society (2022) [121]; (**d**) Schematic of the humidity detection mechanism for the cellulose nanofibres/carbon black composite; reproduced with permission from the American Chemical Society (2022) [121].

Similarly, Tokito et al. reported an ink-based printing, low-cost, fast-fabricated, and flexible humidity sensor [121] (Figure 6c). Conductive ink, which is composed of cellulose nanofibres and carbon black, changes the resistance of the ink according to the change in humidity (Figure 6d). The hygroscopic expansion of nanofibres destroys the conductive path of the carbon black, resulting in an increase in electrical resistance. The porous physical structure of nanofibres makes this hygroscopic ability even better. Additionally, the carbon black content is critical for the conductivity of the ink. Low content prevents the ink from forming sufficient conductive paths, whereas high content reduces the sensitivity of the sensor. In a high humidity environment, the ionic conductivity of the ink is expected to achieve sensitive humidity detection. The sensor realises low-cost and high-efficiency humidity detection and has unique advantages in the disposable product manufacturing industry. It is worth noting that the adhesion between the sensing layer and substrate is insufficient, and it can easily fall off in a humid environment. The combination of this with the non-standard phenolic polymers that have been mentioned in the previous materials section, perhaps with unexpected effects, would be interesting.

Sensors play a significant role in the research on medical devices. The combined use of multiple sensors allows the detection of information regarding the human body from multiple perspectives and provides a more comprehensive understanding of human physiology. The application to different parts of the human body [122], particularly to uneven areas, in a flexible package [123] creates conformal contact with the body. The development of these electronic skins and the design of medical devices have significantly improved the diagnosis and treatment methods of human diseases, opening a 'new window' in the future of smart medicine.

3.2. In Vivo Sensing

The inherent mechanical and biological properties of human tissues vary among people, and implantable devices that achieve immediate detection face major problems [124], such as the following: (1) it is difficult to apply existing electronic devices to organs with 3D structures, which can easily cause rejection by the body; (2) the devices are in a complex environment in the body, and long-term monitoring may lead to device failure; and (3) there are difficulties in achieving a stable transmission using passive wireless reading devices for strain sensors and maintaining the quality factor of the RLC circuit. Additionally, the complex physiological environment inside the human body, the secretion of stomach acid and catabolic enzymes, and the inherent contraction and expansion movements of organs and tissues can cause damage to the device structure and the human body. In vivo sensor devices are often implantable, adaptive, and degradable, allowing for in vivo 'spaced' sensing [125,126], wireless charging [127], and self-powered power devices [128–130], which are harmless and adaptable in vivo. The device is stable in the event of an in vivo organ 'movement' and can be displayed in real-time on a mobile device terminal via Bluetooth [131].

Recently, Lee et al. reported a suture-process-connected, wireless capacitive fibre optic strain sensor (Figure 7a) for connective tissue monitoring [124]. The sensing system consists of a hollow double-helix structure comprising two stretchable conductive fibres combined with a passive RLC circuit. The relationship between the mechanical and electrical properties is derived using an analytical expression. The hollow double-helix structure forms excellent tensile properties. During the stretching process, the two fibres approach each other and are squeezed after contact, resulting in a change in the capacitance. The addition of an induction coil to the wireless system realises the wireless signal-reading capability. The mechanical–electrical property relationship can be adjusted through the conductive fibre length diameter, creating more relaxed conditions for the implantation position, and the system enables wireless transmission without welding [132].

Jeong et al. proposed a wirelessly charged, fully implantable soft optoelectronic system [133] (Figure 7b), which realises rapid adjustment and adaptation in complex environments and is widely used in in vivo neuroscience research. The system performs a light stimulation through optoelectronic nerve probes, realises wireless transmission through coil antenna inductive coupling, receives radio-frequency energy, and achieves wireless charging. The system simultaneously enables remote selective wireless control and closed-loop wireless automatic charging capabilities for multiple animals using an off-the-shelf smartphone. In the preparation of devices, the development of phenolic polymers, which have a strong adsorption force in liquid environment, has made outstanding contributions to the bonding and integration of devices. Phenolic polymers are often used as carriers for drug delivery and have excellent antibacterial and anti-inflammatory properties [134]. This enables long-term in vivo studies without intervention [135], creating a new benchmark for in vivo sensing. However, the system suffers from metal-sensitive interconnects and uses batteries, resulting in a certain degree of incompatibility with MRI machines, narrowing its potential for biomedical applications; however, the design is still good enough.

In another study, Wang et al. reported a biodegradable, self-fed pressure sensor (Figure 7c) that could convert environmental pressure into an electrical signal for post-operative cardiovascular care [136]. The core of the sensor consists of two triboelectric layers, which are in contact under the action of an external force in order to achieve voltage output. An air space structure was introduced in order to ensure that the force–electricity change curve had good linearity. The sensor exhibits good sensitivity and durability by providing kinetic energy through the frictional electrical effect of the degradable material. Additionally, the sensor had 99% antimicrobial properties, a 5-day service time, and an 84-day degradation time, which can effectively prevent wound infections.

Rogers et al. reported a wireless implantable microflow sensing system (Figure 7d) that enables stable and reliable flow sensing with readout on a mobile terminal [137]. The sensor detects the blood flow through a heater and thermistor and is connected through a small Bluetooth module that is attached to the skin for data reading and for visual display on the mobile terminal. The sensing system was applied to surgical flaps and later, to organ transplantation in order to detect bleeding in microvessels using a thermal probe for continuous monitoring. The probe barbs exhibit excellent resistance to temperature and mechanical disturbances, are made of degradable materials, facilitate the mitigation of inflammation in organ tissues, and exhibit strong biocompatibility.

Interestingly, Bao et al. evaluated a stretchable biological interface [138] for neural tissue (Figure 7e), which was prepared by the laser patterning of a metal-complexed polyimide into an interconnected graphene/nanoparticle network that was embedded in an elastomer. The interface is used for a seamless connection between the central nervous system and gastrointestinal tract tissue. The real-time detection of neurotransmitters is realised through different impedances in different stretching states, and autonomous monitoring can be performed without external stimulation. The interface has excellent stretchability and softness and is highly compliant with intestinal tissue, avoiding the unnecessary irritation that is caused by peristalsis. Moreover, the interface has excellent adhesion and easily adheres to tissue membranes. It can also be mounted on endoscopes in order to realise direct sensing of the stomach, which is expected to solve the problem of the correlation between microorganisms and intestinal chemical kinetics [139].

These studies of implantable sensing systems imply that in vivo sensing has unique advantages in biomedicine, enabling more direct information acquisition [140] and health monitoring [141]. Additionally, implantable sensor devices are uniquely positioned to carry drugs [142,143] and can provide promising targeted drug delivery for localised treatment. This near-attrition-free drug therapy significantly enhances the treatment efficiency and is a novel clinical treatment tool.

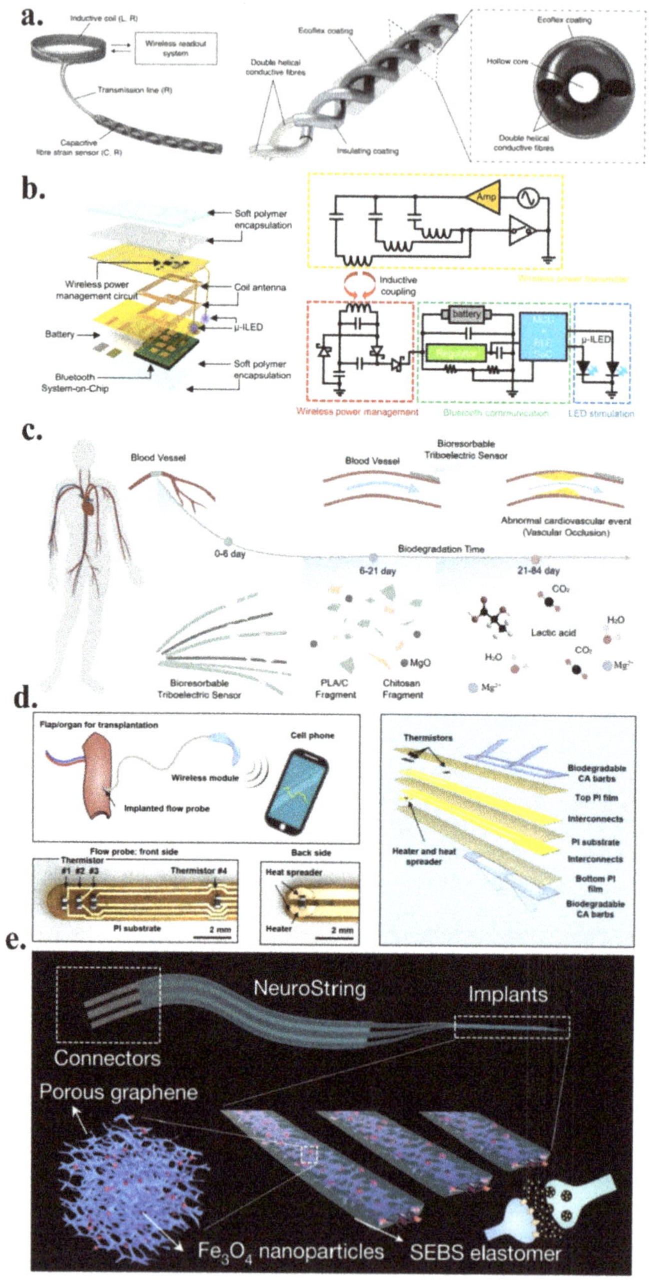

Figure 7. (**a**) Suture–process–connected wireless capacitive fibre optic strain sensor; reproduced with permission from Springer Nature (2021) [124]; (**b**) Wirelessly rechargeable, fully implantable soft optoelectronic system; reproduced with permission from Springer Nature (2021) [133]; (**c**) Biodegradable self–powered pressure sensor; reproduced with permission from Wiley (2021) [136]; (**d**) Wirelessly implantable microflow sensing system; reproduced with permission from Elsevier (2022) [137]; (**e**) Stretchable biological interface for neural tissue; reproduced with permission from Springer Nature (2022) [138].

4. Discussion and Conclusion

Notably, flexible wearable sensors that are used in biomedicine are different from the traditional sensors that are used in signal processing. Owing to the particularity of the working environment of wearable sensors and the miniaturisation of devices, wearable sensors often use integrated back-end circuits in order to analyse signals. Microcontrollers are usually used for signal acquisition and transmission, combined with the C language and Python, in order to develop PC-based applications for data visualisation. The real-time display on the terminal is realised through the interaction with the application program through serial communication. In wireless sensing devices, coil coupling, or optical communication is often used in order to realise data transmission. A real-time display on the mobile terminal can be realised using Bluetooth. The development of vector machine-supervised learning algorithms, deep learning algorithms, and applications is also extremely important in the medical Internet of Things.

Flexible wearable sensors for biomedicine can realise force detection in high-strain soft tissues, including human skin, the detection of temperature and humidity, and the detection of physiological information in complex environments in vivo. Different flexible wearable sensors are prepared using different materials, structures, sensing mechanisms, and communication methods, which can realise the stable and accurate detection of human pressure, strain, multi-angle force, vibration, temperature, humidity, cell tissue, flow, and microorganisms. Therefore, we believe that the application of flexible wearable devices in biomedical applications is feasible.

In this review, we have presented an updated overview of wearable devices that are used in biomedical applications. Starting with materials, we have presented cutting-edge preparation techniques and have analysed the important properties of these materials. By relying on advanced material preparation techniques, we have combined modern sensing technologies and MEMS processes in order to design and fabricate excellent sensor devices. We have discussed the unique sensing modalities, the excellent performance, and some exciting applications of these sensor devices from the perspective of in vivo and in vitro sensing. To date, these wearable devices have exhibited remarkable performance in solving specific problems with a much greater efficiency. This emerging field encompasses considerable opportunities and challenges in terms of the sensing capability, orientation, and adaptability. Flexible wearable sensor devices are increasingly moving toward adaptive, self-feeding, and personalised directions, providing multimodal [144], multifunctional [145], and multidirectional [146] detection capabilities. This vision has been made more relevant by combining it with data-processing and wireless-transmission systems. However, the types of signals that sensors can perceive are still limited. For example, the effect of in vitro detection at the cellular and protein levels are poor. The active detection of sensor targeting in vivo [147,148] remains a problem that modern science urgently needs to overcome. There is considerable room for the optimisation of the compatibility between wearable devices and human tissue, from harmless to beneficial, from short-term to long-term monitoring, and the optimisation of the degradation time. In the future, the development of wearable devices will require the cooperation of various disciplines (materials science, biomedicine, chemistry, microelectronics, and communication technology) in order to break professional barriers. It is worth noting that the development of processing technology and testing technology is equally important in order to realise the leap from theory to reality. Flexible wearables are the result of a multidisciplinary intersection. With the rapid development of various disciplines, it is reasonable to predict that flexible wearables will definitely attract attention in biomedical applications.

5. Summary and Outlook

We have presented the latest research on various flexible wearable sensing devices for biomedical applications, focusing on in vivo and in vitro sensing. As mentioned earlier, the detection of human physiological signals has expanded with the development of new materials and advances in sensing technologies. Flexible wearable sensors have great development potential in the fields of assisted diagnosis and treatment, health monitoring, cell capture, medical prosthetics, human–computer interaction, drug delivery, and targeted therapy.

Future work and product promotion face many obstacles. Whether in vitro or in vivo sensing, the known materials still cannot fully meet the needs of detection, which is closely related to the weakness of human signals. Discovering more emerging materials, performing different degrees of synthesis, doping on known materials, and exploring more suitable materials will be beneficial. For example, combining different groups with MXene and MOF materials, or crosslinking and modifying hydrogels, can achieve better performance. The development of various adaptive and self-powered materials has significantly boosted the development of implantable sensors. In addition, the shortcomings of sensor communication technology restrict the detection of human physiological signals, which is especially obvious in in vivo sensing. There are three reasons for this finding, as follows: First, human tissues have certain obstacles to signal transmission. Second, the complex environment of the body is not conducive for the operation of the sensor. Third, the rejection reaction of the human body itself, as well as the material properties of the sensor, can easily cause discomfort to the human body. Additionally, the development of processing and testing technologies is equally important for the development of new materials, new structures, and more sophisticated wearable devices. The rise in 3D printing technology has led to the refinement of device products and a higher yield. Simultaneously, the development of medicines is extremely important for drug delivery and targeted therapy.

Notably, the detection of human physiological information using wearable devices has great application potential in the medical Internet of Things. Through wearable device protocol communication technology, the IoT gateway is constructed, and the IoT server realises the real-time communication between the users and the medical staff. Medical staff can detect physiological information based on the wearable devices, set detection parameters, and synchronously process the user's physical condition. The data visualisation of the wearable sensor on the mobile terminal through Bluetooth, which has been realised at this stage, has explored the architecture of medical IoT. Whether the wearable sensor comprehensively detects the physiological signal is directly related to whether the medical staff can make an accurate judgment. The architecture of the protocol communication technology, the IoT gateway, and the server must be jointly developed by researchers in many industries.

The application of flexible wearable devices in biomedicine requires joint development in various disciplines. It is believed that with the efforts of various disciplines, the application of flexible wearable devices in biomedicine will become more comprehensive. The transition of these applications from the laboratory to clinical applications and to actual life is something that is worthy of further study.

Author Contributions: Conceptualisation, X.N. and X.W.; methodology, X.W.; software, X.W.; formal analysis, X.N. and X.W.; investigation, X.W.; resources, X.N.; writing—original draft preparation, X.W.; writing—review and editing, X.N., T.K., J.Z., L.D., J.D., P.X. and D.W.; supervision, X.N.; project administration, X.N.; funding acquisition, X.N. All authors have read and agreed to the published version of the manuscript.

Funding: This research was supported by the National Natural Science Foundation of China (No. 61804090).

Conflicts of Interest: The authors declare no conflict of interest.

References

1. Ohm, Y.; Pan, C.; Ford, M.J.; Huang, X.; Liao, J.; Majidi, C. An electrically conductive silver–polyacrylamide–alginate hydrogel composite for soft electronics. *Nat. Electron.* **2021**, *4*, 185–192. [CrossRef]
2. Baumgartner, M.; Hartmann, F.; Drack, M.; Preninger, D.; Wirthl, D.; Gerstmayr, R.; Lehner, L.; Mao, G.; Pruckner, R.; Demchyshyn, S.; et al. Resilient yet entirely degradable gelatin-based biogels for soft robots and electronics. *Nat. Mater.* **2020**, *19*, 1102–1109. [CrossRef] [PubMed]

3. Song, H.; Luo, G.; Ji, Z.; Bo, R.; Xue, Z.; Yan, D.; Zhang, F.; Bai, K.; Liu, J.; Cheng, X.; et al. Highly-integrated, miniaturized, stretchable electronic systems based on stacked multilayer network materials. *Sci. Adv.* **2022**, *8*, eabm3785. [CrossRef] [PubMed]
4. Kai, W.; Denis, T.; Shuang, L.; Renata, S.; Kumar, C.V.; JianPing, W. Giant Magnetoresistance Biosensors in Biomedical Applications. *ACS Appl. Mater. Interfaces* **2022**, *14*, 9945–9969. [CrossRef]
5. Su, Q.; Zou, Q.; Li, Y.; Chen, Y.; Teng, S.-Y.; Kelleher, J.T.; Nith, R.; Cheng, P.; Li, N.; Liu, W.; et al. A stretchable and strain-unperturbed pressure sensor for motion interference–free tactile monitoring on skins. *Sci. Adv.* **2021**, *7*, eabi4563. [CrossRef]
6. Zhu, M.; Sun, Z.; Chen, T.; Lee, C. Low cost exoskeleton manipulator using bidirectional triboelectric sensors enhanced multiple degree of freedom sensory system. *Nat. Commun.* **2021**, *12*, 2692. [CrossRef]
7. Yu, Y.; Li, J.; Solomon, S.A.; Min, J.; Tu, J.; Guo, W.; Xu, C.; Song, Y.; Gao, W. All-printed soft human-machine interface for robotic physicochemical sensing. *Sci. Robot.* **2022**, *7*, eabn0495. [CrossRef]
8. Anwer, A.H.; Khan, N.; Ansari, M.Z.; Baek, S.-S.; Yi, H.; Kim, S.; Noh, S.M.; Jeong, C. Recent Advances in Touch Sensors for Flexible Wearable Devices. *Sensors* **2022**, *22*, 4460. [CrossRef]
9. Guo, K.; Wustoni, S.; Koklu, A.; Díaz-Galicia, E.; Moser, M.; Hama, A.; Alqahtani, A.A.; Ahmad, A.N.; Alhamlan, F.S.; Shuaib, M.; et al. Rapid single-molecule detection of COVID-19 and MERS antigens via nanobody-functionalized organic electrochemical transistors. *Nat. Biomed. Eng.* **2021**, *5*, 666–677. [CrossRef]
10. Chen, C.; Zhao, S.; Pan, C.; Zi, Y.; Wang, F.; Yang, C.; Wang, Z.L. A method for quantitatively separating the piezoelectric component from the as-received "Piezoelectric" signal. *Nat. Commun.* **2022**, *13*, 1391. [CrossRef]
11. Cui, X.; Chen, J.; Wu, W.; Liu, Y.; Li, H.; Xu, Z.; Zhu, Y. Flexible and breathable all-nanofiber iontronic pressure sensors with ultraviolet shielding and antibacterial performances for wearable electronics. *Nano Energy* **2022**, *95*, 107022. [CrossRef]
12. Xu, H.; Zheng, W.; Wang, Y.; Xu, D.; Zhao, N.; Qin, Y.; Yuan, Y.; Fan, Z.; Nan, X.; Duan, Q.; et al. Flexible tensile strain-pressure sensor with an off-axis deformation-insensitivity. *Nano Energy* **2022**, *99*, 107384. [CrossRef]
13. Wang, C.; Gong, X.; Li, J.; Chen, Y.; Li, B.; Zhang, L.; Fu, S. Ultrahigh-sensitivity thermochromic smart fabrics and flexible temperature sensors based on intramolecular proton-coupled electron transfer. *Chem. Eng. J.* **2022**, *446*, 136444. [CrossRef]
14. Wang, Y.; Xu, C.; Yu, X.; Zhang, H.; Han, M. Multilayer flexible electronics: Manufacturing approaches and applications. *Mater. Today Phys.* **2022**, *23*, 100647. [CrossRef]
15. Yu, Z.; Wu, P. Underwater Communication and Optical Camouflage Ionogels. *Adv. Mater.* **2021**, *33*, 2008479. [CrossRef]
16. Cao, K.; Wu, M.; Bai, J.; Wen, Z.; Zhang, J.; Wang, T.; Peng, M.; Liu, T.; Jia, Z.; Liang, Z.; et al. Beyond Skin Pressure Sensing: 3D Printed Laminated Graphene Pressure Sensing Material Combines Extremely Low Detection Limits with Wide Detection Range. *Adv. Funct. Mater.* **2022**, *32*, 2202360. [CrossRef]
17. Lan, L.; Ping, J.; Xiong, J.; Ying, Y. Sustainable Natural Bio-Origin Materials for Future Flexible Devices. *Adv. Sci.* **2022**, *9*, 2200560. [CrossRef]
18. Chen, Y.; Zhang, C.; Yin, R.; Yin, A.; Feng, Q.; Liu, F.; Shao, J.; Su, T.; Wang, H.; Chen, G.; et al. Environmentally adaptive and durable hydrogels toward multi-sensory application. *Chem. Eng. J.* **2022**, *449*, 137907. [CrossRef]
19. Peng, X.; Dong, K.; Zhang, Y.; Wang, L.; Wei, C.; Lv, T.; Wang, Z.L.; Wu, Z. Sweat-Permeable, Biodegradable, Transparent and Self-powered Chitosan-Based Electronic Skin with Ultrathin Elastic Gold Nanofibers. *Adv. Funct. Mater.* **2022**, *32*, 2112241. [CrossRef]
20. Khandelwal, G.; Dahiya, R. Self-Powered Active Sensing Based on Triboelectric Generators. *Adv. Mater.* **2022**, *34*, 2200724. [CrossRef]
21. Kim, J.; Campbell, A.S.; de Ávila, B.E.-F.; Wang, J. Wearable biosensors for healthcare monitoring. *Nat. Biotechnol.* **2019**, *37*, 389–406. [CrossRef] [PubMed]
22. Sharma, A.; Badea, M.; Tiwari, S.; Marty, J. Wearable Biosensors: An Alternative and Practical Approach in Healthcare and Disease Monitoring. *Molecules* **2021**, *26*, 748. [CrossRef] [PubMed]
23. Lu, C.; Gao, Y.; Yu, S.; Zhou, H.; Wang, X.; Li, L. Non-Fluorinated Flexible Superhydrophobic Surface with Excellent Mechanical Durability and Self-Cleaning Performance. *ACS Appl. Mater. Interfaces* **2022**, *14*, 4750–4758. [CrossRef] [PubMed]
24. Lin, S.; Hu, S.; Song, W.; Gu, M.; Liu, J.; Song, J.; Liu, Z.; Li, Z.; Huang, K.; Wu, Y.; et al. An ultralight, flexible, and biocompatible all-fiber motion sensor for artificial intelligence wearable electronics. *NPJ Flex. Electron.* **2022**, *6*, 27. [CrossRef]
25. Swain, N.; Tripathy, A.; Thirumurugan, A.; Saravanakumar, B.; Schmidt-Mende, L.; Ramadoss, A. A brief review on stretchable, compressible, and deformable supercapacitor for smart devices. *Chem. Eng. J.* **2022**, *446*, 136876. [CrossRef]
26. Chen, J.; Wu, T.; Zhang, L.; Tang, C.; Song, H.; Huang, F.; Zuo, C. Flexible ionic-gel strain sensor with double network, high conductivity and high frost-resistance using electrohydrodynamic printing method. *Addit. Manuf.* **2022**, *58*, 103021. [CrossRef]
27. Nie, Y.; Yue, D.; Xiao, W.; Wang, W.; Chen, H.; Bai, L.; Yang, L.; Yang, H.; Wei, D. Anti-freezing and self-healing nanocomposite hydrogels based on poly(vinyl alcohol) for highly sensitive and durable flexible sensors. *Chem. Eng. J.* **2022**, *436*, 135243. [CrossRef]
28. Ji, J.; Zhang, C.; Yang, S.; Liu, Y.; Wang, J.; Shi, Z. High Sensitivity and a Wide Sensing Range Flexible Strain Sensor Based on the V-Groove/Wrinkles Hierarchical Array. *ACS Appl. Mater. Interfaces* **2022**, *14*, 24059–24066. [CrossRef]
29. Wang, J.; Liu, L.; Yang, C.; Zhang, C.; Li, B.; Meng, X.; Ma, G.; Wang, D.; Zhang, J.; Niu, S.; et al. Ultrasensitive, Highly Stable, and Flexible Strain Sensor Inspired by Nature. *ACS Appl. Mater. Interfaces* **2022**, *14*, 16885–16893. [CrossRef]
30. Yu, W.; Gong, K.; Li, Y.; Ding, B.; Li, L.; Xu, Y.; Wang, R.; Li, L.; Zhang, G.; Lin, S. Flexible 2D Materials beyond Graphene: Synthesis, Properties, and Applications. *Small* **2022**, *18*, 2105383. [CrossRef]
31. Zhang, X.; Li, Z.; Du, W.; Zhao, Y.; Wang, W.; Pang, L.; Chen, L.; Yu, A.; Zhai, J. Self-powered triboelectric-mechanoluminescent electronic skin for detecting and differentiating multiple mechanical stimuli. *Nano Energy* **2022**, *96*, 107115. [CrossRef]
32. Qu, X.; Xue, J.; Liu, Y.; Rao, W.; Liu, Z.; Li, Z. Fingerprint-shaped triboelectric tactile sensor. *Nano Energy* **2022**, *98*, 107324. [CrossRef]

33. Zarei, M.; Lee, G.; Lee, S.G.; Cho, K. Advances in Biodegradable Electronic Skin: Material Progress and Recent Applications in Sensing, Robotics, and Human–Machine Interfaces. *Adv. Mater.* **2022**, 2203193. [CrossRef]
34. He, R.; Liu, H.; Niu, Y.; Zhang, H.; Genin, G.M.; Xu, F. Flexible Miniaturized Sensor Technologies for Long-Term Physiological Monitoring. *NPJ Flex. Electron.* **2022**, *6*, 20. [CrossRef]
35. Huang, Y.; Li, H.; Hu, T.; Li, J.; Yiu, C.K.; Zhou, J.; Li, J.; Huang, X.; Yao, K.; Qiu, X.; et al. Implantable Electronic Medicine Enabled by Bioresorbable Microneedles for Wireless Electrotherapy and Drug Delivery. *Nano Lett.* **2022**, *22*, 5944–5963. [CrossRef]
36. Zhang, S.; Zhu, J.; Zhang, Y.; Chen, Z.; Song, C.; Li, J.; Yi, N.; Qiu, D.; Guo, K.; Zhang, C.; et al. Standalone stretchable RF systems based on asymmetric 3D microstrip antennas with on-body wireless communication and energy harvesting. *Nano Energy* **2022**, *96*, 107069. [CrossRef]
37. Zhao, X.; Chen, G.; Zhou, Y.; Nashalian, A.; Xu, J.; Tat, T.; Song, Y.; Libanori, A.; Xu, S.; Li, S.; et al. Giant Magnetoelastic Effect Enabled Stretchable Sensor for Self-Powered Biomonitoring. *ACS Nano* **2022**, *16*, 6013–6022. [CrossRef]
38. Zhao, Z.; Liu, K.; Liu, Y.; Guo, Y.; Liu, Y. Intrinsically flexible displays: Key materials and devices. *Natl. Sci. Rev.* **2022**, *9*, nwac090. [CrossRef]
39. Zhao, C.; Wang, Y.; Tang, G.; Ru, J.; Zhu, Z.; Li, B.; Guo, C.F.; Li, L.; Zhu, D. Ionic Flexible Sensors: Mechanisms, Materials, Structures, and Applications. *Adv. Funct. Mater.* **2022**, *32*, 2110417. [CrossRef]
40. Eivazzadeh-Keihan, R.; Noruzi, E.B.; Chidar, E.; Jafari, M.; Davoodi, F.; Kashtiaray, A.; Gorab, M.G.; Hashemi, S.M.; Javanshir, S.; Cohan, R.A.; et al. Applications of carbon-based conductive nanomaterials in biosensors. *Chem. Eng. J.* **2022**, *442*, 136183. [CrossRef]
41. Wang, Q.; Yu, X.; Chen, X.; Gao, J.; Shi, D.; Shen, Y.; Tang, J.; He, J.; Li, A.; Yu, L.; et al. A Facile Composite Strategy to Prepare a Biodegradable Polymer Based Radiopaque Raw Material for "Visualizable" Biomedical Implants. *ACS Appl. Mater. Interfaces* **2022**, *14*, 24197–24212. [CrossRef] [PubMed]
42. Wu, B.; Feng, E.; Liao, Y.; Liu, H.; Tang, R.; Tan, Y. Brush-Modified Hydrogels: Preparations, Properties, and Applications. *Chem. Mater.* **2022**, *34*, 6210–6231. [CrossRef]
43. Lee, H.J.; Yang, J.C.; Choi, J.; Kim, J.; Lee, G.S.; Sasikala, S.P.; Lee, G.-H.; Park, S.-H.K.; Lee, H.M.; Sim, J.Y.; et al. Hetero-Dimensional 2D $Ti_3C_2T_x$ MXene and 1D Graphene Nanoribbon Hybrids for Machine Learning-Assisted Pressure Sensors. *ACS Nano* **2021**, *15*, 10347–10356. [CrossRef] [PubMed]
44. Zhang, S.; Liu, H.; Yu, J.; Li, B.; Ding, B. Multi-functional flexible 2D carbon nanostructured networks. *Nat. Commun.* **2020**, *11*, 5134. [CrossRef]
45. Yang, J.; Wang, C.; Liu, L.; Zhang, H.; Ma, J. Water-Tolerant MXene Epidermal Sensors with High Sensitivity and Reliability for Healthcare Monitoring. *ACS Appl. Mater. Interfaces* **2022**, *14*, 21253–21262. [CrossRef]
46. Cheng, B.; Yu, J.; Arisawa, T.; Hayashi, K.; Richardson, J.J.; Shibuta, Y.; Ejima, H. Ultrastrong underwater adhesion on diverse substrates using non-canonical phenolic groups. *Nat. Commun.* **2022**, *13*, 1892. [CrossRef]
47. Scalable production of ultrafine polyaniline fibres for tactile organic electrochemical transistors. *Nat. Commun.* **2022**, *13*, 2101. [CrossRef] [PubMed]
48. Lim, C.; Hong, Y.J.; Jung, J.; Shin, Y.; Sunwoo, S.-H.; Baik, S.; Park, O.K.; Choi, S.H.; Hyeon, T.; Kim, J.H.; et al. Tissue-like skin-device interface for wearable bioelectronics by using ultrasoft, mass-permeable, and low-impedance hydrogels. *Sci. Adv.* **2021**, *7*, eabd3716. [CrossRef]
49. Meng, X.; Qiao, Y.; Do, C.; Bras, W.; He, C.; Ke, Y.; Russell, T.P.; Qiu, D. Hysteresis-Free Nanoparticle-Reinforced Hydrogels. *Adv. Mater.* **2021**, *34*, 2108243. [CrossRef]
50. Verma, M.L.; Sukriti; Dhanya, B.S.; Saini, R.; Das, A.; Varma, R.S. Synthesis and application of graphene-based sensors in biology: A review. *Environ. Chem. Lett.* **2022**, *20*, 2189–2212. [CrossRef]
51. Kohls, A.; Ditty, M.M.; Dehghandehnavi, F.; Zheng, S.-Y. Vertically Aligned Carbon Nanotubes as a Unique Material for Biomedical Applications. *ACS Appl. Mater. Interfaces* **2022**, *14*, 6287–6306. [CrossRef]
52. Huang, H.; Dong, C.; Feng, W.; Wang, Y.; Huang, B.; Chen, Y. Biomedical engineering of two-dimensional MXenes. *Adv. Drug Deliv. Rev.* **2022**, *184*, 114178. [CrossRef]
53. Choi, E.; Lee, J.; Kim, Y.-J.; Kim, H.; Kim, M.; Hong, J.; Kang, Y.C.; Koo, C.M.; Kim, D.W.; Kim, S.J. Enhanced stability of Ti3C2Tx MXene enabled by continuous ZIF-8 coating. *Carbon* **2022**, *191*, 593–599. [CrossRef]
54. Gaihre, B.; Potes, M.A.; Serdiuk, V.; Tilton, M.; Liu, X.; Lu, L. Two-dimensional nanomaterials-added dynamism in 3D printing and bioprinting of biomedical platforms: Unique opportunities and challenges. *Biomaterials* **2022**, *284*, 121507. [CrossRef]
55. Zhang, Z.; Lee, Y.; Haque, F.; Leem, J.; Hsieh, E.Y.; Nam, S. Plasmonic sensors based on graphene and graphene hybrid materials. *Nano Converg.* **2022**, *9*, 1–24. [CrossRef]
56. Sharabani, S.Z.; Edelstein-Pardo, N.; Molco, M.; Schwartz, N.B.; Morami, M.; Sivan, A.; Rom, Y.G.; Evental, R.; Flaxer, E.; Sitt, A. Messy or Ordered? Multiscale Mechanics Dictates Shape-Morphing of 2D Networks Hierarchically Assembled of Responsive Microfibers. *Adv. Funct. Mater.* **2022**, *32*, 2111471. [CrossRef]
57. Bai, X.; Guan, J. MXenes for electrocatalysis applications: Modification and hybridization. *Chin. J. Catal.* **2022**, *43*, 2057–2090. [CrossRef]
58. Yao, Y.; Han, Y.; Zhou, M.; Xie, L.; Zhao, X.; Wang, Z.; Barsan, N.; Zhu, Z. $MoO_3/TiO_2/Ti_3C_2T_x$ nanocomposite based gas sensors for highly sensitive and selective isopropanol detection at room temperature. *J. Mater. Chem. A Mater. Energy Sustain.* **2022**, *10*, 8283–8292. [CrossRef]
59. He, T.; Puente-Santiago, A.R.; Xia, S.; Ahsan, A.; Xu, G.; Luque, R. Experimental and Theoretical Advances on Single Atom and Atomic Cluster-Decorated Low-Dimensional Platforms towards Superior Electrocatalysts. *Adv. Energy Mater.* **2022**, *12*, 2200493. [CrossRef]

60. Yang, Z.; He, Y.; Liao, S.; Ma, Y.; Tao, X.; Wang, Y. Renatured hydrogel painting. *Sci. Adv.* **2021**, *7*, eabf9117. [CrossRef]
61. Tang, Z.; He, C.; Tian, H.; Ding, J.; Hsiao, B.S.; Chu, B.; Chen, X. Polymeric nanostructured materials for biomedical applications. *Prog. Polym. Sci.* **2016**, *60*, 86–128. [CrossRef]
62. Wang, Z.; Cui, H.; Li, S.; Feng, X.; Aghassi-Hagmann, J.; Azizian, S.; Levkin, P.A. Facile Approach to Conductive Polymer Microelectrodes for Flexible Electronics. *ACS Appl. Mater. Interfaces* **2021**, *13*, 21661–21668. [CrossRef] [PubMed]
63. Son, S.Y.; Lee, G.; Wang, H.; Samson, S.; Wei, Q.; Zhu, Y.; You, W. Integrating charge mobility, stability and stretchability within conjugated polymer films for stretchable multifunctional sensors. *Nat. Commun.* **2022**, *13*, 2739. [CrossRef]
64. Ma, W.; Zhang, Y.; Pan, S.; Cheng, Y.; Shao, Z.; Xiang, H.; Chen, G.; Zhu, L.; Weng, W.; Bai, H.; et al. Smart fibers for energy conversion and storage. *Chem. Soc. Rev.* **2021**, *50*, 7009–7061. [CrossRef]
65. Wu, G.; Wu, X.; Zhu, X.; Xu, J.; Bao, N. Two-Dimensional Hybrid Nanosheet-Based Supercapacitors: From Building Block Architecture, Fiber Assembly, and Fabric Construction to Wearable Applications. *ACS Nano* **2022**, *16*, 10130–10155. [CrossRef]
66. Dong, K.; Peng, X.; Cheng, R.; Ning, C.; Jiang, Y.; Zhang, Y.; Wang, Z.L. Advances in High-Performance Autonomous Energy and Self-Powered Sensing Textiles with Novel 3D Fabric Structures. *Adv. Mater.* **2022**, *34*, 2109355. [CrossRef]
67. Muir, V.G.; Burdick, J.A. Chemically Modified Biopolymers for the Formation of Biomedical Hydrogels. *Chem. Rev.* **2020**, *121*, 10908–10949. [CrossRef]
68. Yang, J.; An, J.; Sun, Y.; Zhang, J.; Zu, L.; Li, H.; Jiang, T.; Chen, B.; Wang, Z.L. Transparent self-powered triboelectric sensor based on PVA/PA hydrogel for promoting human-machine interaction in nursing and patient safety. *Nano Energy* **2022**, *97*, 107199. [CrossRef]
69. Yan, N.; Sujanani, R.; Kamcev, J.; Jang, E.-S.; Kobayashi, K.; Paul, D.R.; Freeman, B.D. Salt and ion transport in a series of crosslinked AMPS/PEGDA hydrogel membranes. *J. Membr. Sci.* **2022**, *653*, 120549. [CrossRef]
70. Nishat, Z.S.; Hossain, T.; Islam, N.; Phan, H.; Wahab, A.; Moni, M.A.; Salomon, C.; Amin, M.A.; Ibn Sina, A.A.; Hossain, S.A.; et al. Hydrogel Nanoarchitectonics: An Evolving Paradigm for Ultrasensitive Biosensing. *Small* **2022**, *18*, 2107571. [CrossRef]
71. Kwon, J.H.; Kim, Y.M.; Moon, H.C. Porous Ion Gel: A Versatile Ionotronic Sensory Platform for High-Performance, Wearable Ionoskins with Electrical and Optical Dual Output. *ACS Nano* **2021**, *15*, 15132–15141. [CrossRef]
72. Qu, X.; Wang, S.; Zhao, Y.; Huang, H.; Wang, Q.; Shao, J.; Wang, W.; Dong, X. Skin-inspired highly stretchable, tough and adhesive hydrogels for tissue-attached sensor. *Chem. Eng. J.* **2021**, *425*, 131523. [CrossRef]
73. Zhang, X.; Cheng, X.; Si, Y.; Yu, J.; Ding, B. Elastic and highly fatigue resistant ZrO_2-SiO_2 nanofibrous aerogel with low energy dissipation for thermal insulation. *Chem. Eng. J.* **2021**, *433*, 133628. [CrossRef]
74. Kim, J.; Zhang, G.; Shi, M.; Suo, Z. Fracture, fatigue, and friction of polymers in which entanglements greatly outnumber cross-links. *Science* **2021**, *374*, 212–216. [CrossRef]
75. Han, J.; Bin Kim, D.; Kim, J.H.; Kim, S.W.; Ahn, B.U.; Cho, Y.S. Origin of high piezoelectricity in carbon nanotube/halide nanocrystal/P(VDF-TrFE) composite nanofibers designed for bending-energy harvesters and pressure sensors. *Nano Energy* **2022**, *99*, 107421. [CrossRef]
76. Arens, L.; Weißenfeld, F.; Klein, C.O.; Schlag, K.; Wilhelm, M. Osmotic Engine: Translating Osmotic Pressure into Macroscopic Mechanical Force via Poly(Acrylic Acid) Based Hydrogels. *Adv. Sci.* **2017**, *4*, 1700112. [CrossRef]
77. Daniel, M.; Mathew, G.; Anpo, M.; Neppolian, B. MOF based electrochemical sensors for the detection of physiologically relevant biomolecules: An overview. *Coord. Chem. Rev.* **2022**, *468*, 214627. [CrossRef]
78. Cun, J.-E.; Fan, X.; Pan, Q.; Gao, W.; Luo, K.; He, B.; Pu, Y. Copper-based metal–organic frameworks for biomedical applications. *Adv. Colloid Interface Sci.* **2022**, *305*, 102686. [CrossRef]
79. Zhai, Z.; Zhang, X.; Wang, J.; Li, H.; Sun, Y.; Hao, X.; Qin, Y.; Niu, B.; Li, C. Washable and flexible gas sensor based on UiO-66-NH2 nanofibers membrane for highly detecting SO_2. *Chem. Eng. J.* **2022**, *428*, 131720. [CrossRef]
80. Wang, Z.; Sun, Q.; Liu, B.; Kuang, Y.; Gulzar, A.; He, F.; Gai, S.; Yang, P.; Lin, J. Recent advances in porphyrin-based MOFs for cancer therapy and diagnosis therapy. *Coord. Chem. Rev.* **2021**, *439*, 213945. [CrossRef]
81. Huang, D.; Wang, Y.; Wang, X.; Li, H.; Tan, X.; Chen, Y.; Wang, W.; Cheng, Q.; Yi, M.; Han, G.; et al. Rational in situ construction of Fe-modified MXene-derived MOFs as high-performance acetone sensor. *Chem. Eng. J.* **2022**, *444*, 136526. [CrossRef]
82. Donaldson, L. Wearable sweat sensor for healthcare monitoring. *Mater. Today* **2022**, *52*, 1–2. [CrossRef]
83. Ke, Y.; Jia, K.; Zhong, W.; Ming, X.; Jiang, H.; Chen, J.; Ding, X.; Li, M.; Lu, Z.; Wang, D. Wide-range sensitive all-textile piezoresistive sensors assembled with biomimetic core-shell yarn via facile embroidery integration. *Chem. Eng. J.* **2022**, *435*, 135003. [CrossRef]
84. Park, Y.; Yun, I.; Chung, W.G.; Park, W.; Lee, D.H.; Park, J. High-Resolution 3D Printing for Electronics. *Adv. Sci.* **2022**, *9*, 2104623. [CrossRef]
85. Shim, H.; Jang, S.; Yu, C. High-resolution patterning of organic semiconductors toward industrialization of flexible organic electronics. *Matter* **2022**, *5*, 23–25. [CrossRef]
86. Li, J.; Huo, W.; Yu, J.; Liu, X.; Haslinger, M.; Muehlberger, M.; Kulha, P.; Huang, X. Micro and nano materials and processing techniques for printed biodegradable electronics. *Mater. Today Nano* **2022**, *18*, 100201. [CrossRef]
87. Cheng, Y.; Gong, X.; Yang, J.; Zheng, G.; Zheng, Y.; Li, Y.; Xu, Y.; Nie, G.; Xie, X.; Chen, M.; et al. A touch-actuated glucose sensor fully integrated with microneedle array and reverse iontophoresis for diabetes monitoring. *Biosens. Bioelectron.* **2022**, *203*, 114026. [CrossRef]

88. Xiong, H.; Lacin, E.; Ouyang, H.; Naik, A.; Xu, X.; Xie, C.; Youn, J.; Wilson, B.A.; Kumar, K.; Kern, T.; et al. Probing Neuropeptide Volume Transmission In Vivo by Simultaneous Near-Infrared Light-Triggered Release and Optical Sensing. *Angew. Chem. Int. Ed.* **2022**, *61*, e202206122. [CrossRef]
89. Cha, G.D.; Lee, W.H.; Sunwoo, S.-H.; Kang, D.; Kang, T.; Cho, K.W.; Kim, M.; Park, O.K.; Jung, D.; Lee, J.; et al. Multifunctional Injectable Hydrogel for In Vivo Diagnostic and Therapeutic Applications. *ACS Nano* **2022**, *16*, 554–567. [CrossRef]
90. Cornet, B.; Fang, H.; Ngo, H.; Boyer, E.W.; Wang, H. An Overview of Wireless Body Area Networks for Mobile Health Applications. *IEEE Netw.* **2022**, *36*, 76–82. [CrossRef]
91. Liu, L.; Niu, S.; Zhang, J.; Mu, Z.; Li, J.; Li, B.; Meng, X.; Zhang, C.; Wang, Y.; Hou, T.; et al. Bioinspired, Omnidirectional, and Hypersensitive Flexible Strain Sensors. *Adv. Mater.* **2022**, *34*, 2200823. [CrossRef]
92. Pan, L.; Han, L.; Liu, H.; Zhao, J.; Dong, Y.; Wang, X. Flexible sensor based on Hair-like microstructured ionic hydrogel with high sensitivity for pulse wave detection. *Chem. Eng. J.* **2022**, *450*, 137929. [CrossRef]
93. Shu, Q.; Xu, Z.; Liu, S.; Wu, J.; Deng, H.; Gong, X.; Xuan, S. Magnetic flexible sensor with tension and bending discriminating detection. *Chem. Eng. J.* **2022**, *433*, 134424. [CrossRef]
94. Niu, H.; Li, H.; Gao, S.; Li, Y.; Wei, X.; Chen, Y.; Yue, W.; Zhou, W.; Shen, G. Perception-to-Cognition Tactile Sensing Based on Artificial-Intelligence-Motivated Human Full-Skin Bionic Electronic Skin. *Adv. Mater.* **2022**, *34*, 2202622. [CrossRef]
95. Wang, Y.; Liu, D.; Zhang, Y.; Fan, L.; Ren, Q.; Ma, S.; Zhang, M. Stretchable Temperature-Responsive Multimodal Neuromorphic Electronic Skin with Spontaneous Synaptic Plasticity Recovery. *ACS Nano* **2022**, *16*, 8283–8293. [CrossRef]
96. Yang, T.; Deng, W.; Chu, X.; Wang, X.; Hu, Y.; Fan, X.; Song, J.; Gao, Y.; Zhang, B.; Tian, G.; et al. Hierarchically Microstructure-Bioinspired Flexible Piezoresistive Bioelectronics. *ACS Nano* **2021**, *15*, 11555–11563. [CrossRef]
97. Bai, N.; Wang, L.; Xue, Y.; Wang, Y.; Hou, X.; Li, G.; Zhang, Y.; Cai, M.; Zhao, L.; Guan, F.; et al. Graded Interlocks for Iontronic Pressure Sensors with High Sensitivity and High Linearity over a Broad Range. *ACS Nano* **2022**, *16*, 4338–4347. [CrossRef]
98. Chen, Y.; Lei, H.; Gao, Z.; Liu, J.; Zhang, F.; Wen, Z.; Sun, X. Energy autonomous electronic skin with direct temperature-pressure perception. *Nano Energy* **2022**, *98*, 107273. [CrossRef]
99. Wei, S.; Liu, L.; Huang, X.; Zhang, Y.; Liu, F.; Deng, L.; Bilotti, E.; Chen, G. Flexible and Foldable Films of SWCNT Thermoelectric Composites and an S-Shape Thermoelectric Generator with a Vertical Temperature Gradient. *ACS Appl. Mater. Interfaces* **2022**, *14*, 5973–5982. [CrossRef]
100. Zhang, W.; Xi, Y.; Wang, E.; Qu, X.; Yang, Y.; Fan, Y.; Shi, B.; Li, Z. Self-Powered Force Sensors for Multidimensional Tactile Sensing. *ACS Appl. Mater. Interfaces* **2022**, *14*, 20122–20131. [CrossRef]
101. Lee, Y.; Chung, J.W.; Lee, G.H.; Kang, H.; Kim, J.-Y.; Bae, C.; Yoo, H.; Jeong, S.; Cho, H.; Kang, S.-G.; et al. Standalone real-time health monitoring patch based on a stretchable organic optoelectronic system. *Sci. Adv.* **2021**, *7*, eabg9180. [CrossRef] [PubMed]
102. Anwar, A.R.; Sajjad, M.T.; Johar, M.A.; Hernández-Gutiérrez, C.A.; Usman, M.; Łepkowski, S.P. Recent Progress in Micro-LED-Based Display Technologies. *Laser Photon. Rev.* **2022**, *16*, 2100427. [CrossRef]
103. Jia, C.; Xia, Y.; Zhu, Y.; Wu, M.; Zhu, S.; Wang, X. High-Brightness, High-Resolution, and Flexible Triboelectrification-Induced Electroluminescence Skin for Real-Time Imaging and Human–Machine Information Interaction. *Adv. Funct. Mater.* **2022**, *32*, 2201292. [CrossRef]
104. Falcucci, T.; Presley, K.F.; Choi, J.; Fizpatrick, V.; Barry, J.; Sahoo, J.K.; Ly, J.T.; Grusenmeyer, T.A.; Dalton, M.J.; Kaplan, D.L. Degradable Silk-Based Subcutaneous Oxygen Sensors. *Adv. Funct. Mater.* **2022**, *32*, 2202020. [CrossRef]
105. Wei, X.; Wang, B.; Wu, Z.; Wang, Z.L. An Open-Environment Tactile Sensing System: Toward Simple and Efficient Material Identification. *Adv. Mater.* **2022**, *34*, 2203073. [CrossRef]
106. Li, C.; Liu, D.; Xu, C.; Wang, Z.; Shu, S.; Sun, Z.; Tang, W.; Wang, Z.L. Sensing of joint and spinal bending or stretching via a retractable and wearable badge reel. *Nat. Commun.* **2021**, *12*, 2950. [CrossRef]
107. Zhang, Y.; Zhang, T.; Huang, Z.; Yang, J. A New Class of Electronic Devices Based on Flexible Porous Substrates. *Adv. Sci.* **2022**, *9*, 2105084. [CrossRef]
108. Yao, Y.; Chen, Y.; Wang, K.; Turetta, N.; Vitale, S.; Han, B.; Wang, H.; Zhang, L.; Samorì, P. A robust vertical nanoscaffold for recyclable, paintable, and flexible light-emitting devices. *Sci. Adv.* **2022**, *8*, eabn2225. [CrossRef]
109. Liu, D.; Zhang, D.; Sun, Z.; Zhou, S.; Li, W.; Li, C.; Li, W.; Tang, W.; Wang, Z.L. Active-Matrix Sensing Array Assisted with Machine-Learning Approach for Lumbar Degenerative Disease Diagnosis and Postoperative Assessment. *Adv. Funct. Mater.* **2022**, *32*, 2113008. [CrossRef]
110. Zhang, J.; Lai, Z.; Kong, H.; Shen, L. Robust Twin Bounded Support Vector Classifier With Manifold Regularization. *IEEE Trans. Cybern.* **2022**, 1–16. [CrossRef]
111. Roh, Y.; Kim, M.; Won, S.M.; Lim, D.; Hong, I.; Lee, S.; Kim, T.; Kim, C.; Lee, D.; Im, S.; et al. Vital signal sensing and manipulation of a microscale organ with a multifunctional soft gripper. *Sci. Robot.* **2021**, *6*, eabi6774. [CrossRef]
112. Sun, B.; Xu, R.; Han, X.; Xu, J.; Hong, W.; Xu, Y.; Fu, Z.; Zhu, H.; Sun, X.; Wang, J.; et al. Ultra-high temperature tolerant flexible transparent electrode with embedded silver nanowires bundle micromesh for electrical heater. *NPJ Flex. Electron.* **2022**, *6*, 48. [CrossRef]
113. Hu, J.; Qiu, Y.; Wang, X.; Jiang, L.; Lu, X.; Li, M.; Wang, Z.; Pang, K.; Tian, Y.; Zhang, W.; et al. Flexible six-dimensional force sensor inspired by the tenon-and-mortise structure of ancient Chinese architecture for orthodontics. *Nano Energy* **2022**, *96*, 107073. [CrossRef]

114. Pang, Q.; Hu, H.; Zhang, H.; Qiao, B.; Ma, L. Temperature-Responsive Ionic Conductive Hydrogel for Strain and Temperature Sensors. *ACS Appl. Mater. Interfaces* **2022**, *14*, 26536–26547. [CrossRef]
115. Jinxu, Q.; Xigui, Y.; Chenglong, S.; Yu, C.; Yuan, D.; Zhenfeng, Z.; Hang, L.; Chaofan, L.; Yizhe, L.; Chuang, Z.; et al. Carbon nanodot-based humidity sensor for self-powered respiratory monitoring. *Nano Energy* **2022**, *101*, 107549. [CrossRef]
116. Ding, H.; Lv, G.; Cai, X.; Chen, J.; Cheng, Z.; Peng, Y.; Tang, G.; Shi, Z.; Xie, Y.; Fu, X.; et al. An Optoelectronic thermometer based on microscale infrared-to-visible conversion devices. *Light. Sci. Appl.* **2022**, *11*, 130. [CrossRef]
117. Zheng, Q.; Chalise, D.; Jia, M.; Zeng, Y.; Zeng, M.; Saeidi-Javash, M.; Tanvir, A.N.M.; Uahengo, G.; Kaur, S.; Garay, J.E.; et al. Structured illumination with thermal imaging (SI-TI): A dynamically reconfigurable metrology for parallelized thermal transport characterization. *Appl. Phys. Rev.* **2022**, *9*, 021411. [CrossRef]
118. Eichhorn, T.R.; Parker, A.J.; Josten, F.; Müller, C.; Scheuer, J.; Steiner, J.M.; Gierse, M.; Handwerker, J.; Keim, M.; Lucas, S.; et al. Hyperpolarized Solution-State NMR Spectroscopy with Optically Polarized Crystals. *J. Am. Chem. Soc.* **2022**, *144*, 2511–2519. [CrossRef]
119. Wang, N.; Tong, J.; Wang, J.; Wang, Q.; Chen, S.; Sheng, B. Polyimide-Sputtered and Polymerized Films with Ultrahigh Moisture Sensitivity for Respiratory Monitoring and Contactless Sensing. *ACS Appl. Mater. Interfaces* **2022**, *14*, 11842–11853. [CrossRef]
120. Zou, S.; Tao, L.-Q.; Wang, G.; Zhu, C.; Peng, Z.; Sun, H.; Li, Y.; Wei, Y.; Ren, T.-L. Humidity-Based Human–Machine Interaction System for Healthcare Applications. *ACS Appl. Mater. Interfaces* **2022**, *14*, 12606–12616. [CrossRef]
121. Tachibana, S.; Wang, Y.-F.; Sekine, T.; Takeda, Y.; Hong, J.; Yoshida, A.; Abe, M.; Miura, R.; Watanabe, Y.; Kumaki, D.; et al. A Printed Flexible Humidity Sensor with High Sensitivity and Fast Response Using a Cellulose Nanofiber/Carbon Black Composite. *ACS Appl. Mater. Interfaces* **2022**, *14*, 5721–5728. [CrossRef] [PubMed]
122. Xu, J.; Zhang, L.; Lai, X.; Zeng, X.; Li, H. Wearable RGO/MXene Piezoresistive Pressure Sensors with Hierarchical Microspines for Detecting Human Motion. *ACS Appl. Mater. Interfaces* **2022**, *14*, 27262–27273. [CrossRef] [PubMed]
123. Lin, J.-C.; Liatsis, P.; Alexandridis, P. Flexible and Stretchable Electrically Conductive Polymer Materials for Physical Sensing Applications. *Polym. Rev.* **2022**, 1–60. [CrossRef]
124. Lee, J.; Ihle, S.J.; Pellegrino, G.S.; Kim, H.; Yea, J.; Jeon, C.-Y.; Son, H.-C.; Jin, C.; Eberli, D.; Schmid, F.; et al. Stretchable and suturable fibre sensors for wireless monitoring of connective tissue strain. *Nat. Electron.* **2021**, *4*, 291–301. [CrossRef]
125. Nan, K.; Babaee, S.; Chan, W.W.; Kuosmanen, J.L.P.; Feig, V.R.; Luo, Y.; Srinivasan, S.S.; Patterson, C.M.; Jebran, A.M.; Traverso, G. Low-cost gastrointestinal manometry via silicone–liquid-metal pressure transducers resembling a quipu. *Nat. Biomed. Eng.* **2022**. [CrossRef]
126. Qi, A.; Ma, M.; Luo, Y.; Fernandes, G.; Shi, G.; Fan, J.; Qi, Y.; Ma, J. WISe: Wireless Intelligent Sensing for Human- Centric Applications. *IEEE Wirel. Commun.* **2022**, 1–13. [CrossRef]
127. Ryu, H.; Park, H.-M.; Kim, M.-K.; Kim, B.; Myoung, H.S.; Kim, T.Y.; Yoon, H.-J.; Kwak, S.S.; Kim, J.; Hwang, T.H.; et al. Self-rechargeable cardiac pacemaker system with triboelectric nanogenerators. *Nat. Commun.* **2021**, *12*, 4374. [CrossRef]
128. Shi, Y.; Wang, K.; Feng, X.; Li, S.; Liang, H.; Li, G.; Li, W.; Ke, W.; Chen, A.; Yang, X.; et al. Additive Manufactured Self-Powered Mechanoelectric Sensor as the Artificial Nucleus Pulposus for Monitoring Tissue Rehabilitation after Discectomy. *Nano Energy* **2022**, *96*, 107113. [CrossRef]
129. Ma, J.; Shen, L.; Jiang, Y.; Ma, H.; Lv, F.; Liu, J.; Su, Y.; Zhu, N. Wearable Self-Powered Smart Sensors for Portable Nutrition Monitoring. *Anal. Chem.* **2022**, *94*, 2333–2340. [CrossRef]
130. Alagumalai, A.; Shou, W.; Mahian, O.; Aghbashlo, M.; Tabatabaei, M.; Wongwises, S.; Liu, Y.; Zhan, J.; Torralba, A.; Chen, J.; et al. Self-powered sensing systems with learning capability. *Joule* **2022**, *6*, 1475–1500. [CrossRef]
131. Ra, Y.; Kim, J.; Kim, H.; Cho, S.; Lee, D.; Jang, S.; Kam, D.; La, M.; Park, S.J.; Choi, D. Smart conveyor roller system for self-powered product size identification in electrically off-grid condition via hybridization of triboelectric-electromagnetic generators. *Nano Energy* **2022**, *100*, 107447. [CrossRef]
132. Jiang, C.; Li, X.; Lian, S.W.M.; Ying, Y.; Ho, J.S.; Ping, J. Wireless Technologies for Energy Harvesting and Transmission for Ambient Self-Powered Systems. *ACS Nano* **2021**, *15*, 9328–9354. [CrossRef]
133. Kim, C.Y.; Ku, M.J.; Qazi, R.; Nam, H.J.; Park, J.W.; Nam, K.S.; Oh, S.; Kang, I.; Jang, J.-H.; Kim, W.Y.; et al. Soft subdermal implant capable of wireless battery charging and programmable controls for applications in optogenetics. *Nat. Commun.* **2021**, *12*, 535. [CrossRef]
134. Wang, X.; Fan, Y.; Yan, J.; Yang, M. Engineering polyphenol-based polymeric nanoparticles for drug delivery and bioimaging. *Chem. Eng. J.* **2022**, *439*, 135661. [CrossRef]
135. Wang, J.; Wang, L.; Feng, J.; Tang, C.; Sun, X.; Peng, H. Long-term In Vivo Monitoring of Chemicals with Fiber Sensors. *Adv. Fiber Mater.* **2021**, *3*, 47–58. [CrossRef]
136. Ouyang, H.; Li, Z.; Gu, M.; Hu, Y.; Xu, L.; Jiang, D.; Cheng, S.; Zou, Y.; Deng, Y.; Shi, B.; et al. A Bioresorbable Dynamic Pressure Sensor for Cardiovascular Postoperative Care. *Adv. Mater.* **2021**, *33*, 2102302. [CrossRef]
137. Lu, D.; Li, S.; Yang, Q.; Arafa, H.M.; Xu, Y.; Yan, Y.; Ostojich, D.; Bai, W.; Guo, H.; Wu, C.; et al. Implantable, wireless, self-fixing thermal sensors for continuous measurements of microvascular blood flow in flaps and organ grafts. *Biosens. Bioelectron.* **2022**, *206*, 114145. [CrossRef]
138. Li, J.; Liu, Y.; Yuan, L.; Zhang, B.; Bishop, E.S.; Wang, K.; Tang, J.; Zheng, Y.-Q.; Xu, W.; Niu, S.; et al. A tissue-like neurotransmitter sensor for the brain and gut. *Nature* **2022**, *606*, 94–101. [CrossRef]
139. Díaz-Marín, C.D.; Zhang, L.; Lu, Z.; Alshrah, M.; Grossman, J.C.; Wang, E.N. Kinetics of Sorption in Hygroscopic Hydrogels. *Nano Lett.* **2022**, *22*, 1100–1107. [CrossRef]

140. Shi, Z.; Lu, Y.; Shen, S.; Xu, Y.; Shu, C.; Wu, Y.; Lv, J.; Li, X.; Yan, Z.; An, Z.; et al. Wearable battery-free theranostic dental patch for wireless intraoral sensing and drug delivery. *NPJ Flex. Electron.* **2022**, *6*, 49. [CrossRef]
141. Sheng, F.; Zhang, B.; Zhang, Y.; Li, Y.; Cheng, R.; Wei, C.; Ning, C.; Dong, K.; Wang, Z.L. Ultrastretchable Organogel/Silicone Fiber-Helical Sensors for Self-Powered Implantable Ligament Strain Monitoring. *ACS Nano* **2022**, *16*, 10958–10967. [CrossRef] [PubMed]
142. Kar, A.; Ahamad, N.; Dewani, M.; Awasthi, L.; Patil, R.; Banerjee, R. Wearable and implantable devices for drug delivery: Applications and challenges. *Biomaterials* **2022**, *283*, 121435. [CrossRef] [PubMed]
143. Chaochao, Z.; Qin, S.; Hu, L.; Xi, C.; Yuan, X.; Yu, C.; Zhuo, X.; Feng, L.; Jinyan, S.; Jiacheng, L.; et al. Shape Designed Implanted Drug Delivery System for In Situ Hepatocellular Carcinoma Therapy. *ACS Nano* **2022**, *16*, 8493–8503. [CrossRef]
144. Chun, K.; Seo, S.; Han, C. A Wearable All-Gel Multimodal Cutaneous Sensor Enabling Simultaneous Single-Site Monitoring of Cardiac-Related Biophysical Signals. *Adv. Mater.* **2022**, *34*, 2110082. [CrossRef] [PubMed]
145. Jiang, L.; Liu, J.; He, S.; Liu, A.; Zhang, J.; Xu, H.; Shao, W. Flexible wearable sensors based on lignin doped organohydrogels with multi-functionalities. *Chem. Eng. J.* **2022**, *430*, 132653. [CrossRef]
146. Zhang, X.; Ke, L.; Zhang, X.; Xu, F.; Hu, Y.; Lin, H.; Zhu, J. Breathable and Wearable Strain Sensors Based on Synergistic Conductive Carbon Nanotubes/Cotton Fabrics for Multi-directional Motion Detection. *ACS Appl. Mater. Interfaces* **2022**, *14*, 25753–25762. [CrossRef]
147. Sonay, A.Y.; Kalyviotis, K.; Yaganoglu, S.; Unsal, A.; Konantz, M.; Teulon, C.; Lieberwirth, I.; Sieber, S.; Jiang, S.; Behzadi, S.; et al. Biodegradable Harmonophores for Targeted High-Resolution In Vivo Tumor Imaging. *ACS Nano* **2021**, *15*, 4144–4154. [CrossRef]
148. Nakamura, T.; Sato, Y.; Yamada, Y.; Elwakil, M.M.A.; Kimura, S.; Younis, M.A.; Harashima, H. Extrahepatic targeting of lipid nanoparticles in vivo with intracellular targeting for future nanomedicines. *Adv. Drug Deliv. Rev.* **2022**, *188*, 114417. [CrossRef]

micromachines

Article

On-Skin Flexible Pressure Sensor with High Sensitivity for Portable Pulse Monitoring

Weihao Zheng [1], Hongcheng Xu [1], Meng Wang [1], Qikai Duan [2], Yangbo Yuan [1], Weidong Wang [1,*] and Libo Gao [3,*]

1 School of Mechano-Electronic Engineering, Xidian University, Xi'an 710071, China
2 School of Automation and Software Engineering, Shanxi University, Taiyuan 030013, China
3 Department of Mechanical and Electrical Engineering, School of Aerospace Engineering, Xiamen University, Xiamen 361102, China
* Correspondence: wangwd@mail.xidian.edu.cn (W.W.); lbgao@xmu.edu.cn (L.G.)

Abstract: Radial artery pulse pressure contains abundant cardiovascular physiological and pathological information, which plays an important role in clinical diagnosis of traditional Chinese medical science. However, many photoelectric sensors and pressure sensors will lose a large number of waveform features in monitoring pulse, which will make it difficult for doctors to precisely evaluate the patients' health. In this letter, we proposed an on-skin flexible pressure sensor for monitoring radial artery pulse. The sensor consists of the MXene ($Ti_3C_2T_x$)-coated nonwoven fabrics (n-WFs) sensitive layer and laser-engraved interdigital copper electrodes. Benefiting from substantially increased conductive paths between fibers and electrodes during normal compression, the sensor obtains high sensitivity (3.187 kPa^{-1}), fast response time (15 ms), low detection limit (11.1 Pa), and long-term durability (20,000 cycles). Furthermore, a flexible processing circuit was connected with the sensor mounted on wrist radial artery, achieving wirelessly precise monitoring of the pulse on smart phones in real time. Compared with the commercial flexible pressure sensor, our sensor successfully captures weak systolic peak precisely, showing its great clinical potential and commercial value.

Keywords: flexible sensor; pressure; high sensitivity; portable; pulse monitoring

Citation: Zheng, W.; Xu, H.; Wang, M.; Duan, Q.; Yuan, Y.; Wang, W.; Gao, L. On-Skin Flexible Pressure Sensor with High Sensitivity for Portable Pulse Monitoring. *Micromachines* **2022**, *13*, 1390. https://doi.org/10.3390/mi13091390

Academic Editor: Joost Lötters

Received: 9 August 2022
Accepted: 21 August 2022
Published: 25 August 2022

1. Introduction

Radial arterial pulse pressure (RAPP) monitoring is crucial for Chinese medical-aided diagnosis since rich physiological features in pulse waves can help doctors to evaluate and diagnose diseases within the body, instead of scanning devices [1–4]. The light wave difference caused by a light–absorption change in human hemoglobin is used to monitor pulse wave by ordinary photoelectric pulse sensors, which usually lose substantial information of RAPP owing to their instability during near-infrared light transmittance [5,6]. In addition, monitoring weak entire RAPP remains challenging for most reported flexible pressure sensors due to their low sensitivity [7,8]. Hence, it is necessary to develop a highly sensitive pressure sensor to precisely monitor arterial pulse.

Currently, most reported works focused on the novel two-dimensional (2D) material used to being sensitive elements in flexible pressure sensors, such as graphene, carbon nano-tube, and $SnSe_2$, etc. [9]. For example, Jing et al. proposed a flexible piezoresistive pressure sensors based on graphene, which achieved high elasticity of 85% but exhibited a limited sensitivity (0.075 kPa^{-1}) and long response time (120 ms) [10]. Park et al. proposed a flexible sensor based on carbon nanotube thin films, which exhibited a high sensitivity of 278.5 kPa^{-1} but a limited responding range of 0–30 Pa [11]. In addition, the sensor based on $SnSe_2$ nanosheet reported by Tannarana et al. can only work under a pressure beyond 2 kPa, resulting in disability for monitoring RAPP [12]. MXene is intrinsically hydrophilic and yet they have demonstrated higher electrical conductivity than solution processed graphene [13], thereby enhancing sensor's sensitivity greatly by the reduction of contact resistance and possibly enabling sensors to attain more pulse. In addition, their exceptional electrochemical properties were widely used in flexible sensors [14]. Consequently, MXene-based flexible pressure sensor (FPS) is a potential candidate to solving problems like low sensitivity, long response time, and a small working range of forementioned sensors.

Here, an integrated FPS composed of laser-engraved interdigital copper electrodes and the MXene/n-WFs (nonwoven fabrics) sensing layer was proposed. A key sensitive element was made of MXene-coated n-WFs. The sensor was connected to an on-skin wirelessly flexible processing circuit to successfully monitor RAPP on a smart phone. Owing to substantially increased conductive paths between fibers and electrodes during compression, the FPS achieved a high sensitivity of 3.187 kPa^{-1} within 6 kPa and 1.059 kPa^{-1} in a larger pressure region of 100 kPa, as well as a low-pressure detected limit down to 11.1 Pa. Furthermore, FPS is able to monitor RAPP precisely compared with a commercial pressure sensor and can be used to help for diagnosis of cardiovascular diseases.

2. Experimental Details

2.1. Structure Design and Mechanism

During arterial constriction and relaxation, a weak blood pressure was produced by the process periodically forming a pulse waveform (Figure 1a). By converting this mechanical force into an electrical output, a thin and flexible pressure sensor is applied to the wrist skin to capture RAPP in real time and enable long-term monitoring on a smartphone via a back-end processing circuit on the skin (Figure 1b). As shown in Figure 1c, the designed FPS is applied to the wrist skin with the processing circuit. To connect closely with the FPS and meet the wearable needs, we designed the signal processing circuit to be flexible The FPS mainly consists of a PU encapsulation layer at the top, an MXene/n-WFs sensitive layer, interdigital electrodes, and a supporting polyimide (PI) layer (Figure 1d). Among them, the interdigital electrodes are designed to sense local weak signals to improve the sensitivity of the sensor on uneven skin. The two pieces of 3M tape (3M3300LSE-9495LSE, Linxing Company, Shenzhen, Guangdong province, China) are the key to improving the sensitivity of FPS, which is used to increase the initial compression space of the sensitive layer, leading to a high initial contact resistance and the ability of detecting a weak pressure signal. The MXene/n-WFs become contacted with each other when the sensor is squeezed by a weak pulse force, thereby reducing the contact resistance between the fibers. In addition, another reduction in contact resistance is attributed to the contact between sensitive fibers and electrodes under compression. Both of these reasons are formative mechanisms for the sensor to capture external stimuli.

2.2. Materials and Fabrication

The material of a sensitive layer determines the property of the sensor. $Ti_3C_2T_x$ MXene is one of the 2D transition metal carbides/nitrides that possess good metallic conductivity. Figure 2a shows the scanning electron microscope (SEM) image of $Ti_3C_2T_x$ nanosheet (5 mg·mL^{-1}, Xiyan New Materials Company, Nanjing, Jiangsu province China) in which the insert exhibits a Tyndall effect in solution. Because of the existence of a large number of hydroxyl and fluorine groups on the surface, $Ti_3C_2T_x$ nanosheets can be evenly distributed in solution, which also shows that the metallic conductivity and rich surface functionalities can coexist without mutual interference [15,16]. Benefiting from the irregular microfiber structure and high porosity of n-WFs (250 mm * 380 mm * 500 pieces, Order Flagship Store), MXene can be dipped and coated on the fabric surface as the sensitive layer. To obtain the sensitive layer, the n-WFs were first washed with ethanol and deionized water and dried. Next, the dried n-WFs were immersed in aqueous $Ti_3C_2T_x$ solution (5 mg mL^{-1}) for one minute and dried at 50 °C, and this operation was repeated several times. Figure 2b shows the morphology (inserts) and square resistance of the n-WFs coated for 3, 5, 7, and 9 times. The square resistance decreased from 14.8 Ω cm^{-2} to 5.2 Ω cm^{-2} as the coating number increased. However, the $Ti_3C_2T_x$ nanosheets on the fibers were severely aggregated when the coating reached nine times, so we used MXene/n-WFs modified by seven times as the sensitive layer in the following study. The fabricated MXene/n-WFs samples showed good mechanical stability in the uniaxial compression test (Figure 2c). When the compressive strain reached 80%, the sensitive layer still exhibited almost uniform loading and offloading processes, which was attributed to the mechanical stability of n-WFs. Compared to pure n-WFs, MXene/n-WFs exhibited almost the same tensile fracture strength of 75 MPa, indicating that the modification of MXene had no effect on its mechanical behavior (Figure 2d). Figure 2e illustrates the fabrication process of the FPS. The interdigital electrodes were engraved on PI films using a laser ablation. The sensitive layer was aligned with the interdigital electrodes, and the prepared sensor was packed in a PU tape on PET film.

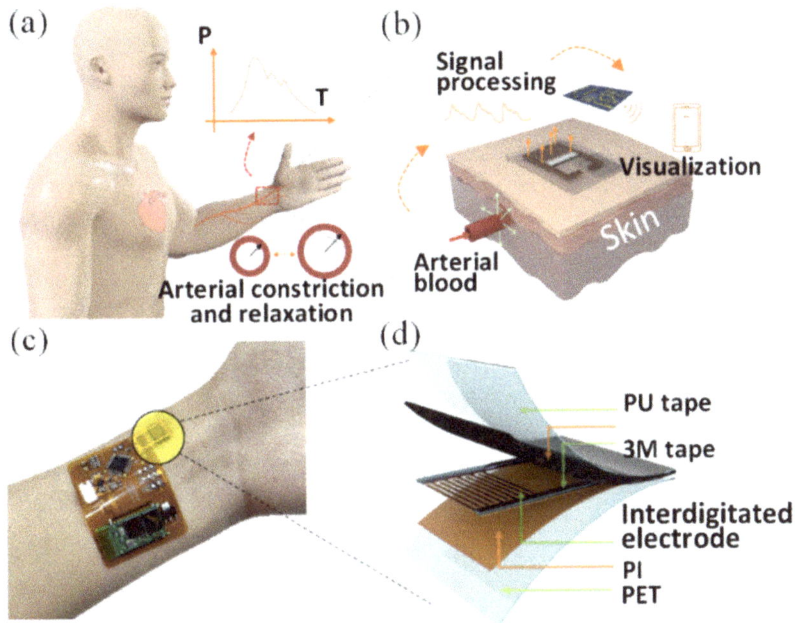

Figure 1. (**a**) The beating of the human heart causes the radial artery vasodilatation and vasodilatation, resulting in a slight pressure change on the skin surface; (**b**) collecting the pressure response signals by flexible sensor and sending signals to the mobile phone for display by a signal processing circuit; (**c**) optical image of the FPS connected with a processing circuit mounted on a participant's wrist; (**d**) schematic diagram of the structure of FPS.

3. Results and Discussion

3.1. Test Platform

The test platform is shown in Figure 3a, which comprises a force gauge (Zhiqu, ZQ-990B, Zhiqv Company, Dongguan, Guangdong province, China) and an electrochemical workstation (CHI760E, Chenhua Company, Shanghai, China) and their respective upper computers. The fabricated sensor was placed on the platform of the force gauge with its electrodes connected to the electrochemical workstation. By applying different pressure to FPS using force gauge, the performance of the sensor is characterized through the variation of current recorded by electrochemical workstation under 1 V voltage.

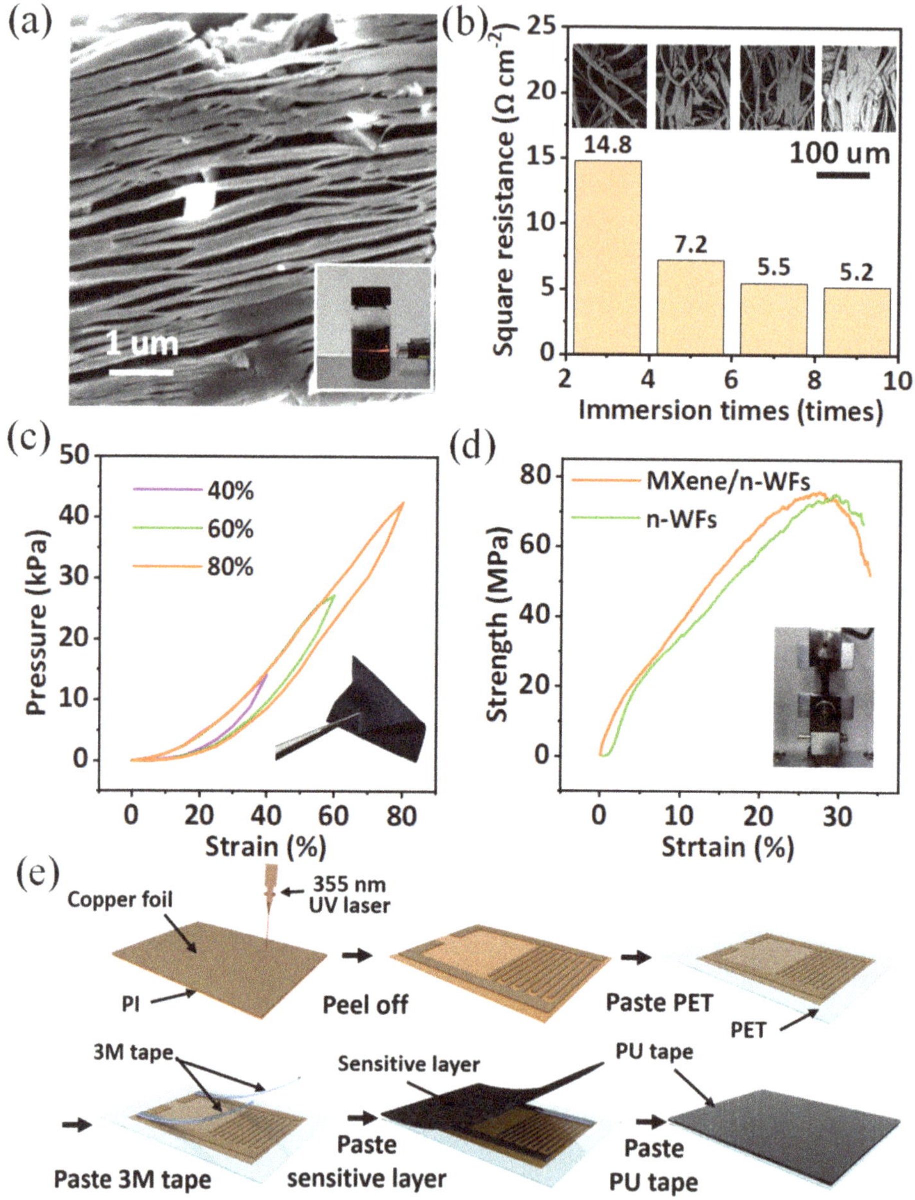

Figure 2. (**a**) SEM image of $Ti_3C_2T_x$ nanosheet and Tyndall effect of $Ti_3C_2T_x$ nanosheets solution; (**b**) morphology and square resistance of the n-WFs coated for 3, 5, 7, and 9 times; (**c**) compression test's comparison under various strains; (**d**) tensile test of the n-WFs and MXene/n-WFs; (**e**) fabricated process of FPS.

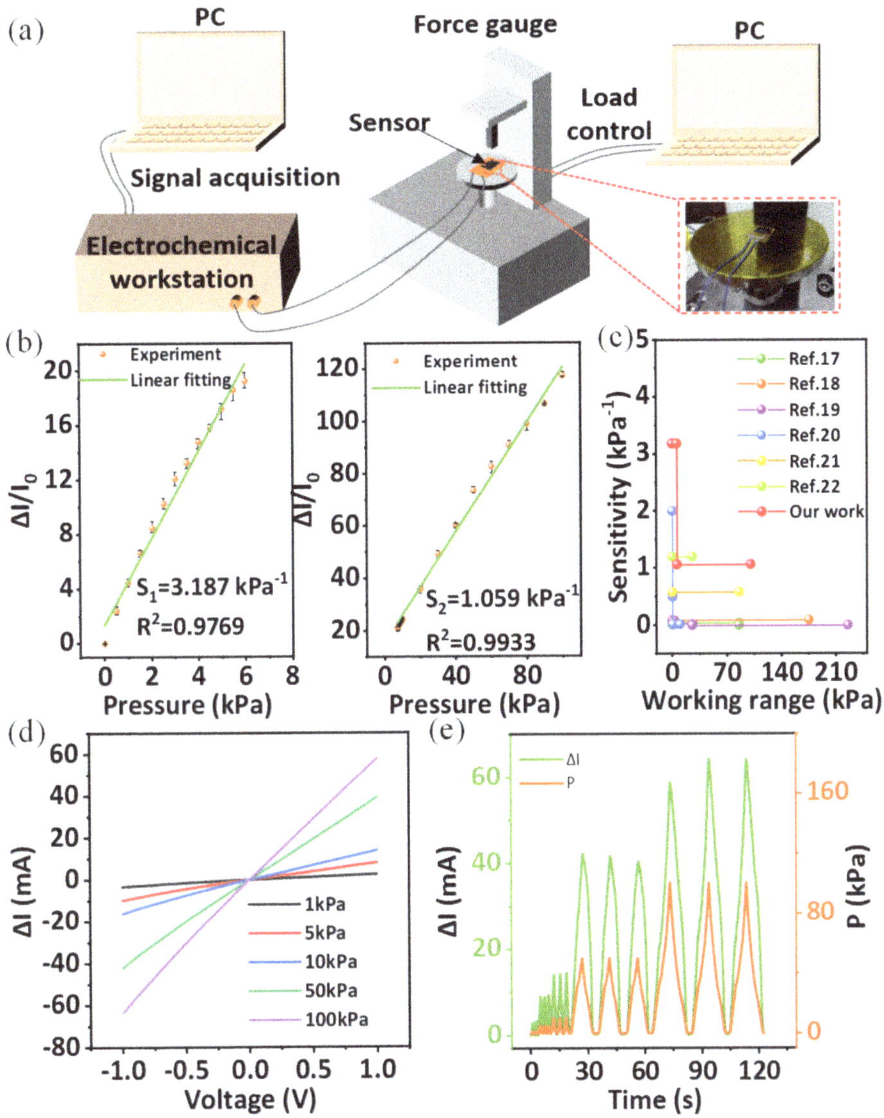

Figure 3. (**a**) Schematic illustration of the experimental setup; (**b**) the sensitivity of FPS, showing a high sensitivity of 3.187 kPa^{-1} within 6 kPa and 1.059 kPa^{-1} in a larger pressure region up to 100 kPa; (**c**) comparison of the fabricated sensor with other flexible sensors; (**d**) I–V curves of the sensor under various applied pressures; (**e**) current response under increasing pressures.

3.2. Device Characterization

Figure 3b shows the normalized current variation of the sensors under different pressures. The FPS shows a high sensitivity of 3.187 kPa^{-1} within 6 kPa and 1.059 kPa^{-1} in a larger pressure region up to 100 kPa, where the sensitivity is calculated by the formula:

$$S = \frac{\Delta I / I_0}{\Delta P} \tag{1}$$

where the I_0 is the initial current, ΔI is the variation of current under different pressure, and ΔP is the amount of the pressure change from I_0 to I. It can be seen that the FPS has a good linearity and sensitivity in a working range, especially under the pressure lower than 6 kPa, which performs the advantage of detecting weak pressure. Compared to compression under a larger force, the sensitive layer can form a larger compressed strain at a low pressure range, which is determined by the compression response as shown in Figure 2c. Larger strain in the modified n-WFs can excite more contact between inner conductive fibers to achieve higher resistance decrease, thus to form higher sensitivity at low pressure. Compared to previous works, our sensor still exhibits eminent sensitivity and working range as shown in Figure 3c [17–22]. The I–V curves of the FPS show a good ohmic contact between sensitive layer and electrodes that the voltage ranges from −1 to 1 V (Figure 3d). Figure 3e shows a small hysteresis between current change and applied pressure, verifying the good response capability of the sensor to the applied pressure. In addition, the sensor exhibits fast response and recovery times of 15 ms and 36 ms, respectively, at a continuous pressure of 10 kPa (Figure 4a). In addition, our sensor can sense weak pressures down to 11.1 Pa, as shown in Figure 4b, demonstrating the potential of the sensor for weak RAPP monitoring. To evaluate the recoverability of the sensor, the FPS remains stable for 120 s under different continuous forces and recovers immediately after offloading, as shown in Figure 4c. It is important to verify the response of FPS to the force under different curvature because the curvature of the skin at radial arteries of people are different. We attach the FPS to cylinders with different radii and apply different pressure to it, and the results are shown in Figure 4d. It can be seen that the response of sensors with different curvature is approximately the same under the same pressure, demonstrating the ability of the detection of RAPP of different people. In addition, the FPS maintains a constant output without significant signal degradation after 6000 cycles under loading-unloading pressure of 1 kPa (Figure 4d), verifying the stability of FPS to monitor RAPP. Furthermore, we apply a much bigger pressure of 50 kPa to FPS for more than 14,000 cycles, and the output is still stable, showing good durability.

3.3. Applications

The fabricated sensor has high sensitivity and linearity, leading to a wide range of applications. As shown in Figure 5a, a balloon is used for a blowing test, and the results show that the FPS is able to monitor human breath. The responses caused by finger touch also prove the potential application of FPS in robotic tactile sensation in complex environments (Figure 5b). As shown in Figure 5c, we fix the sensor to the second joint of the index finger with medical tape and bend the joint into 30°, 60°, and 90° three times, respectively. The increasing current variation proves the ability of FPS to monitor finger movements, so it can be used to collect signals for human finger rehabilitation training. Furthermore, by placing the sensor on the human carotid artery, we successfully collected the human carotid pulse and swallow signals, demonstrating the application of the monitoring of human larynx health.

Considering the significance of RAPP in medical diagnosis and the performance of the FPS, we apply the sensor to long-term remote wireless monitoring of RAPP. To achieve real-time monitoring, a flexible wireless signal processing circuit was designed as shown in Figure 6a. A Wheatstone bridge is used to convert the resistance change of the FPS into a voltage change, and then the voltage signal is transmitted to an amplifier. Next, the analog voltage signal is converted to a digital signal by an Analog to Digital Converter (ADC) and filtered by digital software in the Microcontroller Units (MCU). Finally, the processed signal is sent to the cell phone via a low-power Bluetooth module for remote data transmission and visualization. As shown in Figure 6b, the fabricated thin FPS can be tightly attached to the volunteer's wrist skin and integrated with the flexible circuit. The real-time monitoring of the RAPP on a smartphone further demonstrates the feasibility of our sensor system in physical scenarios.

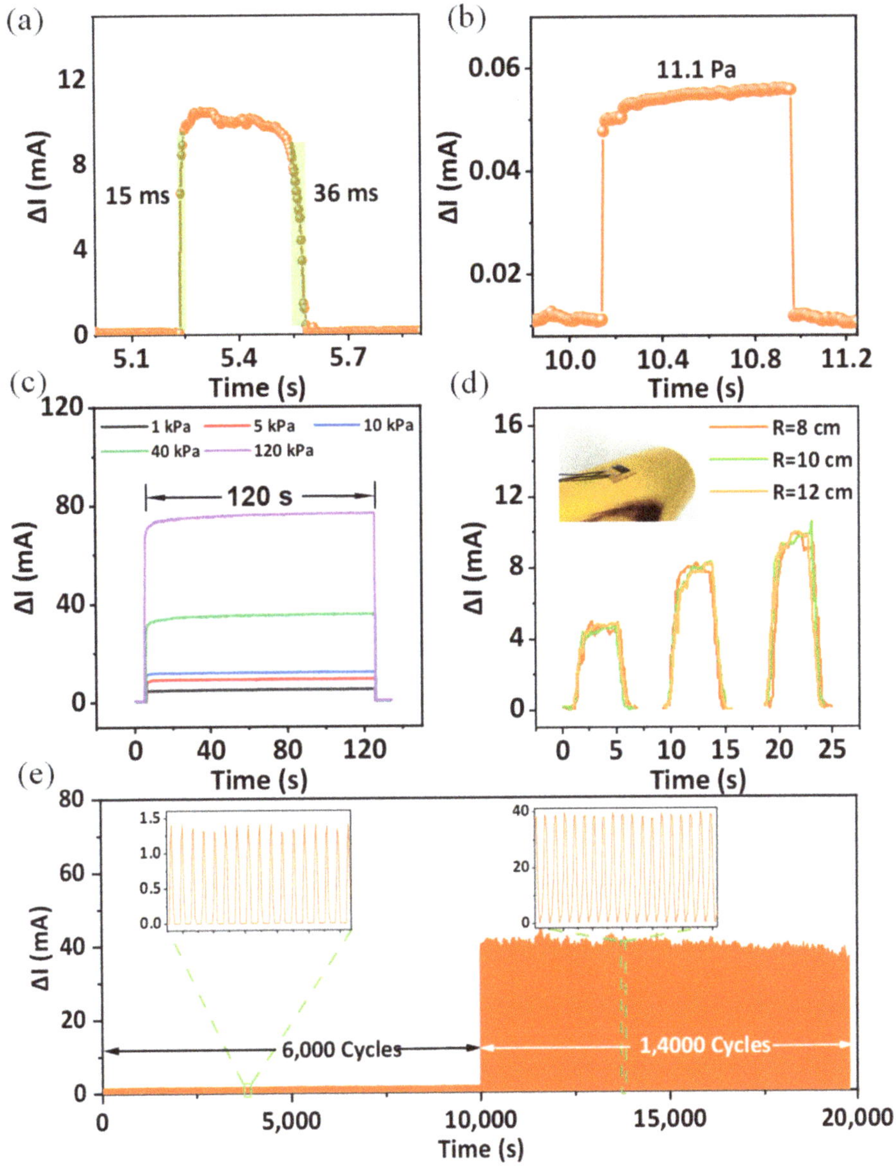

Figure 4. (**a**) Outlined response and relaxation time under a pressure of 10 kPa; (**b**) minimum detection pressure is 11.1 Pa; (**c**) response of FPS under different pressure for 120 s; (**d**) The FPS were tested on the sides of cylinders with different radii; (**e**) long-term durability over 6000 cycles under a pressure of 1 kPa and 14,000 cycles under a pressure of 50 kPa.

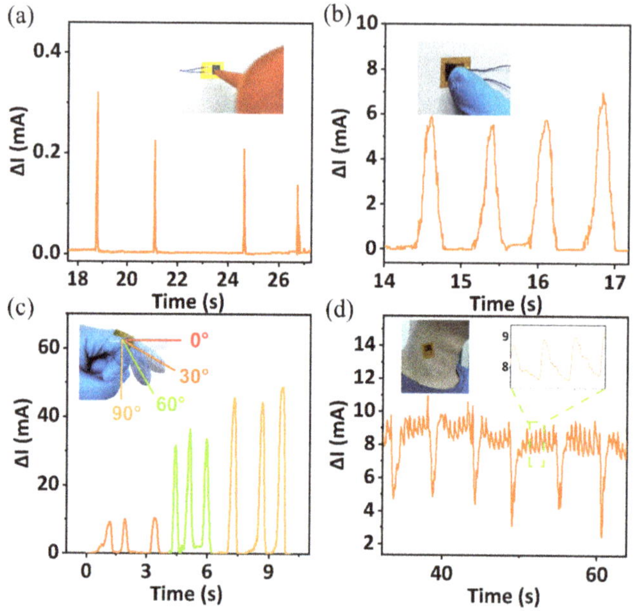

Figure 5. (**a**) The current response output of gas leak monitoring, and the image is shown in the inset; (**b**) the response of touching; (**c**) the response of bending the finger; (**d**) the monitoring of swallowing and carotid pulse.

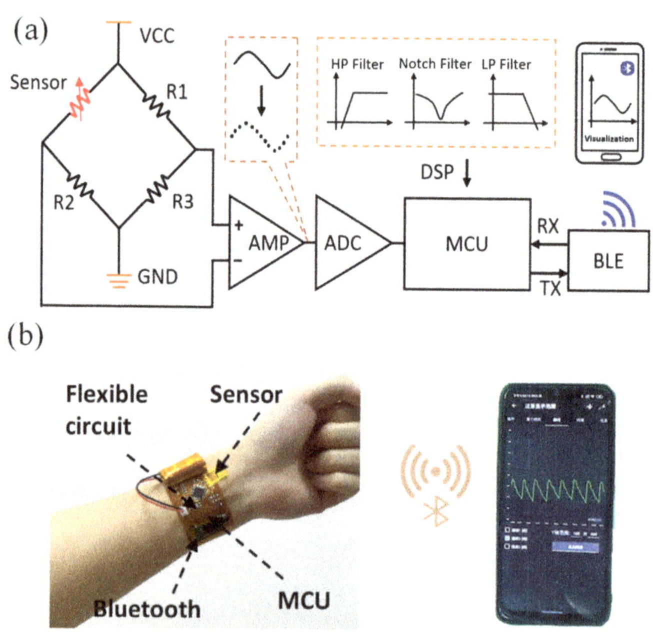

Figure 6. (**a**) Functional block diagram of pulse pressure signal acquisition; (**b**) The FPS is attached to the skin together with the flexible processing circuit. By connecting to the Bluetooth of mobile phone, the remote monitoring of RAPP was achieved.

Stable and consistent waveform can reflect a health condition of the tested subject. Clinically, pulse waveform is a key indicator of many cardiovascular diseases, and it mainly includes a systolic peak (P_1), a reflected systolic peak (P_2), a dicrotic notch (P_3), a diastolic peak (P_4), and a diastolic valley (P_5) [23,24]. These peaks can be analyzed and processed to determine the cardiovascular health patients. To evaluate the superiority of our FPS, we compared it with a commercial pulse sensor (RP-C7.6-ST-LF2, brought from LEGACT). During the test, we fix the FPS and commercial sensor with medical tape to a 25-year-old male volunteer's radial artery successively. It is important to reduce the impact of external environment on the results. Therefore, in the process of fixing the sensors, we try to ensure the consistent position of two sensors, and let the tape exert a pressure of about 1 kPa on the sensors to reduce the impact of external pressure. The results obtained are normalized and compared, and the RAPP peaks are marked, as shown in Figure 7a. It can be seen that there are obvious differences of the two pulse waveforms. The values of P_2 and P_5 of two curves are basically the same, but the commercial sensor hardly detects the characteristics of P_2 peak, which leads to the difference of P_3 peak compared to FPS. Hence, compared to a commercial sensor, our FPS can keep more characteristics of P_2 and P_3. The values of P_2 and P_3 can be used to evaluate important physiological and pathological indexes of the human body, such as the radial artery augmentation index ($AI_r = P_2/P_1$) and the radial diastolic augmentation ($DAI = P_3/P_1$) [25,26]. Hence, the RAPP waveform monitored by FPS can be observed to predict diseases such as arteriosclerosis and hypertension. Furthermore, according to the pulse wave theory [25,27,28], the mean arterial pressure P_m in the radial artery is calculated as follows:

$$P_m = \frac{1}{T}\int_0^T p(t)\mathrm{d}t = P_5 + K(P_1 - P_5) \tag{2}$$

where $p(t)$ is the value of RAPP waveform, K is the waveform coefficient of RAPP and a parameter indicating the degree of arteriosclerosis, and we can obtain that:

$$K = \frac{P_m - P_5}{P_1 - P_5} \tag{3}$$

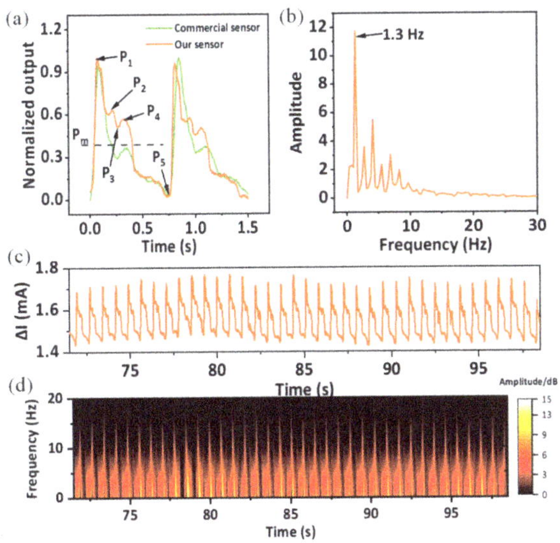

Figure 7. (**a**) Analysis and of RAPP waveform; (**b**) spectrum of (**a**); (**c**) long-term RAPP monitoring; (**d**) the STFE image of (**c**).

Consequently, the K is just related to the shape of pulse pressure waveform [27]. It corresponds to the ratio of the mean value and the peak value of RAPP. Vascular resistance and arterial elasticity of healthy young people are low with a K value of about 0.32–0.39. The middle-aged and elderly people have higher K value, which is about 0.4 owing to blood viscosity increase. The K value of patients with severe hypertension and atherosclerosis is about 0.45–0.5. The K (0.344) we obtained from our sensor reflects the healthy cardiovascular status of an adult male. The RAPP waveform can also be used to monitor the heart rate. Figure 7b shows the spectrum of the monitored RAPP waveform, where the volunteer's heart rate is 1.3 Hz. Furthermore, we monitored the RAPP for a period of time, and the waveform remained stable, indicating the stability of FPS (Figure 7c). Figure 7d shows the short-time Fourier transform (STFT) signals of Figure 7c. Judging from the frequency component at each time, we can obtain the human cardiovascular activity.

4. Conclusions

In this work, an on-skin flexible pulse pressure sensor was proposed and connected with a flexible processing circuit to successfully monitor radial artery pulse wirelessly. MXene-modified n-WFs as the sensing layer markedly improve the sensitivity compared to many previous works. Furthermore, a flexible functional circuit was connected with the sensor and provides an epidermally real-time and wireless monitoring for pulse wave on the wrist. The sensor also can distinguish clear pulse wave peaks related to a commercial sensor. By rationally building sensing material into a thin sensor and integrating with an on-skin system instead of merely developing sensors, we hope this design will facilitate the development of portable clinics in the future.

Author Contributions: Conceptualization, L.G. and M.W.; methodology, W.Z.; software, H.X.; writing—original draft preparation, W.Z.; writing—review and editing, L.G., H.X., Q.D. and Y.Y.; supervision, W.W. All authors have read and agreed to the published version of the manuscript.

Funding: This work was supported in part by the State Key Laboratory of Mechanics and Control of Mechanical Structures (Nanjing University of Aeronautics and Astronautics) (Grant No. MCMS-E-0422G03), the National Natural Science Foundation of China (No. 61904141), the Fundamental Research Funds for the Central Universities (JB210407), the Key Research and Development Program of Shaanxi (Program No. 2021GY-277), the Shenzhen-Hong Kong-Macau Technology Research Program (Type C, 202011033000145), and the Innovation Fund of Xidian University.

Data Availability Statement: The data presented in this study are available on request from the corresponding author.

Conflicts of Interest: The authors declare no conflict of interest.

References

1. Ghasemzadeh, N.; Zafari, A.M. A brief journey into the history of the arterial pulse. *Cardiol. Res. Pract.* **2011**, *2011*, 164832. [CrossRef] [PubMed]
2. Kelly, R.; Hayward, C.; Avolio, A.; O'Rourke, M. Noninvasive determination of age-related changes in the human arterial pulse. *Circulation* **1989**, *80*, 1652–1659. [CrossRef] [PubMed]
3. Campbell, B. Arterial waveforms: Monitoring changes in configuration. *Heart Lung* **1997**, *26*, 204–214. [CrossRef]
4. Chemla, D.; Hebert, J.L.; Coirault, C.; Zamani, K.; Suard, I.; Colin, P.; Lecarpentier, Y. Total arterial compliance estimated by stroke volume-to-aortic pulse pressure ratio in humans. *Am. J. Physiol.* **1998**, *274*, H500–H505. [CrossRef]
5. Dimin, W.; Zhang, D.; Guangming, L. A Novel Multichannel Wrist Pulse System with Different Sensor Arrays. *IEEE Trans. Instrum. Meas.* **2015**, *64*, 2020–2034. [CrossRef]
6. Nitzan, M.; Taitelbaum, H. The measurement of oxygen saturation in arterial and venous blood. *IEEE Instrum. Meas. Mag.* **2008**, *11*, 9–15. [CrossRef]
7. Thouti, E.; Nagaraju, A.; Chandran, A.; Prakash, P.V.B.S.S.; Shivanarayanamurthy, P.; Lal, B.; Kumar, P.; Kothari, P.; Panwar, D. Tunable flexible capacitive pressure sensors using arrangement of polydimethylsiloxane micro-pyramids for bio-signal monitoring. *Sens. Actuators A Phys.* **2020**, *314*, 112251. [CrossRef]
8. Mohammad Haniff, M.A.; Muhammad Hafiz, S.; Wahid, K.A.; Endut, Z.; Wah Lee, H.; Bien, D.C.; Abdul Azid, I.; Abdullah, M.Z.; Ming Huang, N.; Abdul Rahman, S. Piezoresistive effects in controllable defective HFTCVD graphene-based flexible pressure sensor. *Sci. Rep.* **2015**, *5*, 14751. [CrossRef]

9. Li, J.; Bao, R.; Tao, J.; Peng, Y.; Pan, C. Recent progress in flexible pressure sensor arrays: From design to applications. *J. Mater. Chem. C* **2018**, *6*, 11878–11892. [CrossRef]
10. Jing, Z.; Zhang, Q.; Cheng, Y.; Ji, C.; Zhao, D.; Liu, Y.; Jia, W.; Pan, S.; Sang, S. Highly sensitive, reliable and flexible piezoresistive pressure sensors based on graphene-PDMS @ sponge. *J. Micromech. Microeng.* **2020**, *30*, 085012. [CrossRef]
11. Park, S.-J.; Kim, J.; Chu, M.; Khine, M. Flexible Piezoresistive Pressure Sensor Using Wrinkled Carbon Nanotube Thin Films for Human Physiological Signals. *Adv. Mater. Technol.* **2018**, *3*, 1700158. [CrossRef]
12. Tannarana, M.; Solanki, G.K.; Bhakhar, S.A.; Patel, K.D.; Pathak, V.M.; Pataniya, P.M. 2D-SnSe2 Nanosheet Functionalized Piezo-resistive Flexible Sensor for Pressure and Human Breath Monitoring. *ACS Sustain. Chem. Eng.* **2020**, *8*, 7741–7749. [CrossRef]
13. Xu, H.; Gao, L.; Zhao, H.; Huang, H.; Wang, Y.; Chen, G.; Qin, Y.; Zhao, N.; Xu, D.; Duan, L.; et al. Stretchable and anti-impact iontronic pressure sensor with an ultrabroad linear range for biophysical monitoring and deep learning-aided knee rehabilitation. *Microsyst. Nanoeng.* **2021**, *7*, 92. [CrossRef]
14. Li, Y.; Huang, S.; Wei, C.; Wu, C.; Mochalin, V.N. Adhesion of two-dimensional titanium carbides (MXenes) and graphene to silicon. *Nat. Commun.* **2019**, *10*, 3014. [CrossRef]
15. Kim, S.J.; Koh, H.J.; Ren, C.E.; Kwon, O.; Maleski, K.; Cho, S.Y.; Anasori, B.; Kim, C.K.; Choi, Y.K.; Kim, J.; et al. Metallic Ti3C2Tx MXene Gas Sensors with Ultrahigh Signal-to-Noise Ratio. *ACS Nano* **2018**, *12*, 986–993. [CrossRef]
16. Gao, L.; Wang, M.; Wang, W.; Xu, H.; Wang, Y.; Zhao, H.; Cao, K.; Xu, D.; Li, L. Highly Sensitive Pseudocapacitive Iontronic Pressure Sensor with Broad Sensing Range. *Nanomicro. Lett.* **2021**, *13*, 140. [CrossRef]
17. Kannichankandy, D.; Pataniya, P.M.; Narayan, S.; Patel, V.; Sumesh, C.K.; Patel, K.D.; Solanki, G.K.; Pathak, V.M. Flexible piezo-resistive pressure sensor based on conducting PANI on paper substrate. *Synth. Met.* **2021**, *273*, 116697. [CrossRef]
18. Tang, Z.; Jia, S.; Zhou, C.; Li, B. 3D Printing of Highly Sensitive and Large-Measurement-Range Flexible Pressure Sensors with a Positive Piezoresistive Effect. *ACS Appl. Mater. Interfaces* **2020**, *12*, 28669–28680. [CrossRef]
19. Cheng, H.; Wang, B.; Tan, Y.; Yin, Y.; Wang, C. Low-Cost, Highly Sensitive, and Flexible Piezoresistive Pressure Sensor Characterized by Low-Temperature Interfacial Polymerization of Polypyrrole on Latex Sponge. *Macromol. Mater. Eng.* **2021**, *306*, 2000772. [CrossRef]
20. Li, S.; Li, R.; González, O.G.; Chen, T.; Xiao, X. Highly sensitive and flexible piezoresistive sensor based on c-MWCNTs decorated TPU electrospun fibrous network for human motion detection. *Compos. Sci. Technol.* **2021**, *203*, 108617. [CrossRef]
21. Luo, N.; Dai, W.; Li, C.; Zhou, Z.; Lu, L.; Poon, C.C.Y.; Chen, S.-C.; Zhang, Y.; Zhao, N. Flexible Piezoresistive Sensor Patch Enabling Ultralow Power Cuffless Blood Pressure Measurement. *Adv. Funct. Mater.* **2016**, *26*, 1178–1187. [CrossRef]
22. Shi, J.; Wang, L.; Dai, Z.; Zhao, L.; Du, M.; Li, H.; Fang, Y. Multiscale Hierarchical Design of a Flexible Piezoresistive Pressure Sensor with High Sensitivity and Wide Linearity Range. *Small* **2018**, *14*, e1800819. [CrossRef]
23. Cheng, W.; Wang, J.; Ma, Z.; Yan, K.; Wang, Y.; Wang, H.; Li, S.; Li, Y.; Pan, L.; Shi, Y. Flexible Pressure Sensor with High Sensitivity and Low Hysteresis Based on a Hierarchically Microstructured Electrode. *IEEE Electron. Device Lett.* **2018**, *39*, 288–291. [CrossRef]
24. Dagdeviren, C.; Su, Y.; Joe, P.; Yona, R.; Liu, Y.; Kim, Y.S.; Huang, Y.; Damadoran, A.R.; Xia, J.; Martin, L.W.; et al. Conformable amplified lead zirconate titanate sensors with enhanced piezoelectric response for cutaneous pressure monitoring. *Nat. Commun.* **2014**, *5*, 4496. [CrossRef] [PubMed]
25. Park, J.; Kim, M.; Lee, Y.; Lee, H.S.; Ko, H. Fingertip skin-inspired microstructured ferroelectric skins discriminate static/dynamic pressure and temperature stimuli. *Sci. Adv.* **2015**, *1*, e1500661. [CrossRef]
26. Yi, Z.; Liu, Z.; Li, W.; Ruan, T.; Chen, X.; Liu, J.; Yang, B.; Zhang, W. Piezoelectric Dynamics of Arterial Pulse for Wearable Continuous Blood Pressure Monitoring. *Adv. Mater.* **2022**, *34*, e2110291. [CrossRef]
27. Song, J.; Zhang, S.; Qiao, Y.; Luo, Z.; Zhang, J.; Zeng, Y.; Wang, L. Predicting pregnancy-induced hypertension with dynamic hemodynamics. *Eur. J. Obstet. Gynecol. Reprod. Biol.* **2004**, *117*, 162–168. [CrossRef]
28. Meng, K.; Chen, J.; Li, X.; Wu, Y.; Fan, W.; Zhou, Z.; He, Q.; Wang, X.; Fan, X.; Zhang, Y.; et al. Flexible Weaving Constructed Self-Powered Pressure Sensor Enabling Continuous Diagnosis of Cardiovascular Disease and Measurement of Cuffless Blood Pressure. *Adv. Funct. Mater.* **2018**, *29*, 1806388. [CrossRef]

micromachines

Article

Ramie Fabric Treated with Carboxymethylcellulose and Laser Engraved for Strain and Humidity Sensing

Shangxuan Shi [1,2], Jiao Liang [1,2], Chenkai Qu [1,2], Shangbi Chen [3] and Bin Sheng [1,2,*]

1 School of Optical Electrical and Computer Engineering, University of Shanghai for Science and Technology, Shanghai 200093, China
2 Shanghai Key Laboratory of Modern Optical Systems, Engineering Research Center of Optical Instruments and Systems, Shanghai 200093, China
3 Shanghai Aerospace Control Technology Institute, Shanghai 200233, China
* Correspondence: bsheng@usst.edu.cn

Abstract: Wearable fabric sensors have attracted enormous attention due to their huge potential in human health and activity monitoring, human–machine interaction and the Internet of Things (IoT). Among natural fabrics, bast fabric has the advantage of high strength, good resilience and excellent permeability. Laser engraving, as a high throughput, patternable and mask-free method, was demonstrated to fabricate fabric sensors. In this work, we developed a simplified, cost-effective and environmentally friendly method for engraving ramie fabric (a kind of bast fabric) directly by laser under an ambient atmosphere to prepare strain and humidity sensors. We used carboxymethylcellulose (CMC) to pretreat ramie fabric before laser engraving and gained laser-carbonized ramie fabrics (LCRF) with high conductivity (65 Ω sq^{-1}) and good permeability. The strain and humidity sensors had high sensitivity and good flexibility, which can be used for human health and activity monitoring.

Keywords: ramie fabric; laser engraving; CMC; strain sensor; humidity sensor

Citation: Shi, S.; Liang, J.; Qu, C.; Chen, S.; Sheng, B. Ramie Fabric Treated with Carboxymethylcellulose and Laser Engraved for Strain and Humidity Sensing. *Micromachines* **2022**, *13*, 1309. https://doi.org/10.3390/mi13081309

Academic Editors: Libo Gao and Zhuoqing Yang

Received: 21 July 2022
Accepted: 11 August 2022
Published: 13 August 2022

1. Introduction

Electronic textiles (E-textiles) have been widely used in recent years owing to their functionality in sensing [1–5], energy harvesting [1,6–8] and wireless transmission [9–11]. In terms of efficient signal detecting, the fabric sensor plays a significant role and attracts enormous attention due to its ubiquitous nature of fabrics such as low cost, high toughness, good wearability, non-invasive manner of sensing and excellent permeability [12]. Nowadays, the fabric sensor is applied in various fields, such as human health and activity monitoring [13–16], human–machine interaction [17,18] and the Internet of Things (IoT) [19,20].

To date, the major technology of functionalizing fabric to fabricate fabric sensors includes screen printing [21,22], inkjet printing [23,24], physical vapor deposition (PVD) [25], pyrolysis [26,27], electrophoretic deposition [28], and others [29]. Inkjet printing has several advantages including mask-free fabrication, high print resolution, and scalability from table-top devices to big press units. The screen-printing technique has the capability to print electronic devices at a low cost, with very little or no material wastage [30]. Conductive metals (e.g., Cu, Au, Pt, Ag) can be deposited on fabrics by PVD, making it possible to produce flexible and lightweight electroconductive textiles. Pyrolysis is a simple and effective method to carbonize the fabric in an oven and prepare highly flexible strain sensors [31,32]. Electrophoretic deposition is performed from an aqueous dispersion at ambient temperatures under a direct current (DC) electric field, which is rapid, repeatable, and environmentally benign. However, these aforementioned methods for preparing fabric sensors still have certain limitations. In inkjet printing, it takes time-consuming steps to obtain ink with appropriate viscosity and surface tension to prevent nozzle blockage. The screen-printing process provides high wet film thickness, resulting in high spreading of ink and low resolution if not cured instantly. PVD can only be carried out in a high vacuum environment and pyrolysis needs the conditions of high temperature and a protective atmosphere. Electrophoretic deposition takes time to disperse and the formation of gases due to electrolysis creating bubbles, leading to non-uniform deposition on the fabrics. As a high throughput, patternable, and mask-free method, laser engraving was demonstrated to fabricate carbonized fabric for preparing fabric sensors [33–37]. Some chemically synthesized fabrics, such as PI [35] and Kevlar fabric [34], can be engraved directly by laser to synthesize graphene under an ambient atmosphere, without any pretreatment.

However, when some natural materials (cotton, bast, wood and silk) were engraved by laser under an ambient atmosphere, they are ablated and become fragile due to violent oxidation, resulting in poor conductivity and inapplicability to the preparation of flexible sensors. In order to solve the above problem, many methods [36,38–40] had been developed in different ways. One direct method is to use a controlled atmosphere chamber to allow Ar/H_2 to flow through the chamber in order to isolate oxygen and avoid natural materials experiencing ablation [38]. Another strategy of multiple lasing and defocusing the laser has also proved to be an effective method [39]. Moreover, using fire retardant to treat samples seems to be a simpler and more economical method [36,39]. Although there are the aforementioned effective solutions, these laser engraving methods on natural materials still have certain limitations. The controlled atmosphere chamber required for isolating oxygen increases the cost and multiple lasing and defocusing of the laser involves complicated preparation routes. The addition of commercial flame retardants containing phosphorus or boron may harm the environment to a certain extent. As a result, exploring a low-cost, simplified and environmentally friendly method to engrave natural fabrics by laser and fabricate a fabric sensor is necessary.

Among natural fabrics, bast fabric, which is made from the fiber of natural bast plants (rich in cellulose and lignin), mainly includes jute, ramie, flax, and hemp fabrics, which have the advantage of high strength, good resilience and excellent permeability [41]. Previously, Liang heated linen fabric with graphene oxide at 900 °C under the protection of nitrogen and then integrated the fabric with silver nanowires to prepare a wearable strain sensor [42]. Liu carbonized the hemp fabrics to 800 °C with a nitrogen atmosphere and fabricated the permeable pressure sensors [43]. In Liu's other work, a high-performance stretchable strain sensor based on linen fabric was developed through a similar carbonization method and polymer-assisted copper deposition [44]. However, the carbonization of bast fabric can only be carried out under the conditions of high temperature and a protective atmosphere, and can not be patterned and functionalized for some sensing applications. As a result, it is a very attractive challenge to engrave bast fabric by laser under an ambient atmosphere for preparing the bast fabric sensor.

In this work, a simplified, cost-effective and environmentally friendly method was developed for engraving patterns on ramie fabric (a kind of bast fabric) directly under ambient atmosphere to fabricate strain and humidity sensors based on laser-carbonized ramie fabrics (LCRF). We used carboxymethylcellulose (CMC) to pretreat ramie fabric before laser engraving and gained LCRF with high conductivity (65 Ω sq^{-1}) and good flexibility. We carried out a series of comparative tests and proved that CMC pretreatment has a good effect on the flame retardancy of fabrics during laser engraving. In order to demonstrate the ability of this method in fabricating flexible sensors on ramie fabric, wearable strain sensors and humidity sensors based on LCRF were fabricated. Compared with the traditional high-temperature carbonizing fabric in the oven, this method has the advantages of low equipment requirements, no protective gas atmosphere, low cost, less time-consuming, and customization.

2. Materials and Methods

2.1. Materials

Ramie fabric (100% ramie, ≈700 μm of thickness) was made from the stem fiber of natural ramie plant, which was purchased from Huidian company (Guangzhou, China). CMC powder (which contains 9.9 wt% of sodium) was used to treat ramie fabric, which was purchased from Sinopharma Chemical Reagent Co., Ltd. (Shanghai, China). Silver conductive glue used to connect copper foil with LCRF was purchased from Shenzhen Ausbond Co., Ltd. (Guangdong, China). NaCl, $LiCl_2$, MgCl, NaBr, KCl and K_2SO_4 for preparation of saturated solution were purchased from Sinopharma Chemical Reagent Co., Ltd. (Shanghai, China).

2.2. LCRF Fabrication by Laser Engraving

LCRF was fabricated following the workflow shown in Figure 1a. We first fully soaked a piece of ramie fabric (≈4 × 4 cm^2, Figure 1b) in CMC solution (range from 0.5 to 3.9 wt% in deionized water) for one hour and dried it in a vacuum oven at 80 °C for two hours. After drying, we directly engraved ramie fabric via irradiation using a semiconductor laser (450 nm wavelength, 0.30 mm spot width, from DAJA, Dongguan, China) under ambient atmosphere. The laser beam is a Gaussian beam, and the area far from the center of the laser receives less energy in each scan. The width of the overlapping area in consecutive passes is about 0.14 mm, which ensures that the area far from the center of the laser can also obtain enough energy in consecutive passes. The surface resistance value of carbonized fabric in this paper was measured through a four-point probe meter (HPS2526, Changzhou, China). In each surface resistance measurement, we chose the average surface resistance of five samples as the surface resistance value under the same condition. The error distributions of the samples tested in each case ranged from 2% to 4%. Figure 1c,d show that the laser-engraved area turned black and the generated LCRF has high integrity and conductivity (surface resistance is around 10^2 Ω sq^{-1}). We also observed untreated ramie fabric engraved by laser and the generated LCRF had low-degree carbonation and poor conductivity (surface resistance is around 10^6 Ω sq^{-1}, Figure 1e,f).

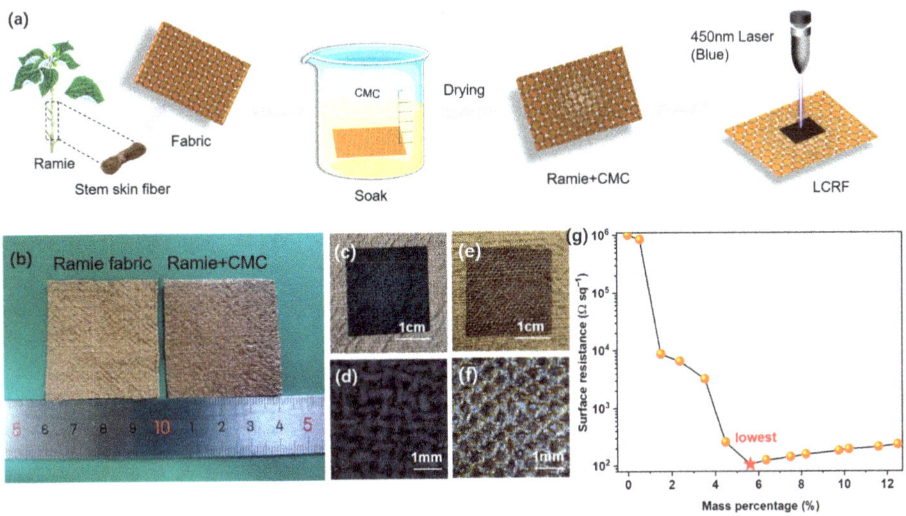

Figure 1. (**a**) Schematic illustration of the fabrication for LCRF. (**b**) photographs of untreated ramie fabric and ramie fabric treated with CMC. (**c**,**d**) surface morphology of carbonized ramie fabric treated with CMC. (**e**,**f**) surface morphology of carbonized untreated ramie fabric. (**g**) surface resistance of LCRF adding CMC with different mass percentages.

Then, we weighed the ramie fabric before and after soaking to obtain the mass change and tested the effect of adding CMC with different mass percentages on the conductivity of LCRF under fixed laser scan speed and power (37% maximum laser power of 15 W and scan speed = 36 mm s^{-1}, Figure 1g). When the mass percentage of CMC on ramie fabric initially increased, the conductivity of LCRF significantly improved. When the mass percentage of CMC is greater than 4.48%, the surface resistance can be stably maintained at the level of 10^2 Ω sq^{-1}. We found that when the mass percentage of CMC was 5.62% (gained from 2.1 wt% CMC solution), the surface resistance had a minimum value (105 Ω sq^{-1}), which was about 10^4 times lower than carbonized untreated ramie fabric. Considering the operability of the performance tests, all further characterization and experiments were carried out when the ramie fabric was soaked in 2.1 wt% CMC solution for one hour.

2.3. Laser Operation Parameter Optimization for LCRF Fabrication

The LCRF with the lowest surface resistance can be obtained by adjusting the laser power percentage and laser scan speed. Laser power is varied from 31% to 41% of maximum operational value (4.65–6.15 W, respectively), with the laser scan speed varying from 12–48 mm s^{-1}. Figure 2a,b show the optical photographs of samples (1×1 cm^2) representing different combinations of parameters and their surface resistance values, respectively. At 33% low power, the carbonization of fabric is insufficient and there are areas of the fabric that are not carbonized (Figure 2(c1)). High laser power of up to 40% ensures that there is enough energy for the conversion of cellulose and lignin fibers into LCRF with good conductivity (surface resistance ≈ 75 Ω sq^{-1}, Figure 2(c2)), which can also be achieved at low scan speed. Moreover, there appears to be a limit above which the energy imparted on the fabric begins to damage its fibers significantly (Figure 2(c3)). When the scanning speed is as low as 12 mm s^{-1} or 24 mm s^{-1}, increasing the laser power is more likely to lead to ablation. The samples will be ablated at 37% or 38% power, and the lowest surface resistance is about 241 Ω sq^{-1}. When the scanning speed increases to 36 mm s^{-1} or 48 mm s^{-1}, the samples will be ablated at 40% power. As a result, appropriately increasing the scanning speed will help to obtain more samples at higher power and reduce the surface resistance of the samples. The lowest surface resistance of all samples (≈65 Ω sq^{-1}) is obtained when the power = 39% and scan speed = 36 mm s^{-1}.

2.4. Laser Engraving of Strain and Humidity Sensor on Ramie Fabric

We firstly used laser to engrave a rectangular pattern (2×6 cm^2) and an interdigital pattern (2×2 cm^2) on the ramie fabric and obtained the LCRF. The patterning of LCRF for strain and humidity sensors was controlled by a translational platform (voidmicro, DAJA). The laser power percentage and scanning speed for the laser engraving were 39% and 36 mm s^{-1}, respectively. After laser engraving, the two ends of rectangular patterned LCRF were connected with copper foil and silver glue. Then, the sample was dried at 80 °C for 2 h and we gained the strain sensor. We applied silver glue to bottom of the interdigital patterned LCRF, dried it at 80 °C for 2 h as well and gained the humidity sensor. Finally, nine humidity sensors engraved on the ramie fabric could form a 3×3 humidity sensing array.

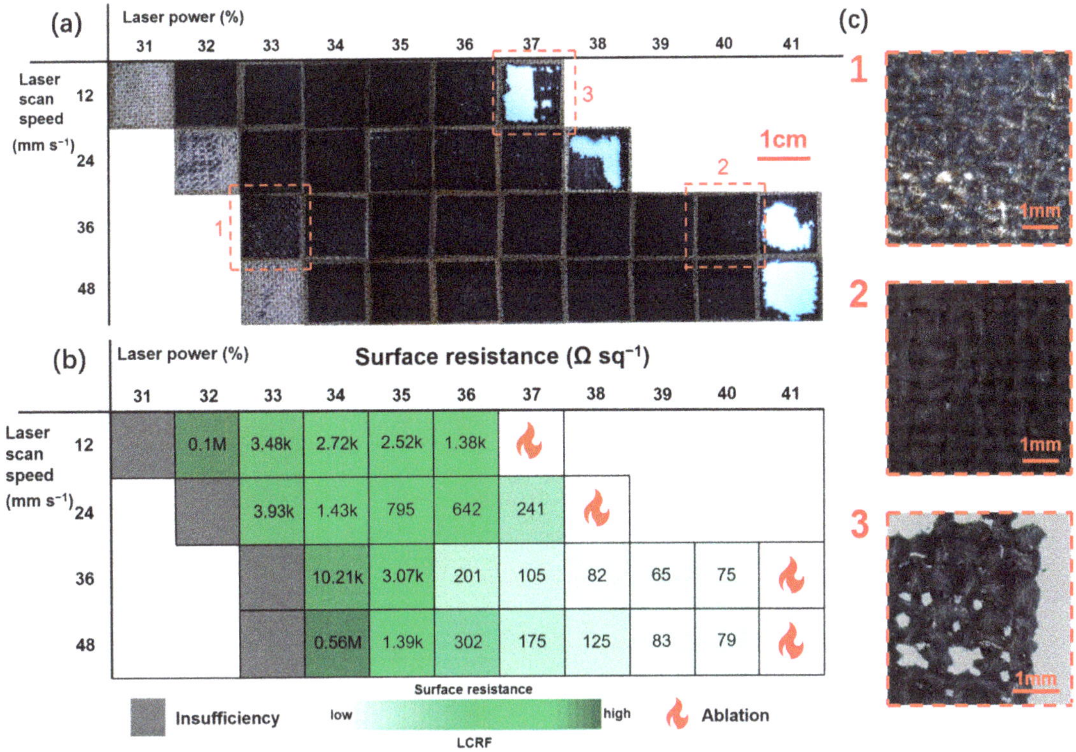

Figure 2. (**a**) LCRF engraved by different laser scan speed and power and (**b**) their surface resistance. (**c**) different LCRF for three different laser scan speeds and power.

2.5. Performance Test of the Strain and Humidity Sensor

The resistance of the strain sensor was measured using a desktop digital multimeter (DMM6500, Tek Keithley Co., Ltd., Cleveland, OH, USA). The gauge factor (GF) of the sensor was calculated according to the equation:

$$GF = (\Delta R/R_0)/\varepsilon \tag{1}$$

In which ΔR, R_0, and ε are the resistance variations, original resistance, and bending strains applied, respectively. According to the reported literature [45,46], the ε was determined based on the radius of curvature and thickness of the sensor under tension or compression. A high-precision single-axis electrodynamic force tester (ZQ-990B, Zhiqu Precision Instrument Co., Ltd., Dongguan, China) was used to bend and release the strain sensor repeatedly.

The capacitance of the humidity sensor was measured at room temperature (25 °C) using a desktop digital bridge (VC4092A, Xi'an Shengli Instrument Co., Ltd., Xi'an, China) with a frequency of 1 kHz, an AC voltage of 3.0 V and a recording interval of 1 s. According to the literature [47–49], six saturated salt solutions with specific relative humidity (RH), which were $LiCl_2$ (11% RH), MgCl (32% RH), NaBr (57% RH), NaCl (75% RH), KCl (84% RH) and K_2SO_4 (97% RH), respectively were prepared as the performance test environment of the humidity sensor. The sensor's sensitivity S is defined as follows:

$$S = (C - C_0)/\Delta RH \tag{2}$$

where S is the sensitivity, C_0 and C are the sensor's capacitance in 11% RH and humidity environments, respectively, and ΔRH is the variation in RH. A high-precision single-axis electrodynamic force tester was used to bend and release the humidity sensor repeatedly (same as the strain sensor). The response and recovery times are defined as the time required to reach 90% of the change of sensor capacitance.

3. Results and Discussion

3.1. The Effect of CMC Treatment on Ramie Fabric

We further evaluated the effect of CMC treatment on ramie fabric under laser engraving. CMC, as a water-soluble derivative of cellulose, can be synthesized from some plant-based precursors (such as corn cobs, banana pseudo-stem and pineapple peel, etc.) and some waste materials (such as wastepaper, waste textiles and cotton gin wastes, etc.) [50]. Thermogravimetric analysis carried out in the air showed that after high-temperature treatment, the ramie fabric treated with CMC has a higher residual weight than that of untreated ramie fabric (Figure 3). The weight of the untreated ramie fabric was reduced to less than 1.0% by 460 °C. As a comparison, ramie fabric treated with CMC retained approximately 8.6% of its original weight even at 600 °C. The residual weight of the ramie fabric treated with CMC was considerably higher than the weight of Na^+ (approximately 0.56 wt%) in the fabric, which might indicate the existence of carbonaceous materials. These results showed that the CMC treatment was conducive to avoiding ablation for ramie fabrics exposed to the air during laser engraving, which was consistent with the conclusion that the surface resistance of the ramie fabric significantly reduced more than 10^4 times after the CMC treatment.

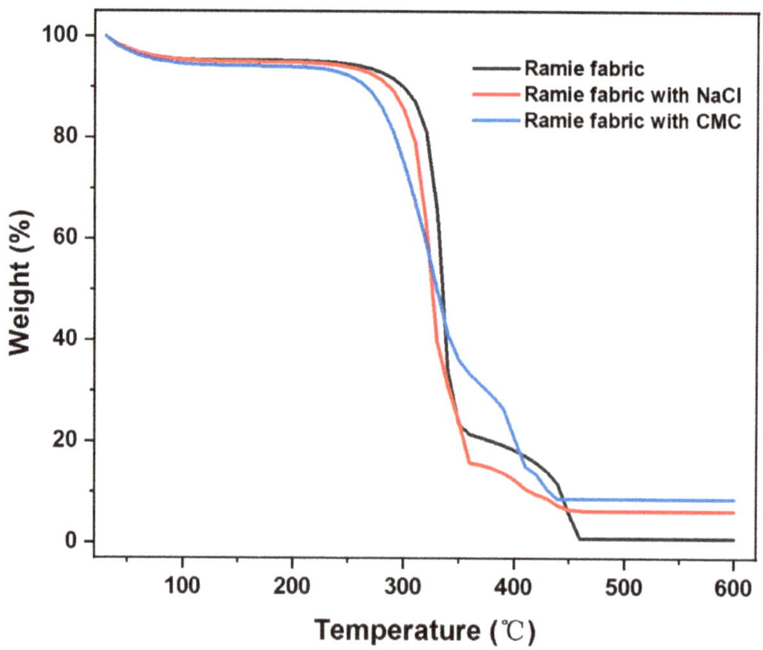

Figure 3. Thermogravimetric curves of different ramie fabrics in air.

According to previous literature [51,52], Na^+ can lower the activation energy of the dehydration step and help the growth of carbonaceous materials at high temperatures. The good effect on flame retardancy of CMC might be attributed to the presence of Na^+ in CMC. In order to further research the effect of Na^+ in CMC on fabric during the laser-engraving, a controlled experiment using NaCl to treat fabric was designed. We soaked the ramie fabric with NaCl solution and the mass percentage of Na^+ on the ramie fabric was ~0.56 wt%, which was the same as the ramie fabric treated with CMC. Then, we engraved ramie fabric treated with NaCl in the same way as we did with the ramie fabric treated with CMC (power = 39% and scan speed = 36 mm s^{-1}). The surface resistance of the carbonized ramie fabric treated with NaCl was ~4128 Ω sq^{-1}, more than 240 times smaller than that of the carbonized untreated ramie fabric. Thermogravimetric analysis showed that the residual weight of ramie fabric treated with NaCl was approximately 6.3% at 600 °C, which was more than six times higher than that of the untreated ramie fabric. These results seemed to indicate that the addition of Na^+ could avoid ablation and reduce the surface resistance of the carbonized ramie fabric. However, compared with the surface resistance of the carbonized ramie fabric treated with NaCl, the surface resistance of the carbonized ramie fabric treated with CMC was more than 63 times smaller, confirming that the CMC treatment had a better effect in reducing the surface resistance of the carbonized ramie fabric. Moreover, compared with NaCl, carbon and Na^+ in CMC can synergistically play a better role in the flame retardant effect, which could explain why the residual weight of the ramie fabric treated with CMC is higher than the ramie fabric treated with NaCl. As a result, using CMC to pretreat ramie fabric is an effective method to avoid fabric ablation and gain LCRF with high conductivity.

3.2. LCRF Characterization

A thorough physicochemical characterization of LCRF was performed to retrieve information about its chemical and morphological properties. Taking the previous work of adjusting laser parameters as a reference, the LCRF fabricated by laser was selected (power = 39% and scan speed = mm s^{-1}). The surface morphology of pristine ramie fabric and LCRF was observed by a scanning electron microscope (TESCAN MIRA3). A scanning electron microscope (SEM) image of the pristine ramie fabric and LCRF is shown in Figures 4a–c and 4d–f, respectively. Figure 4a,d illustrate that the pristine fabric experiences slight shrinkage after carbonization, which would be attributed to the release of CO_2, CO and H_2O during the laser engraving. It can be seen from Figure 4e that some fibers seem to stick together, which is distinguished from the bast fibers carbonized in the oven [43] and may be related to the wrapping of CMC on ramie fibers.

Figure 4. (**a**–**c**) SEM images of carbonized pristine ramie fabric. (**d**–**f**) SEM images of LCRF.

Figure 5 showed the Raman spectrum and X-ray photoelectron spectroscopy (XPS) spectra of pristine ramie fabric and LCRF. Raman spectrum was measured with a Raman spectroscopy (WITec, Apyron, Beijing, China) equipped with a 532 nm laser wavelength. After laser engraving, the typical G-band at ~1334 cm^{-1}, and D-band at ~1582 cm^{-1} could be seen clearly in Figure 5a, which could possibly indicate the graphite structure [43,44]. The XPS test (Thermo Fisher Scientific K-Alpha, Waltham, MA, USA) was conducted to analyze the element information in detail (Figure 5b). The peaks of C_{1s} and O_{1s} are found in the pristine ramie fabric while C_{1s}, O_{1s} and Na_{1s} are found in the LCRF due to the addition of CMC. Further, the C/O atomic ratio of LCRF (2.81) is higher than that of the pristine ramie fabric (2.04), confirming the efficient loss of O atoms from the ramie fabric and thus the successful carbonization of the laser-engraved area [53], which is consistent with the conclusion of the Raman analysis. For further distinction, the analysis of the C_{1s} XPS spectra revealed four dominant peaks: sp^2 C–C bond at ~284.5 eV, C–O bond at ~286.0 eV, C=O bond at ~288.5 eV, and COO bond at ~289.5 eV for the two samples [54], as illustrated in Figure 5c,d. It can be seen clearly that the content of the C–C bond increased from 18.57% to 69.43% and the content of the C–O bond decreased from 66.67% to 13.93%. The sp^2 C–C peak became dominant in LCRF, and the carbonation degree was comparable to that in the laser-carbonized materials (e.g., wood [38,39], nanocellulose [52], lignin/PEO film [55]). As a result, the ramie fabric with CMC treatment can be carbonized to a high degree during laser engraving.

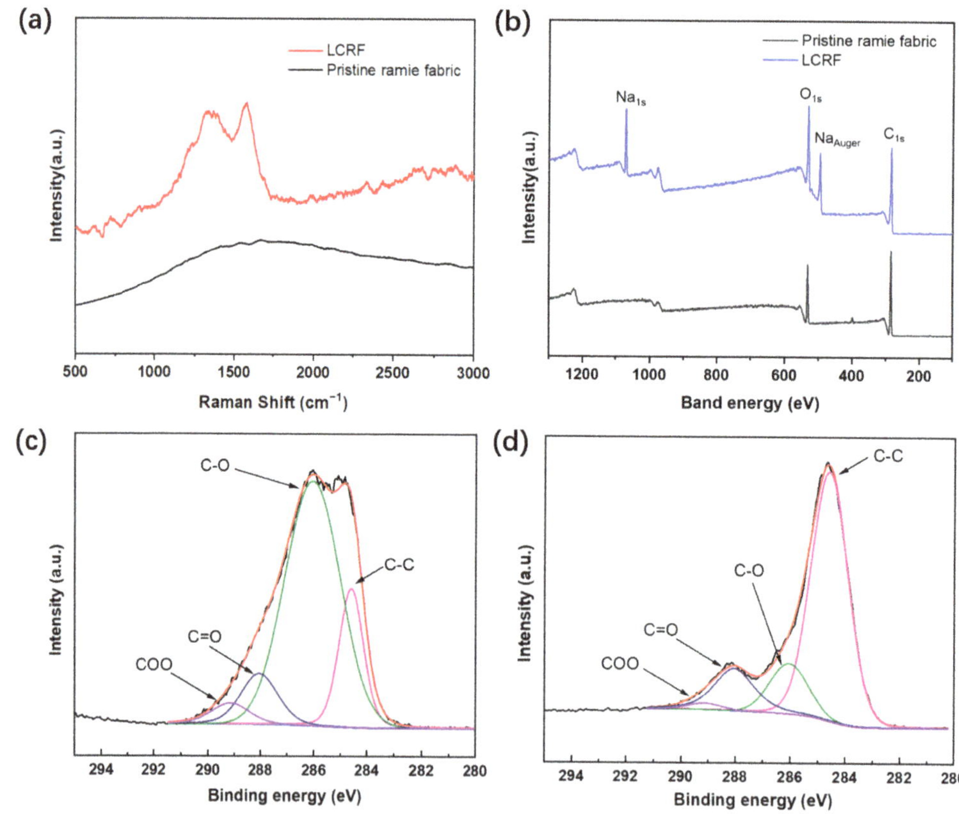

Figure 5. (**a**) Raman spectrum of pristine ramie fabric and LCRF. (**b**) XPS spectra of pristine ramie fabric and LCRF. (**c**) The detailed element information of the C_{1s} XPS spectra for pristine ramie fabric. (**d**) The detailed element information of the C_{1s} XPS spectra for LCRF.

3.3. LCRF-Based Sensors

3.3.1. Strain Sensor

In order to investigate the strain sensitivity, we applied two kinds of strains, compressing and tensioning, to the LCRF sensor and observed that the resistance of the strain sensor decreased or increased accordingly. This phenomenon can be explained by the compression and tension of carbonized fiber with cracks in LCRF (Figure 6a) [46,47]. By bending the inner surface, the sensor is under compressive stress, and the crack gap in the fibers narrows, which results in decreasing the resistance of the strain sensor. On the contrary, by bending the outer surface, the sensor was under tensile stress, and the crack gap in the fibers widens, which resulted in increasing the resistance of the strain sensor. The following experimental phenomena of the pre-bending test can be explained by this mechanism of cracks on the fibers. Figure 6b shows the relationships between the number of pre-bending cycles (radius of curvature was about 2.15 cm, representing 1.5% strain) and the normalized resistance change. Due to the cracks generated in the fibers, the resistance of the sensor in the natural state increased at first. After about 50 pre-bending cycles, the resistance of the sensor in its natural state increased by more than 31% and remained stable, so the LCRF-based strain sensors were all trained for 50 cycles.

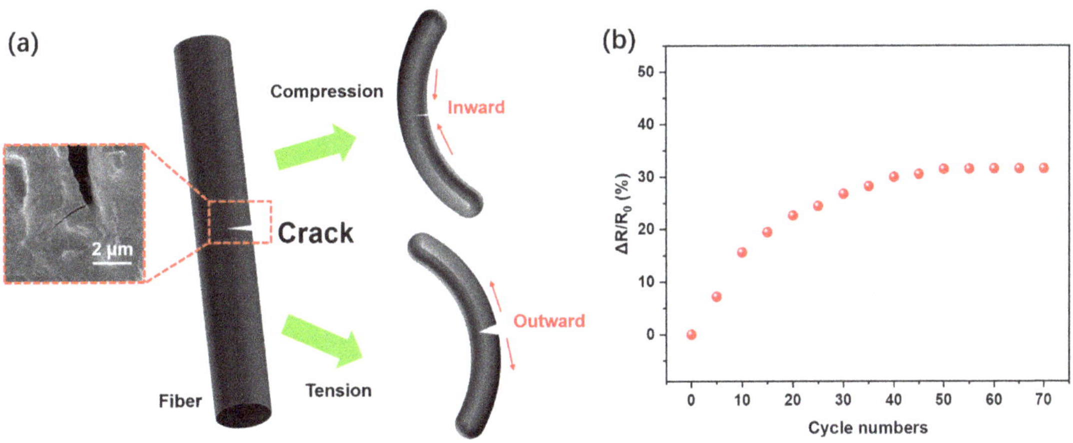

Figure 6. (**a**) Sensing mechanism of LCRF-based strain sensor. (**b**) Normalized resistance changes of sensor in the 70 pre-bending cycles.

As illustrated in Figure 7a,b, the values of GF for compression or tension are 128.9 or 136.3, respectively, higher than the reported bast fabric strain sensors carbonized in the oven [42,44]. Figure 7c exhibits that the sensor generates regular resistance change for bending strain and the response time was approximately 0.22 s or 0.27 s for tension or compression, respectively, which can meet the general sensing requirements of many daily motions. To evaluate the mechanical robustness and the reliability of the strain sensors, the resistance of the sensor was recorded during 5000 bending–unbending cycles (radius of curvature was about 2.93 cm, representing 1.1% strain, Figure 7d). After bending, the change of resistance still maintained stability, suggesting that the sensor is reliable for strain detections.

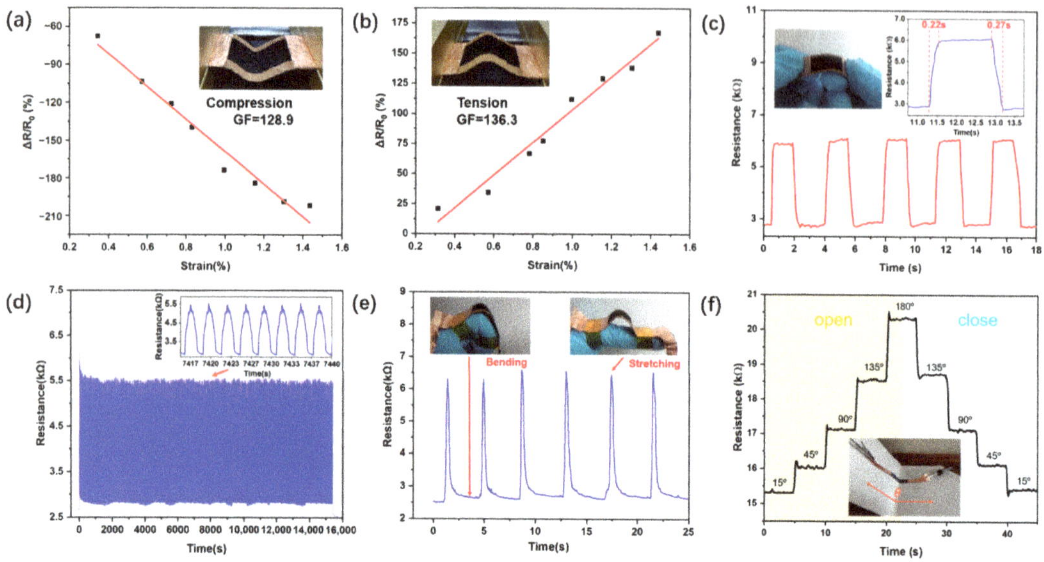

Figure 7. Performance of the sensor for detecting bending strains. Relative resistance changes as a function of applied strain. (**a**,**b**) are in accordance with the different bending directions (inward deflection (compression) or outward deflection (tension), respectively). (**c**) Response time of the sensor under bending strain. (**d**) Resistance change of the strain sensor during application of 5000 bending–unbending cycles. (**e**) Relative resistance changes during the finger bending tests. (**f**) Resistance change of the sensor under opening (orange area) and closing (blue area) the book states.

We further investigated the application potential of the flexible strain sensor in the field of human-motion detection. As fingers are able to perform many complex human actions, we attached the strain sensor to an index finger for the potential application in wearable electronics (Figure 7e). When the finger bent and stretched rapidly, the sensor could respond quickly and produce a short pulse signal, indicating good performance in rapid monitoring of human motion. Moreover, we laid another strain sensor on the folding line of a book as well in order to investigate it as a folding-type sensing device (Figure 7f). By measuring the resistance at the different angles θ, the electromechanical properties of sensors were investigated. We observed the resistance variation of the folding-type sensor by opening the book at 15°, 45°, 90°, 135°, and 180° step by step (each step was held for about five seconds) and then closing gradually from 180° to 135°, 90°, 45°, and 15°. When we opened the book from 15° to 180°, the resistance of the sensor increased from ~15.3 kΩ to ~20.3 kΩ. Then, the book was closed from 180° to 15°, while the resistance of the sensor decreased from ~20.3 kΩ to ~15.3 kΩ. At each turning angle, the folding-type sensor could respond rapidly and the resistance of the sensor could be maintained under the same angle. Similarly, we can also use this folding-type sensor to measure the changes in mechanical arm angle, suggesting an attractive perspective application in robots.

3.3.2. Humidity Sensor

Humidity sensors are widely used in industrial instruments and automation, agriculture, medical care, and climate monitoring [56–58]. Taking advantage of the good water absorption of ramie fabric, the humidity sensor is composed of a laser-carbonized interdigital electrode and uncarbonized ramie fabric, as shown in Figure 8a. The capacitance value variations of the humidity sensor are recorded and depicted when the RH was increased from 11% to 97%. When RH = 11%, the capacitance value was 12.2 pF, and when

RH = 97%, the capacitance value rose to 303.8 pF. The values of S for RH ranging from 0–57% or 57–100% are 5.17 or 55.26, respectively, higher than some reported natural fabric (cotton and wool fabric) humidity sensors (0.34–20.12) [59–61]. The interdigital electrode on ramie fabric had good flexibility and in order to evaluate the mechanical robustness and the reliability of the electrode, the capacitance value of the electrode was recorded during 5000 bending–unbending cycles (radius of curvature was about 1.52 cm, Figure 8b). The result showed that the capacitance value of the electrode only decreased by less than 4.4%, suggesting that the electrode had good reliability. Moreover, when the humidity is alternately converted from 45%RH (in the air) to 84%RH, the sensor generates a regular capacitance signal, as shown in Figure 8c. When the relative humidity in the test container is 84%RH, the capacitance needs about 900 s to reach equilibrium, and it takes about 1000 s to recover in the air. As a result, the humidity sensor's response/recovery time between 45% and 84% RH is about 900 s/1000 s.

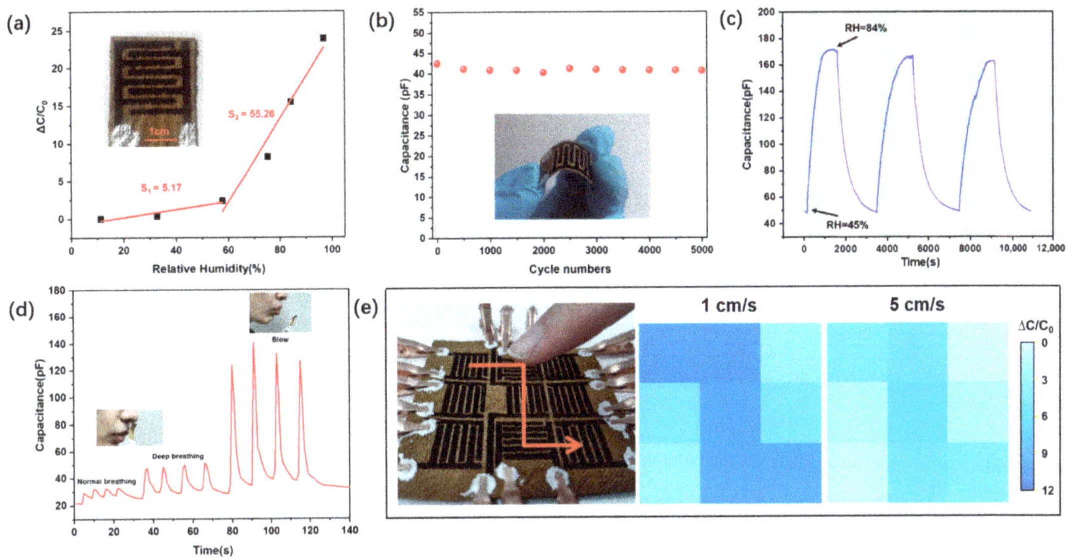

Figure 8. Performance of the sensor for detecting humidity. (**a**) Laser engraved interdigital electrode pattern and its relative capacitance changes as a function of relative humidity. (**b**) Capacitance change of the electrode during application of 5000 bending–unbending cycles. (**c**) Capacitance change of the humidity sensor under alternative humidity change from 45% to 84%. (**d**) Capacitance change of humidity sensor under human blowing and breathing. (**e**) Speed detection of a moving finger using a 3 × 3 humidity sensor array.

For real-time humidity detections, human breathing and blowing were monitored. As shown in Figure 8d, the humidity sensor was respectively close to the nose and mouth of the human body to monitor the water vapor generated by the human body in real-time. When monitoring human breathing, the humidity sensor can distinguish between normal breathing (about 12 times per minute) and deep breathing (about six times per minute) according to the change of capacitance. Furthermore, when blowing air (about five times per minute) onto the humidity sensor, the capacitance value will increase in about 3 s and recover in about 10 s. These results suggest the potential applicability for human-health monitoring. In order to use the patterning advantages of this laser-engraving method, a 3 × 3 humidity sensor array was fabricated to detect moving wet objects (Figure 8e). When a bared finger (3 cm from the array) is passed over the sensor array at different speeds, the capacitance value of each sensor on the array changes differently. When the finger speed was 1 cm/s, more water vapor entered the sensor, resulting in a greater capacitance increase and when the finger speed upgraded to 5 cm/s, less water vapor entered the sensor, resulting in less capacitance increase. We also used a finger with a rubber glove to pass over the array and the capacitance value of the humidity sensor did not change. As a result, these humidity sensors can provide the potential for human-health monitoring and speed detection of wet objects in real-time.

4. Conclusions

In summary, we successfully developed a simplified, cost-effective and environmentally friendly method for engraving ramie fabric directly by laser under an ambient atmosphere to prepare strain and humidity sensors. The addition of CMC to ramie fabric could effectively avoid ablation during laser engraving and the gained LCRF had good conductivity (65 Ω sq^{-1}). The laser-engraving process allows for patterning on the fabric, and thus the fabrication of strain and humidity sensors. The fabric strain sensors feature a high GF of $\approx$128.9 (compressive strain), and $\approx$136.3 (tensile strain) and remarkable reliability and durability (>5000 cycles). The fabric humidity sensors have good flexibility and can detect the moving speed of wet objects in real-time. This low-cost laser-engraving carbonizing method can be used in fabricating electronic components on natural bast materials on a large scale.

Author Contributions: Conceptualization, S.S. and B.S.; methodology, S.S.; software, S.S. and J.L.; validation, S.S., J.L. and B.S.; formal analysis, S.S. and C.Q.; investigation, S.S. and J.L.; resources, B.S.; data curation, C.Q.; writing—original draft preparation, S.S. and B.S.; writing—review and editing, S.S. and B.S.; visualization, S.C. and J.L.; supervision, B.S.; project administration, B.S.; funding acquisition, B.S. All authors have read and agreed to the published version of the manuscript.

Funding: This research was funded by the Natural Science Foundation of Shanghai (19ZR1436100) and the National Natural Science Foundation of China (11105149).

Acknowledgments: The authors express gratitude to the editors and the reviewers for their constructive and helpful review comments.

Conflicts of Interest: The authors declare no conflict of interest.

References

1. Wu, R.; Ma, L.; Patil, A.; Meng, Z.; Liu, S.; Hou, C.; Zhang, Y.; Yu, W.; Guo, W.; Liu, X.Y. Graphene Decorated Carbonized Cellulose Fabric for Physiological Signal Monitoring and Energy Harvesting. *J. Mater. Chem. A* **2020**, *8*, 12665–12673. [CrossRef]
2. Ge, J.; Sun, L.; Zhang, F.-R.; Zhang, Y.; Shi, L.-A.; Zhao, H.-Y.; Zhu, H.-W.; Jiang, H.-L.; Yu, S.-H. A Stretchable Electronic Fabric Artificial Skin with Pressure-, Lateral Strain-, and Flexion-Sensitive Properties. *Adv. Mater.* **2016**, *28*, 722–728. [CrossRef] [PubMed]
3. Rauf, S.; Vijjapu, M.T.; Andrés, M.A.; Gascón, I.; Roubeau, O.; Eddaoudi, M.; Salama, K.N. Highly Selective Metal–Organic Framework Textile Humidity Sensor. *ACS Appl. Mater. Interfaces* **2020**, *12*, 29999–30006. [CrossRef] [PubMed]
4. Wu, S.; Liu, P.; Zhang, Y.; Zhang, H.; Qin, X. Flexible and Conductive Nanofiber-Structured Single Yarn Sensor for Smart Wearable Devices. *Sens. Actuators B Chem.* **2017**, *252*, 697–705. [CrossRef]
5. Xiong, Y.; Xiao, J.; Chen, J.; Xu, D.; Zhao, S.; Chen, S.; Sheng, B. A Multifunctional Hollow TPU Fiber Filled with Liquid Metal Exhibiting Fast Electrothermal Deformation and Recovery. *Soft Matter* **2021**, *17*, 10016–10024. [CrossRef] [PubMed]
6. Zeng, W.; Tao, X.-M.; Chen, S.; Shang, S.; Chan, H.L.W.; Choy, S.H. Highly Durable All-Fiber Nanogenerator for Mechanical Energy Harvesting. *Energy Environ. Sci.* **2013**, *6*, 2631–2638. [CrossRef]
7. Sun, X.; Jin, H.; Qu, W. Lignin-Derived 3D Porous Graphene on Carbon Cloth for Flexible Supercapacitors. *RSC Adv.* **2021**, *11*, 19695–19704. [CrossRef]
8. Fu, X.-Y.; Chen, Z.-D.; Zhang, Y.-L.; Han, D.-D.; Ma, J.-N.; Wang, W.; Zhang, Z.-R.; Xia, H.; Sun, H.-B. Direct Laser Writing of Flexible Planar Supercapacitors Based on GO and Black Phosphorus Quantum Dot Nanocomposites. *Nanoscale* **2019**, *11*, 9133–9140. [CrossRef]
9. Paul, D.L.; Giddens, H.; Paterson, M.; Hilton, G.S.; McGeehan, J.P. Impact of Body and Clothing on a Wearable Textile Dual Band Antenna at Digital Television and Wireless Communications Bands. *IEEE Trans. Antennas Propag.* **2013**, *61*, 2188–2194. [CrossRef]
10. Boutry, C.M.; Beker, L.; Kaizawa, Y.; Vassos, C.; Tran, H.; Hinckley, A.C.; Pfattner, R.; Niu, S.; Li, J.; Claverie, J.; et al. Biodegradable and Flexible Arterial-Pulse Sensor for the Wireless Monitoring of Blood Flow. *Nat. Biomed. Eng.* **2019**, *3*, 47–57. [CrossRef] [PubMed]
11. Ma, L.; Wu, R.; Patil, A.; Zhu, S.; Meng, Z.; Meng, H.; Hou, C.; Zhang, Y.; Liu, Q.; Yu, R.; et al. Full-Textile Wireless Flexible Humidity Sensor for Human Physiological Monitoring. *Adv. Funct. Mater.* **2019**, *29*, 1904549. [CrossRef]

12. Hu, Q.; Nag, A.; Xu, Y.; Han, T.; Zhang, L. Use of Graphene-Based Fabric Sensors for Monitoring Human Activities. *Sens. Actuators A Phys.* **2021**, *332*, 113172. [CrossRef]
13. Fu, Q.; Cui, C.; Meng, L.; Hao, S.; Dai, R.; Yang, J. Emerging Cellulose-Derived Materials: A Promising Platform for the Design of Flexible Wearable Sensors toward Health and Environment Monitoring. *Mater. Chem. Front.* **2021**, *5*, 2051–2091. [CrossRef]
14. Xiao, J.; Xiong, Y.; Chen, J.; Zhao, S.; Chen, S.; Xu, B.; Sheng, B. Ultrasensitive and Highly Stretchable Fibers with Dual Conductive Microstructural Sheaths for Human Motion and Micro Vibration Sensing. *Nanoscale* **2022**, *14*, 1962–1970. [CrossRef] [PubMed]
15. Wang, D.; Sheng, B.; Peng, L.; Huang, Y.; Ni, Z. Flexible and Optical Fiber Sensors Composited by Graphene and PDMS for Motion Detection. *Polymers* **2019**, *11*, 1433. [CrossRef]
16. Ma, L.; Wu, R.; Miao, H.; Fan, X.; Kong, L.; Patil, A.; Liu, X.Y.; Wang, J. All-in-One Fibrous Capacitive Humidity Sensor for Human Breath Monitoring. *Text. Res. J.* **2021**, *91*, 398–405. [CrossRef]
17. Gao, Y.; Lu, C.; Guohui, Y.; Sha, J.; Tan, J.; Xuan, F. Laser Micro-Structured Pressure Sensor with Modulated Sensitivity for Electronic Skins. *Nanotechnology* **2019**, *30*, 325502. [CrossRef]
18. Bai, M.; Zhai, Y.; Liu, F.; Wang, Y.; Luo, S. Stretchable Graphene Thin Film Enabled Yarn Sensors with Tunable Piezoresistivity for Human Motion Monitoring. *Sci. Rep.* **2019**, *9*, 18644. [CrossRef]
19. Afroj, S.; Karim, N.; Wang, Z.; Tan, S.; He, P.; Holwill, M.; Ghazaryan, D.; Fernando, A.; Novoselov, K.S. Engineering Graphene Flakes for Wearable Textile Sensors via Highly Scalable and Ultrafast Yarn Dyeing Technique. *ACS Nano* **2019**, *13*, 3847–3857. [CrossRef]
20. Xu, L.; Liu, Z.; Zhai, H.; Chen, X.; Sun, R.; Lyu, S.; Fan, Y.; Yi, Y.; Chen, Z.; Jin, L.; et al. Moisture-Resilient Graphene-Dyed Wool Fabric for Strain Sensing. *ACS Appl. Mater. Interfaces* **2020**, *12*, 13265–13274. [CrossRef] [PubMed]
21. Yoon, S.; Kim, H.-K. Cost-Effective Stretchable Ag Nanoparticles Electrodes Fabrication by Screen Printing for Wearable Strain Sensors. *Surf. Coat. Technol.* **2020**, *384*, 125308. [CrossRef]
22. Lee, C.G.; Kwon, B.S.; Nam, H.M.; Seo, D.M.; Park, J.; Hwangbo, H.; Park, L.S.; Nam, S.Y. Stretchable Strain Sensors Fabricated by Screen Printing of Silver Paste on the Surface Modified Transparent Elastomeric Polyurethane Films. *Mater. Sci. Appl.* **2018**, *9*, 1008–1020. [CrossRef]
23. Andò, B.; Baglio, S.; Bulsara, A.R.; Emery, T.; Marletta, V.; Pistorio, A. Low-Cost Inkjet Printing Technology for the Rapid Prototyping of Transducers. *Sensors* **2017**, *17*, 748. [CrossRef] [PubMed]
24. Duy Dam, L.; Nguyen, T.; Doan, T.; Dang, T.; Dang, C. Fabrication of Interdigitated Electrodes by Inkjet Printing Technology for Apllication in Ammonia Sensing. *Adv. Nat. Sci. Nanosci. Nanotechnol.* **2016**, *7*, 025002. [CrossRef]
25. Pawlak, R.; Korzeniewska, E.; Frydrysiak, M.; Zięba, J.; Tęsiorowski, Ł.; Gniotek, K.; Stempien, Z.; Tokarska, M. Using Vacuum Deposition Technology for the Manufacturing of Electro-Conductive Layers on the Surface of Textiles. *Fibres Text. East. Eur.* **2012**, *91*, 68–72.
26. Wang, C.; Li, X.; Gao, E.; Jian, M.; Xia, K.; Wang, Q.; Xu, Z.; Ren, T.; Zhang, Y. Carbonized Silk Fabric for Ultrastretchable, Highly Sensitive, and Wearable Strain Sensors. *Adv. Mater.* **2016**, *28*, 6640–6648. [CrossRef]
27. Zhang, M.; Wang, C.; Wang, H.; Jian, M.; Hao, X.; Zhang, Y. Carbonized Cotton Fabric for High-Performance Wearable Strain Sensors. *Adv. Funct. Mater.* **2017**, *27*, 1604795. [CrossRef]
28. Doshi, S.M.; Murray, C.; Chaudhari, A.; Sung, D.H.; Thostenson, E.T. Ultrahigh Sensitivity Wearable Sensors Enabled by Electrophoretic Deposition of Carbon Nanostructured Composites onto Everyday Fabrics. *J. Mater. Chem. C* **2022**, *10*, 1617–1624. [CrossRef]
29. Wang, L.; Xu, T.; Zhang, X. Multifunctional Conductive Hydrogel-Based Flexible Wearable Sensors. *TrAC Trends Anal. Chem.* **2021**, *134*, 116130. [CrossRef]
30. Maddipatla, D.; Narakathu, B.B.; Atashbar, M. Recent Progress in Manufacturing Techniques of Printed and Flexible Sensors: A Review. *Biosensors* **2020**, *10*, 199. [CrossRef]
31. Li, Y.; Samad, Y.A.; Taha, T.; Cai, G.; Fu, S.-Y.; Liao, K. Highly Flexible Strain Sensor from Tissue Paper for Wearable Electronics. *ACS Sustain. Chem. Eng.* **2016**, *4*, 4288–4295. [CrossRef]
32. Li, Y.-Q.; Zhu, W.-B.; Yu, X.-G.; Huang, P.; Fu, S.-Y.; Hu, N.; Liao, K. Multifunctional Wearable Device Based on Flexible and Conductive Carbon Sponge/Polydimethylsiloxane Composite. *ACS Appl. Mater. Interfaces* **2016**, *8*, 33189–33196. [CrossRef] [PubMed]
33. Miao, Y.; Wan, L.; Ling, X.; Chen, B.; Pan, L.; Gao, Y. Mask-Free Preparation of Patterned Carbonized Carboxymethyl Cellulose on Fabrics for Flexible Electronics. *ACS Appl. Electron. Mater.* **2020**, *2*, 855–862. [CrossRef]
34. Wang, H.; Wang, H.; Wang, Y.; Su, X.; Wang, C.; Zhang, M.; Jian, M.; Xia, K.; Liang, X.; Lu, H.; et al. Laser Writing of Janus Graphene/Kevlar Textile for Intelligent Protective Clothing. *ACS Nano* **2020**, *14*, 3219–3226. [CrossRef] [PubMed]
35. Liu, W.; Huang, Y.; Peng, Y.; Walczak, M.; Wang, D.; Chen, Q.; Liu, Z.; Li, L. Stable Wearable Strain Sensors on Textiles by Direct Laser Writing of Graphene. *ACS Appl. Nano Mater.* **2020**, *3*, 283–293. [CrossRef]
36. Li, Z.; Lu, L.; Xie, Y.; Wang, W.; Zhiran, L.; Tang, B.; Lin, N. Preparation of Laser-Induced Graphene Fabric from Silk and Its Application Examples for Flexible Sensor. *Adv. Eng. Mater.* **2021**, *23*, 2100195. [CrossRef]
37. Xu, H.; Gao, L.; Wang, Y.; Cao, K.; Hu, X.; Wang, L.; Mu, M.; Liu, M.; Zhang, H.; Wang, W.; et al. Flexible Waterproof Piezoresistive Pressure Sensors with Wide Linear Working Range Based on Conductive Fabrics. *Nano-Micro Lett.* **2020**, *12*, 159. [CrossRef]
38. Ye, R.; Chyan, Y.; Zhang, J.; Li, Y.; Han, X.; Kittrell, C.; Tour, J.M. Laser-Induced Graphene Formation on Wood. *Adv. Mater.* **2017**, *29*, 1702211. [CrossRef]

39. Chyan, Y.; Ye, R.; Li, Y.; Singh, S.P.; Arnusch, C.J.; Tour, J.M. Laser-Induced Graphene by Multiple Lasing: Toward Electronics on Cloth, Paper, and Food. *ACS Nano* **2018**, *12*, 2176–2183. [CrossRef]
40. Han, X.; Ye, R.; Chyan, Y.; Wang, T.; Zhang, C.; Shi, L.; Zhang, T.; Zhao, Y.; Tour, J. Laser-Induced Graphene from Wood Impregnated with Metal Salts and Use in Electrocatalysis. *ACS Appl. Nano Mater.* **2018**, *1*, 5053–5061. [CrossRef]
41. Dhand, V.; Mittal, G.; Rhee, K.Y.; Park, S.-J.; Hui, D. A Short Review on Basalt Fiber Reinforced Polymer Composites. *Compos. Part. B Eng.* **2015**, *73*, 166–180. [CrossRef]
42. Liang, J.; Zhao, Z.; Tang, Y.; Liang, Z.; Sun, L.; Pan, X.; Wang, X.; Qiu, J. A Wearable Strain Sensor Based on Carbon Derived from Linen Fabrics. *New Carbon Mater.* **2020**, *35*, 522–530. [CrossRef]
43. Liu, Z.; Chen, K.; Fernando, A.; Gao, Y.; Li, G.; Jin, L.; Zhai, H.; Yi, Y.; Xu, L.; Zheng, Y.; et al. Permeable Graphited Hemp Fabrics-Based, Wearing-Comfortable Pressure Sensors for Monitoring Human Activities. *Chem. Eng. J.* **2021**, *403*, 126191. [CrossRef]
44. Liu, Z.; Li, Z.; Zhai, H.; Jin, L.; Chen, K.; Yi, Y.; Gao, Y.; Xu, L.; Zheng, Y.; Yao, S.; et al. A Highly Sensitive Stretchable Strain Sensor Based on Multi-Functionalized Fabric for Respiration Monitoring and Identification. *Chem. Eng. J.* **2021**, *426*, 130869. [CrossRef]
45. Liu, H.; Xiang, H.; Ma, Y.; Li, Z.; Meng, Q.; Jiang, H.; Wu, H.; Li, P.; Zhou, H.; Huang, W. Flexible, Degradable, and Cost-Effective Strain Sensor Fabricated by a Scalable Papermaking Procedure. *ACS Sustain. Chem. Eng.* **2018**, *6*, 15749–15755. [CrossRef]
46. Liao, X.; Liao, Q.; Yan, X.; Liang, Q.; Si, H.; Li, M.; Wu, H.; Cao, S.; Zhang, Y. Flexible and Highly Sensitive Strain Sensors Fabricated by Pencil Drawn for Wearable Monitor. *Adv. Funct. Mater.* **2015**, *25*, 2395–2401. [CrossRef]
47. Zhao, J.; Yu, Z.; Tu, Z.; Bian, H. Influence of Electrode Structure on Performance of Laser Direct Writing Cu-PI Flexible Humidity Sensor. *Micromachines* **2022**, *13*, 992. [CrossRef] [PubMed]
48. Liu, H.; Xiang, H.; Wang, Y.; Li, Z.; Qian, L.; Li, P.; Ma, Y.; Zhou, H.; Huang, W. A Flexible Multimodal Sensor That Detects Strain, Humidity, Temperature, and Pressure with Carbon Black and Reduced Graphene Oxide Hierarchical Composite on Paper. *ACS Appl Mater. Interfaces* **2019**, *11*, 40613–40619. [CrossRef]
49. Wang, N.; Tong, J.; Wang, J.; Wang, Q.; Chen, S.; Sheng, B. Polyimide-Sputtered and Polymerized Films with Ultrahigh Moisture Sensitivity for Respiratory Monitoring and Contactless Sensing. *ACS Appl. Mater. Interfaces* **2022**, *14*, 11842–11853. [CrossRef]
50. Rahman, M.S.; Hasan, M.S.; Nitai, A.S.; Nam, S.; Karmakar, A.K.; Ahsan, M.S.; Shiddiky, M.J.A.; Ahmed, M.B. Recent Developments of Carboxymethyl Cellulose. *Polymers* **2021**, *13*, 1345. [CrossRef]
51. Shi, R.; Tan, L.; Zong, L.; Ji, Q.; Li, X.; Zhang, K.; Cheng, L.; Xia, Y. Influence of Na+ and Ca2+ on Flame Retardancy, Thermal Degradation, and Pyrolysis Behavior of Cellulose Fibers. *Carbohydr. Polym.* **2017**, *157*, 1594–1603. [CrossRef]
52. Lee, S.; Jeon, S. Laser-Induced Graphitization of Cellulose Nanofiber Substrates under Ambient Conditions. *ACS Sustain. Chem. Eng.* **2019**, *7*, 2270–2275. [CrossRef]
53. Le, T.-S.D.; Park, S.; An, J.; Lee, P.S.; Kim, Y.-J. Ultrafast Laser Pulses Enable One-Step Graphene Patterning on Woods and Leaves for Green Electronics. *Adv. Funct. Mater.* **2019**, *29*, 1902771. [CrossRef]
54. Johansson, L.-S.; Campbell, J.M.; Rojas, O.J. Cellulose as the in Situ Reference for Organic XPS. Why? Because It Works. *Surf. Interface Anal.* **2020**, *52*, 1134–1138. [CrossRef]
55. Mahmood, F.; Zhang, C.; Xie, Y.; Stalla, D.; Lin, J.; Wan, C. Transforming Lignin into Porous Graphene via Direct Laser Writing for Solid-State Supercapacitors. *RSC Adv.* **2019**, *9*, 22713–22720. [CrossRef]
56. Li, W.; Xu, F.; Sun, L.; Liu, W.; Qiu, Y. A Novel Flexible Humidity Switch Material Based on Multi-Walled Carbon Nanotube/Polyvinyl Alcohol Composite Yarn. *Sens. Actuators B Chem.* **2016**, *230*, 528–535. [CrossRef]
57. Tong, J.; Wang, N.; Wang, Q.; Chen, S.; Sheng, B. Improved Sensitive Conductive Sponge Sensors with Tunnel-Crack Broadening for Pressure, Humidity and Temperature Sensing Applications. *Sens. Actuators B Chem.* **2022**, *358*, 131497. [CrossRef]
58. Wang, Q.; Tong, J.; Wang, N.; Chen, S.; Sheng, B. Humidity Sensor of Tunnel-Cracked Nickel@polyurethane Sponge for Respiratory and Perspiration Sensing. *Sens. Actuators B Chem.* **2021**, *330*, 129322. [CrossRef]
59. Xu, L.; Zhai, H.; Chen, X.; Liu, Y.; Wang, M.; Liu, Z.; Umar, M.; Ji, C.; Chen, Z.; Jin, L.; et al. Coolmax/Graphene-Oxide Functionalized Textile Humidity Sensor with Ultrafast Response for Human Activities Monitoring. *Chem. Eng. J.* **2021**, *412*, 128639. [CrossRef]
60. Allison, L.K.; Rostaminia, S.; Kiaghadi, A.; Ganesan, D.; Andrew, T.L. Enabling Longitudinal Respiration Monitoring Using Vapor-Coated Conducting Textiles. *ACS Omega* **2021**, *6*, 31869–31875. [CrossRef]
61. de Oliveira, C.R.S.; Batistella, M.A.; de Souza, S.M.D.A.G.U.; de Souza, A.A.U. Development of Flexible Sensors Using Knit Fabrics with Conductive Polyaniline Coating and Graphite Electrodes. *J. Appl. Polym. Sci.* **2017**, *134*, 18. [CrossRef]

micromachines

Article

High Sensitivity and Wide Range Biomimetic Tactile-Pressure Sensor Based on 2D Graphene Film and 3D Graphene Foam

Baolin Sha [1], Xiaozhou Lü [2,*] and La Jiang [2]

[1] The 41st Institute of the Fourth Academy of CASC, Xi'an 710025, China; 13709257563@126.com
[2] School of Aerospace Science and Technology, Xidian University, Xi'an 710071, China; j15603746663@163.com
* Correspondence: xzlu@xidian.edu.cn

Abstract: Bionic electronic skin is a system that simulates human skin and has multiple perceptions. For pressure sensors, high measurement accuracy and wide measurement range restrict each other, and it is difficult to achieve high measurement accuracy and wide measurement range simultaneously. Therefore, the research and application of bionic tactile-pressure sensors are limited due to the mutual constraints of measurement accuracy and range. In this work, a flexible graphene piezoresistive tactile sensor based on a biomimetic structure that utilizes the piezoresistive properties of graphene was reported. The novel tactile-pressure sensor consists of a 2D graphene film tactile sensor and a 3D graphene foam pressure sensor that could achieve high accuracy and a wide-range measurement simultaneously. The testing results show that the measurement range of this sensor was in two intervals of 0–2 N and 2–40 N. For the 0–2 N measurement range, the sensitivity was 472.2 Ω/kPa, the force resolution was 0.01 N, and the response time was less than 40 ms. For the 2–40 N measurement range, the sensitivity was 5.05 kΩ/kPa, the force resolution was 1 N, and the response time was less than 20 ms. The new sensor can realize high-precision and large-scale force measurements and shows great application value in the field of medical instruments and artificial limbs.

Keywords: 2D graphene film; 3D graphene foam; tactile sensor; pressure sensor

Citation: Sha, B.; Lü, X.; Jiang, L. High Sensitivity and Wide Range Biomimetic Tactile-Pressure Sensor Based on 2D Graphene Film and 3D Graphene Foam. *Micromachines* **2022**, *13*, 1150. https://doi.org/10.3390/mi13071150

Academic Editor: Alberto Tagliaferro

Received: 9 June 2022
Accepted: 15 July 2022
Published: 21 July 2022

1. Introduction

With the development of bionic robots and medical devices, it is significant to simulate human perception measurement. As one of the five human senses (vision, touch, hearing, smell, and taste), touch is an important parameter for perceiving the outside world [1]. The pressure sensor can realize the measurement of human touch [2–4]. Its working principle can be divided into four categories: capacitive [5–7], piezoresistive [8–10], piezoelectric [11–14], and triboelectronic [15–20]. Capacitive sensors usually have high sensitivity. Cui et al. fabricated a flexible capacitive tactile sensor based on silver electrodes and Polydimethylsiloxane (PDMS), with a sensitivity of up to 19.8 kPa^{-1} [21]. Wan et al. proposed an ultrasensitive pressure sensor with graphene as an electrode. The response time of the sensor can reach 100 ms, and the sensitivity of the sensor can reach 0.8 kPa^{-1} [22]. Ntagios et al. developed a soft tactile sensor by using the 3D print method, and the soft tactile sensor exhibited a stable response with a sensitivity of 0.00348 kPa^{-1} for pressure <10 kPa and 0.00134 kPa^{-1} for higher pressure [23]. Yang et al. demonstrated a 3D microconformal graphene electrode for a flexible capacitive pressure sensor and the developed capacitive pressure sensor with a high sensitivity of 3.19 kPa^{-1}, a fast response time of 30 ms, and an ultralow detection limit of 1 mg [24]. Kang et al. proposed a highly sensitive capacitive pressure sensor based on a porous structure of PDMS, and the developed sensor resulted in high-performance pressure sensors with high sensitivity of 0.63 kPa^{-1} and extremely low-pressure detection of 2.42 Pa [25]. From the above cases, the capacitive sensor has a high sensitivity but a small measurement range. At the same time, the capacitive sensor has some shortcomings, such as capacitive sensors have large impedances, which lead to them being sensitive to the parasitic capacitance of coaxial cables, and the measurement error is large [26]. Piezoelectric self-sensing is a promising technique that has captured a broad range of attention since its origin. The piezoelectric sensor has good dynamic performance, but its static performance is poor and is prone to drift [27,28]. Spanu et al. proposed a high-sensitivity tactile sensor based on a floating gate organic transistor, an organic charge modulation polyethylene terephthalate (FET) coupled with polyvinylidene difluoride (PVDF), and the lowest pressure can be measured at 300 Pa [29]. The piezoresistive sensor has a simple structure, stable working state, and relatively low cost, so it is widely used. Rinaldi et al. studied PDMS foam sensors coated with multilayer graphene nanosheets, and pressure of about 70 kPa can be measured, corresponding to a sensitivity of 0.23 kPa^{-1} [30].

Traditional tactile sensors are all made of a single pressure sensor. For example, the pressure sensor proposed by Wan et al. has a sensitivity of up to 0.8 kPa^{-1}, high accuracy, and a measurement range of up to 1.2 kPa [22]. Huang et al. incorporated highly conductive polyaniline (PANI) polymers, graphene nanoplates (GNPs), and a small amount of silicone rubber (SR) onto elastic Lycra fabrics by spin coating to fabricate strain sensors [31]. The strain range is up to 40%, but its accuracy is relatively low [31]. It can be seen from the above analysis that traditional sensors are contradictory in terms of measurement range and accuracy, and it is difficult to maintain a large measurement range while satisfying high precision at the same time. Moreover, the haptic function of the human body can not only measure the high-precision micro pressure but also realize the perception of a large measurement range. Human tactile sensation differs by 2 to 3 orders of magnitude in terms of measurement range and accuracy, so it is difficult for a single sensor to meet the above requirements.

Graphene is a star material as a two-dimensional "aromatic" monolayer of carbon atoms with sp^2 atomic configuration that exhibits exceptionally physical properties [32,33] and is widely used in the sensing area [34–36]. In this work, to overcome the contradiction between the measurement range and accuracy of the pressure sensor, this project proposes a graphene-based multi-sensitive cell tactile sensor. The sensor utilizes the piezoresistive properties of graphene. It combines two sensors based on the bionic structure of the human body, which can meet the requirements of simultaneous measurement with high precision and a wide measurement range. This sensor used 2D graphene/PET film as a small-range high-precision tactile module and graphene/polyurethane (PU) sponge foam as a wide-range pressure module. The results show that the measurement range of this sensor was in two intervals of 0–2 N and 2–40 N. For the 0–2 N measurement range, the sensitivity was 472.2 Ω/kPa, the force resolution was 0.01 N, and the response time was less than 40 ms. For the 2–40 N measurement range, the sensitivity was 5.05 kΩ/kPa, the force resolution was 1 N, and the response time was less than 20 ms. The new sensor can realize high-precision and large-scale force measurements and shows great application value in the field of medical instruments and artificial limbs.

2. Methods and Materials

2.1. Structure Design and Working Principle of the Sensor

The sensor proposed in this paper was a biomimetic flexible tactile sensor inspired by human skin, so the sensor should conform to human bionic parameters as much as possible under the premise of ensuring flexibility. The previous research shows that the tactile perceived by the skin of the human hand should have a force measurement sensitivity of 0.01 N, a force measurement range of 0~10 kPa, and a geometric resolution of about 2 mm [37]. Pressure perceived by the skin of the human hand should have a sensitivity of 1 N, a force measurement range of 0~220 kPa, and a 1 cm geometric resolution. Figure 1a was the schematic diagram of the sensor structure. The sensor consists of upper and lower sensor units. The upper sensing units were a layer of 2D graphene film tactile sensors, and the lower sensing units were a 3D graphene foam pressure sensor. The tactile sensor consists of four 2D graphene/PET films with a size of 3 mm × 3 mm and Flexible Printed

Circuit Board (FPCB) electrodes to form four tactile sensing units. Each tactile sensing unit corresponds to a group of electrodes as output, the unit spacing is 3 mm, and the spatial resolution of the tactile sensing unit is 3 mm. The pressure sensor consisted of a top electrode, a PDMS support structure, a 3D graphene foam, and a bottom electrode. The field-shaped PDMS support structure has 4 square hollow grooves, and each hollow groove is embedded with graphene foam with a size of 3 mm × 3 mm × 5 mm.

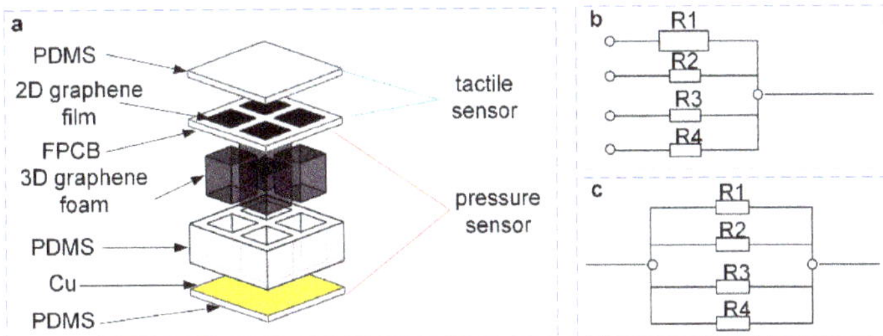

Figure 1. The structure of the designed sensor and measurement circuit schematic. (**a**) The structure of the designed sensor. (**b**) The measurement circuit schematic of the tactile sensor. (**c**) The measurement circuit schematic of the pressure sensor.

The measurement circuit schematic of the sensor is shown in Figure 1b,c, which consists of two parts. Since the four graphene film sensors of the upper tactile sensor were connected to independent electrode outputs, they can be regarded as four independent variable resistors, as shown in Figure 1b. The lower layer is a pressure sensor, due to the "sandwich" structure. The upper and lower electrodes are connected to 4 electrodes separated by PDMS, as shown in Figure 1c. The working principles of tactile and pressure sensors both use the piezoresistive effect of graphene. That is, the C-C bond of the graphene structure is broken when subjected to pressure, resulting in a change in the resistivity of the 2D graphene film and 3D graphene foam, which in turn causes graphene changes in resistance of films and foams.

2.2. Fabrication of the Tactile Sensor

Figure 2 shows the fabrication processes of the sensitive layer of the tactile sensor. Firstly, copper foil (Cu) with a thickness of 50 µm was selected as the substrate, and graphene was grown on its surface by the chemical vapor deposition (CVD) method. Thus, the Cu with graphene thin film(Cu/Graphene) was obtained, as shown in Figure 2a-I. Secondly, the polymethyl methacrylate (PMMA) solution was spin-coated on the surface of the Cu/Graphene, and the sandwich structure of Cu/Graphene/PMMA was obtained, as shown in Figure 2a-II. Here, the PMMA supported the two-dimensional structure of graphene film, which facilitates the subsequent copper foil etching and graphene transfer operations. Thirdly, the copper of the sandwich structure of Cu/Graphene/PMMA was etched using HCL and $CuSO_4$ solution, and the Graphene/PMMA was obtained, as shown in Figure 2a-III. Finally, PET with a thickness of 100 µm was selected as the substrate, graphene film was transferred to the surface of PET using the wetting transfer method, and PMMA film was removed by cleaning with acetone; thus, Graphene/PET film was obtained, as shown in Figure 2a-IV. Figure 2b shows the physical preparation processes of the graphene film. The plane view image of the prepared graphene film is shown in Figure 2c.

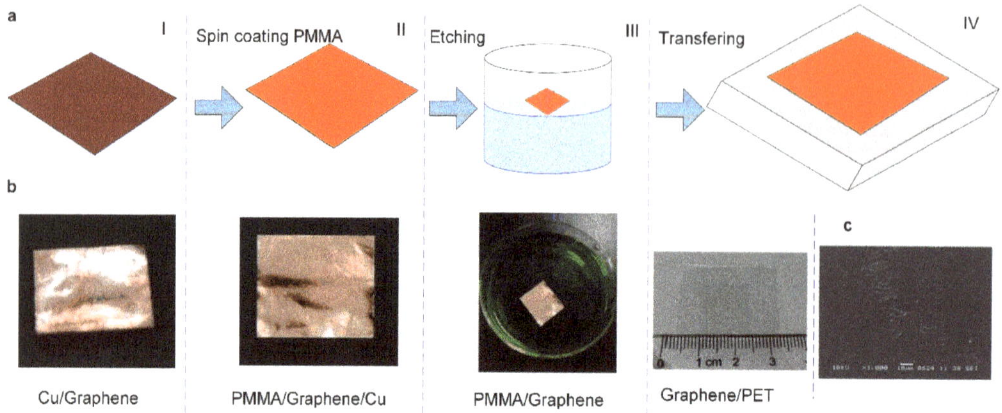

Figure 2. The preparation processes of graphene film. (**a**) The schematic of the fabrication processes of graphene film. (**b**) The physical preparation processes of the graphene film. (**c**) The plane view SEM image of the prepared graphene film.

Figure 3 shows the fabrication processes of the tactile sensor. Firstly, the obtained graphene/PET film was cut into 3 mm × 5 mm, as shown in Figure 3a(I–II). Secondly, the cut graphene/PET film of 3 mm × 5 mm was attached to the electrode surface using silver glue, and the graphene was in contact with the electrodes. The left and right sides of each graphene film were attached to electrodes with a width of 1 mm; therefore, the final size of a single tactile cell unit was 3 mm × 3 mm, as shown in Figure 3a(III–IV). Finally, when the conductive glue is well cured in a vacuum drying oven at a temperature of 80 °C, a layer of PDMS film with a thickness of 200 μm was attached to the top surface of the entire tactile sensor array for protecting the sensitive cells. Figure 3b is the picture of the fabricated tactile sensor. It can be seen that the sensor is small in size and has strong overall flexibility. Since the PDMS film was used as the packing material, the graphene film was robust and not easy to fall off when pressure was applied to the surface of the device.

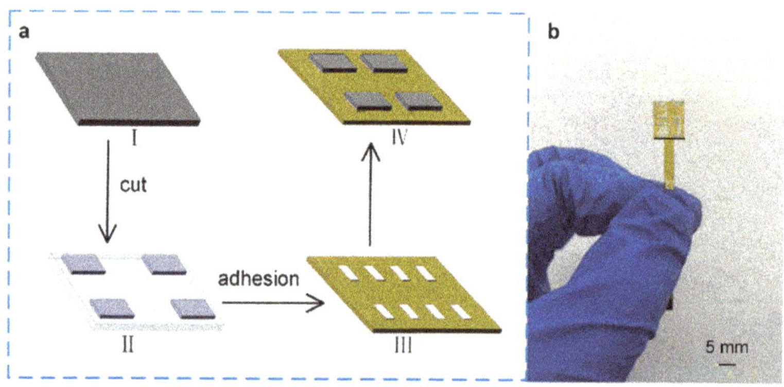

Figure 3. The fabrication processes of the tactile sensor and the photo of the tactile sensor. (**a**) The fabrication processes of the tactile sensor. (**b**) The photo of the fabricated tactile sensor.

2.3. *Fabrication Processes of the Pressure Sensor*

Figure 4 shows the fabrication processes of the pressure sensor. Firstly, the PU foam with a hardness of 27 was cut into a size of 3 mm × 3 mm × 5 mm and washed with absolute ethanol and deionized water sequently, then put into a vacuum drying box to dry for 15 min. Secondly, the graphene oxide (GO) suspension was prepared using graphene oxide and deionized water with a ratio of 2 mol/mL. The clean PU foam was immersed into the GO suspension, and the GO solution was adequately absorbed and then dried at a low temperature of 60 °C for 45 min. Thirdly, the Pu foam that fully absorbed GO (GO/PU) was immersed in the vitamin C (VC) solution at a high temperature of 80 °C to reduce the GO/PU foam to obtain the rGO/PU foam as shown in Figure 4a. Finally, the obtained rGO/PU foam was washed and dried and compressed several times to obtain a stable internal structure, as shown in Figure 4b. The plane view SEM image of the prepared 3D graphene foam is shown in Figure 4c.

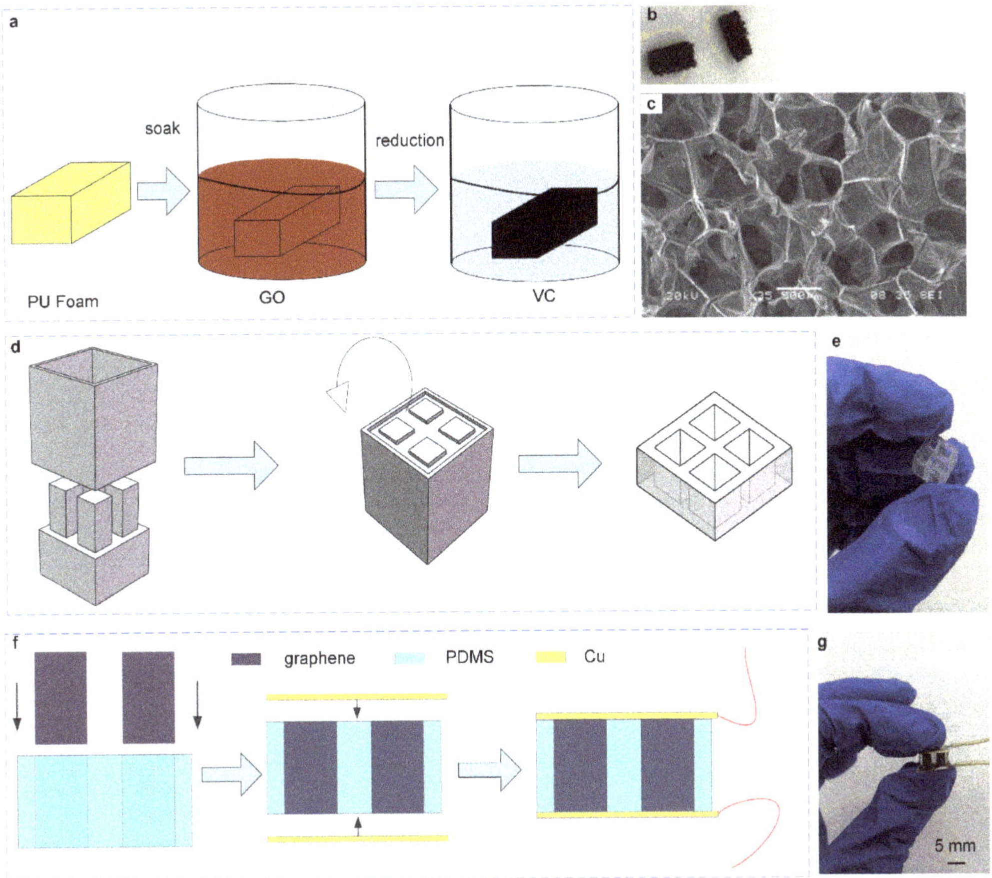

Figure 4. The fabrication processes of the pressure sensor. (**a**) The fabrication processes of the graphene foam. (**b**) The photo of fabricated graphene foam. (**c**) The plane view SEM image of prepared 3D graphene foam. (**d**) The fabrication processes of the support structure of the pressure sensor. (**e**) The photo of the fabricated support structure. (**f**) The fabricated processes of the pressure sensor. (**g**) The photo of the fabricated pressure sensor.

Our previous experiment results show that the modulus of the PDMS with a ratio of 25:1 was close to that of graphene/PU foam and was suitable as a support structure for the foam. The PDMS support structure mold was fabricated using a 3D printer machine, and the design of the mold is shown in Figure 4d. The model was designed with four cuboid columns on the base, and the size of each column was 5 mm, 3 mm, and 3 mm in height, long and wide, respectively. The column was 1 mm away from the boundary, and the distance between each column was 2 mm. The preparation processes of PDMS were as follows. PDMS prepolymer with a ratio of matrix material and curing agent of 25:1 was adequately stirred and degassed in a vacuum dry oven and then injected into the prepared mold. The mold with PDMS mixed solution was put into a vacuum oven and cured at a high temperature of 80 °C for 45 min. Finally, the PDMS was peeled from the mold, and the PDMS support structure was obtained, as shown in Figure 4d. Figure 4e is the picture of the fabricated PDMS structure.

Figure 4f shows the fabrication processes of the pressure sensor. Firstly, the prepared 3 mm × 3 mm × 5 mm graphene/PU foam was inserted into the "+" shaped PDSM support structure with a tweezer and adjusted so that there was no gap between the graphene foam and the inner wall of the support structure. Then, two pieces of copper foil were cut into a size of 10 mm × 12 mm and conductive silver glue was applied to the exposed ends of the graphene foam, and a little adhesive was applied to the surface of the support structure and attached to the copper foil to it. Afterward, the sensor was put into a vacuum oven at a high temperature of 80 °C to cure the conductive glue and adhesive, and then two copper wires were connected to the copper foil; thus, the pressure sensor was successfully fabricated. The actual picture of the pressure sensor is shown in Figure 4g.

2.4. Assemble the Tactile-Pressure Sensor

Figure 5 shows the flow chart and physical diagram of the preparation principle of the tactile-pressure sensor. The preparation process was as follows. Firstly, the bottom of the FPCB of the tactile sensor based on the 2D graphene film was coated with PDMS. Then, a little adhesive was coated on the top of the copper foil of the pressure sensor based on 3D graphene foam. Secondly, the two sensors were pasted together and put into a vacuum drying oven to cure the adhesive at a high temperature of 80 °C. While curing the adhesive, a heavy object of 2 N was placed on the top of the sensor so that the two sensors could fit tightly. After the PDSM was cured, a flexible bionic tactile-pressure sensor based on 2D/3D graphene was obtained, and the actual sensor picture is shown in Figure 5c.

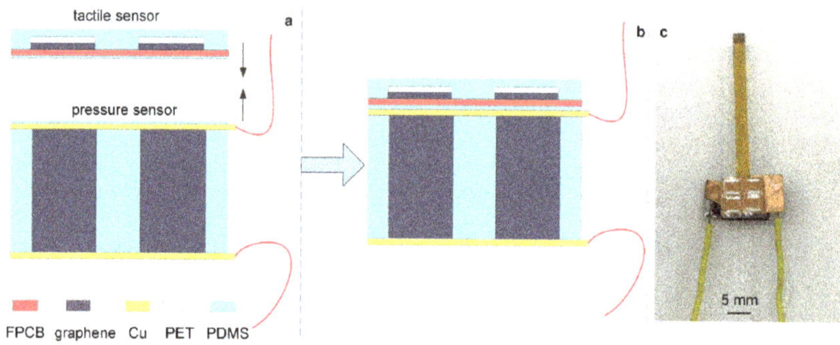

Figure 5. The flow chart and physical diagram of the assembling processes of the tactile-pressure sensor. (**a**) The assemble processes of the tactile-pressure sensor. (**b**) The schematic of the assembled tactile-pressure sensor. (**c**) The photo of the fabricated tactile-pressure sensor.

2.5. Experimental

Figure 6 shows the experimental platform used in this work, a digital pressure gauge and a digital SourceMeter were used to measure the changing of pressure and resistance, respectively. The tactile-pressure sensor unit was placed on the three-dimensional force experiment platform and the digital pressure gauge on the longitudinal pressure arm by applying longitudinal pressure to it. The other end of the tactile-pressure sensor was connected to the digital SourceMeter (Keithley, 2450). The output resistance of the corresponding tactile-pressure sensor was obtained by recording the readings of the SourceMeter in real-time. Figure 6a was the illustration schematic of the testing platform, and Figure 6b was the system's photo.

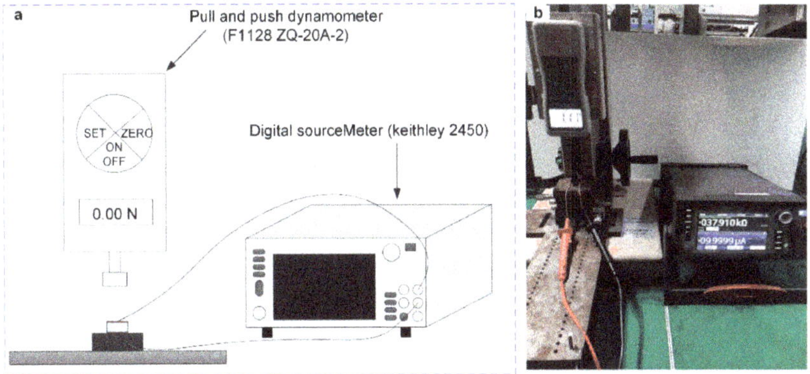

Figure 6. The testing platform. (**a**) The schematic diagram of the force-resistance testing platform. (**b**) The photo of the testing platform.

3. Results and Discussion

Figure 7 shows the output properties of the flexible bionic tactile-pressure sensor proposed in this study. The resistance of the tactile sensor increased with the applied force and the pressure sensor exhibited an opposite changing trend. The sensing range of the tactile and pressure sensor was 0–2 N and 2–40 N, respectively, as shown in Figure 7a,b. The least-squares method was used to fitting the relationship between the input force and the output resistance of the sensors, as shown in Figure 7c,d. The least-squares fitting results of the tactile and pressure sensor, as shown in Equations (1) and (2), were as follows:

$$R = 0.00407\ \mathrm{F} + 1.9337 \tag{1}$$

$$R = -0.45622\ \mathrm{F} + 26.2219 \tag{2}$$

The sensitivity was the ratio of the output increment of the sensor to the input increment during measurement, which was expressed in Equation (3), as follows:

$$S = \frac{\Delta R}{\Delta F} \tag{3}$$

The sensitivity of a linear measurement system can be calculated from the slope of the static characteristic curve. The sensitivity of the tactile and pressure sensor can be calculated by using Equation (3), which was 472.2 Ω/kPa and 5.50 kΩ/kPa, respectively. The demonstrative experiment of the sensor was shown in Video S1.

Figure 8 shows the repeatability and hysteresis properties of the tactile and pressure sensor, and 1#, 2# and 3# represent the first, second and thired sensors. The input force and output resistance relationship of the sensor was recorded and graphed by repeating the measurements on the tactile and pressure sensor three times, and the data are shown in Figure 8a,b. The repeatability refers to the degree to which the output curve deviates from multiple measurements of the sensor's full scale and is an important feature for a sensor. The repeatability error δ_R was usually calculated with the standard deviation. The calculation results show that the repeatability error of the tactile and pressure sensor proposed in this study were 16.55 and 10.48%, respectively, and the tactile-pressure sensor exhibited a good repeatability performance.

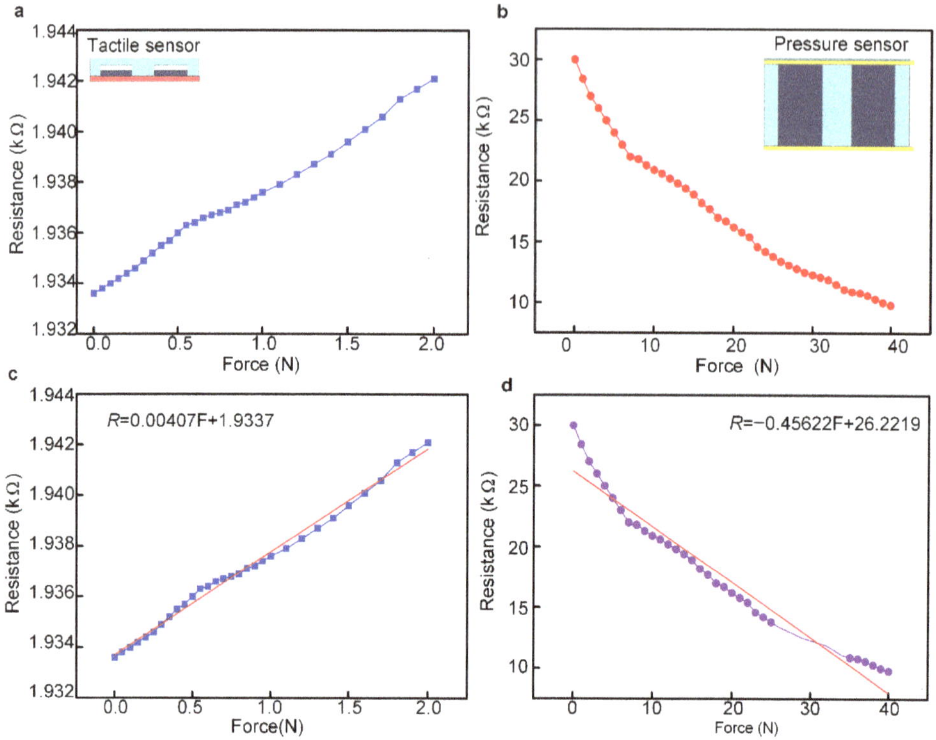

Figure 7. The output properties of the flexible bionic tactile-pressure sensor. (**a**) The relationship between the force and output resistance of the tactile sensor. (**b**) The relationship between the force and output resistance of the pressure sensor. (**c**) The fitting result of the force and output resistance of the tactile sensor by using the least-squares method. (**d**) The fitting result of the force and output resistance of the pressure sensor by using the least-squares method.

The hysteresis of a sensor refers to the non-overlapping phenomenon of the output characteristic curve when the sensor is in the forward (increase in input) and reverse (increase in input) stroke. The hysteresis curves of the tactile and pressure sensor in this study are shown in Figure 8c,d, respectively. The maximum hysteresis was used to evaluate the hysteresis of a sensor, which is expressed in Equation (4) as follows:

$$\delta_H = \frac{\Delta H_{\max}}{Y_{FS}} \times 100\% \quad (4)$$

δ_H is the maximum hysteresis reference error, $\Delta H_{\max}$ is the output maximum absolute deviation, and Y_{FS} is the full-scale value of the sensor. The hysteresis of the tactile and pressure sensor proposed in this study was 4.7% and 7.23% by using Equation (4), respectively.

The dynamic response property refers to the response time to changes in the sensor's input and represents the speed at which the measurement system tracked the input variable parameters. The response time was regarded by many researchers as the most important indicator of dynamic performance and can be used to accurately measure the delay relationship between output and input. Figure 9 shows the dynamic response property of the tactile and pressure sensor in this study. The acquisition frequency of the output resistance was 100 Hz, and it can be obtained from the data in Figure 9a that the response time of the tactile sensor was about 30 ms when the force was applied and about 40 ms when the force was released. The response time of the pressure sensor was less than 20 ms, which can be seen in Figure 9b.

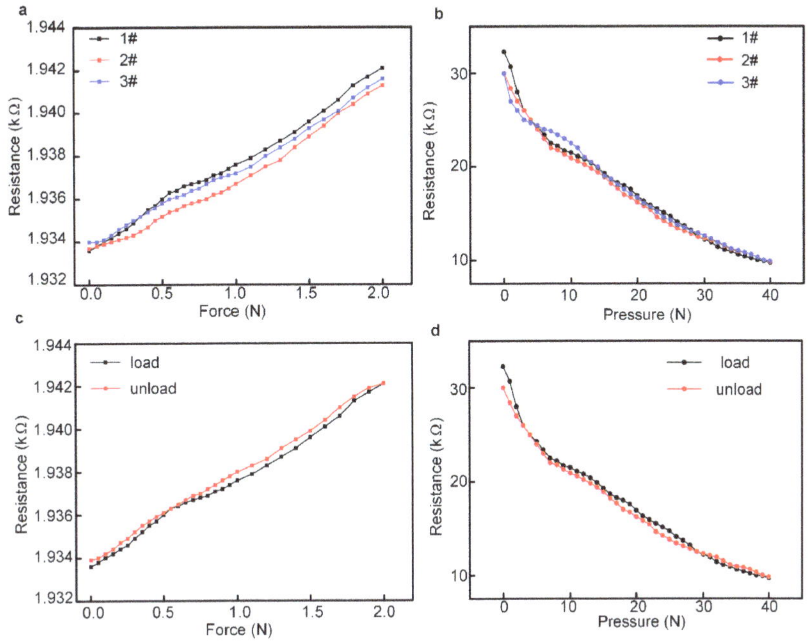

Figure 8. The repeatability and hysteresis properties of the tactile and pressure sensor, and 1#, 2# and 3# represent the first, second and thired sensors. means the (**a**) The repeatability property of the tactile sensor. (**b**) The repeatability property of the pressure sensor. (**c**) The hysteresis property of the tactile sensor. (**d**) The hysteresis property of the pressure sensor.

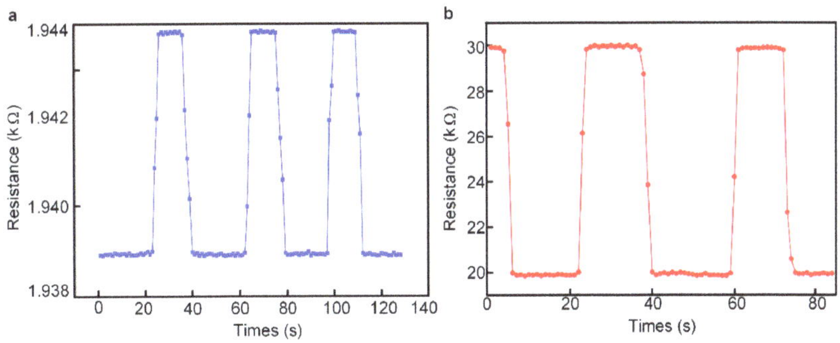

Figure 9. The dynamic response property of the tactile and pressure sensor. (**a**) The response time of the tactile sensor. (**b**) The response time of the pressure sensor.

Figure 10 shows the working mechanism of the tactile-pressure sensor. When an external force was applied to the tactile sensor, the C-C bonds were broken, and the two-dimensional structure of graphene was destroyed. Thereby, the conductive patch was reduced, the resistance was increased, and the resistance change was linearly related to the applied external force to a certain extent, as shown in Figure 10a. For the pressure sensor, when no external force was applied, a layer of graphene was attached to the inner wall and surface of PU foam and had a stable structure. When an external force was applied, the graphene patches on the inner wall of the foam were damaged. While continuously applying an external force to the pressure sensor, the collapsed parts of the foam will come into contact, thereby increasing the number of conductive patches, and the resistance was decreased, as shown in Figure 10b.

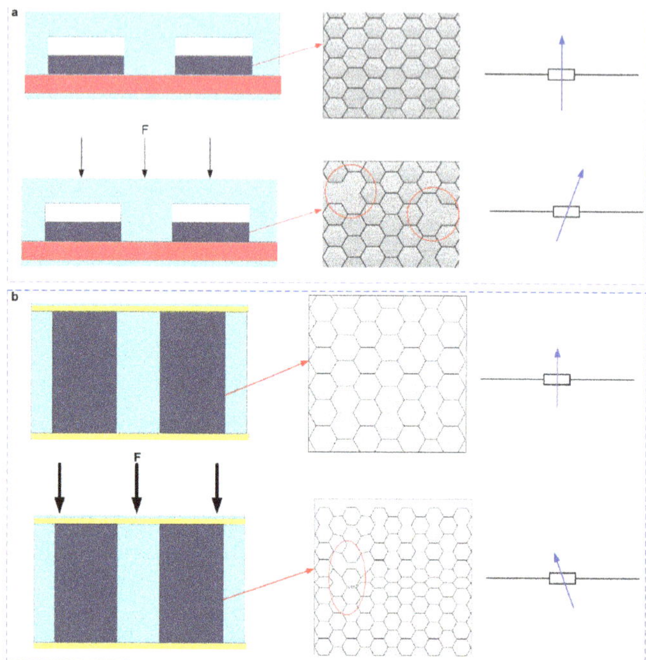

Figure 10. The working mechanism of the tactile-pressure sensor. (**a**) The working mechanism of the tactile sensor. (**b**) The working mechanism of the pressure sensor.

4. Conclusions

In this paper, we present a flexible biomimetic tactile-pressure sensor with high precision and a large measurement range by using 2D graphene film and 3D graphene/PU foam. The tactile-pressure sensor consisted of two layers of structures, and the size was 10 mm × 10 mm × 6 mm in longitude, width, and thickness. The fabrication process, output properties, repeatability, and hysteresis properties were systematically studied. The working mechanism of the tactile-pressure sensor was proposed and discussed. The results exhibit a sensitivity of 472.2 Ω/kPa in the range of 0–2 N, a force resolution of 0.01 N, and a response time of less than 40 ms for the tactile sensor. The sensitivity can reach 5.05 kΩ/kPa, a force resolution of 1 N. The response time was less than 20 ms for the pressure sensor in the measurement range of 2–40 N. The sensor spaned a measurement range of 4 orders of magnitude. These experimental results show that the flexible bionic tactile-pressure sensor designed in this paper can be used to measure both small pressure and large pressure. This has certain reference value for fields such as intelligent robots, electronic skin, and biomedicine.

Supplementary Materials: The following supporting information can be downloaded at: https://www.mdpi.com/article/10.3390/mi13071150/s1, Video S1: The demonstrative experiment of the sensor.

Author Contributions: X.L. and B.S. conceived the idea, analyzed the data, and wrote the paper. X.L., B.S. and L.J. designed the experiments. B.S. and L.J. conducted the experiment. All authors have read and agreed to the published version of the manuscript.

Funding: This research received no external funding.

Institutional Review Board Statement: Not applicable.

Informed Consent Statement: Not applicable.

Data Availability Statement: Not applicable.

Conflicts of Interest: The authors declare no conflict of interest.

References

1. Ponticorvo, M.; Fuccio, R.D.; Ferrara, F.; Rega, A.; Miglino, O. Multisensory educational materials: Five senses to learn. In Proceedings of the International Conference in Methodologies and Intelligent Systems for Technology Enhanced Learning, Ávila, Spain, 26–28 June 2019; Springer: Cham, Switzerland, 2018; pp. 45–52.
2. Lee, S.; Franklin, S.; Hassani, F.A.; Yokota, T.; Nayeem, M.O.G.; Wang, Y.; Leib, R.; Cheng, G.; Franklin, D.W.; Someya, T. Nanomesh pressure sensor for monitoring finger manipulation without sensory interference. *Science* **2020**, *370*, 966–970. [CrossRef] [PubMed]
3. Selamneni, V.; Barya, P.; Deshpande, N.; Sahatiya, P. Low-cost, disposable, flexible, and smartphone enabled pressure sensor for monitoring drug dosage in smart medicine applications. *IEEE Sens. J.* **2019**, *19*, 11255–11261. [CrossRef]
4. Gao, Y.; Ota, H.; Schaler, E.W.; Chen, K.; Zhao, A.; Gao, W.; Fahad, H.M.; Leng, Y.; Zheng, A.; Xiong, F.; et al. Wearable Microfluidic Diaphragm Pressure Sensor for Health and Tactile Touch Monitoring. *Adv. Mater.* **2017**, *29*, 1701985. [CrossRef]
5. Chen, S.; Xin, S.; Yang, L.; Guo, Y.; Zhang, W.; Sun, K. Multi-sized planar capacitive pressure sensor with ultra-high sensitivity. *Nano Energy* **2021**, *87*, 106178. [CrossRef]
6. Xiong, Y.; Shen, Y.; Tian, L.; Hu, Y.; Zhu, P.; Sun, R.; Wong, C.P. A flexible, ultra-highly sensitive and stable capacitive pressure sensor with convex microarrays for motion and health monitoring. *Nano Energy* **2020**, *70*, 104436. [CrossRef]
7. Li, R.; Zhou, Q.; Bi, Y.; Cao, S.; Xia, X.; Yang, A.; Li, S.; Xiao, X. Research progress of flexible capacitive pressure sensor for sensitivity enhancement approaches. *Sens. Actuators A Phys.* **2021**, *321*, 112425. [CrossRef]
8. Shi, J.; Wang, L.; Dai, Z.; Zhao, L.; Du, M.; Li, H.; Fang, Y. Multiscale hierarchical design of a flexible piezoresistive pressure sensor with high sensitivity and wide linearity range. *Small* **2018**, *14*, 1800819. [CrossRef]
9. Gao, L.; Zhu, C.; Li, L.; Zhang, C.; Liu, J.; Yu, H.D.; Huang, W. All paper-based flexible and wearable piezoresistive pressure sensor. *ACS Appl. Mater. Interfaces* **2019**, *11*, 25034–25042. [CrossRef]
10. Liu, W.; Liu, N.; Yue, Y.; Rao, J.; Cheng, F.; Su, J.; Liu, Z.; Gao, Y. Piezoresistive pressure sensor based on synergistical innerconnect polyvinyl alcohol nanowires/wrinkled graphene film. *Small* **2018**, *14*, 1704149. [CrossRef]
11. Yogeswaran, N.; Navaraj, W.T.; Gupta, S.; Liu, F.; Vinciguerra, V.; Lorenzelli, L.; Dahiya, R. Piezoelectric graphene field effect transistor pressure sensors for tactile sensing. *Appl. Phys. Lett.* **2018**, *113*, 014102. [CrossRef]
12. Wang, F.; Jiang, J.; Liu, Q.; Zhang, Y.; Wang, J.; Wang, S.; Han, L.; Liu, H.; Sang, Y. Piezopotential gated two-dimensional InSe field-effect transistor for designing a pressure sensor based on piezotronic effect. *Nano Energy* **2020**, *70*, 104457. [CrossRef]
13. Hosseini, E.S.; Manjakkal, L.; Shakthivel, D.; Dahiya, R. Glycine–chitosan-based flexible biodegradable piezoelectric pressure sensor. *ACS Appl. Mater. Interfaces* **2020**, *12*, 9008–9016. [CrossRef]
14. Chamankar, N.; Khajavi, R.; Yousefi, A.A.; Rashidi, A.; Golestanifard, F. A flexible piezoelectric pressure sensor based on PVDF nanocomposite fibers doped with PZT particles for energy harvesting applications. *Ceram. Int.* **2020**, *46*, 19669–19681. [CrossRef]
15. Zhang, W.; Deng, L.; Yang, L.; Yang, P.; Diao, D.; Wang, P.; Wang, Z.L. Multilanguage-handwriting self-powered recognition based on triboelectric nanogenerator enabled machine learning. *Nano Energy* **2020**, *77*, 105174. [CrossRef]
16. Yang, D.; Guo, H.; Chen, X.; Wang, L.; Jiang, P.; Zhang, W.; Zhang, L.; Wang, Z.L. A flexible and wide pressure range triboelectric sensor array for real-time pressure detection and distribution mapping. *J. Mater. Chem. A* **2020**, *8*, 23827–23833. [CrossRef]
17. Zhang, W.; Wang, P.; Sun, K.; Wang, C.; Diao, D. Intelligently detecting and identifying liquids leakage combining triboelectric nanogenerator based self-powered sensor with machine learning. *Nano Energy* **2019**, *56*, 277–285. [CrossRef]
18. Jin, L.; Zhang, B.; Zhang, L.; Yang, W. Nanogenerator as new energy technology for self-powered intelligent transportation system. *Nano Energy* **2019**, *66*, 104086. [CrossRef]
19. Shi, K.; Chai, B.; Zou, H.; Min, D.; Li, S.; Jiang, P.; Huang, X. Dielectric manipulated charge dynamics in contact electrification. *Research* **2022**, 9862980. [CrossRef]
20. Ren, Z.; Wu, L.; Pang, Y.; Zhang, W.; Yang, R. Strategies for effectively harvesting wind energy based on triboelectric nanogenerators. *Nano Energy* **2022**, *100*, 107522. [CrossRef]
21. Cui, J.; Zhang, B.; Duan, J.; Guo, H.; Tang, J. Flexible pressure sensor with Ag wrinkled electrodes based on PDMS substrate. *Sensors* **2016**, *16*, 2131. [CrossRef]
22. Wan, S.; Bi, H.; Zhou, Y.; Xie, X.; Su, S.; Yin, K.; Sun, L. Graphene oxide as high-performance dielectric materials for capacitive pressure sensors. *Carbon* **2017**, *114*, 209–216. [CrossRef]
23. Ntagios, M.; Nassar, H.; Pullanchiyodan, A.; Navaraj, W.T.; Dahiya, R. Robotic hands with intrinsic tactile sensing via 3D printed soft pressure sensors. *Adv. Intell. Syst.* **2020**, *2*, 1900080. [CrossRef]
24. Yang, J.; Luo, S.; Zhou, X.; Li, J.; Fu, J.; Yang, W.; Wei, D. Flexible, tunable, and ultrasensitive capacitive pressure sensor with microconformal graphene electrodes. *ACS Appl. Mater. Interfaces* **2019**, *11*, 14997–15006. [CrossRef] [PubMed]

25. Kang, S.; Lee, J.; Lee, S.; Kim, S.; Kim, J.K.; Algadi, H.; Al-Sayari, S.; Kim, D.E.; Kim, D.; Lee, T. Highly sensitive pressure sensor based on bioinspired porous structure for real-time tactile sensing. *Adv. Electron. Mater.* **2016**, *2*, 1600356. [CrossRef]
26. Gettings, C.; Speake, C. A method for reducing the adverse effects of stray-capacitance on capacitive sensor circuits. *Rev. Sci. Instrum.* **2019**, *90*, 025004. [CrossRef]
27. Yang, C.; Verbeek, N.; Xia, F.; Wang, Y.; Youcef-Toumi, K. Statically stable charge sensing method for precise displacement estimation of piezoelectric stack-based nanopositioning. *IEEE Trans. Ind. Electron.* **2021**, *68*, 8550–8560. [CrossRef]
28. Hopkins, M.; Vaidyanathan, R.; McGregor, H. Examination of the performance characteristics of veloster as an in-socket pressure sensor. *IEEE Sens. J.* **2020**, *20*, 6992–7000. [CrossRef]
29. Spanu, A.; Pinna, L.; Viola, F.; Seminara, L.; Valle, M.; Bonfiglio, A.; Cosseddu, P. A high-sensitivity tactile sensor based on piezoelectric polymer PVDF coupled to an ultra-low voltage organic transistor. *Org. Electron.* **2016**, *36*, 57–60. [CrossRef]
30. Rinaldi, A.; Tamburrano, A.; Fortunato, M.; Sarto, M.S. A flexible and highly sensitive pressure sensor based on a PDMS foam coated with graphene nanoplatelets. *Sensors* **2016**, *16*, 2148. [CrossRef]
31. Huang, Y.; Gao, L.; Zhao, Y.; Guo, X.; Liu, C.; Liu, P. Highly flexible fabric strain sensor based on graphene nanoplatelet–polyaniline nanocomposites for human gesture recognition. *J. Appl. Polym. Sci.* **2017**, *134*, 45340. [CrossRef]
32. Geim, A.; Novoselov, K. The rise of graphene. *Nat. Mater.* **2007**, *6*, 183–191. [CrossRef] [PubMed]
33. Zhang, W.; Diao, D.; Sun, K.; Fan, X.; Wang, P. Study on friction-electrification coupling in sliding-mode triboelectric nanogenerator. *Nano Energy* **2018**, *48*, 456–463. [CrossRef]
34. Chen, H.; Chen, Z.; Yang, H.; Wen, L.; Yi, Z.; Zhou, Z.; Dai, B.; Zhang, J.; Wu, X.; Wu, P. Multi-mode surface plasmon resonance absorber based on dart-type single-layer graphene. *RSC Adv.* **2022**, *12*, 7821. [CrossRef]
35. Li, J.; Jiang, J.; Zhao, D.; Xu, Z.; Liu, M.; Liu, X.; Tong, H.; Qian, D. Novel hierarchical sea urchin-like Prussian blue@palladium core-shell heterostructures supported on nitrogen-doped reduced graphene oxide: Facile synthesis and excellent guanine sensing performance. *Electrochim. Acta* **2020**, *330*, 135196. [CrossRef]
36. Li, J.; Jiang, J.; Xu, Z.; Liu, M.; Tang, S.; Yang, C.; Qian, D. Facile synthesis of Ag@Cu_2O heterogeneous nanocrystals decorated N-doped reduced graphene oxide with enhanced electrocatalytic activity for ultrasensitive detection of H_2O_2. *Sens. Actuators B Chem.* **2018**, *260*, 529–540. [CrossRef]
37. Lu, X.; Xie, X.; Gao, Q.; Hu, H.; Yang, J.; Wang, H.; Wang, S.; Chen, R. Design of biomimetic human-skin-like tactile flexible sensor. *Sens. Rev.* **2019**, *39*, 397–406. [CrossRef]

micromachines

Article

A Direct-Reading MEMS Conductivity Sensor with a Parallel-Symmetric Four-Electrode Configuration

Zhiwei Liao [1], Junmin Jing [1], Rui Gao [1], Yuzhen Guo [1], Bin Yao [1], Huiyu Zhang [1], Zhou Zhao [1], Wenjun Zhang [2], Yonghua Wang [1], Zengxing Zhang [1,*] and Chenyang Xue [1]

1 State Key Laboratory of Dynamic Measurement Technology, North University of China, Taiyuan 030051, China; s2006174@st.nuc.edu.cn (Z.L.); junmin-jing@outlook.com (J.J.); 18406583750@163.com (R.G.); gyz20000113@163.com (Y.G.); yao_bin2021@163.com (B.Y.); zshy980828@163.com (H.Z.); zzzzhou95@163.com (Z.Z.); wangyonghua@nuc.edu.cn (Y.W.); xuechenyang@nuc.edu.cn (C.X.)

2 School of Aerospace Engineering, Xiamen University, Xiamen 361102, China; 13485461170@163.com

* Correspondence: zhangzengxing@nuc.edu.cn

Abstract: This work proposes a design for a direct-reading conductivity sensor with a parallel symmetrical four-electrode structure, which integrates a silicon-based platinum thin-film strip electrode and a serpentine temperature compensation electrode. The optimal structural parameters of the electrode were determined by finite element simulations performed via COMSOL Multiphysics. Next, the designed conductivity sensor chip was fabricated using MEMS technology, and subsequently, the conductivity measurement circuit was designed to test the fabricated sensor's performance. In laboratory tests, the optimal AC excitation frequency was observed to be 1.067 kHz, while the maximum measurement range was 0–107.41 mS/cm and the measurement precision in low concentration range (0–76.422 mS/cm) was $\pm$0.1 mS/cm. Furthermore, the maximum measurement error of the sensor evaluated using the National Center of Ocean Standards and Metrology was $\pm$0.073 mS/cm. The designed sensor possesses the characteristics of high accuracy, high range, and miniaturization, and enables real-time reading of conductivity value and temperature compensation, which is of great significance for the on-site observation of the physical parameters of marine environment.

Keywords: conductivity sensor; polarization effect; temperature compensation; high accuracy

Citation: Liao, Z.; Jing, J.; Gao, R.; Guo, Y.; Yao, B.; Zhang, H.; Zhao, Z.; Zhang, W.; Wang, Y.; Zhang, Z.; et al. A Direct-Reading MEMS Conductivity Sensor with a Parallel-Symmetric Four-Electrode Configuration. *Micromachines* **2022**, *13*, 1153. https://doi.org/10.3390/mi13071153

Academic Editor: Matteo Rinaldi

Received: 29 June 2022
Accepted: 16 July 2022
Published: 21 July 2022

1. Introduction

The oceans cover about 70.8% of the Earth's surface, and there are various abundant resources in the vast ocean field. The measurement of ocean temperature and salinity is crucial in studying the marine environment. Through these measurements, the ocean circulation, marine ecological environment, marine biodiversity, and marine energy development can be monitored and studied [1,2]. In 1974, the "The Practical Salinity Scale" in UNESCO defined salinity to be calculated based on seawater's conductivity, temperature, and pressure, where pressure has a relatively minor effect on the salinity [3]. Likewise, measuring salinity by conductivity, temperature, and pressure has many advantages, such as high accuracy, fast measurement speed, high reliability, and easy on-site measurement [4]. In this regard, there exist a variety of commercial CTD (Conductivity, Temperature, Depth) sensors used in high-precision marine development, such as the Seabird series, which occupies a leading position in the global market [5]. However, its large size, high cost, and energy consumption limit its promotion in the marine ranching and three-dimensional marine environment monitoring [6–8].

From 2005 to 2007, Broadbent, a scholar at University of South Florida in the United States, focused on using the liquid crystal polymer (LCP) with low hygroscopicity and permeability to manufacture a substrate and resistive temperature sensor (RTD) [9]. In that work, the electroplated nickel-gold-platinum alloy was used as a flat film four-electrode conductivity cell to develop a miniaturized CTD system. Relevant experiments concluded that the temperature and conductivity measurement accuracies were ±0.546 °C and ±0.882 mS/cm, respectively. In 2005, S. Bhansali et al. designed a MEMS-based CTD sensor [10], in which conductivity sensors generally use a parallel-plate capacitor structure, where most of the electric field is confined between the two parallel plates. However, some of the electric field is still distributed at the edges of the plates. To prevent the influence of external electric field on the measurement accuracy, installing guard rings at both ends of a board or adding electrodes and auxiliary circuits is adopted.

In 2011, X. Huang et al. from University of Southampton, UK, developed a measurement system based on the MEMS process for a seven-electrode conductivity sensor and a platinum temperature resistance sensor [11]. In that work, a 500 μm thick glass substrate was coated with 100 nm thick platinum layer serving as electrodes, wires, and pads. The platinum layer was then covered with a 25 μm epoxy laminate (SY320) insulation. The conductivity and temperature measurement accuracies were up to ±0.03 mS/cm and ±0.005 °C, respectively. Meanwhile, the temperature sensor drifted by 0.1 °C while the conductivity drifted by about 5.003 mS/cm, after five weeks. In 2013, Myounggon Kim et al. from the School of Mechatronics, Gwangju Institute of Science and Technology proposed an integrated microfluidic-based sensor module for real-time monitoring of reverse osmosis (RO) that measures temperature, conductivity, and salinity. The microfluidic device was constructed from a thin metal film and a microfluidic channel that was fabricated using the microelectromechanical system (MEMS) technology [12]. Recently, in 2020, Wu Chaonan et al. of Ningbo University also proposed a conductivity and temperature sensor fabricated using MEMS technology, where the sensor chip size was about 12 mm × 12 mm, and 34 chips could be fabricated simultaneously on a 4-inch silicon substrate [13]. The developed chip has high sensitivity, fast response time, and a good repeatability of temperature measurement. The experimental results revealed that the MEMS-based CT sensor has a temperature sensitivity of 0.0619 °C/Ω, a cell constant of 2.559 cm^{-1}, and conductivity and temperature measurement accuracies of ±0.08 mS/cm and ±0.05 °C, respectively, providing valuable experimental data for ocean measurements.

This paper proposes a conductivity sensor that integrates temperature-compensated electrodes with a parallel four-electrode structure. The optimal parameters of the sensor structure were determined by finite element simulation, and the design of the sensor package structure was completed. Simultaneously, the hardware circuit and data acquisition algorithm were designed. The sensor's range, precision, accuracy, and consistency were tested in the laboratory and third-party testing institutions. Furthermore, the sensor can read the conductivity through the master computer.

2. Working Principle of Sensor

Mainly, there are two types of conductivity sensors for marine environment monitoring applications: electrode conductivity sensors [14–17] and inductive conductivity sensors [18–20]. The inductive conductivity sensor is electrodeless, and the metal part of the sensor is not in direct contact with seawater. Moreover, the non-metallic shell is not easy to corrode, and the influence of the polarization effect is also avoided. Therefore, an inductive sensor is more suitable for the field measurements with high stability in harsh environments. However, it is susceptible to the proximity effect, and consequently, the sensor's electric field is easily disturbed or distorted by the surrounding objects. The inductive conductivity sensor uses a toroidal transformer, which inherently results in a large size of the sensor, thereby limiting its integration with other miniaturized MEMS sensors.

Alternatively, electrode conductivity sensors measure the conductivity through a conductivity cell whose parameters are closely related to the position and shape of electrodes. In addition, the electrodes are divided into excitation electrodes and measurement electrodes. The excitation circuit provides a constant AC signal to the excitation electrodes, and a stable electric field is generated in the conductivity cell. Here, the measurement electrodes detect the potentials in different areas inside the conductivity cell and output them after the signal processing.

The conductivity σ of seawater is a physical quantity that describes its current transport capability, essentially reflecting the level of electrolyte concentration in seawater. Conductance G is inversely proportional to resistance R, and is given by the ratio of current I and voltage V:

$$G = \frac{1}{R} = \frac{I}{V} \tag{1}$$

The formula for calculating conductivity can then be expressed as:

$$\sigma = \kappa \cdot G \tag{2}$$

where κ is the cell constant, which can be expressed as [13]:

$$\kappa = \frac{2}{\pi l}\text{arccosh}\frac{d}{2a} \tag{3}$$

Among them, l is the length of electrode, d is the distance between two inner electrodes, and a is the width of voltage electrode. Meanwhile, the four electrodes are axially symmetrically distributed.

When an electrode conductivity sensor operates in seawater, electrochemical reactions occur at the contact surface between seawater and metal electrodes, which cause electron transfer between seawater and electrode. Notably, the polarization effect is defined as the change in ion concentration around the electrode due to the electric field generated by electrode. At the junction of electrode and seawater, the molecules and ions partially dissolved in seawater will be adsorbed on the surface of electrode in the form of chemical bonds, to form an inner layer. Here, the electric center trajectory of these ions is called the inner Helmholtz plane (IHP). Furthermore, the cations in seawater are attracted by the Coulomb force of the anions on electrodes, and the electric center trajectories of these cations are called the outer Helmholtz plane (OHP). In addition, some ions and molecules are not adsorbed, and instead, they are distributed in the diffusion layer due to the influence of electric field force and thermal motion, as demonstrated in Figure 1a.

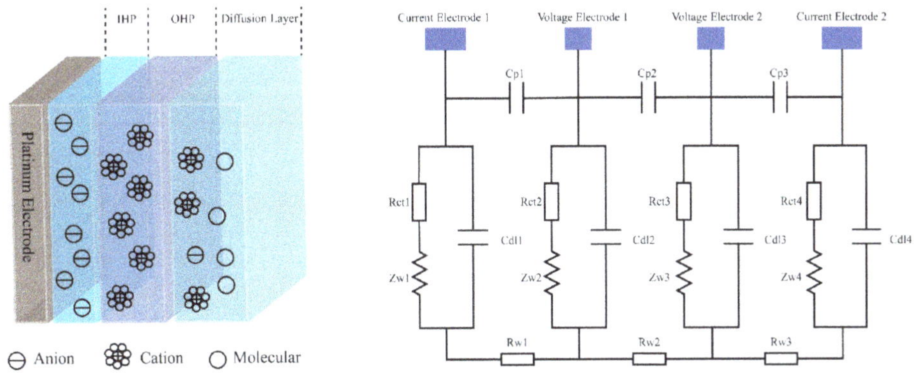

(a) (b)

Figure 1. (**a**) A diagram demonstrating the structure of the electrode–electrolyte interface; (**b**) the equivalent circuit diagram of a four-electrode conductivity sensor with AC excitation applied to current electrode and voltage drop measured by voltage electrode.

The contact surface between electrode and seawater plays a capacitance role when there is no charge transfer at the electrode–seawater interface. This capacitance is called as the double-layer capacitance C_{dl}, formed by the Helmholtz capacitance C_H in series with the equivalent capacitance C_G of diffusion layer. When a voltage is applied to the electrodes, charge transfer occurs between the seawater and electrodes, and such form of charge transfer is similar to a leakage current through a double-layer capacitor, which can be in fact seen as an impedance in parallel with the double-layer capacitor—the Faraday impedance [21]. Accordingly, the Faraday impedance can be expressed as a series connection of charge transfer impedance R_{ct} and Warburg impedance Z_W [22], where Z_W is expressed as:

$$Z_W = \frac{\sqrt{2}A_W}{\sqrt{j\omega}} \tag{4}$$

where j is an imaginary number, ω is the angular frequency of excitation signal, and A_W is the Warburg coefficient [23].

When an AC excitation is applied to the excitation electrode, the contact surface between electrode and seawater can be represented by the double-layer capacitance C_{dl}, charge transfer impedance R_{ct}, and Warburg impedance Z_W. Combined with the stray capacitance C_P and seawater equivalent resistance R_W, the equivalent circuit of the four-electrode conductivity sensor can be represented as shown in Figure 1b. When an AC excitation is applied between the two current electrodes, there will be a voltage drop across the equivalent resistance R_{W2} of the solution between the voltage electrodes. This voltage drop can be measured by a voltage electrode with high input impedance. The current through the solution can be measured by the current electrode, so that the conductivity of the solution can be measured.

In addition, as the temperature of solution changes, the mobility of ions in the solution is also affected, thereby resulting in a temperature-dependent change in the conductivity. Therefore, to overcome the influence of temperature and make the conductivity of different solutions comparable at different temperatures, the conductivity should be temperature compensated. The temperature compensation formula is provided in Equation (5) as:

$$\sigma_{15} = \frac{\sigma_t}{1 + \beta(t - 15)} \tag{5}$$

where σ_t is the conductivity of solution at temperature t °C, σ_{15} is the conductivity of solution at 15 °C, and β is the temperature coefficient of solution conductivity.

3. Design and Fabrication

3.1. Structure and Package Design

In this work, a 525 µm thick 4-inch P-type <100> silicon wafer with high accuracy, good consistency, and low cost is used as the substrate. Essentially, electrode design is the most crucial part in the design of silicon-based thin film conductivity sensor. In designing a multi-electrode conductivity sensor, each pair of electrodes has strict spacing requirements. To reduce the parasitic capacitance, positions of current and voltage electrodes should be strictly symmetrical. As shown in Figure 2, the four-electrode design separates the voltage electrode from the current electrode, which can further weaken the electrode polarization phenomenon. The proposed sensor integrates four parallel strip electrodes, serving as current and voltage electrodes. Furthermore, a temperature electrode with a twisted alignment is also designed between the two voltage electrodes. In addition, platinum metal with good chemical stability and corrosion resistance is used as the material for the electrodes. Its resistance value varies linearly with temperature, and hence, it is used as a temperature compensating electrode. Moreover, silicon nitride film is used to insulate and protect the temperature compensating electrode. Silicon nitride film has good mechanical properties, high dielectric strength, chemical stability, and low film stress. Thus, it can be an excellent protective and insulating layer for the temperature electrode.

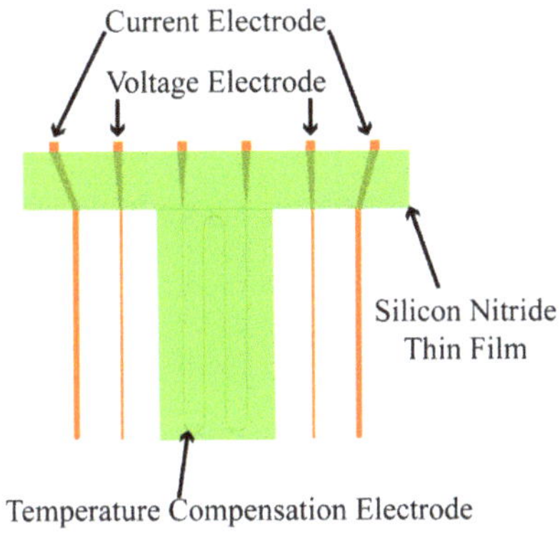

Figure 2. The figure is a schematic diagram of the chip structure.

The dimensions of the electrodes were determined using COMSOL FEM simulations. The chip was surrounded by seawater, and Figure 3a shows the potential distribution of the chip, where a clear potential difference between the two voltage electrodes can be observed. Notably, when the current density of conductivity sensor is too high, the electrodes may get damaged. In addition, increasing the electric field strength can improve the measurement accuracy. Therefore, when designing the electrode, the length of electrode should be appropriately selected to reduce the current density. While keeping the spacing and width of all electrodes constant, the effect of electrode length on the current density and electric field strength is illustrated in Figure 3b. As the electrode length increases, the current density decreases, while the electric field strength increases. Nevertheless, the current density should not be too small, otherwise, it will affect the measurement accuracy in the low range. With these considerations, the electrode length is determined to be 14 mm for this work. Additionally, for fixed electrode length and electrode spacing, the variation of current density with the width of current electrode is shown in Figure 3c. It can be seen from the results that when the current electrode width is 100 μm, the current density is the smallest, and likewise, the width of the current electrode is designed to be 100 μm. Since no current flows through the voltage electrodes, they can be as narrow as the manufacturing process allows. Correspondingly, the width of voltage electrode is set to 10 μm. When the potential difference between the voltage electrodes of conductivity sensor is higher, the sensor's sensitivity will be higher. Evidently, the potential difference between the voltage electrodes varies with the electrode spacing, while keeping the electrode length and width unchanged, as elaborated in Figure 3d. Likewise, when the current electrode spacing is constant, the potential difference increases as the voltage electrode spacing increases. For a voltage electrode spacing of 4 mm, the potential difference is observed to be the largest, therefore the selected voltage electrode spacing is 4 mm. On the other hand, for a fixed voltage electrode spacing, the potential difference increases with the decreasing current electrode spacing. Accordingly, the current electrode spacing is set to 4.5 mm following the selection of voltage electrode spacing, while considering the manufacturing capabilities.

As shown in Figure 4b, epoxy resin is used here to encapsulate the chip adhered to the PCB, which can effectively prevent the exposed pads on the chip from corroding. Furthermore, the size of acquisition circuit board is designed to be 57.5 mm × 18 mm × 1.5 mm, as shown in Figure 4a. The protective cover made of acrylic material can prevent the probe from crashing, and the segmented package design allows a quick replacement of the probe. Moreover, the acquisition circuit board is installed inside the tube shell, and the assembled conductivity sensor is shown in Figure 4c. This package design can transmit data in real-time through cables or realize self-capacitive data storage through TF cards.

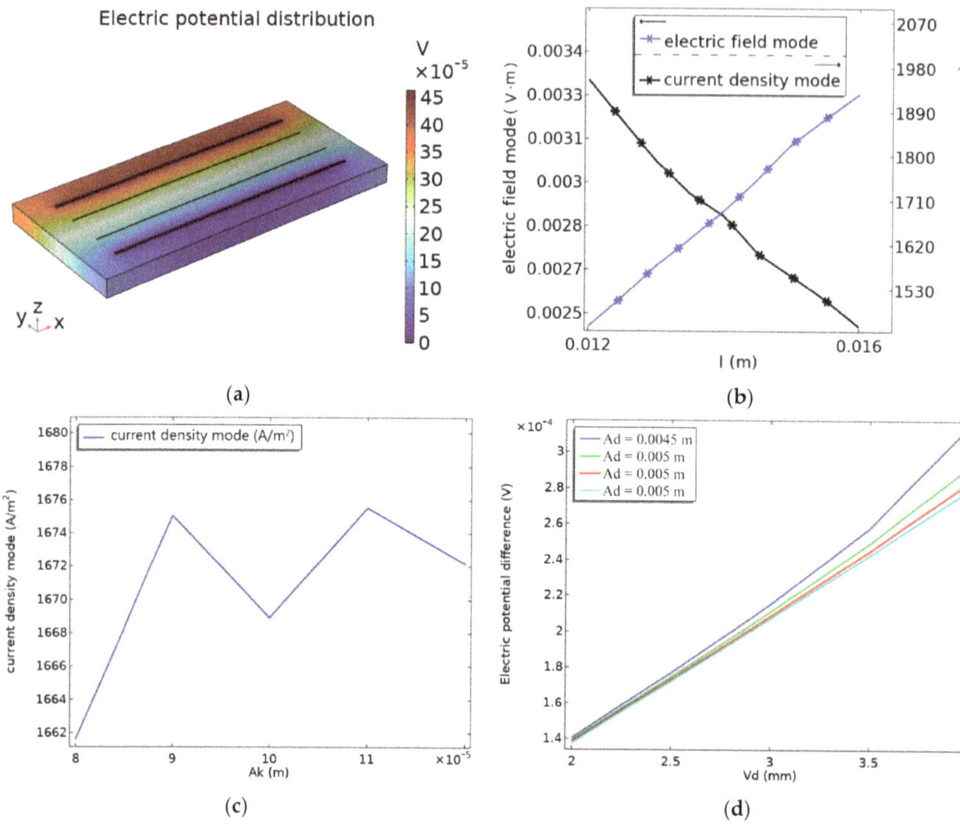

Figure 3. (**a**) The distribution of electric potential; (**b**) the effect of electrode length on current density and electric field strength; (**c**) the effect of current electrode width on current density; (**d**) the relationship between electrode spacing and voltage electrode potential.

3.2. Structure and Package Design

A single chip integrates four parallel electrodes and a serpentine-shaped platinum thin-film resistor with a size of 17 mm × 7.5 mm × 0.5 mm, as shown in Figure 5b, and 61 such chips can be simultaneously fabricated on a 4-inch silicon wafer substrate. The fabrication process of the silicon-based thin-film platinum conductivity sensor chip is explained in Figure 5a. Initially, a 4-inch P-type <100> single-side polished silicon wafer with a 1-micron silicon dioxide film deposited on the surface was cleaned, and a low-stress silicon nitride film was grown on it by chemical vapor deposition (PECVD). Next, the photoresist was spin-coated, and the conductivity electrodes and temperature compensation electrode patterns were photo-etched on the photoresist by ultraviolet lithography (UVL). Then, conductivity electrodes and temperature compensation electrodes with a thickness of 300 nm were fabricated by electron beam evaporation (EBE) followed by the lift-off process. To reduce the influence of temperature compensation electrode on the measurement of seawater conductivity and prevent the temperature compensation electrode from deforming with the temperature changes, a silicon nitride film was deposited again on the temperature compensation electrode for protection and insulation. Since chemical vapor deposition deposits the silicon nitride over the whole surface, it is also necessary to use UVL and reactive ion etching (RIE) to pattern the silicon nitride protective layer.

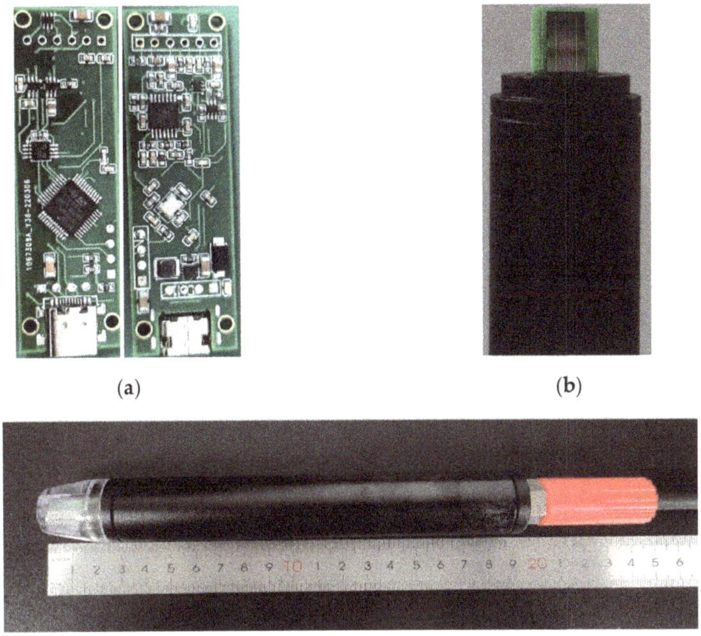

(**a**) (**b**)

(**c**)

Figure 4. (**a**) The size of circuit board with double-sided patch is 57.5 mm × 18 mm × 1.5 mm; (**b**) a corrosion-resistant epoxy potted probe; (**c**) the assembled conductivity sensor with a length of 20 cm.

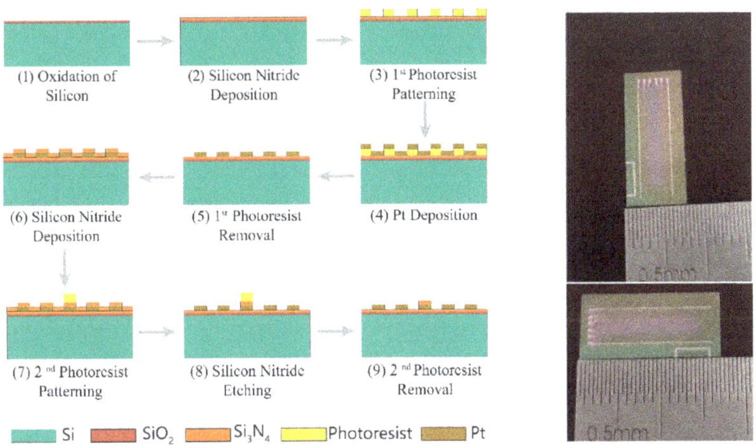

(**a**) (**b**)

Figure 5. (**a**) The fabrication process of the conductivity sensor chip; (**b**) sixty-one chips, each with a size of 17 mm × 7.5 mm × 0.5 mm, can be fabricated on a 4-inch silicon wafer.

3.3. Measurement Hardware and Algorithm

The design scheme for the processing circuit of conductivity sensor is shown in Figure 6. According to the measurement principle of four-electrode conductivity sensor, the master control chip generates and supplies a triangular wave AC signal to current electrode of the conductivity sensor. Meanwhile, a differential operational amplifier measures the voltage difference between the voltage electrodes with high input impedance. Next, the differential amplified output voltage is input to the master control chip for the integral operation through ADC collection. In addition, the current flowing through the other current electrode is input to the operational amplifier after being collected by the ADC, and also input to the master control chip for integral operation. In the master chip, the ratio of current and voltage obtained from the integration process is output through the serial port, and the output value is the original signal of the conductivity sensor.

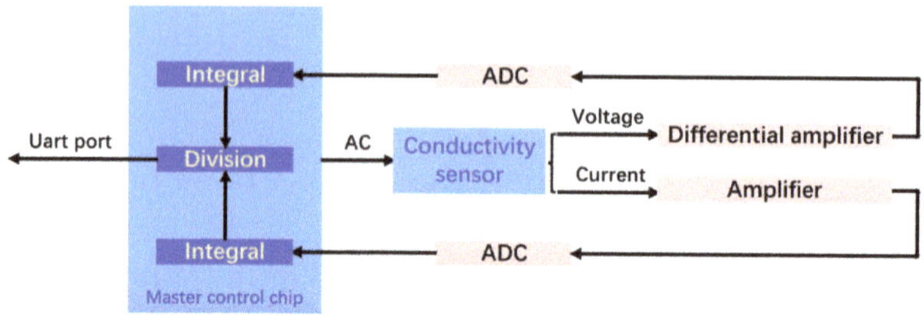

Figure 6. The designed circuit scheme of the four-electrode conductivity sensor. The signal is collected by ADC and processed by main control chip.

4. Experimental Method

4.1. Conductivity Sensor Calibration

A triangular wave is used as an excitation to overcome the influence of double-layer capacitance and polarization effect. As shown in Figure 1b, when the excitation frequency is low, the impedance of C_P is big, and most of the current goes through R_W. However, because of the voltage drop across the double-layer capacitances (C_{dl}), the measured impedance magnitude will be higher than R_W. This effect diminishes at increasing frequencies. However, at higher frequencies, the impedance of C_P will decrease, so that a part of the injected current will go through C_P. Therefore, the impedance magnitude will be lower than R_w because of stray capacitance. Therefore, selection of an appropriate excitation frequency is critical in improving the measurement accuracy. To determine the proper excitation frequency, potassium chloride solutions of 0.35 mol/L, 0.40 mol/L, and 0.6 mol/L were prepared, and these three solutions were placed in a constant temperature bath with a temperature fluctuation of ±0.005 °C. Next, conductivity tests were conducted on the three solutions, and the changes in the circuit output with the concentration under different excitation frequencies are shown in Figure 7. It can be seen from the test results that when the excitation frequency is higher than 1500 Hz or lower than 937 Hz, the output of the circuit does not change significantly with the concentration. When the frequency is 1067 Hz, the output of the circuit is the largest, and the change with the concentration is the most obvious. Therefore, 1067 Hz is selected as the excitation frequency.

4.1.1. The Laboratory Calibration

The output of conductivity sensor is the ratio of current and voltage after the integral and mean value processing, thus the conductivity sensor needs to be calibrated. During the laboratory tests, the sensor was preliminarily calibrated to take into account the influence of experimental environment. Different standard conductivity solutions were prepared with dried potassium chloride and deionized water, and the sensor's output in these solutions was tested at 25 °C. As shown in Figure 8, from 0 to 77.08 ms/cm, the output of conductivity sensor changes linearly with the standard conductivity, and R^2 = 0.99968. According to the fitting curve, the relationship between the output X_O of conductivity sensor and the standard conductivity σ is:

$$\sigma = 8.184 \cdot X_O - 0.265 \tag{6}$$

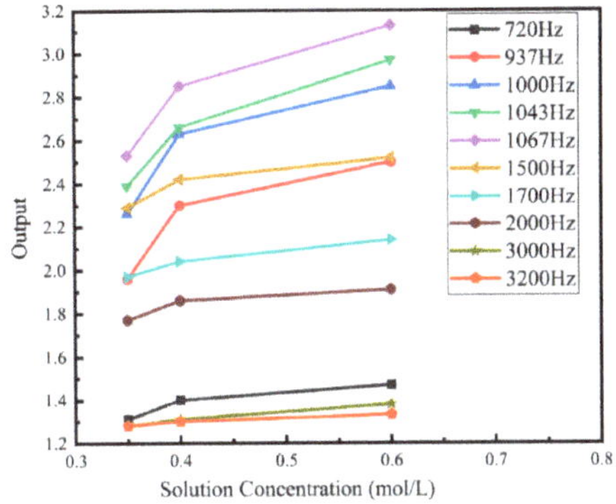

Figure 7. The relationship between concentration and the circuit output at different frequencies. From 720 Hz to 3200 Hz, the frequency is adjusted by dichotomy to determine the optimal excitation frequency.

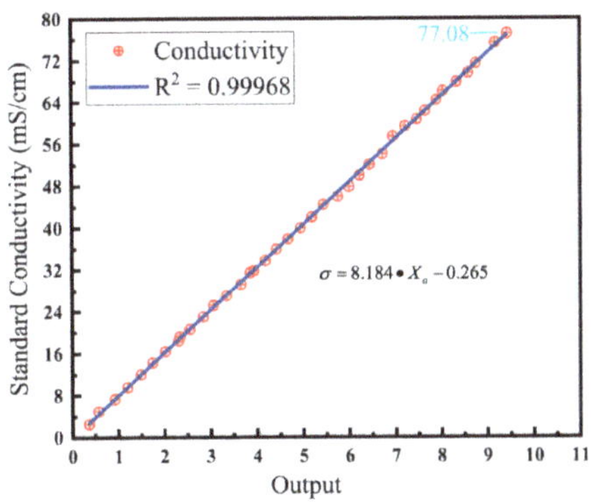

Figure 8. Fitted curve calibrated in the laboratory, calibrated by standard conductivity solutions with different concentrations.

4.1.2. Third Party Mechanism Calibration

To overcome the impact of experimental environment, a calibration at a third-party organization was carried out, at the National Center of Ocean Standards and Metrology in China. Considering the temperature-dependent characteristics of conductivity of standard seawater, the conductivity sensor was placed in a pool of standard seawater, and data are collected at 5 °C intervals between 0 and 35 °C. The corresponding relationship between the output of the conductivity sensor circuit and the temperature is shown in Figure 9a. According to the calibration protocol, the conductivity of air is considered to be 0 mS/cm. Therefore, the fitted curve between the output of the circuit and the standard conductivity is shown in Figure 9b. The fitted relationship between the conductivity y and the output x of circuit is described in Equation (7).

$$y = a_0 + a_1 \cdot x + a_2 \cdot x^2 + a_3 \cdot x^3 + a_4 \cdot x^4 + a_5 \cdot x^5 \tag{7}$$

$a_0 = -0.02189, a_1 = 3.23449, a_2 = 5.94996, a_3 = -2.03875, a_4 = 0.29531, a_5 = -0.01522$

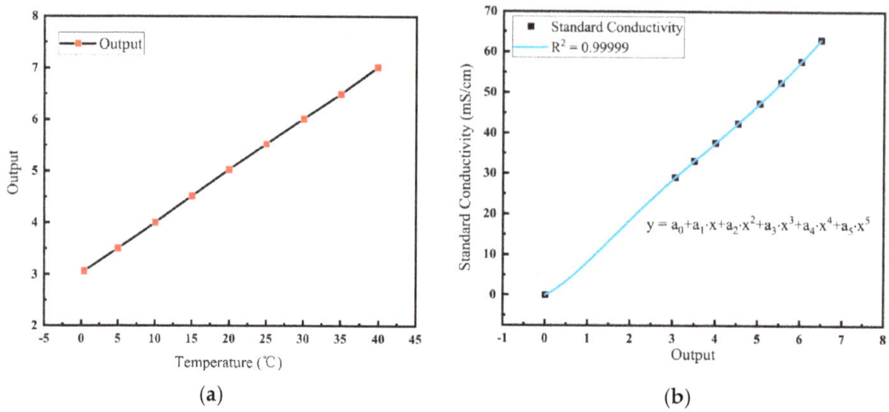

Figure 9. (**a**) Correspondence between temperature and circuit output. (**b**) Fitting curve calibrated by a third party, which is calibrated by changing the temperature of standard seawater.

4.2. Temperature Calibration

Fundamentally, the resistance of thin-film platinum resistors designed with serpentine structure can reach up to 13 kΩ. Moreover, the resistance of thin-film platinum resistors changes linearly with temperature. According to the eight-point calibration method, the sensor was placed in a constant temperature bath with a temperature fluctuation of ±0.005 °C. Next, eight temperature points were set between 5–45 °C, YOWEXA's YET-710 temperature sensor was used as the standard instrument, and the relationship between the resistance of thin-film platinum and the actual temperature was measured, as shown in Figure 10. The linear fitting relationship between the resistance of thin-film platinum resistance and the actual temperature is given as:

$$T = 0.02548 \cdot X - 309.0508 \quad (8)$$

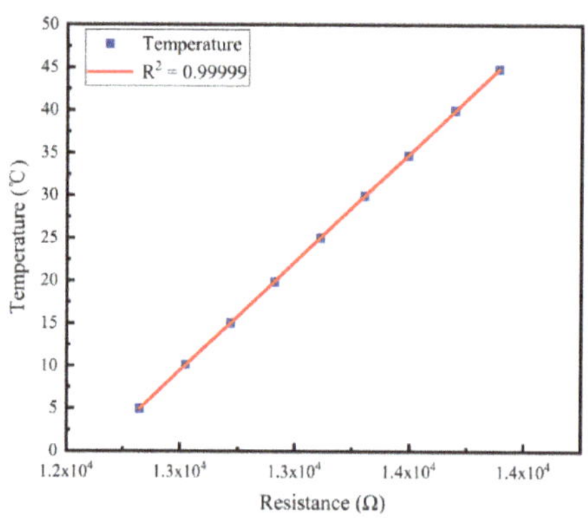

Figure 10. Variation in the resistance of platinum film resistor with temperature.

The R^2 of the fitted curve is 0.99999, demonstrating that the indication value of thin-film platinum resistance possesses an excellent linear relationship with the temperature.

5. Results and Discussion

5.1. Range and Precision of Sensor

The conductivity of potassium chloride solution with different concentrations was tested using the developed silicon-based thin film conductivity sensor. As evident from Figure 11a, the sensor has a maximum range of 107.41 mS/cm. Each concentration of KCl solution was tested 60 times, and the corresponding standard deviation of the measured conductivity value is shown in Figure 11b. The measurement precision between 0 and 76.422 mS/cm ranges from ±0.005 mS/cm at 5.939 mS/cm to ±0.165 mS/cm at 76.422 mS/cm. Furthermore, the measurement precision between 81.879 mS/cm~107.41 mS/cm ranges from ±0.229 mS/cm to ±0.401 mS/cm.

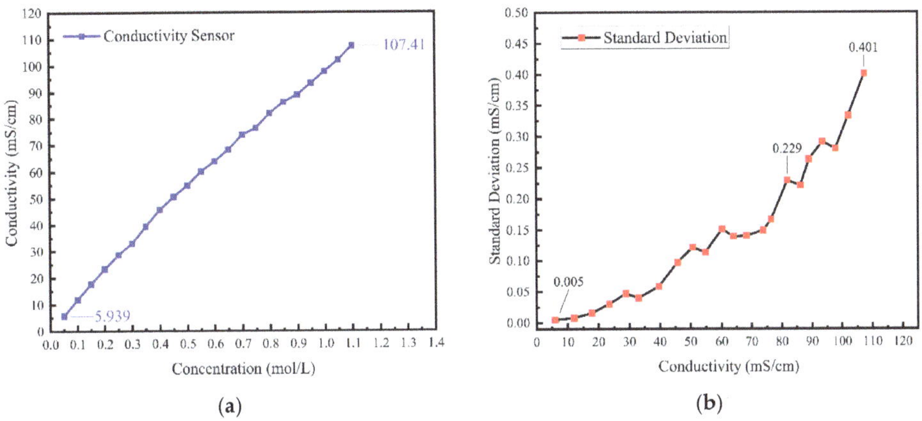

Figure 11. (**a**) The measurement range of our sensor. The maximum range of our sensor can reach 107.41 mS/cm. (**b**) The precision of our sensor, and the measurement precision in low and medium range is about ±0.1 mS/cm.

5.2. Performance Consistency

Conductivity tests were carried out on potassium chloride solutions with different concentrations using three distinct silicon-based thin-film conductivity sensors. As shown in Figure 12, when the concentration is 0.1 mol/L, the error in the test results of three sensors is ±0.086 mS/cm. When the concentration is 0.3 mol/L, the error between Sensor 1 and Sensor 2 is ±0.012 mS/cm, while the measurement error of Sensor 3 is ±1.697 mS/cm. The measurement error is because the temperature of thermostatic bath had not stabilized before the test was started. Therefore, it can be safely stated that the output of silicon-based thin-film conductivity sensor is consistent within the effective range.

5.3. Performance of Temperature Compensation and Salinity Testing

To reduce the effect of temperature on the conductivity test, the conductivity must be temperature compensated. The conductivity of standard seawater was tested from 0 °C to 40 °C. The conductivity values obtained at different temperatures were all converted to conductivity at 15 °C according to Equation (5). Figure 13 shows the actual conductivity of seawater at different temperatures and the conductivity after temperature compensation, with a maximum error of 0.1917 mS/cm after the compensation. Conductivity and temperature can be measured simultaneously by the sensor and displayed directly on the master computer. The temperature, conductivity, and salinity indications can be read out in real time through the master computer, and the sampling time and sampling frequency can also be set through the master computer. When the test was performed in the laboratory, the sensor was fixed about 10 cm below the water surface. Here, the pressure depends on the depth, hence the salinity of the solution, can be calculated directly in the master computer. The developed sensor essentially enables the integrated testing of conductivity, temperature, and salinity.

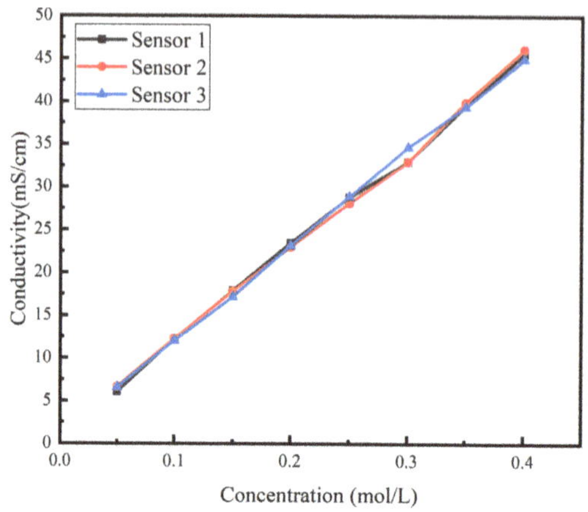

Figure 12. This figure shows the consistent performance of the proposed silicon-based four-electrode conductivity sensor based on platinum film. In different standard conductivity solutions, the test results of the three sensors are basically identical.

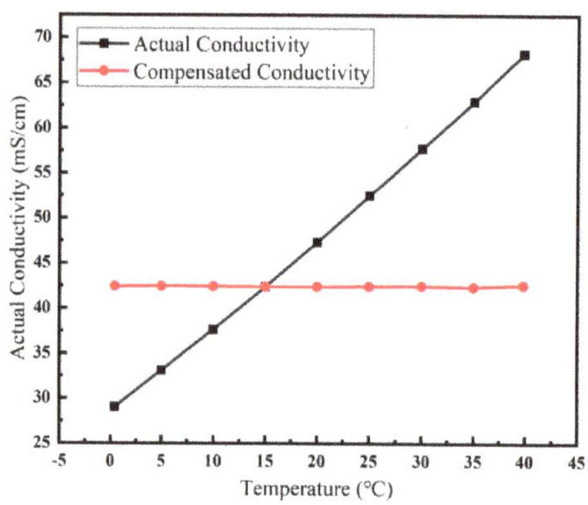

Figure 13. Measured values of the sensor at different temperatures compared to the temperature compensated values.

5.4. Sensor Accuracy Test and Performance Comparison

The accuracy of conductivity sensor was tested at the National Center of Ocean Standards and Metrology. In the experiment, the conductivity of air and the conductivity of standard seawater at 0 °C, 5 °C, 10 °C, 15 °C, 20 °C, 25 °C, 30 °C, 35 °C, and 40 °C were tested. In comparison with the standard conductivity, the maximum measurement error of our sensor was ±0.073 mS/cm at 57.6772 mS/cm, as illustrated in Figure 14.

The performance comparison of the proposed sensor and other conductivity sensors using MEMS technology for marine measurement is presented in Table 1. It can be seen that the accuracy of our sensor is better than that of Hyldgrad multi-sensor system and Chaonan Wu's CT sensor. The measurement range of the proposed conductivity sensor is larger than the other sensors. The chip size is also smaller than Broadbent's PCB MEMS CTD size.

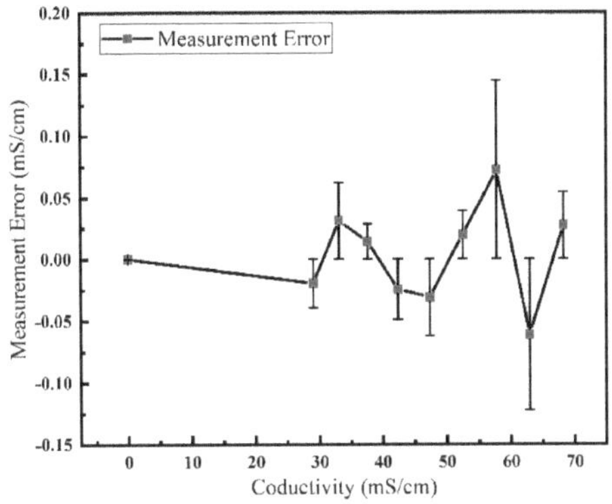

Figure 14. The graph shows the measurement error of the sensor.

Table 1. Comparison of our conductivity sensor and other sensors.

Sensor	Accuracy	Range	Chip size
Hyldgrad multi-sensor system [23]	±0.6 mS/cm	-	4 mm × 4 mm
Broadbent PCB MEMS CTD [24]	-	2–70 mS/cm	18 mm × 28 mm
Huangxi CT sensor [11]	±0.03 mS/cm	25–55 mS/cm	10 mm × 20 mm
Chaonan Wu CT sensor [13]	±0.08 mS/cm	0–101 mS/cm	12 mm × 12 mm
Our conductivity sensor	±0.073 mS/cm (0–70 mS/cm)	0–107.41 mS/cm	17 mm × 7.5 mm

6. Conclusions

This paper proposes a direct-reading MEMS conductivity sensor with four parallel electrodes, integrating temperature-compensated electrodes for real-time temperature compensation. The sensor can directly read out the measured conductivity, temperature, and salinity through the master computer. The developed sensor exhibited good consistency with 61 chips successfully fabricated on a 4-inch silicon wafer. Furthermore, the measurement circuit and algorithm of conductivity sensor were also developed, and an integrated package of sensor probe and circuit was realized. The maximum measurement range of the sensor in the laboratory was 107.41 mS/cm. Most importantly, a third-party standardized calibration was also carried out, and the accuracy of the conductivity sensor was better than ±0.073 mS/cm over the range of 0–70 mS/cm. Future research will focus on improving the measurement accuracy at a high range and the long-term stability of the sensor.

Author Contributions: Conceptualization, Z.Z. (Zengxing Zhang) and C.X.; methodology, Z.Z. (Zengxing Zhang) and Y.W.; software, Z.L. and B.Y.; validation, Z.L., J.J. and Y.G.; formal analysis, W.Z. and R.G.; investigation, Y.G., H.Z. and Z.Z. (Zhou Zhao); resources, Y.W.; data curation, Z.Z. (Zengxing Zhang); writing—original draft preparation, Z.L.; writing—review and editing, Z.L. and Z.Z. (Zengxing Zhang). All authors have read and agreed to the published version of the manuscript.

Funding: This work was supported by the National Natural Science Foundation of China (Grant No. 62101511, 61727806), Applied basic research project of Shanxi Province (Grant No. 20210302124539), Shanxi '1311 project' Key Subject Construction (1331 KSC), and Tan Kah Kee Innovation Laboratory (RD2022020511).

Data Availability Statement: The data presented in this study are available on request from the corresponding author.

Conflicts of Interest: The authors declare no conflict of interest.

References

1. Nag, A.; Mukhopadhyay, S.C.; Kosel, J. Sensing system for salinity testing using laser-induced graphene sensors. *Sens. Actuators A Phys.* **2017**, *264*, 107–116. [CrossRef]
2. Pawlowicz, R.; Feistel, R.; McDougall, T.J.; Ridout, P.; Seitz, S.; Wolf, H. Metrological challenges for measurements of key climatological observables Part 2: Oceanic salinity. *Metrologia* **2016**, *53*, R12–R25. [CrossRef]
3. Lewis, E. The practical salinity scale 1978 and its antecedents. *IEEE J. Ocean. Eng.* **1980**, *5*, 3–8. [CrossRef]
4. Jijesh, J.J.; Shivashankar; Susmitha, M.; Bhanu, M.; Sindhanakeri, P. Development of a CTD Sensor Subsystem for Oceanographic Application. In Proceedings of the 2nd IEEE International Conference on Recent Trends in Electronics, Information and Communication Technology (RTEICT), Bangalore, India, 19–20 May 2017; pp. 1487–1492.
5. Venkatesan, R.; Ramesh, K.; Arul Muthiah, M.; Thirumurugan, K.; Atmanand, M.A. Analysis of drift characteristic in conductivity and temperature sensors used in Moored buoy system. *Ocean Eng.* **2019**, *171*, 151–156. [CrossRef]
6. Broadbent, H.A.; Ketterl, T.P.; Silverman, A.M.; Torres, J.J. Development of a CTD biotag: Challenges and pitfalls. *Deep Sea Res. Part II: Top. Stud. Oceanogr.* **2013**, *88–89*, 131–136. [CrossRef]
7. Marcelli, M.; Piermattei, V.; Madonia, A.; Mainardi, U. Design and application of new low-cost instruments for marine environmental research. *Sensors* **2014**, *14*, 23348–23364. [CrossRef] [PubMed]
8. Oiler, J.; Shock, E.; Hartnett, H.; Dombard, A.J.; Yu, H. Harsh Environment Sensor Array-Enabled Hot Spring Mapping. *IEEE Sens. J.* **2014**, *14*, 3418–3425. [CrossRef]
9. Broadbent, H.A.; Ivanov, S.Z.; Fries, D.P. A miniature, low cost CTD system for coastal salinity measurements. *Meas. Sci. Technol.* **2007**, *18*, 3295–3302. [CrossRef]
10. Aravamudhan, S.; Bhat, S.; Bethala, B.; Bhansali, S.; Langebrake, L. MEMS based Conductivity-Temperature-Depth (CTD) sensor for harsh oceanic environment. In Proceedings of the Oceans 2005 Conference, Washington, DC, USA, 17–23 September 2005; pp. 1785–1789.
11. Huang, X.; Pascal, R.W.; Chamberlain, K.; Banks, C.J.; Mowlem, M.; Morgan, H. A Miniature, High Precision Conductivity and Temperature Sensor System for Ocean Monitoring. *IEEE Sens. J.* **2011**, *11*, 3246–3252. [CrossRef]
12. Kim, M.; Choi, W.; Lim, H.; Yang, S. Integrated microfluidic-based sensor module for real-time measurement of temperature, conductivity, and salinity to monitor reverse osmosis. *Desalination* **2013**, *317*, 166–174. [CrossRef]
13. Wu, C.; Gao, W.; Zou, J.; Jin, Q.; Jian, J. Design and Batch Microfabrication of a High Precision Conductivity and Temperature Sensor for Marine Measurement. *IEEE Sens. J.* **2020**, *20*, 10179–10186. [CrossRef]
14. Gong, W.D.; Mowlem, M.; Kraft, M.; Morgan, H. Oceanographic Sensor for in-situ temperature and conductivity monitoring. In Proceedings of the International Conference OCEANS 2008 and MTS/IEEE Kobe Techno-Ocean, Kobe, Japan, 8–11 April 2008; pp. 42–47.
15. Li, X.; Meijer, G.C.M. A Low-Cost and Accurate Interface for Four-Electrode Conductivity Sensors. *IEEE Trans. Instrum. Meas.* **2005**, *54*, 2433–2437. [CrossRef]
16. Ramos, P.M.; Pereira, J.M.D.; Ramos, H.M.G.; Ribeiro, A.L. A Four-Terminal Water-Quality-Monitoring Conductivity Sensor. *IEEE Trans. Instrum. Meas.* **2008**, *57*, 577–583. [CrossRef]
17. Horiuchi, T.; Wolk, F. Long-term stability of a new conductivity-temperature sensor tested on the VENUS cabled observatory. In Proceedings of the OCEANS'10 IEEE SYDNEY, Sydney, Australia, 24–27 May 2010; pp. 1–4. [CrossRef]
18. Fougere, A.J. New non-external field inductive conductivity sensor (NXIC) for long term deployments in biologically active regions. In Proceedings of the OCEANS 2000 MTS/IEEE Conference and Exhibition. Conference Proceedings (Cat. No.00CH37158), Providence, RI, USA, 11–14 September 2000; Volume 621, pp. 623–630. [CrossRef]
19. Natarajan, S.P.; Huffman, J.; Weller, T.M.; Fries, D.P. Contact-less toroidal fluid conductivity sensor based on RF detection. In Proceedings of the SENSORS, 2004 IEEE, Vienna, Austria, 24–27 October 2004; Volume 301, pp. 304–307. [CrossRef]
20. Striggow, K.; Dankert, R. The exact theory of inductive conductivity sensors for oceanographic application. *IEEE J. Ocean. Eng.* **1985**, *10*, 179. [CrossRef]
21. Aikens, D.A. Electrochemical methods, fundamentals and applications. *J. Chem. Educ.* **1983**, *60*, A25. [CrossRef]
22. Barbero, G.; Lelidis, I. Analysis of Warburg's impedance and its equivalent electric circuits. *Phys. Chem. Chem. Phys.* **2017**, *19*, 24934–24944. [CrossRef] [PubMed]
23. Hyldgård, A.; Mortensen, D.; Birkelund, K.; Hansen, O.; Thomsen, E.V. Autonomous multi-sensor micro-system for measurement of ocean water salinity. *Sens. Actuators A Phys.* **2008**, *147*, 474–484. [CrossRef]
24. Broadbent, H.A.; Ketterl, T.P.; Reid, C.S. A miniature rigid/flex salinity measurement device fabricated using printed circuit processing techniques. *J. Micromech. Microeng.* **2010**, *20*, 085008. [CrossRef]

micromachines

Article

Design, Fabrication, and Dynamic Environmental Test of a Piezoresistive Pressure Sensor

Rui Gao [1], Wenjun Zhang [2], Junmin Jing [1], Zhiwei Liao [1], Zhou Zhao [1], Bin Yao [1], Huiyu Zhang [1], Yuzhen Guo [1], Yanbo Xu [1], Yonghua Wang [1], Zengxing Zhang [2,*], Zhidong Zhang [1,*] and Chenyang Xue [1]

[1] State Key Laboratory of Dynamic Measurement Technology, North University of China, Taiyuan 030051, China; 18406583750@163.com (R.G.); junmin-jing@outlook.com (J.J.); s2006174@st.nuc.edu.cn (Z.L.); Zzzzhou95@163.com (Z.Z.); yao_bin2021@163.com (B.Y.); zshy980828@163.com (H.Z.); gyz20000113@163.com (Y.G.); xuyanbo_2001@163.com (Y.X.); wangyonghua@nuc.edu.cn (Y.W.); xuechenyang@nuc.edu.cn (C.X.)

[2] School of Aerospace Engineering, Xiamen University, Xiamen 361102, China; 13485461170@163.com

* Correspondence: zhangzengxing@nuc.edu.cn (Z.Z.); zdzhang@nuc.edu.cn (Z.Z.)

Abstract: Microelectromechanical system (MEMS) pressure sensors have a wide range of applications based on the advantages of mature technology and easy integration. Among them, piezoresistive sensors have attracted great attention with the advantage of simple back-end processing circuits. However, less research has been reported on the performance of piezoresistive pressure sensors in dynamic environments, especially considering the vibrations and shocks frequently encountered during the application of the sensors. To address these issues, this paper proposes a design method for a MEMS piezoresistive pressure sensor, and the fabricated sensor is evaluated in a series of systematic dynamic environmental adaptability tests. After testing, the output sensitivity of the sensor chip was 9.21 mV·bar^{-1}, while the nonlinearity was 0.069% FSS. The sensor overreacts to rapidly changing pressure environments and can withstand acceleration shocks of up to 20× g. In addition, the sensor is capable of providing normal output over the vibration frequency range of 0–5000 Hz with a temperature coefficient sensitivity of −0.30% FSS °C^{-1} over the temperature range of 0–80 °C. Our proposed sensor can play a key role in applications with wide pressure ranges, high-frequency vibrations, and high acceleration shocks, as well as guide MEMS-based pressure sensors in high pressure ranges and complex environmental adaptability in their design.

Keywords: piezoresistive pressure sensor; silicon-on-insulator (SOI) structure; dynamic environment test

Citation: Gao, R.; Zhang, W.; Jing, J.; Liao, Z.; Zhao, Z.; Yao, B.; Zhang, H.; Guo, Y.; Xu, Y.; Wang, Y.; et al. Design, Fabrication, and Dynamic Environmental Test of a Piezoresistive Pressure Sensor. *Micromachines* **2022**, *13*, 1142. https://doi.org/10.3390/mi13071142

Academic Editor: Libo Gao

Received: 24 June 2022
Accepted: 16 July 2022
Published: 19 July 2022

1. Introduction

Microelectromechanical system (MEMS)-based piezoresistive pressure sensors have wide applications in industrial construction [1–4], healthcare monitoring [5–11], and ocean depth detection due to their high process maturity, easy sensor integration, and easy signal processing [12,13]. However, systematic research of the sensor performance in harsh environments is lacking, although many studies have been reported on the design and manufacturing of such piezoresistive pressure sensors. Various interference factors such as temperature, vibration, shock, and strong light irradiation may cause the abnormal output of the sensitive unit of the piezoresistive pressure sensor when in practical application.

Enabled by smart-cut and other technologies, silicon-on-insulator (SOI) wafers have been made commercially available for CMOS (Complementary Metal Oxide Semiconductor) and MEMS. In addition to its significant advantages in manufacturing high-performance CMOS circuits, SOI has been demonstrated as a perfect option for MEMS fabrication, owing to its excellent electrical isolation. Thanks to SOI technology, the pressure sensor's fabrication process is becoming easier, and the sensor's temperature preference can be enhanced significantly.

The sensor materials play an important role in enhancing the stability of the MEMS pressure sensor in a harsh environment. The drift of carriers in silicon material makes the piezoresistive pressure sensor unstable in high-temperature environments. In 2015, Sainan Li et al. achieved stable operation of a silicon-based piezoresistive pressure sensor at 350 °C by using the method of accurately calculating the heavy doping dose [14]. However, due to the inherently narrow forbidden band of silicon materials, pressure testing above 300 °C is still a difficult problem for silicon-based piezoresistive pressure sensors. Therefore, the breakthrough of high-temperature-resistant materials has become a research hotspot. Over the years, sensitive materials for piezoresistive pressure sensors have progressed, and silicon carbide (SiC) corresponds to one of the current candidates. Notably, SiC is one of the most promising materials for applications in harsh environments, thanks to its excellent electrical, mechanical, and chemical properties [15,16].

The sensor structure also plays an imperative role in sensor performance in harsh environments [17–19]. Various new film structures have been developed for increasing the sensor sensitivity while maintaining linearity, such as E-type film [20], hollow enhanced film [21], circular groove film [22], and peninsula structure film [23]. Using the optimized design of the diaphragm structure [20–23], the sensitivity of the sensor indeed increases, compared with the traditional C-type structure. However, as shown in Figure 1, etching grooves on the upper surface of the pressure-bearing diaphragm or adding a beam structure to the lower surface of the diaphragm will certainly have an impact on the natural frequency of the sensor. In practical applications of the sensor, it may experience high-frequency and low-frequency vibrations, as well as high-acceleration shocks, suggesting that it cannot be employed in a harsh environment.

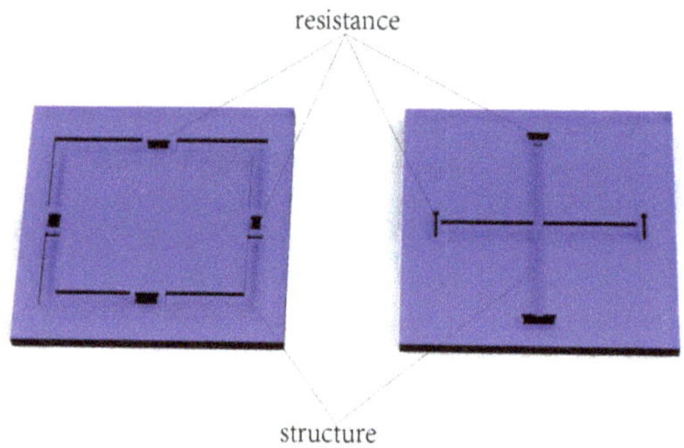

Figure 1. Etching a groove on the film structure to improve sensitivity might lead to instability in a harsh environment.

Due to the piezoresistive effect of the semiconductor, its resistance changes in the stressed state. Subsequently, the resistance change in the stress concentration area of the diaphragm is sensed using the Wheatstone bridge circuit, which then converts the resistance change into a voltage change.

This study proposes a piezoresistive pressure sensor on an SOI substrate, which adopts the traditional C-type film structure, with a design range of 0–40 bar. The optimal parameters of the film structure are extracted through extensive finite element simulations. In addition, the relationship among the film size L, maximum stress size S, and film thickness h was established using a linear fitting method. To validate the design, the designed sensor chip was fabricated, and the sensor's performance was tested in a harsh environment using a series of dynamic environmental tests including (a) a rapid pressure shock test, (b) a temperature test, (c) a high-frequency vibration test, and (d) a high-acceleration shock test.

Following the testing, the full-scale output of the pressure sensor prepared at room temperature reached 368.8 mV, while the temperature coefficient sensitivity was −0.3% FSS/°C.

Furthermore, the dynamic environment test showed that the sensor could maintain a low output drift under 5000 Hz high-frequency vibration and 20× g acceleration shock. Our proposed sensor can play an important role in pressure measurement under high-frequency vibration and a high-acceleration shock environment, as well as guide the design of MEMS-based pressure sensors with high environmental adaptability. The performance comparison of the fabricated sensors is shown in the Table 1.

Table 1. Comparison of performance with other pressure sensors at room temperature.

Sensor	Sensitivity (mV/V/bar)	Pressure Nonlinearity (%FSS)	Full Range Pressure
Proposed sensor	9.21	0.069%	40 bar
Sensor in [24]	6.03	3.89%	30 bar
Sensor in [24]	6.58	0.33%	10 bar
Sensor in [25]	5.18	0.02%	30 bar
Sensor in [25]	3.69	0.011%	30 bar

2. Structure Design and Simulation

This work designs an absolute pressure sensor with a sensing range of 0–40 bar using a typical C-type film structure. The structural parameters are further optimized by the small deflection theory and finite element analysis. Due to the sandwiched structure of the SOI substrate, the fabrication of the sensitive diaphragm usually employs a buried oxide layer as an etch stop layer for silicon etching. The thicknesses of the buried oxide layer and device layer in the SOI substrate determine the thickness of the sensitive diaphragm, as shown in Figure 2.

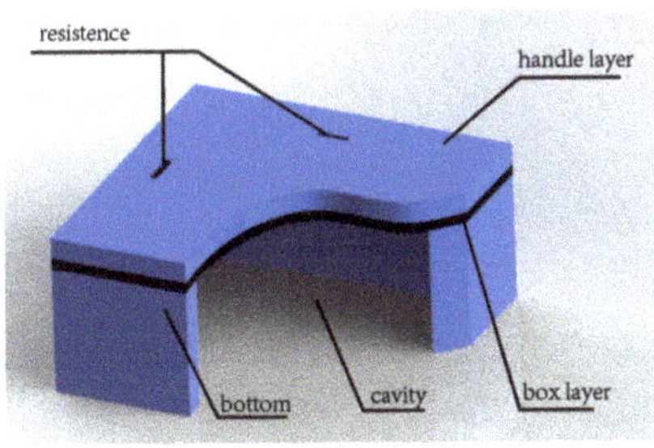

Figure 2. Diagram of the piezoresistive pressure sensor with back cavity. The thickness of the device layer is the sum of the thickness of the handle layer and the thickness of the box layer.

We selected a diaphragm with various thicknesses to ensure that the size value reached the optimal linear stress under the maximum pressure. To find the optimal solution, the finite element analysis was used, which took the square film as the structure in the solid mechanics simulation. Film thicknesses of h = 5, 10, 20, 30, 40, and 50 μm were established. Next, the small deflection theory was used as the boundary condition to simulate and calculate the diaphragm size with a deflection of 1/5 h [26,27], under the maximum pressure. The finite element simulation results are shown in Figure 3a. Meanwhile, the relationship between film thickness and optimal film length is demonstrated in Figure 3b. It can be seen from Figure 3b that the optimal film length is proportional to the film thickness, and the fitting equation is given as follows:

$$L = 27.8\,h. \quad (1)$$

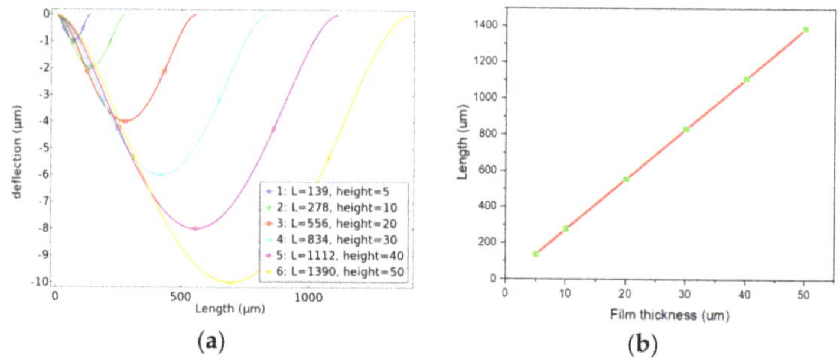

Figure 3. (**a**) Optimum film size for different film thicknesses; (**b**) relationship between film thickness and optimal film length.

To effectively sense the stress-induced resistant change in the piezoresistive diaphragm, the Wheatstone bridge consisting of four resistors should be placed as close as possible in the stress-concentrated area. The stress distribution of the diaphragm is shown in Figure 4a. Moreover, the centerline stress distribution and the maximum length of the stress concentration area S of the diaphragm are shown in Figure 4b. The maximum stress σ_{max} on the diaphragm is near the edge, starting from the edge of the diaphragm and moving inward along the centerline until the second point of equal stress is encountered. The distance S between the two points can be used as a reference for the design dimensions of the sensor resistance. *S* should be the focus when laying out the Wheatstone bridge and sizing the sensitive resistors, i.e., the optimum stress length.

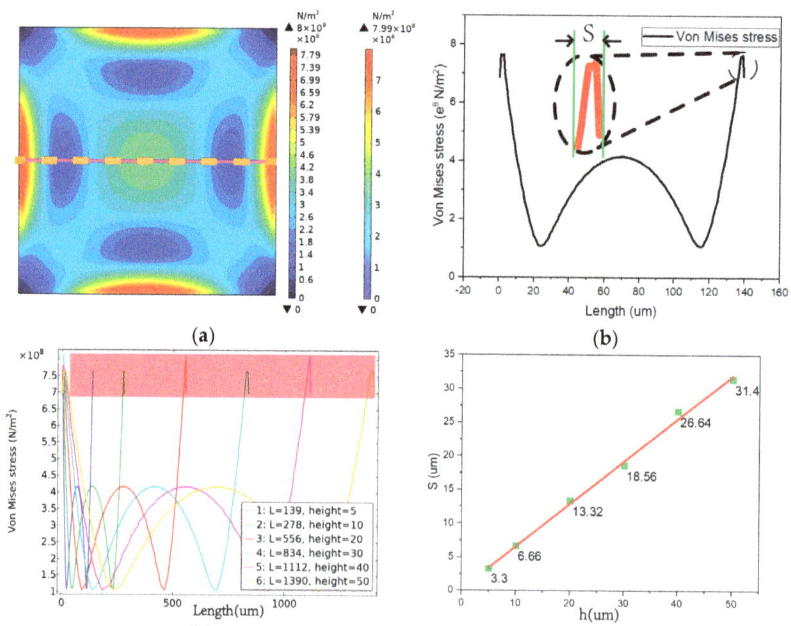

Figure 4. (**a**) Sensitive diaphragm stress distribution; (**b**) optimal stress length *S*; (**c**) centerline stress distribution; (**d**) relationship between the thickness of the sensitive diaphragm and the length S of the stress region.

Different optimal film sizes, the sensitive film under maximum pressure, and the centerline stress distribution are shown in Figure 4c. The maximum stress values of the diaphragms with different optimal sizes tended to be consistent under full-scale pressure. Evidently, the optimal stress region length S increased with the size of the diaphragm, and the variation trend of S with the film thickness is illustrated in Figure 4d.

The finite element simulation results revealed that S increased with the increase in film thickness h, with a large value of S indicating a larger space for the arrangement of resistors. This actually indicates that, in the design process, the thickness of the sensor sensitive diaphragm should be appropriately increased, and the fitting equation of S and film thickness h can be given as follows:

$$S = 0.632329\, h + 0.311507. \tag{2}$$

The thickness of the designed sensitive diaphragm in this work was 11 μm. Calculated according to Equation (1), the side length of the square diaphragm was L = 305.8 μm, while the optimum stress length calculated using Equation (2) was S = 7.267 μm. The structural parameters of the sensor are shown in Table 2.

Table 2. Sizes of structural parameters of the sensor chip.

Parameter	Diaphragm Length	Diaphragm Thickness	Resistor Length	Resistor Width	Cavity Height	Chip Size
Value (μm)	306	11	80	8	389	2500*2500

3. Fabrication and Measurement

The cross-sectional schematic diagrams of the fabrication process for the pressure sensor are shown in Figure 5a. A double-side polished, 400 μm ± 5 um thick, 4 inch (100 ± 0.5 mm) P-type <100> SOI wafer with a resistance of more than 10,000 Ω·cm was used as the substrate, where the thicknesses of the handle layer and buried oxide layer were 10 μm and 1 μm, respectively. The fabrication process is summarized below.

Initially, silicon dioxide (SiO_2) was grown on both sides of the SOI wafer by thermal oxidation, which served as the protection layer. Then, lithography was performed on the front side of SOI substrate to pattern the area for piezoresistors, followed by boron ion implantation to form the piezoresistors. Next, the ohmic contact area at the end of the resistive strip was patterned by photolithography, and a high-concentration boron ion implantation was performed on the front side of the SOI wafer, to form a highly doped region near the sensitive cell. Then, the SOI wafer was rapidly annealed at 1000 °C for 20 s to activate the impurities. Later, SiO_2 deposited by PECVD (plasma-enhanced chemical vapor deposition) was adapted as a passivation layer to protect the piezoresistor, while the ohmic contact region was exposed by wet etching of SiO_2. Then, Ti/Au bonding pads were patterned to connect the resistors, to form a Wheatstone bridge. Finally, the Bosch etching process and anodic bonding created a vacuum back cavity to form an absolute pressure sensor [28,29]. An appropriate etching rate can ensure the verticality during the etching process to ensure the accuracy of the diaphragm size (Bosch etching rate of 1.6 μm per minute). The fabricated sensor chip was observed using SEM (scanning electron microscope), as shown in Figure 5c. The high verticality of the rear cavity ensured good diaphragm consistency and a good stress response. The actual chip manufactured in this work is shown in Figure 5b, indicating that the manufacturing was in line with the design expectations.

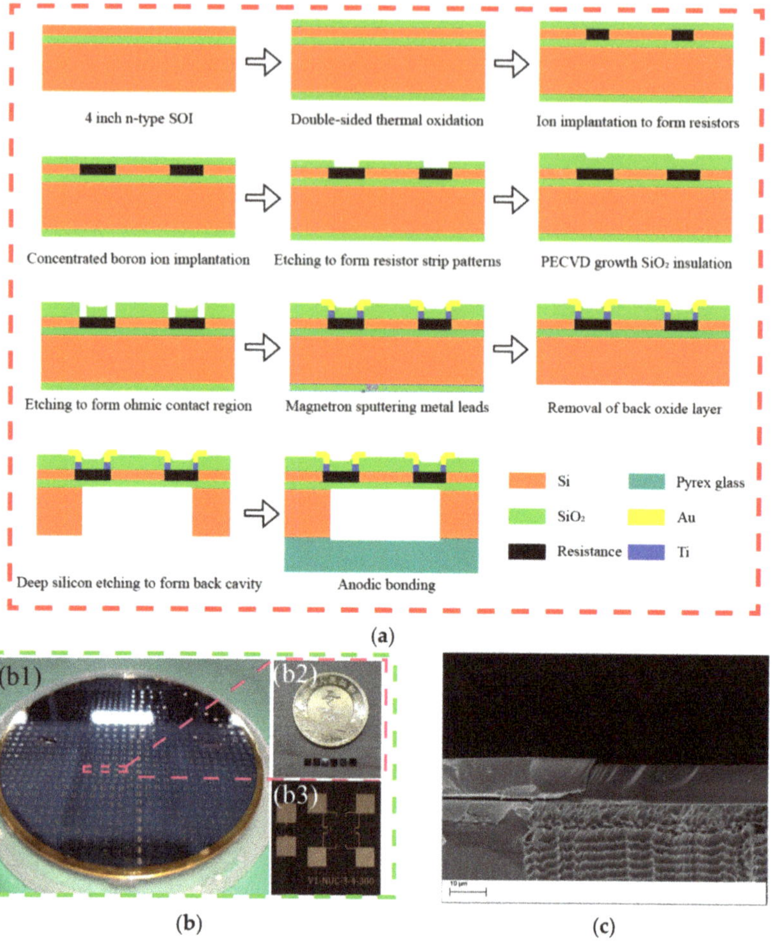

Figure 5. (**a**) Diagram illustrating chip manufacturing process; (**b1**) diagram of fabricated sensor wafers; (**b2**) diagram of several individual pressure sensor chips; (**b3**) diagram of the top surface of the chip under the microscope; (**c**) diagram of cross-section of chip under SEM (scanning electron microscope).

The basic testing process of the sensor is described below. A semiconductor analyzer and a four-probe table were utilized to test the bridge resistance, as shown in Figure 6. A standard 6 MPa pressure test pump (Const 118) was used to pressurize the sensor, and the minimum control unit of the pump was 0.1 kPa, which could accurately control the pressure. A constant 5 V voltage source was used to power the sensor, and the output voltage signal of the sensor was monitored using a Keithley 2450 SourceMeter SMU. During the test, the pressure was changed with a step of 4.75 bar, and a total of nine points were selected. The pressure applied to the sensor was increased to a high value of 40 bar, and then gradually reduced from 40 bar to 2 bar. At each transition point, data were recorded only after the sensor output was stable, and then the basic index of the sensor was calculated. The basic test procedure of the chip is elaborated in Figure 7a,b.

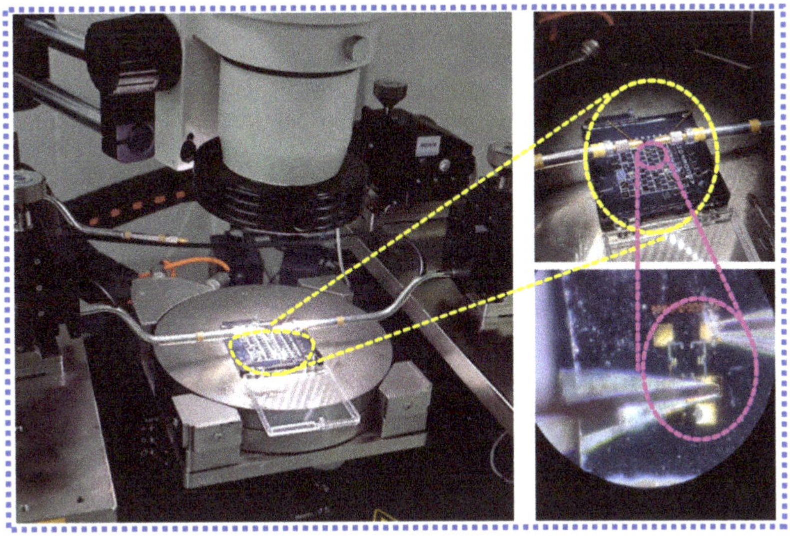

Figure 6. Diagram of semiconductor analyzer and a four-probe table.

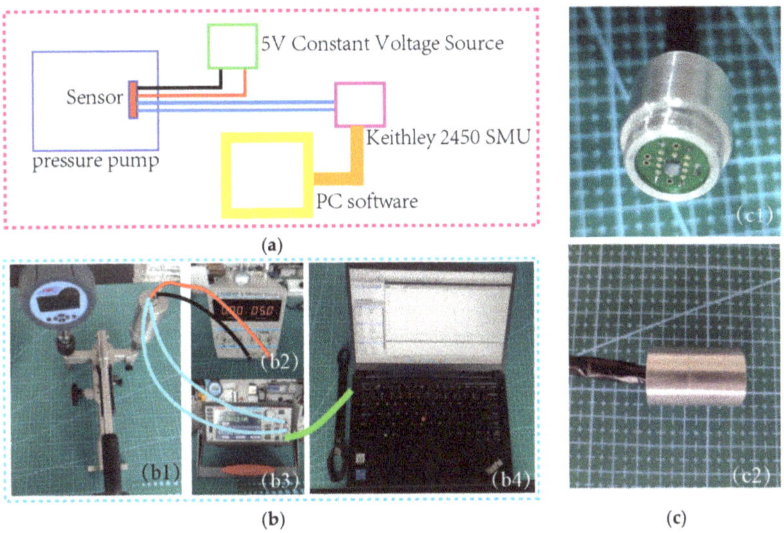

Figure 7. (**a**) Diagram of basic test process; (**b1**) diagram of pressure pump; (**b2**) diagram of DC stabilized power supply; (**b3**) Keithley 2450 SourceMeter SMU; (**b4**) diagram of PC software; (**c1**) diagram of axonometric view of packaged sensor; (**c2**) side view of packaged sensor.

The packaged chip is shown in Figure 7c. As shown in the figure, aluminum alloy was used as the shell of the sensor package, and the back of the sensor was potted with epoxy resin. In addition, an aluminum alloy protective cover was added to the front end of the package shell to protect the chip.

We also designed a series of experiments to verify the responsiveness and reliability of the fabricated pressure sensor while facing complex environments. Pressure is a continuous curve over time, and, when the sensor is used in a complex environment, it will encounter sudden changes in temperature, pressure, vibration, acceleration shock, etc. Likewise, these actual conditions can bring significant challenges to the sensor chip. The question is whether the sensor can respond correctly to these rapidly changing environmental conditions. In view of the above problems, the designed experiments were as follows: (a) six reference points were selected from 10 bar to 35 bar (10 bar, 15 bar, 20 bar, 25 bar, 30 bar, and 35 bar), and a pressure shock was applied at each reference point to conduct a rapid response test of the sensor; (b) a temperature experiment was conducted on the sensor, and a pressure scan was performed with the sensor at different temperatures; (c) the sensor was placed on the vibration table, with the sensor chip film perpendicular to the vibration direction, 10–5000 Hz frequency vibration was applied to the sensor, and the sensor output was observed; (d) 5×, 10×, 15×, and 20× *g* gravitational acceleration shocks were applied to the sensor, and the experiment was repeated five times to verify the reliability of the sensor.

4. Results and Discussion

The sensor chip was tested by a semiconductor analyzer, where it was found that the sensor units on the edge of the SOI chip possessed an abnormal Wheatstone bridge resistance, and the volt–ampere characteristic curve showed nonlinearity, which may have been due to the uneven implantation of boron ions making Schottky contacts in some areas of the SOI wafer; subsequently, the sensor chip with a resistance value closest to the design value was selected as the test sample.

The measurement indicators of the sensor included zero output, sensitivity, nonlinearity, hysteresis, and maximum overload. The sensor's full-scale pressure scan and sensor nonlinearity are shown in Figure 8a, while the positive and negative stroke pressure scans of the sensor are shown in Figure 8b. First, the sensor was kept under a full-scale pressure environment for 5 min, and the corresponding output of the sensor chip is shown in Figure 8c. Next, the sensor's nonlinearity, repeatability, temperature coefficient sensitivity (TCS), and hysteresis were calculated [30]. The detailed data of the sensor at room temperature are shown in Table 3. The sensor had a sensitivity of 9.21 mV/V/bar and a nonlinear error of 0.069% FSS.

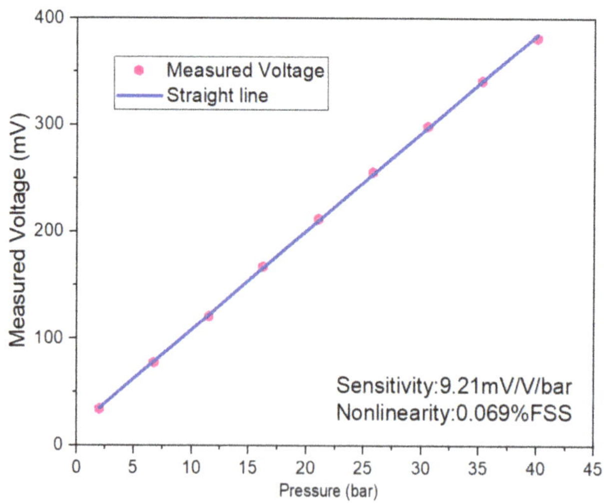

(a)

Figure 8. *Cont.*

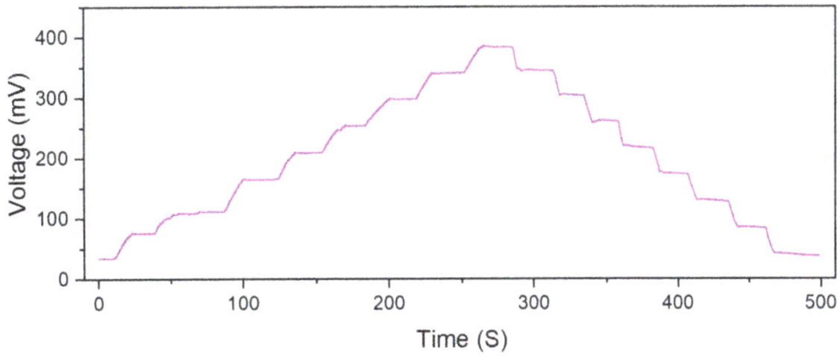

(b)

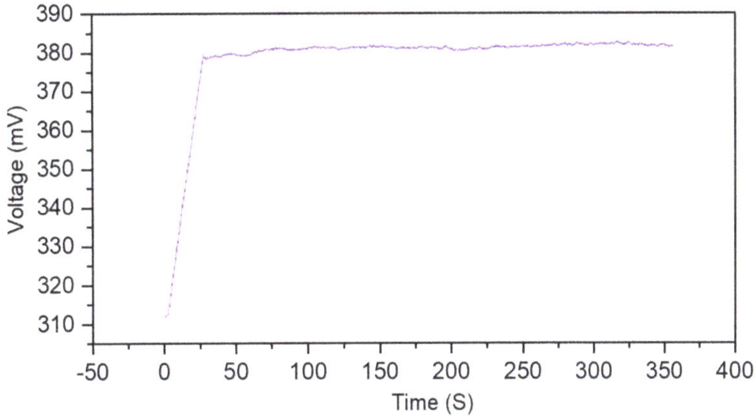

(c)

Figure 8. (**a**) Output voltage of the sensor; (**b**) forward and reverse step pressure test; (**c**) diagram of the output of sensor chip at full-scale pressure.

Table 3. Technical data of the sensor at room temperature.

Parameter	Value	Parameter	Value
Resistance (kΩ)	1.9	Nonlinearity (%FSS)	0.069
Zero output (mV)	16.5	TCS (%FSS/°C)	−0.030
Sensitivity (mV/V/bar)	9.21	Full-range time drift (mV/V/min)	0.553
Repeatability (%FSS)	0.49	Hysteresis (%FSS)	2.6

To validate the dynamic performance and environmental stability of the sensor, a dynamic environmental adaptability test of the sensor was carried out. First, the sensor was tested for a rapid pressure increase. The test results under a rapid pressure increase are shown in Figure 9. Pressure values of 10, 15, 20, 25, 30, and 35 bar were selected as the base points, and a 0.4 bar instantaneous shock was applied to the sensor within 180 ms. The growth rate of the pressure shock was 2.22 bar/s. When the pressure sensor was impacted, the sensor's output had a prominent peak, and it subsequently tended to be stable.

The shock was 0.4 bar, and the response time of the pressure pump was 10 ms. The difference between the voltage peak output of the sensor and the average value is shown in Table 4 Evidently, the maximum instantaneous shock error was 0.53% FSS. It can be seen from Figure 8 that the rapid increase in pressure caused the sensor to have a peak error. However, 2 s after the output peaked, the sensor output tended to a stable value, and the sensor's error after stabilization reached 0.27%. Therefore, we can conclude that, when the pressure changes rapidly, the sensor will output excessive voltage, and the sensor output spike errors must be considered.

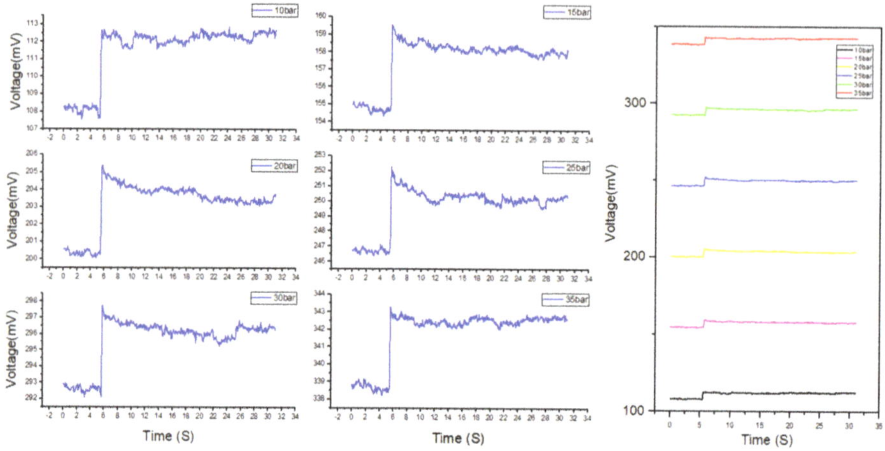

Figure 9. Diagram of sensor output during the rapid shock; the left picture is the separate output of the six sampling points, and the right picture shows a comparison of the six-point output. All pressure shocks were 0.4 bar.

Table 4. Sensor spike shock data.

Pressure/bar	10	15	20	25	30	35
Average voltage (mV)	112.25	158.21	203.80	250.30	296.27	342.51
Peak voltage (mV)	112.87	159.55	205.39	252.27	297.74	343.29
ΔPeak voltage (mV)	0.611	1.340	1.590	1.968	1.470	0.771

Figure 10a indicates the sensor's output voltage for temperature ranging from 0 to 80 °C. As the temperature increased, the zero-point output voltage of the sensor increased, and the slope of fitting curve of the sensor decreased, indicating that the sensitivity of the sensor diminished with the growth of the temperature. Here, the data fitting for the sensor sensitivity was carried out at different temperatures. Taking the sensor sensitivity at 20 °C as the benchmark, the temperature sensitivity coefficient (TCS) of the sensor was −0.03% FSS/°C. The sensitivity of the sensor dropped by 19.6% at 80 °C compared to the output at 20 °C. Nevertheless, the minimum resolution of the sensor could still reach 0.337% FSS. Moreover, the chip's sensitivity at different temperatures is shown in Figure 10b.

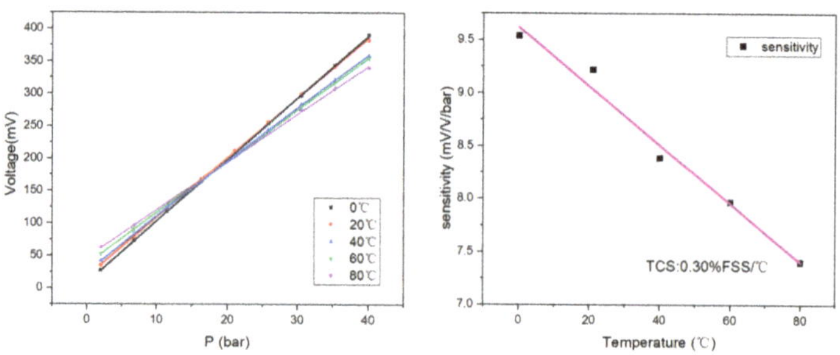

(a) **(b)**

Figure 10. (**a**) Output of the sensor at different temperatures; (**b**) sensitivity drift at different temperatures.

It can be observed that, as the temperature increased, the sensitivity of the sensor decreased, while the zero-output increased (for the convenience of comparison, this paper regards the output of the pressure sensor as the zero output at 2 bar). This is because the carrier mobility of silicon decreases with the increasing temperature; thus, the piezoresistive coefficient of the sensor decreases [31], resulting in a decrease in the sensor sensitivity. The ideal sensor chip has a uniform doping concentration and four identical resistors (R1 = R2 = R3 = R4). As shown in Figure 11, according to the voltage output formula of the Wheatstone bridge [32], the zero output of the sensor chip should be 0 mV; however, the output of the fabricated sensor chip was 16.5 mV. As the temperature increases, the ratio of ΔA to ΔB changes; hence, the zero output of the sensor changes. This phenomenon might be caused by various reasons, among which the consistency of bridge resistance and the residual stress of the chip are two critical influencing factors. The boron ion doping error can also easily produce a resistance bar error. Under the supply voltage of 5 V, a resistance difference of 10 Ω can lead to a zero-point output drift of 13.1 mV. Notably, the SiO_2 insulating layer may also cause this phenomenon after high-temperature annealing at 1000 °C, due to residual tensile stress [33–35].

$$V_{out} = V_{in} \times \frac{R2 \times R4 - R3 \times R1}{(R1 + R2)(R3 + R4)}. \quad (3)$$

$$A = R2 \times R4 - R3 \times R1. \quad (4)$$

$$B = (R1 + R2)(R3 + R4). \quad (5)$$

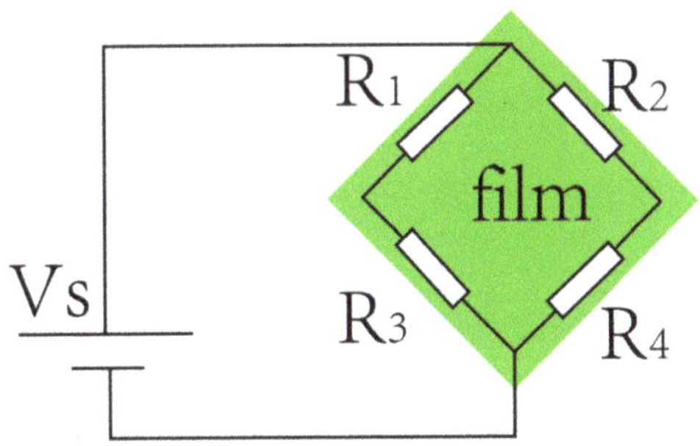

Figure 11. Schematic diagram of the Wheatstone bridge circuit.

A finite element simulation of the characteristic frequency of the sensor was carried out, and the results are reported in Figure 12a. It is concluded that the smallest characteristic frequency of the sensor was 1,777,100 Hz [36] when the normal atmospheric pressure (92.5 kPa) was 20 °C. At ambient pressure (92.5 kPa) and temperature (20 °C), the sensor was subjected to 0–5000 Hz vibration scanning. When the direction of sensor-sensitive film was parallel to the vibration direction, the output during the vibration sweep was as shown in Figure 12(b1). The data were subjected to Fourier transform [37] to study the frequency-domain characteristics, as shown in Figure 12(b2). It can be observed that the zero-point output of the sensor was within the test range of 0–5000 Hz, and the sensor chip did not resonate, as expected by the simulation. The output drift of the sensor was 0.37% FSS, indicating that the sensor had a good anti-vibration performance, and that it can be applied in a high-frequency vibration environment. It is worth mentioning that, during the test, the package shell of the sensor caused resonance at the vibration frequency of 700–750 Hz; the short-term resonance did not cause any abnormal sensor output. However, long-term resonance is very likely to damage the chip; thus, in a high-frequency vibration environment, the natural frequency of the sensor package structure must be considered as one of the critical indicators in the package design.

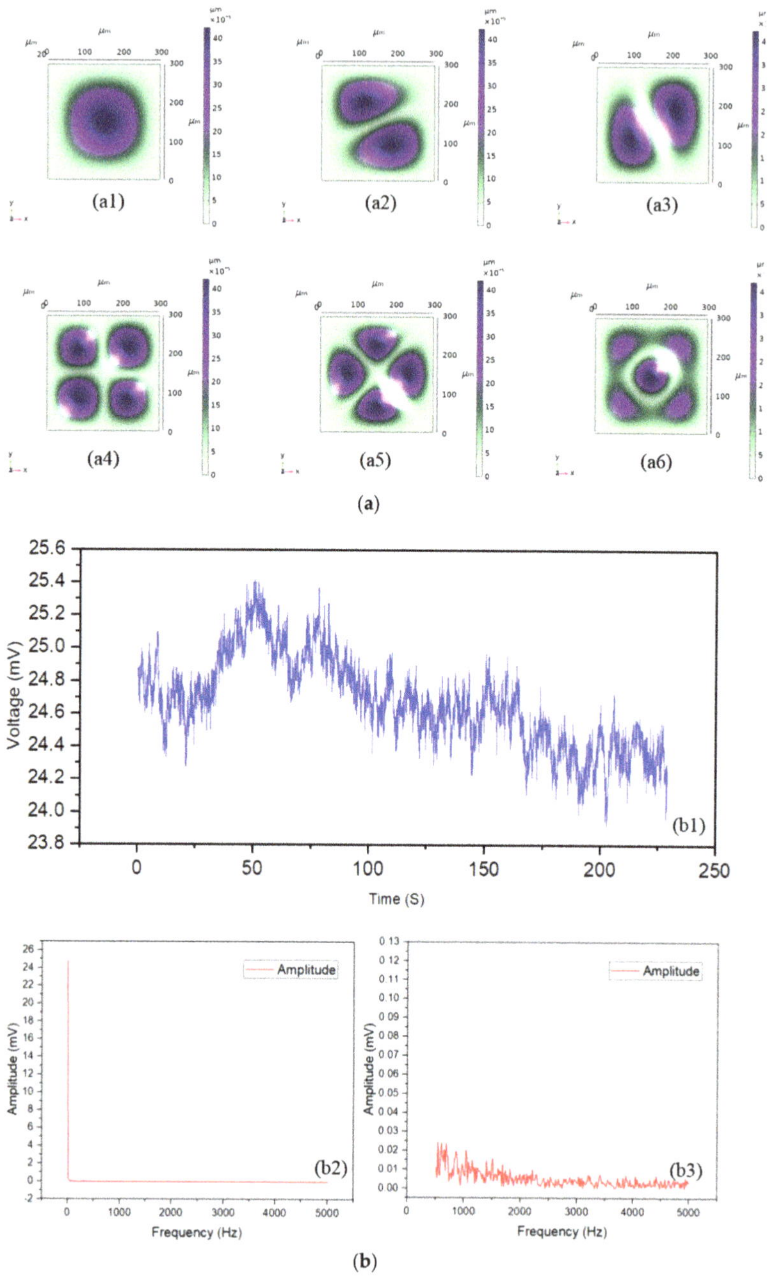

Figure 12. (**a**) Sensor eigenfrequency simulation. The diaphragm had six eigenfrequencies corresponding to six vibration modes: (**a1**) 1,777,100 Hz; (**a2**) 3,584,700 Hz; (**a3**) 3,583,800 Hz; (**a4**) 5,232,400 Hz; (**a5**) 6,328,600 Hz; (**a6**) 6,364,100 Hz. (**b1**) Output during the vibration sweep; (**b2**) Fourier-transform spectrogram based on b1; (**b3**) frequency-domain characteristics of sensor output during 500–5000 Hz vibration, the max amplitude is 0.025 mV.

After the sensor was tested by the above-simulated impact test, it was tested in the pressure pump. The result shows that the output of the sensor was normal, as shown in Figure 13.

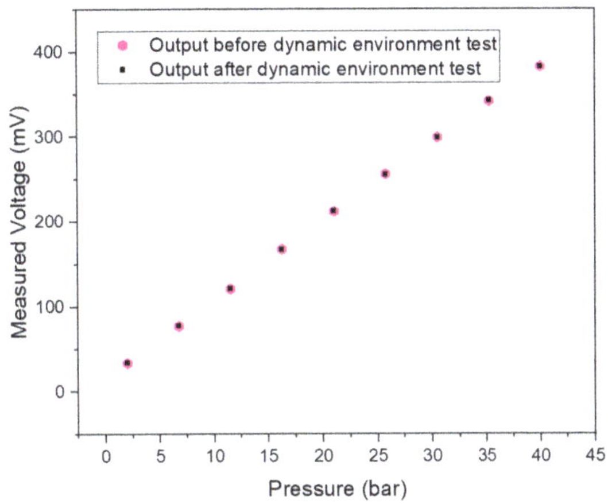

Figure 13. Sensor output before and after dynamic test.

5. Conclusions

This paper proposed an optimal design method for a C-structure piezoresistive pressure sensor and presented the optimal sensor parameter equation through finite element simulation. The designed sensor was manufactured and tested, and the test results validated the accuracy of the finite element simulation results. In addition, the fabricated sensor had high sensitivity and linearity, with a full-scale output of 368.8 mV and a nonlinear error of 0.069% over a pressure range of 40 bar. Meanwhile, the TCS of the sensor at 0–80 °C was −0.3% FSS/°C. Moreover, the sensor was capable of resisting a high-frequency vibration of 100–5000 Hz and an acceleration up to 20× g. In addition, according to the dynamic environment simulation test, we can conclude that the sensor's output error became more significant when the sensor faced extremely rapid pressure pulses. It was also concluded that the anti-vibration performance of the sensor mainly depended on the packaging structure of the sensor. Therefore, to achieve a high sensor accuracy in a complex application environment, the influence of the environment should be one of the main reference factors in the design.

6. Patents

On the basis of this research, the team applied for a patent titled: An MEMS piezoresistive pressure sensor (Patent number: ZL 2021 2 2838774,X).

Author Contributions: Conceptualization, C.X. and Z.Z. (Zengxing Zhang); methodology, R.G.; software, W.Z. and R.G.; validation, B.Y., Y.G. and Z.Z. (Zhou Zhao); investigation, J.J. and Z.L.; writing—manuscript preparation, R.G.; writing—review and editing, R.G., Y.X. and Z.Z. (Zhidong Zhang); project management, H.Z. and Y.W. All authors have read and agreed to the published version of the manuscript.

Funding: This work was supported by the National Natural Science Foundation of China (Grant Nos. 62101511 and 61727806), the Applied Basic Research Project of Shanxi Province (Grant No. 20210302124539), the Shanxi '1311 project' Key Subject Construction (1331 KSC), and the Tan Kah Kee Innovation Laboratory (RD2022020511).

Data Availability Statement: Data are contained within the article.

Conflicts of Interest: The authors declare no conflict of interest.

References

1. Nguyen, T.; Dinh, T.; Phan, H.P.; Pham, T.A.; Dau, V.T.; Nguyen, N.T.; Dao, D.V. Advances in ultrasensitive piezoresistive sensors: From conventional to flexible and stretchable applications. *Mater. Horiz.* **2021**, *8*, 2123–2150. [CrossRef] [PubMed]
2. Gupta, Y.M. Stress measurements using piezoresistance gauges: Modeling the gauge as an elastic-plastic inclusion. *J. Appl. Phys.* **1983**, *54*, 6256–6266. [CrossRef]
3. Jing, Z.; Zhang, G.Y.; Shi, D.X. Review of graphene-based strain sensors. *Chin. Phys. B* **2013**, *22*, 35–43.
4. Cranny, A.; Cotton, D.; Chappell, P.H.; Beeby, S.P.; White, N.M. Thick-film force and slip sensors for a prosthetic hand. *Sens. Actuators A Phys.* **2005**, *123–124*, 162–171. [CrossRef]
5. Zhou, C.; Zang, J.; Xue, C.; Ma, Y.; Hua, X.; Gao, R.; Zhang, Z.; Li, B.; Zhang, Z. Design of a Novel Medical Acoustic Sensor Based on MEMS Bionic Fish Ear Structure. *Micromachines* **2022**, *13*, 163. [CrossRef]
6. Lipomi, D.J.; Vosgueritchian, M.; Tee, B.C.; Hellstrom, S.L.; Lee, J.A.; Fox, C.H.; Bao, Z. Skin-like pressure and strain sensors based on transparent elastic films of carbon nanotubes. *Nat. Nanotechnol.* **2011**, *6*, 788–792. [CrossRef]
7. Amjadi, M.; Kyung, K.U.; Park, I.; Sitti, M. Stretchable, Skin-Mountable, and Wearable Strain Sensors and Their Potential Applications: A Review. *Adv. Funct. Mater.* **2016**, *26*, 1678–1698. [CrossRef]
8. Gao, L.; Wang, Y.; Hu, X.; Zhou, W.; Lu, Y. Cellular Carbon-Film-Based Flexible Sensor and Waterproof Supercapacitors. *ACS Appl. Mater. Interfaces* **2019**, *11*, 26288–26297. [CrossRef]
9. Balderrama, V.S.; Leon-Gil, J.A.; Fernández-Benavides, D.A.; Ponce-Hernández, J.; Bandala-Sánchez, M. MEMS Piezoresistive Pressure Sensor Based on Flexible PET Thin-Film for Applications in Gaseous-Environments. *IEEE Sens. J.* **2022**, *22*, 1939–1947. [CrossRef]
10. Meena, K.V.; Sankar, A.R. Biomedical catheters with integrated miniature piezoresistive pressure sensors: A Review. *IEEE Sens. J.* **2021**, *21*, 10241–10290. [CrossRef]
11. Kumar, K.S.; Xu, Z.; Kalairaj, M.S.; Ponraj, G.; Huang, H.; Ng, C.-F.; Wu, Q.H.; Ren, H. Stretchable Capacitive Pressure Sensing Sleeve Deployable onto Catheter Balloons towards Continuous Intra-Abdominal Pressure Monitoring. *Biosensors* **2021**, *11*, 156. [CrossRef] [PubMed]
12. Zhong, G.-Y.; Liu, Y.; Song, J.; Zhao, Q.; Li, Y.-S.; Li, F.-Y. Strong piezoresistance effect of small molecule organic semiconductors. *J. Phys. D-Appl. Phys.* **2008**, *41*, 205106. [CrossRef]
13. Komati, B.; Agnus, J.L.; Clévy, C.; Lutz, P. Prototyping of a highly performant and integrated piezoresistive force sensor for microscale applications. *J. Micromech. Microeng.* **2014**, *24*, 035018. [CrossRef]
14. Li, S.; Liang, T.; Wang, W.; Hong, Y.; Zheng, T. A novel SOI pressure sensor for high temperature application. *J. Semicond.* **2015**, *36*, 014014. [CrossRef]
15. Barlian, A.A.; Park, W.-T.; Mallon, J.R., Jr.; Rastegar, A.J.; Pruitt, B.L. Review: Semiconductor Piezoresistance for Microsystems. *Proc. IEEE* **2009**, *97*, 513–552. [CrossRef]
16. Phan, H.-P.; Dao, D.V.; Nakamura, K.; Dimitrijev, S.; Nguyen, N.-T. The Piezoresistive Effect of SiC for MEMS Sensors at High Temperatures: A Review. *J. Microelectromech. Syst.* **2015**, *24*, 1663–1677. [CrossRef]
17. Medjahdi, N.; Benmoussa, N.; Benyoucef, B. Modeling, Simulation and Optimization of the Mechanical Response of Micromechanical Silicon Cantilever: Application to Piezoresistive Force Sensor. *Phys. Procedia* **2014**, *55*, 348–355. [CrossRef]
18. Pramanik, C.; Saha, H. Piezoresistive pressure sensing by porous silicon membrane. *IEEE Sens. J.* **2006**, *6*, 301–309. [CrossRef]
19. Wang, W.; He, H.; Bian, Y.; Yang, Y. A MEMS High Temperature Pressure Sensor. *Micronanoelectron. Technol.* **2016**, *53*, 387–393. [CrossRef]
20. Shimazoe, M.; Matsuoka, Y.; Yasukawa, A.; Tanabe, M. A Special Silicon Diaphragm Pressure Sensor with High Output and High-Accuracy. *Sens. Actuators* **1982**, *2*, 275–282. [CrossRef]
21. Kinnell, P.K.; King, J.; Lester, M.; Craddock, R. A Hollow Stiffening Structure for Low Pressure Sensors. In Proceedings of the 23rd Eurosensors Conference, Lausanne, Switzerland, 6–9 September 2009.
22. Zhang, S.; Wang, T.; Lou, L.; Tsang, W.M.; Sawada, R.; Kwong, D.-L.; Lee, C. Annularly Grooved Diaphragm Pressure Sensor with Embedded Silicon Nanowires for Low Pressure Application. *J. Microelectromech. Syst.* **2014**, *23*, 1396–1407. [CrossRef]
23. Huang, X.; Zhang, D. A high sensitivity and high linearity pressure sensor based on a peninsula-structured diaphragm for low-pressure ranges. *Sens. Actuators A-Phys.* **2014**, *216*, 176–189. [CrossRef]
24. Kumar, S.S.; Pant, B.D. Fabrication and characterization of pressure sensor, and enhancement of output characteristics by modification of operating pressure range. In Proceedings of the 2015 19th International Symposium on VLSI Design and Test (VDAT), Ahmedabad, India, 26–29 June 2015.
25. Kumar, S.; Pant, B. Design of piezoresistive MEMS absolute pressure sensor. In Proceedings of the 16th International Workshop on Physics of Semiconductor Devices in Kanpur, India, Kanpur, India, 14–17 December 2012; SPIE: Bellingham, WA, USA; pp. 308–317.
26. Leondes, C.T. *MEMSNEMS: Handbook Techniques and Applications*; Springer: New York, NY, USA, 2006.
27. Elgamel, H.E. Closed-form expressions for the relationships between stress, diaphragm deflection, and resistance change with pressure in silicon piezoresistive pressure sensors. *Sens. Actuators A Phys.* **1995**, *50*, 17–22. [CrossRef]
28. Zhang, Z.; Liu, G.; Wang, K. Fabrication of Needle-Like Silicon Nanowires by Using a Nanoparticles-Assisted Bosch Process for Both High Hydrophobicity and Anti-Reflection. *Micromachines* **2021**, *12*, 1009. [CrossRef] [PubMed]
29. Knowles, K.M.; van Helvoort, A.T.J. Anodic bonding. *Int. Mater. Rev.* **2006**, *51*, 273–311. [CrossRef]

30. Nag, M.; Singh, J.; Kumar, A.; Alvi, P.A.; Singh, K. Sensitivity enhancement and temperature compatibility of graphene piezoresistive MEMS pressure sensor. *Microsyst. Technol. -Micro-Nanosyst. -Inf. Storage Processing Syst.* **2019**, *25*, 3977–3982. [CrossRef]
31. Zheng, P.; Rougieux, F.E.; Macdonald, D.; Cuevas, A. Measurement and Parameterization of Carrier Mobility Sum in Silicon as a Function of Doping, Temperature and Injection Level. *IEEE J. Photovolt.* **2014**, *4*, 560–565. [CrossRef]
32. Castro, H.F.; Correia, V.; Pereira, N.; Costab, P.; Oliveiraa, J.; Lanceros-Mendez, S. Printed Wheatstone bridge with embedded polymer based piezoresistive sensors for strain sensing applications. *Addit. Manuf.* **2018**, *20*, 119–125. [CrossRef]
33. Anh Vang, T.; Zhang, X.; Zhu, B. Effects of Temperature and Residual Stresses on the Output Characteristics of a Piezoresistive Pressure Sensor. *IEEE Access* **2019**, *7*, 27668–27676. [CrossRef]
34. Tian, B.; Shang, H.; Wang, W. Research on Temperature Zero Drift of SiC Piezoresistive Pressure Sensor Based on Asymmetric Wheatstone Bridge. *Silicon* **2021**, *14*, 5445–5451. [CrossRef]
35. Hsueh, C.H.; Luttrell, C.R.; Cui, T. Thermal stress analyses of multilayered films on substrates and cantilever beams for micro sensors and actuators. *J. Micromech. Microeng.* **2006**, *16*, 2509–2515. [CrossRef]
36. Bower, A.F. *Applied Mechanics of Solids*; CRC Press: Boca Raton, FL, USA, 2009.
37. Lin, H.-C.; Ye, Y.-C. Reviews of bearing vibration measurement using fast Fourier transform and enhanced fast Fourier transform algorithms. *Adv. Mech. Eng.* **2019**, *11*, 1–12. [CrossRef]

micromachines

Article

Sound Source Localization Based on Multi-Channel Cross-Correlation Weighted Beamforming

Mengran Liu, Junhao Hu, Qiang Zeng, Zeming Jian * and Lei Nie

Hubei Key Laboratory of Modern Manufacturing Quantity Engineering, School of Mechanical Engineering, Hubei University of Technology, Wuhan 430068, China; mrliu@hbut.edu.cn (M.L.); hujunhao199802@163.com (J.H.); zengqiang199907@163.com (Q.Z.); leinie@hbut.edu.cn (L.N.)
* Correspondence: jianzemingx@163.com

Abstract: Beamforming and its applications in steered-response power (SRP) technology, such as steered-response power delay and sum (SRP-DAS) and steered-response power phase transform (SRP-PHAT), are widely used in sound source localization. However, their resolution and accuracy still need improvement. A novel beamforming method combining SRP and multi-channel cross-correlation coefficient (MCCC), SRP-MCCC, is proposed in this paper to improve the accuracy of direction of arrival (DOA). Directional weight (DW) is obtained by calculating the MCCC. Based on DW, suppressed the non-incoming wave direction and gained the incoming wave direction to improve the beamforming capabilities. Then, sound source localizations based on the dual linear array under different conditions were simulated. Compared with SRP-PHAT, SRP-MCCC has the advantages of high positioning accuracy, strong spatial directivity and robustness under the different signal–noise ratios (SNRs). When the SNR is −10 dB, the average positioning error of the single-frequency sound source at different coordinates decreases by 5.69%, and that of the mixed frequency sound sources at the same coordinate decreases by 5.77%. Finally, the experimental verification was carried out. The results show that the average error of SRP-MCCC has been reduced by 8.14% and the positioning accuracy has been significantly improved, which is consistent with the simulation results. This research provides a new idea for further engineering applications of sound source localization based on beamforming.

Keywords: sound source localization; multi-channel cross-correlation coefficient; microphone array; beamforming

Citation: Liu, M.; Hu, J.; Zeng, Q.; Jian, Z.; Nie, L. Sound Source Localization Based on Multi-Channel Cross-Correlation Weighted Beamforming. *Micromachines* **2022**, *13*, 1010. https://doi.org/10.3390/mi13071010

Academic Editors: Libo Gao and Zhuoqing Yang

Received: 20 May 2022
Accepted: 23 June 2022
Published: 26 June 2022

1. Introduction

Sound source localization techniques have a wide range of application prospects in civil and military systems, such as intelligent medical systems, security monitoring and sonar detection [1–5]. Existing sound source localization techniques can be divided into the subspace, time delay estimation and beamforming. The subspace approach uses the orthogonality between the signal and noise subspaces to determine the waveform direction, including multiple signal classification (MUSIC) and estimating signal parameters via rotational invariance techniques (ESPRIT) [6,7]. Their performance is heavily dependent on the covariance matrix estimation, which is influenced by the signal-to-noise ratio (SNR). Hu et al. proposed an improved MUSIC algorithm to calculate the spatial spectrum and achieve azimuth estimation [8]. Herzog et al. developed a novel EB-ESPRIT method to estimate the incoming wave direction of a sound source [9]. The direction estimation accuracy of studies [8] and [9] is higher than that of MUSIC and ESPRIT, respectively. However, the accuracy is still significantly reduced at a lower SNR. The time delay estimation method achieves the source location based on the arrival time difference [10]. In [11], sound source localization in an indoor environment is completed by using two dual-element arrays to estimate time delay. However, the time delay estimation method is susceptible to noise.

When the number of array elements is increased, there is redundant information in the signals. The multi-channel cross-correlation coefficient (MCCC) method in [12] can improve the robustness of delay estimation through multi-element array signals. Beamforming is to obtain the direction of the sound source by summing weighted array signals, and is classified into frequency domain beamforming and time domain beamforming [13,14]. In frequency domain beamforming, many methods have been proposed to improve its performance [15,16]. Reference [15] improved the robustness by adding norm constraints and spatial smoothing techniques. Reference [16] proposed nested arrays to improve beamforming performance. Time domain beamforming is compared with frequency domain beamforming in [17]. The two beamforming methods have similar performance. Time domain beamforming is a natural broadband method which is suitable for single-frequency and multi-frequency signals, and it does not require repeated multiple frequency processing. In time domain beamforming, the steered-response power (SRP) is commonly used. Steered-response power delay and sum (SRP-DAS) is used for direction estimation based on microphone arrays. The steered-response power phase transform (SRP-PHAT) algorithm is an optimization of the SRP-DAS, which is easy to implement and has stronger robustness than SRP-DAS by whitening the signals [18,19]. The SRP-PHAT algorithm as time domain beamforming has been widely used in target tracking and distributed localization [20–22]. Salvati et al. reported the SRP-PHAT modification algorithm, which can speed up the operation [23]. However, the directivity based on SRP-PHAT is not outstanding in azimuth estimation, and the localization accuracy still needs further improvement.

Therefore, combining the advantages of MCCC and SRP, a new beamforming method (SRP-MCCC) is proposed in this paper. In this method, the wave direction weight (DW) is calculated by the MCCC, which adjusts the SRP value to enhance the directivity of microphone arrays and improve spatial resolution. In this paper, Section 2 describes the positioning principle of the proposed method. Then, the sound source localization simulation is reviewed in Section 3. Section 4 verifies the performance of the proposed method under experimental conditions. Finally, conclusions are given in Section 5.

2. Positioning Principle

Figure 1 shows the calculating flow of the position. Suppose the numbers of arrays and each array elements are M and N, respectively. For each array, the MCCC is evaluated by the N-elements' signals. The DW is constructed by using redundant information of MCCC. After obtaining the DW, the weighted beamforming is performed, and finally, the relative direction $(\theta_I, \theta_{II}, \ldots, \theta_M)$ is found. When $M \geq 2$, the source position can be calculated from $(\theta_I, \theta_{II}, \ldots, \theta_M)$.

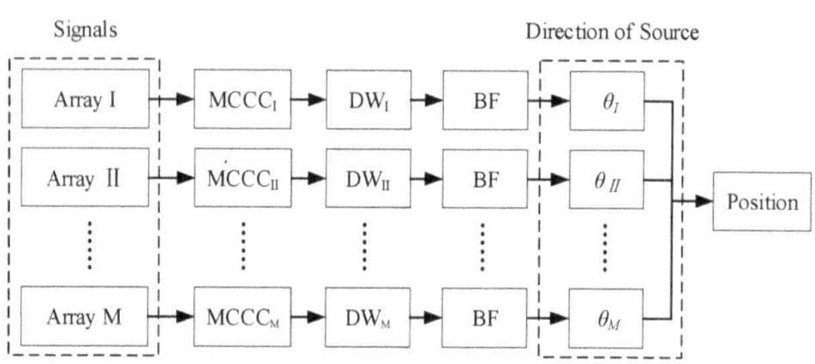

Figure 1. Positioning flowchart.

2.1. Signal Model

The far-field signal propagation and the N-element linear microphone array model are shown in Figure 2. θ_0 represents the direction from the far-field source to the array, and d_i $(i = 1, 2 \ldots \ldots N-1)$ denotes the spacing between the two array elements in the array.

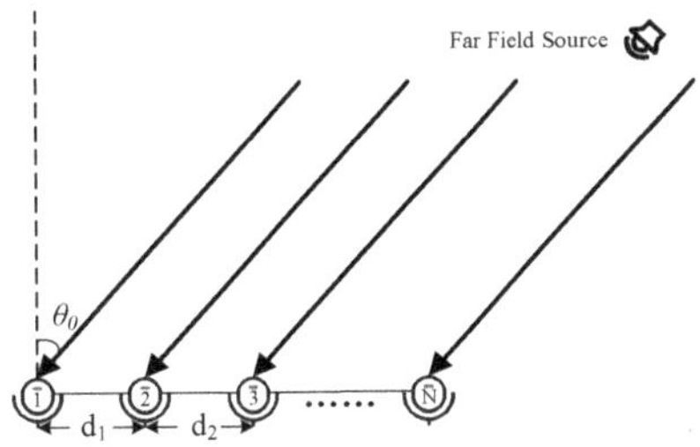

Figure 2. Signal and array model.

Assume that the source signal is $s(n)$, which is a time series. The signals received by the N-element linear microphone array can be expressed as $X(n)$.

$$X(n) = \begin{bmatrix} x_1(n) \\ x_2(n) \\ \cdots \\ x_{N-1}(n) \\ x_N(n) \end{bmatrix} = \begin{bmatrix} \alpha_1 s(n-t_1) + v_1(n) \\ \alpha_2 s(n-t_2) + v_2(n) \\ \cdots \\ \alpha_{N-1} s(n-t_{N-1}) + v_{N-1}(n) \\ \alpha_N s(n-t_N) + v_N(n) \end{bmatrix} \tag{1}$$

where $x_i(n)(i = 1,\ 2,\ \ldots,\ N)$ represents the signal received by the i-th microphone, α_i is the attenuation coefficient of the signal received by the i-th microphone, t_i is the propagation time from the sound source localization to the i-th microphone and $v_i(n)$ is the noise signal received by the i-th microphone.

Taking the first microphone as the reference, the aligned signal $y(n,\ \tau)$ can be written as:

$$\begin{aligned} y(n,\tau) &= [\ \alpha'_1 x_1(n-\tau_1) \quad \alpha'_2 x_2(n-\tau_2) \quad \ldots \quad \alpha'_{N-1} x_{N-1}(n-\tau_{N-1}) \quad \alpha'_N x_N(n-\tau_N)\ ]^T \\ &= [\ \alpha'_1 s'_1(n-\tau_1) \quad \alpha'_2 s'_2(n-\tau_2) \quad \ldots \quad \alpha'_{N-1} s'_{N-1}(n-\tau_{N-1}) \quad \alpha'_N s'_N(n-\tau_N)\ ]^T \\ &+ [\ v_1(n-\tau_1) \quad v_2(n-\tau_2) \quad \ldots \quad v_{N-1}(n-\tau_{N-1}) \quad v_N(n-\tau_N)\ ]^T \end{aligned} \tag{2}$$

where $s'_i(n-\tau_i)$ represents the i-th signal after alignment. τ_i $(i = 1,\ 2,\ \ldots,\ n)$ is the time delay from the i-th microphone to the first microphone, and its value is related to the position of the sound source and the microphone array structure, where $\tau_1 = 0$. α'_i is the relative attenuation coefficient, and it is calculated by

$$\alpha'_i = \frac{\alpha_i}{\alpha_1} \quad (i = 1, 2, \ldots, N) \tag{3}$$

For a linear array, the following relationship exists between τ_i and d_n (the distance between the n-th and (n + 1)-th microphone):

$$\tau_i = \begin{cases} 0 \quad (i = 1) \\ \frac{\sum_{n=1}^{i-1} d_n}{d_1} \tau_2 (i \geq 2) \end{cases} \tag{4}$$

2.2. Direction Weight

The Pearson coefficient of the normalized signals is used to replace the value of the correlation function [24,25]. The correlation coefficient matrix of different time delays is $\rho(\tau)$:

$$\begin{aligned}\rho(\tau) &= Py(n,\tau)y(n,\tau)^T P \\ &= \begin{bmatrix} \rho_{11}(\tau_1,\tau_1) & \rho_{12}(\tau_1,\tau_2) & \dots & \rho_{1N}(\tau_1,\tau_N) \\ \rho_{21}(\tau_2,\tau_1) & \rho_{22}(\tau_2,\tau_2) & \dots & \rho_{2N}(\tau_2,\tau_N) \\ \dots & \dots & \dots & \dots \\ \rho_{N1}(\tau_N,\tau_1) & \rho_{N2}(\tau_N,\tau_2) & \dots & \rho_{NN}(\tau_N,\tau_N) \end{bmatrix}\end{aligned} \tag{5}$$

where

$$P = \begin{bmatrix} \frac{1}{\sigma_1} & & & \\ & \frac{1}{\sigma_2} & & \\ & & \dots & \\ & & & \frac{1}{\sigma_N} \end{bmatrix} \tag{6}$$

$\rho_{ij}(\tau)$ is the correlation coefficient between the i-th and the j-th alignment signal, which can be expressed as:

$$\begin{aligned}\rho_{ij}(\tau_i,\tau_j) &= \frac{x_i(n-\tau_i)\cdot x_j(n-\tau_j)}{\sigma_i\sigma_j} \\ &= \frac{\alpha'_i s'_i(n-\tau_i)\cdot\alpha'_j s_j(n-\tau_j)}{\sigma_i\sigma_j} + \frac{\alpha'_i s'_i(n-\tau_i)\cdot v_j(n-\tau_j)}{\sigma_i\sigma_j} \\ &+ \frac{\alpha'_j s'_j(n-\tau_i)\cdot v_i(n-\tau_i)}{\sigma_i\sigma_j} + \frac{v'_i(n-\tau_i)\cdot v_j(n-\tau_j)}{\sigma_i\sigma_j} \\ &= r_{ij} + r_{iv_j} + r_{jv_i} + r_{v_iv_j}\end{aligned} \tag{7}$$

where $\sigma_i = \sum x_i^2$ is the energy of the i-th signal and $\cdot$ indicates the inner product. Observing Equation (7), it can be deduced that r_{ij} is the correlation coefficient between the i-th signal and the j-th signal. r_{iv_j} represents the correlation coefficient between the i-th signal and the j-th noise, and r_{jv_i} is the correlation coefficient between the j-th signal and the i-th noise. $r_{v_iv_j}$ is expressed as the correlation coefficient between the i-th noise and the j-th noise. It can be seen that when $s'_i(n-\tau_i) = s'_j(n-\tau_j)$, r_{ij} has the maximum value. The $h_{DW}(\tau)$ is constructed by using redundant information of the MCCC. $h_{DW}(\tau)$ is calculated from the correlation coefficient matrix, which is directly related to the signal itself, and it can be represented as:

$$h_{DW}(\tau) = \prod_{i=1}^{N}\prod_{j=1}^{i}\rho_{ij}(\tau_i,\tau_j) \tag{8}$$

2.3. Weighted Beamforming

The weighted beamforming process of a single array after obtaining the $h_{DW}(\tau)$ is shown in Figure 3. First, the compensation time delay τ'_i (i = 1, 2 … … N) is performed on the signals $x_i(t)$ received by the array i_A (i_A = 1, 2 … … M) to obtain the compensation signals $x_i(t-\tau'_i)$. Then, the weighted signals $h_{DW}(\tau)x_i(t-\tau'_i)$ are calculated by multiplying $x_i(t-\tau'_i)$ and $h_{DW}(\tau)$. Subsequently, the spatial integration of $h_{DW}(\tau)x_i(t-\tau'_i)$ at the same instant is performed, and the time integration of all signals at different times is operated. Finally, $D_{i_A}(\theta)$ (beam output of array i_A) is calculated. Since $h_{DW}(\tau)$ contains spatial information, it can enhance the incoming wave direction and suppress the non-incoming wave direction, which can improve the directional ability of $D_{i_A}(\theta)$.

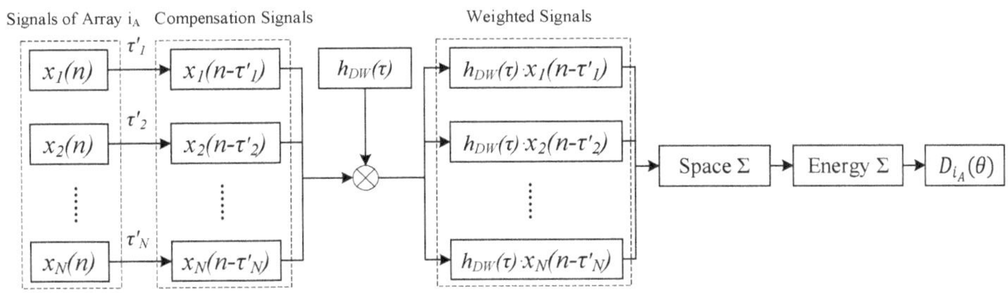

Figure 3. Weighted beamforming.

$D_{i_A}(\theta)$ can be expressed as:

$$D_{i_A}(\theta) = \sum \left| h^T y(n,\tau) \right|^2 \tag{9}$$

where $h^T = h_{DW}(\tau)[1 \quad 1 \quad \ldots \quad 1]$, $\tau = \frac{d_1 \sin\theta}{c}$ and c is the speed of sound in the environment. When $\theta = \theta_0$, $D_{i_A}(\theta)$ has a maximum value; at this time, θ_0 is the direction of the sound source.

The frequency domain expression can be written as:

$$D_{i_A}(\theta) = \frac{1}{2\pi} \prod_{j=1}^{N} \prod_{i=1}^{j} \int_{-\pi}^{\pi} \rho_{ij}(\tau_i, \tau_j) X_i(e^{j\omega}) X_j^*(e^{j\omega}) e^{j\omega(\tau_i - \tau_j)} d\omega \tag{10}$$

where $X_i(e^{j\omega})$ and $X_j(e^{j\omega})$ represent the Fourier transforms of the signals $x_i(t)$ and $x_j(t)$, respectively.

As the number of arrays is M, the coordinates (x, y) of the sound source can be obtained by the following equation:

$$\begin{gathered} D_{all} = \sum_{i_A=1}^{M} D_{i_A}(f_{i_A}(x,y)) \\ (x,y) = \underset{(x,y)}{\operatorname{argmax}} D_{all} \end{gathered} \tag{11}$$

where $f_{i_A}(x, y) = \theta_{i_A}$ represents the direction of the source coordinate (x, y) relative to the normal direction of the array i_A. $f_{i_A}(x, y)$ represents a mapping from coordinate space to direction space. When D_{all} is maximized, the corresponding (x_0, y_0) is the maximum probability of the sound source.

3. Simulation Analysis

We used two sets of equally spaced linear arrays to simulate the sound source localization. The coordinates of the i-th microphone of the array I are $((i-1) \times \mathrm{d}, 0)$ (i = 1, 2, 3...16), and the coordinates of the i-th microphone of the array II are $((i-1) \times \mathrm{d}, 10)$. d is the distance between two adjacent microphones and d = 0.043 m. D is the distance between the array I and array II angle and D = 10 m. The sound source and array locations are shown in Figure 4. θ_1 and θ_2 are sound source directions relative to array I and array II, respectively. Assuming that the normal array is the beginning edge and the incoming wave direction is the end edge, θ_1 and θ_2 are positive counterclockwise and are negative clockwise. Moreover, θ_1 and $\theta_2 \in [-90°, 90°)$. The center of the array is used as a reference point, and the sound source coordinates (x, y) can be calculated by the following:

$$\begin{gathered} \frac{y}{x-7.5d} = \tan(\theta_1 + 90°) \\ \frac{y-D}{x-7.5d} = \tan(\theta_2 - 90°) \end{gathered} \tag{12}$$

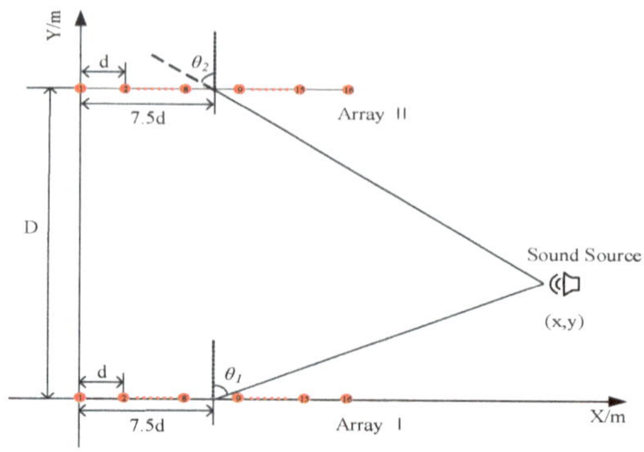

Figure 4. Sound source and microphone array simulation location.

Figure 5 shows the direction estimation results of the SRP-PHAT and the SRP-MCCC when the source signal is a 600 Hz sine signal with source coordinates (14, 4) and the SNR is −5 dB. The brightness of a point represents the probability that the point is the location of the sound source. Figure 6 shows the location estimation results of the SRP-PHAT and the SRP-MCCC when the SNR is −5 dB.

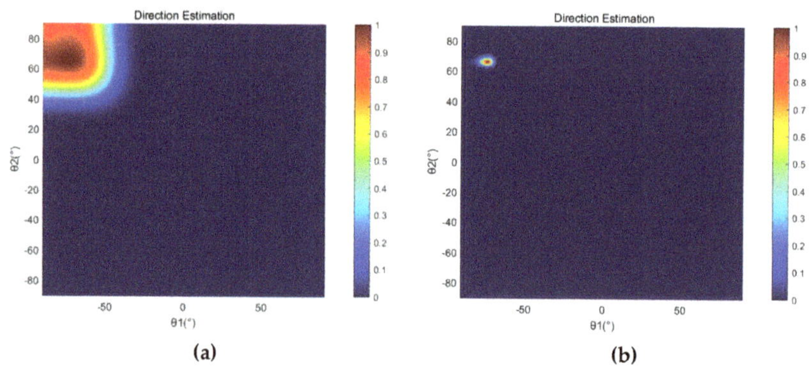

Figure 5. The direction estimation simulation results of (**a**) SRP-PHAT and (**b**) SRP-MCCC when the SNR is −5 dB.

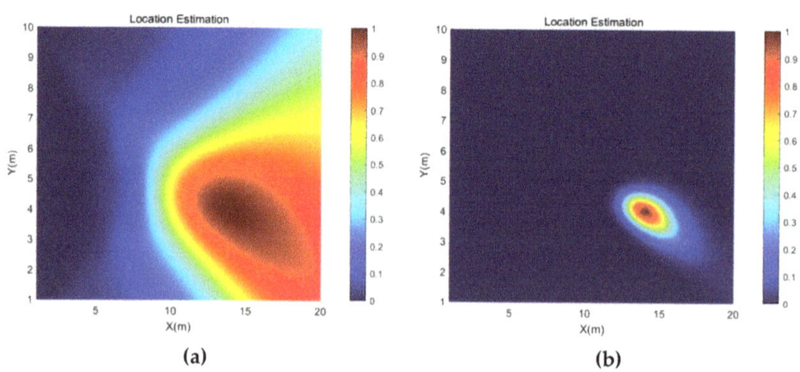

Figure 6. The location estimation simulation results of (**a**) SRP-PHAT and (**b**) SRP-MCCC when the SNR is −5 dB.

From Figures 5 and 6, we can observe that the SRP-MCCC has a superior localization convergence capability and can improve the localization accuracy by aggregating the localization results into smaller areas and improving spatial pointing. To further explore the advantages of this method, simulation analyses of SRP-PHAT and the SRP-MCCC were conducted under different SNR conditions, and the results are shown in Table 1.

Table 1. Simulation results for different SNR conditions (600 Hz, (14, 4)).

SNR/dB	SRP-PHAT			SRP-MCCC		
	Position/m	Directional Deviation/°	Distance Error/%	Position/m	Directional Deviation/°	Distance Error/%
10	(14.25, 3.94)	(−0.36, 0.12)	1.77	(13.96, 4.00)	(0.04, −0.08)	0.27
0	(14.27, 3.86)	(−0.46, −0.28)	2.09	(14.19, 3.96)	(−0.26, −0.03)	1.33
−5	(15.24, 3.84)	(−1.76, 1.12)	8.59	(13.78, 3.90)	(−0.11, −0.83)	1.66
−10	(12.59, 4.10)	(1.59, −2.03)	9.71	(14.31, 4.13)	(0.24, 0.87)	2.31
−15	(15.61, 4.06)	(−1.33, 2.29)	11.06	(13.48, 4.10)	(1.01, −0.46)	3.64
−20	(16.14, 4.07)	(−1.76, 3.02)	14.71	(14.57, 4.26)	(0.49, 1.57)	4.30

As shown in Table 1, when the SNR is high, the localization errors of both methods are comparatively insignificant. As the SNR decreases, the localization error of the SRP-PHAT increases significantly, while that of the SRP-MCCC remains at a low level. When the SNR is reduced from 10 dB to −20 dB, the position error of the SRP-PHAT increases from 1.77% to 14.71%, and the error of the SRP-MCCC only increases from 0.27% to 4.30%. Simulation experiments show that the SRP-MCCC is feasible. The localization effect is excellent, and the robustness is outstanding.

To further explore the method proposed in this paper, the single-frequency signal at different source positions and the mixed-frequency signals at the same coordinate were simulated when SNR = −10 dB. The results are shown in Tables 2 and 3, respectively. We can infer that the average error of the single-frequency signal at different coordinates decreased by 5.69%, and that of the mixed-frequency signals decreased by 5.77% at the same coordinate.

Table 2. Simulation results for different source locations (SNR = −10 dB, 600 Hz).

Source Location/m	SRP-PHAT			SRP-MCCC		
	Position/m	Directional Deviation/°	Distance Error/%	Position/m	Directional Deviation/°	Distance Error/%
(12, 7)	(10.96, 6.62)	(1.28, −3.53)	8.81	(12.22, 6.79)	(−0.96, −0.84)	2.19
(15, 6)	(16.26, 6.18)	(1.41, 1.11)	7.88	(15.52, 6.03)	(−0.46, 0.51)	3.22
(17, 4.5)	(18.27, 4.04)	(−2.34, −0.23)	7.51	(17.47, 4.60)	(−0.02, 0.85)	2.73
(19, 6.5)	(20.99, 6.84)	(0.78, 1.87)	10.05	(19.81, 6.92)	(0.46, 1.58)	4.05
(21, 5)	(21.21, 4.15)	(1.08, −0.63)	8.83	(21.52, 4.90)	(−0.62, 0.50)	2.45

Table 3. Simulation results for different frequencies (SNR = −10 dB, (15, 5)).

Frequency/Hz	SRP-PHAT			SRP-MCCC		
	Position/m	Directional Deviation/°	Distance Error/%	Position/m	Directional Deviation/°	Distance Error/%
600	(13.59, 5.09)	(2.21, −1.52)	8.93	(14.61, 5.09)	(0.81, −0.16)	2.53
600, 900	(16.17, 5.21)	(−0.62, 2.00)	7.52	(15.41, 5.13)	(−0.03, 0.92)	2.72
600, 900, 1500	(13.68, 5.08)	(2.03, −1.42)	8.36	(14.73, 5.23)	(1.15, 0.50)	2.24

4. Experiment

To verify the feasibility of the SRP-MCCC, a field localization experiment was conducted, as shown in Figure 7. The array structure and the sound source frequency are consistent with the simulation during the experiment, as shown in Figure 4. This paper presents an experimental comparison between the SRP-PHAT and the SRP-MCCC.

Figure 7. (**a**) Experimental scene. (**b**) Experimental equipment.

The direction estimation and position estimation results of SRP-PHAT and SRP-MCCC when the sound source coordinate is (10, 5) are shown in Figures 8 and 9, respectively. From Figures 8 and 9, it can be found that the spot size of the SRP-MCCC, in both angle and position, is smaller than SRP-PHAT. Therefore, the SRP-MCCC has a stronger angular resolution and can control the beam width in a smaller range, which can lead to more accurate direction and position estimation.

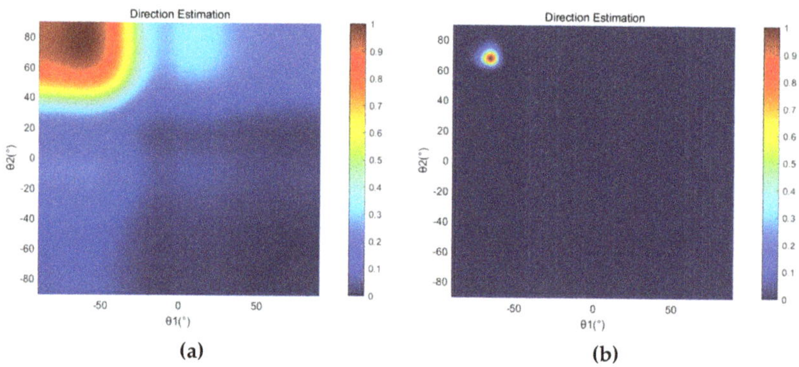

Figure 8. The direction estimation experimental results of (**a**) SRP-PHAT and (**b**) SRP-MCCC when the sound source coordinate is (10, 5).

Subsequently, localization experiments were conducted for the sound source coordinates of (11, 2) and (12, 7.5), and the experimental results are shown in Table 4. Compared with the SRP-PHAT, the SRP-MCCC has higher accuracy in both direction and position estimation, and the average error is reduced by 8.14%. Therefore, SRP-MCCC has a better localization effect, which is in line with the expected results and verifies the feasibility and superiority of the SRP-MCCC.

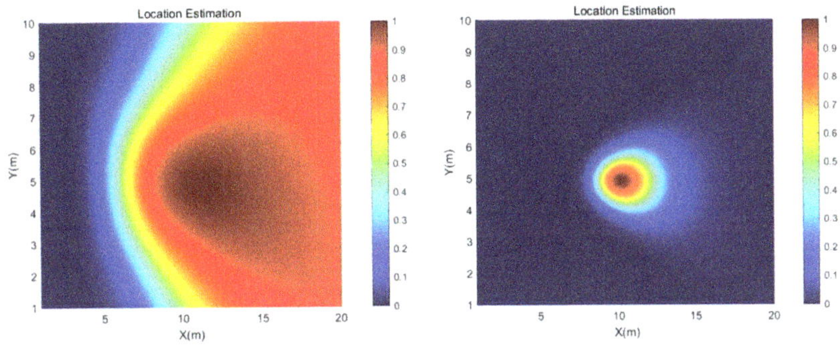

(a) (b)

Figure 9. The location estimation experimental results of (**a**) SRP-PHAT and (**b**) SRP-MCCC when the sound source coordinate is (10, 5).

Table 4. Localization results of SRP-PHAT and MCCC-SRP at different coordinates.

Source Location/m	$(\theta_1, \theta_2)/^\circ$	SRP-PHAT			SRP-MCCC		
		Direction/°	Position/m	Error/%	Direction/°	Position/m	Error/%
(10, 5)	(−64.16, 64.16)	(−65.60, 66.08)	(11.14, 4.80)	10.35	(−64.99, 64.18)	(10.20, 4.91)	1.96
(11, 2)	(−79.98, 54.76)	(−81.94, 55.73)	(11.83, 1.72)	7.83	(−79.11, 54.75)	(10.81, 2.13)	2.06
(12, 7.5)	(−58.67, 78.53)	(−61.77, 79.12)	(13.39, 7.36)	12.62	(−59.52, 78.26)	(12.24, 7.39)	2.36

5. Conclusions

A new high-precision beamforming algorithm (SRP-MCCC) is proposed to improve the positioning accuracy by combining SPR and MCCC. A detailed theoretical analysis of the method is presented here, and the simulations and experiments verify its feasibility. The results show that the method has the advantages of strong robustness and high localization accuracy. Furthermore, the SRP-MCCC has better spatial resolution and localization capability than the SRP-PHAT. Both the simulations and experiments verify the effectiveness of the method. These results provide a new idea for the weighted beamforming algorithm, which is essential for researching high-precision sound source localization in complex environments.

Author Contributions: Conceptualization, M.L. and Z.J. ; methodology, M.L. and J.H.; validation, M.L., Z.J. and J.H.; formal analysis, Z.J.; investigation, J.H. and Q.Z.; data curation, J.H. and Q.Z.; writing—original draft preparation, J.H. and Q.Z.; writing—review and editing J.H. and Q.Z.; visualization, J.H. and Q.Z.; supervision, Z.J. and L.N.; project administration; M.L., Z.J. and L.N.; funding acquisition, L.N. All authors have read and agreed to the published version of the manuscript.

Funding: This research was funded by the National Natural Science Foundation of China [Grant Nos. 51805154 and 51975191], Green Industry Technology Leading Project of Hubei University of Technology [XJ2021004901], and the Scientific Research Foundation of Hubei University of Technology [GCRC 2020010].

Institutional Review Board Statement: Not applicable.

Informed Consent Statement: Not applicable.

Data Availability Statement: The data used to support the findings of this study are available from the corresponding author upon request.

Conflicts of Interest: The authors declare no conflict of interest.

References

1. Aich, A.; Palanisamy, P. On-grid DOA estimation method using orthogonal matching pursuit. In Proceedings of the International Conference on Signal Processing and Communications (ICSPC) IEEE, Coimbatore, India, 28–29 July 2017; pp. 483–487. [CrossRef]
2. Argentieri, S.; Danès, P.; Souères, P. A survey on sound source localization in robotics: From binaural to array processing methods. *Comput. Speech Lang.* **2015**, *34*, 87–112. [CrossRef]
3. Pradhan, D.; Bera, R. Direction of Arrival Estimation via ESPRIT Algorithm for Smart Antenna System. *Int. J. Comput. Appl.* **2015**, *118*, 5–7. [CrossRef]
4. Manamperi, W.; Abhayapala, T.D.; Zhang, J.; Samarasinghe, P.N. Drone audition: Sound source localization using on-board microphones. *IEEE/ACM Trans. Audio Speech Lang. Process.* **2022**, *30*, 508–519. [CrossRef]
5. Kong, D.F.N.; Shen, C.; Tian, C.; Zhang, K. A New Low-Cost Acoustic Beamforming Architecture for Real-Time Marine Sensing: Evaluation and Design. *J. Mar. Sci. Eng.* **2021**, *9*, 868. [CrossRef]
6. Schmidt, R.O. Multiple emitter location and signal parameter estimation. *IEEE Trans. Antennas Propag.* **1986**, *34*, 276–280. [CrossRef]
7. Roy, R.; Kailath, T. ESPRIT-estimation of signal parameters via rotational invariance techniques. *IEEE Trans. Acoust. Speech Signal Process.* **1989**, *37*, 984–995. [CrossRef]
8. Hu, Y.; Abhayapala, T.D.; Samarasinghe, P.N. Multiple source direction of arrival estimations using relative sound pressure based music. *IEEE/ACM Trans. Audio Speech Lang. Process.* **2020**, *29*, 253–264. [CrossRef]
9. Herzog, A.; Habets, E.A.P. Eigenbeam-ESPRIT for DOA-Vector Estimation. *IEEE Signal Process. Lett.* **2019**, *26*, 572–576. [CrossRef]
10. Dvorkind, T.G.; Gannot, S. Time difference of arrival estimation of speech source in a noisy and reverberant environment. *Signal Process.* **2005**, *85*, 177–204. [CrossRef]
11. Chung, M.-A.; Chou, H.-C.; Lin, C.-W. Sound Localization Based on Acoustic Source Using Multiple Microphone Array in an Indoor Environment. *Electronics* **2022**, *11*, 890. [CrossRef]
12. He, H.; Wu, L.; Lu, J.; Qiu, X.; Chen, J. Time Difference of Arrival Estimation Exploiting Multichannel Spatio-Temporal Prediction. *IEEE Trans. Audio Speech, Lang. Process.* **2012**, *21*, 463–475. [CrossRef]
13. Chiariotti, P.; Martarelli, M.; Castellini, P. Acoustic beamforming for noise source localization—Reviews, methodology and applications. *Mech. Syst. Signal Process.* **2018**, *120*, 422–448. [CrossRef]
14. Yang, Y.; Zhigang, C. A Review of High-performance Beamforming Methods for Acoustic Source Identification. *Chin. J. Mech. Eng.* **2021**, *57*, 166–183. [CrossRef]
15. Zhao, Y.; Jensen, J.R.; Jensen, T.L.; Chen, J.; Christensen, M.G. Experimental study of robust acoustic beamforming for speech acquisition in reverberant and noisy environments. *Appl. Acoust.* **2020**, *170*, 107531. [CrossRef]
16. Yang, J.; Liao, G.; Li, J. Robust adaptive beamforming in nested array. *Signal Process.* **2015**, *114*, 143–149. [CrossRef]
17. Padois, T.; Fischer, J.; Doolan, C.; Doutres, O. Acoustic imaging with conventional frequency domain beamforming and generalized cross correlation: A comparison study. *Appl. Acoust.* **2021**, *177*, 107914. [CrossRef]
18. Cobos, M.; Marti, A.; Lopez, J.J. A Modified SRP-PHAT Functional for Robust Real-Time Sound Source Localization With Scalable Spatial Sampling. *IEEE Signal Process. Lett.* **2010**, *18*, 71–74. [CrossRef]
19. Boora, R.; Dhull, S.K. Performance Evaluation of Iterative SRP-PHAT Techniques for Acoustic Source Localization. In Proceedings of the First International Conference on Computational Electronics for Wireless Communication, Singapore, 3 January 2022; Springer: Berlin/Heidelberg, Germany; pp. 403–418. [CrossRef]
20. Wu, K.; Khong, A.W.H. Acoustic source tracking in reverberant environment using regional steered response power measurement. In Proceedings of the Asia-Pacific Signal and Information Processing Association Annual Summit and Conference, Kaohsiung, Taiwan, 29 October–1 November 2013; pp. 1–6. [CrossRef]
21. Huang, J.; Jin, N.; Wang, L.; Chen, X.; Sheng, X.; Yang, S.; Zhao, X.; Sun, L.; Zhu, M. Acoustic Source Localization with Distributed Smartphone Arrays. In Proceedings of the 2015 IEEE Global Com-munications Conference (GLOBECOM), San Diego, CA, USA, 6–10 December 2015; pp. 1–6. [CrossRef]
22. Diaz-Guerra, D.; Miguel, A.; Beltran, J.R. Robust Sound Source Tracking Using SRP-PHAT and 3D Convolutional Neural Networks. *IEEE/ACM Trans. Audio Speech Lang. Process.* **2020**, *29*, 300–311. [CrossRef]
23. Salvati, D.; Drioli, C.; Foresti, G.L. Sensitivity-based region selection in the steered response power algorithm. *Signal Process.* **2018**, *153*, 1–10. [CrossRef]
24. Dunn, O.J.; Clark, V.A. *Applied Statistics: Analysis of Variance and Regression*; John Wiley and Sons, Inc.: Hoboken, NJ, USA, 1986. [CrossRef]
25. Rezayee, A.; Gazor, S. An adaptive KLT approach for speech enhancement. *IEEE Trans. Speech Audio Process.* **2001**, *9*, 87–95. [CrossRef]

micromachines

MDPI

Article

Enhancing Part-to-Part Repeatability of Force-Sensing Resistors Using a Lean Six Sigma Approach

Andrés O. Garzón-Posada [1,2], Leonel Paredes-Madrid [1,*], Angela Peña [3], Victor M. Fontalvo [1] and Carlos Palacio [4]

1 Faculty of Engineering, Universidad Católica de Colombia, Carrera 13 # 47-30, Bogota 110221, Colombia; aogarzon@ucatolica.edu.co (A.O.G.-P.); vmfontalvo@ucatolica.edu.co (V.M.F.)

2 Department of Applied Physics, Materials and Surface Lab (Nanotechnology Unit), Faculty of Sciences, Universidad de Málaga, ES29071 Malaga, Spain

3 Faculty of Mechanical, Biomedical and Electronic Engineering, Universidad Antonio Nariño, Carrera 7 # 21-84, Tunja 150001, Colombia; angelapena@uan.edu.co

4 GIFAM Group, Faculty of Sciences, Universidad Antonio Nariño, Carrera 7 # 21-84, Tunja 150001, Colombia; carlospalacio@uan.edu.co

* Correspondence: ljparedes@ucatolica.edu.co; Tel.: +57-(1)-327-7300

Abstract: Polymer nanocomposites have found wide acceptance in research applications as pressure sensors under the designation of force-sensing resistors (FSRs). However, given the random dispersion of conductive nanoparticles in the polymer matrix, the sensitivity of FSRs notably differs from one specimen to another; this condition has precluded the use of FSRs in industrial applications that require large part-to-part repeatability. Six Sigma methodology provides a standard framework to reduce the process variability regarding a critical variable. The Six Sigma core is the DMAIC cycle (Define, Measure, Analyze, Improve, and Control). In this study, we have deployed the DMAIC cycle to reduce the process variability of sensor sensitivity, where sensitivity was defined by the rate of change in the output voltage in response to the applied force. It was found that sensor sensitivity could be trimmed by changing their input (driving) voltage. The whole process comprised: characterization of FSR sensitivity, followed by physical modeling that let us identify the underlying physics of FSR variability, and ultimately, a mechanism to reduce it; this process let us enhance the sensors' part-to-part repeatability from an industrial standpoint. Two mechanisms were explored to reduce the variability in FSR sensitivity. (i) It was found that the output voltage at null force can be used to discard noncompliant sensors that exhibit either too high or too low sensitivity; this observation is a novel contribution from this research. (ii) An alternative method was also proposed and validated that let us trim the sensitivity of FSRs by means of changing the input voltage. This study was carried out from 64 specimens of Interlink FSR402 sensors.

Keywords: nanocomposites; tactile sensors; force sensors; pressure sensors; sensor phenomena and characterization; regression analysis; Gaussian distribution

Citation: Garzón-Posada, A.O.; Paredes-Madrid, L.; Peña, A.; Fontalvo, V.M.; Palacio, C. Enhancing Part-to-Part Repeatability of Force-Sensing Resistors Using a Lean Six Sigma Approach. *Micromachines* **2022**, *13*, 840. https://doi.org/10.3390/mi13060840

Academic Editor: Kazunori Hoshino

Received: 25 February 2022
Accepted: 22 May 2022
Published: 27 May 2022

1. Introduction

Force-sensing resistors (FSRs) are typically manufactured from a blend of an insulating polymer with conductive nanoparticles ([1–6]. The resulting nanocomposite exhibits a piezoresistive response that can be used to either measure compressive forces [6] or tensile loads [3,4,7]. Given the low profile and low cost of FSRs, their usage in research and industrial applications is currently increasing [8]. Multiple studies related to gait analysis [9–11], robotics [12,13], and other disciplines have reported the usage of polymer nanocomposites to perform strain/stress measurements [14–17].

In recent years, there has been an explosion in the number of studies that have developed novel techniques for manufacturing conductive polymers composites (CPCs). Recent studies have incorporated conductive polymers as a replacement for conductive particles [18–20]; this is a desirable characteristic given the toxicity of graphene, carbon nanotubes (CNTs), and carbon black (CB) [21]. Similarly, the introduction of self-healing properties to the nanocomposite has been also studied by multiple authors [22–24]. Previous authors' works have mainly focused on: first, improving the performance of FSRs by studying the effect of source voltage in the repeatability of measurements [25], and second, developing tailored driving circuits that can help us to minimize time drift and electrical hysteresis [26].

Despite the previously mentioned efforts, the overall performance of FSRs and CPCs still lags behind that of load cells performance in multiple ways. By comparing commercial FSRs [27–29] with an LCHD-5 load cell from Omega Engineering (Norwalk, CT, USA) [30], a difference of up to two orders of magnitude is evident in the metrics of hysteresis and accuracy. Fortunately, studies from Urban et al. [31] and Nguyen and Chauban [32] have helped to narrow the performance gap between both sensing technologies. However, one of the most important drawbacks of FSRs, which has not been yet addressed by specific literature, is the inability to know sensor sensitivity a priori. Given the random dispersion of conductive nanoparticles along the insulating polymer matrix [33], it is not possible to determine the resulting sensitivity of a given nanocomposite, i.e., every specimen has a different sensitivity. This characteristic limits the extensive usage of FSRs since individual sensor calibration is required before use. This condition is not a major concern when only a few sensors are required in the final application, but when multiple sensors are required, sensor characterization is a time-consuming task; ultimately, a more suitable sensing technology may be preferred instead. Robotic skins and tactile pads are representative examples of applications requiring multiple sensing points. In these types of deployments, sensor arrays with multiple tactels are employed to detect shapes and force profiles [34].

As pointed out by Castellanos-Ramos et al., the characterization of piezoresistive tactile pads required a complex test bench with an air compressor and tailored hardware; this was performed to match the specific dimensions of the tactile pad [34]. The aim of this study is to develop techniques that save time and resources by avoiding individual sensor calibration. In this research, we address such a concern by using the Lean Six Sigma Methodology (LSSM) to a group of 64 specimens of commercial FSRs, manufactured by Interlink Electronics, Inc. (Westlake Village, CA, USA) [27].

It must be stated that the application of the LSSM to FSRs represents a novelty. By looking up the following keyword combinations in the Scopus search engine without year constraints: Six Sigma and FSR, Six Sigma and polymer composite, Six Sigma and piezoresistive sensor, a total of 68 entries were found. Most of the entries found were inaccurate because Sigma is used to designate sensors' sensitivity. Only six studies truly reported the use of LSSM [35–40], but most of them reported calibration procedures for pressure-sensing equipment in automotive applications.

The rest of this paper is organized as follows: Section 2 briefly describes the theoretical foundations of the Lean Six Sigma methodology and the physical modeling of FSRs, the experimental setup for gathering sensor data is described in Section 3, followed by the application of the LSSM in Section 4. Conclusions are addressed in Section 5.

2. Theoretical Foundations

2.1. Application of the Six Sigma Methodology

Readers may refer to Appendix A for a theoretical description of the Lean Six Sigma methodology (LSSM). In this section, we describe the application of the LSSM to our study case. The goal of this research is to reduce the sensitivity dispersion of FSRs by adjusting their driving voltage and/or by discarding noncompliant sensors. As previously mentioned, the reduction in sensitivity dispersion avoids individual sensor calibration, which ultimately saves time and resources.

In order to reduce sensitivity dispersion, we have deployed the core of the LSSM, i.e., the DMAIC cycle (Define, Measure, Analyze, Improve, and Control). The Six Sigma methodology can be implemented during any phase of product development, manufacturing, and later deployment into the final application [41]. The current research only considered the application of the LSSM to reduce FSR variability, as measured from their sensitivity; this is known in industry as the enhancing of part-to-part repeatability [27]. In this study, we only considered compensation techniques into the last stage of final application deployment. However, several authors have explored different methods during sensor fabrication, such as addition of surfactants [42] and the application of magnetic fields to reduce the percolation threshold [43]. Given the fact that we only applied compensation techniques into the last stage of final application deployment, the ultimate goal of Six Sigma reduction was partially achieved.

In order to obtain a Six Sigma reduction in sensitivity dispersion, it is required to apply the DMAIC cycle from the reception of raw materials, followed by rigorous control of sensor manufacturing and assembly; this whole process, although possible, would require separate research.

In this study, the DMAIC was deployed by measuring the output voltage at null force (U_o^{0N}) for each sensor as received from the manufacturer. Later, we correlated U_o^{0N} with sensor sensitivity and developed a statistical model to trim the driving voltage of the FSRs; this process required a thorough understanding of the sensing mechanism of FSRs, which are discussed in Section 2.2. Specific details of the DMAIC cycle are later addressed in Section 4.

2.2. Physical Modeling of Force-Sensing Resistors

Multiple authors have studied the underlying physics of CPCs under different mechanical and electrical conditions, such as compressed/uncompressed operation [44–46], sourcing at low/high voltages [47,48], and finite element analysis by considering changes in particle dimension and spatial distribution [49,50]. The aforementioned studies agree on the fact that piezoresistivity mainly originates from two phenomena: quantum tunneling occurring among adjacent conductive particles separated by the insulating polymer, and constriction resistance occurring at clusters of multiple particles. Each phenomenon is subsequently described.

2.2.1. Quantum Tunneling as a Source of Piezoresistivity

This conduction mechanism can be explained from widely known equations by Simmons that describe the tunneling current between electrodes separated by a thin insulating film [51]. When operating at voltages near zero, a tunneling barrier of thickness (s), with an input applied voltage (U), exhibits a current density (J) equal to:

$$J(U,s) = \frac{3e^2\sqrt{2mV_a}}{2h^2s}U\exp\left(-\frac{4\pi s}{h}\sqrt{2mV_a}\right) \tag{1}$$

where (h) is the Planck constant, (V_a) is the height of the insulating potential barrier and (m), (e) are the electron mass and charge, respectively. However, if $U >> V_a/e$, the current density can be obtained from the following expression:

$$J(U,s) = \frac{2.2e^3U^2}{8\pi hV_as^2}\exp\left(-\frac{8\pi s}{2.96heU}\sqrt{2mV_a{}^3}\right) \tag{2}$$

For the sake of this paper, the full set of equations for the intermediate voltage ranges are not presented since they are not required, but they can be found in the original study from Simmons [51]. Note that regardless of the applied voltage, U, current density changes in a negative exponential fashion with the interparticle separation, i.e., this observation also holds for the intermediate voltage equations not included in this study. Similarly, when operating at high input voltages, a change in U impacts current density (and also the sensor's sensitivity) in a nonlinear fashion. Under a compression regime, it is possible to relate s with the external applied force (F) as next:

$$s(F) = s_o(1 - F/(AM)) \quad (3)$$

where (s_o) is the uncompressed interparticle separation, (A) is the sensor area, and (M) is the compressive modulus of the nanocomposite. It is possible to substitute (3) into (1) and (2) to obtain unified equations that relate current density with the external applied force.

Figure 1 shows the tunneling phenomenon occurring in multiple spots along the nanocomposite. In practice, the net tunneling resistance (R_{tun}) is originated from the multiple parallel and series connections that occur in the 3D polymer matrix. Unfortunately, an explicit expression for R_{tun} can only be found for voltages near zero using (1), but when $U >> V_a/e$, an explicit relationship for U/J cannot be found since U appears as part of the argument in the exponential function, see (2).

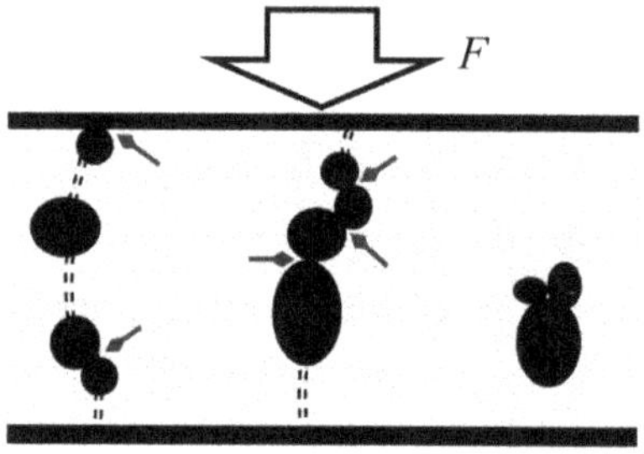

Figure 1. Sketch of a nanocomposite comprising randomly spaced conductive particles in an insulating polymer matrix; the material is sandwiched between metallic electrodes and subjected to an external compressive force (F). Quantum tunneling conduction is shown as double dashed line paths; they connect isolated particles, thus creating the tunneling resistance. Greyed diamond arrows mark the constriction resistance occurring between adjacent particles (as well as between electrode and particles in contact); they create the constriction resistance (R_c). Particles located too far from each other fail to create a conduction path.

For most CPCs involving conductive nanoparticles, the height of the potential barrier is, at most, 0.57 eV for a Sn–Pb/PS nanocomposite [52], where PS stands for polystyrene. Later, in Section 4, we demonstrate that the optimal operating range for the Interlink sensors is accessible for voltages around 3 V. By taking the largest case of V_a = 0.57 eV, we can straightforwardly discard (1) as a valid model since it only holds for voltages near 0 V. On the other hand, Equations (2) and (3) are better suited to modeling the piezoresistive response of Interlink sensors when predominantly operating under quantum tunneling regime; this occurs because (2) holds when $U >> Va/e$, which is the working case since U is around 3 V and V_a = 0.57 eV.

Finally, it must be clarified that current density can be converted to current by considering the effective tunneling area. Nonetheless, the effective tunneling area is not the same as the sensor area, A, because electrons flow only through some regions of the polymer with high particle concentration; see Figure 1. A comprehensive discussion of Simmons's equations for modeling piezoresistive sensors was performed by the authors in a previous study [25]. Such study experimentally determined the effective tunneling area, as well as the whole set of parameters considered in (1)–(3).

2.2.2. Constriction Resistance as a Source of Piezoresistivity

The constriction resistance (R_c) originates at two different spots: first, at the incomplete percolation paths located along the polymer matrix; these are the particle-particle interactions, and second, at the sensor boundary where electrode-particle interactions occur; both cases are shown in Figure 1. For particles with diameters ranging from a hundred nanometers to tens of micrometers, the contact size is comparable with the mean free path of electrons, thus causing a restriction to free electron motion [53]. According to the model developed by Mikrajuddin et al. [54], when operating under compression regime, the constriction resistance changes in an inversely proportional fashion with the applied force:

$$R_c(F) \propto R_o/F \tag{4}$$

The constant (R_o) depends on the particles physical dimensions, Poisson ratio of the material, and elasticity modulus. For simplicity purposes, the exact expression is not presented here as it comprises a piecewise function for the elastic and inelastic interaction occurring at the interface; such an explanation falls out the scope of this article. However, we must emphasize that the constriction resistance is a *voltage-independent* phenomenon.

By recalling Figure 1, we note that R_c originates at multiple spots along the nanocomposite, therefore, we can only measure the net contribution of the multiple series-connected and parallel-connected constriction resistances, i.e., R_c forms an intricate network of resistances. The net contribution of the multiple constriction resistances, R_c, is henceforth designated as the contact resistance (R_{con}). We will no longer use R_c in this manuscript.

2.2.3. Combining Tunneling and Contact Resistances

Sensor resistance (R_{FSR}) is calculated by summing the contribution of the tunneling and the contact resistance as follows:

$$R_{FSR} = R_{tun} + R_{con} \tag{5}$$

Equation (5) was initially proposed by Kalantari et al. [45], and was later embraced by the authors [25].

As demonstrated in the next section, the random dispersion of conductive nanoparticles creates a specific sensor response in which either R_{tun} or R_{con} dominates for a given specimen. If particles are grouped in clusters separated by the insulating polymer, then R_{tun} dominates, but if particles are in direct contact (forming percolation paths), then R_{con} dominates. A discussion regarding the influence of U over R_{tun} and R_{con} can be found in previous works [25].

3. Materials and Methods

In previous work [55], the experimental setup was thoroughly described; therefore, only a brief description is presented here.

3.1. Mechanical Setup

The mechanical setup comprised a tailored test bench capable of handling up to 16 Interlink FSR402 sensors simultaneously; this let us speed up the characterization process by avoiding single-sensor measurement [55]. Forces were applied from a linear motor and a spring. An LCHD-5 load cell was deployed to close the force loop [30].

3.2. Electrical Setup

An amplifier in inverting configuration was used as the interface circuit; see Figure 2. This setup was preferably chosen over a voltage divider because the amplifier let us control the voltage across the sensor at any time. Analog multiplexers were deployed to enable time-multiplexed readings of the 16 FSRs. From the amplifier output voltage (U_o), we could determine sensor resistance, R_{FSR}, given the amplifier model:

$$U_o = -(R_f/R_{FSR}) \cdot U \tag{6}$$

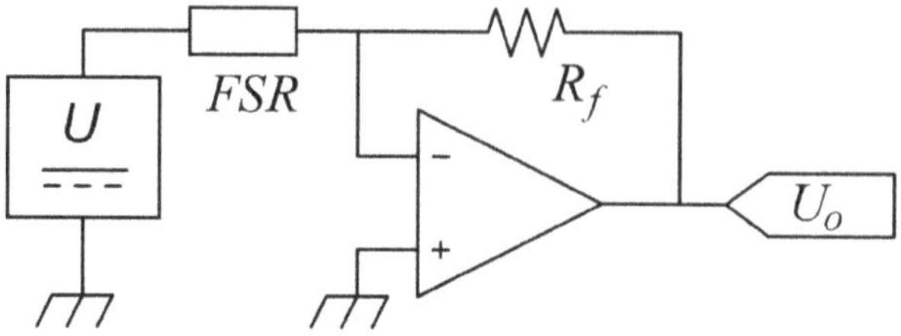

Figure 2. Electrical setup for driving the FSRs. The input voltage (U) was implemented from a digital-to-analog converter (DAC) to enable sensor characterization at multiple voltages.

From (6), we can either use R_{FSR} or U_o as the variable to measure. However, U_o was preferably chosen over R_{FSR}, because the former exhibited a linear relationship with the applied force, F. The feedback resistor (R_f) was set to 560 Ω for all the experimental tests in this study. The acquisition system comprised a 16-bit analog-to-digital converter (ADC) model NI-9205 and a 16-bit digital-to-analog converter (DAC) model NI-9263; the former was required for measuring U_o, whereas the latter was employed to generate multiple U values. The system controller was the CRIO-9035 running LabVIEW Real Time.

3.3. Methods

Each step of the DMAIC cycle is briefly described in Table 1. A thorough description of the process is presented in the rest of the manuscript.

Table 1. Summary of the DMAIC cycle with a brief description of each step.

Step of the Cycle	Description
Define	Sensitivity (m) of 64 specimens of commercial FSRs, model Interlink FSR402 [27]. A total of 48 sensors were considered for the DMAI stages and 16 for the C stage.
Measure	Sensitivity was measured in force steps of 1 N, starting at 0 N up to 20 N. A total of 19 input voltages (U) were considered: 0.25 V, 0.5 V, 0.75 V, and 1 V. Above 1 V, voltage increments of 0.5 V were applied up to 8.5 V.
Analyze	Evaluation of the experimental data in perspective of the underlying physics of FSRs. Four claims were stated to ease the analysis and to derive conclusions.
Improve	The improve step comprised two stages: finding the optimal input voltage that minimizes dispersion in sensitivity, proposing and test two different methods to reduce the dispersion in sensitivity.
Control	Validate the two methods developed in the improve stage using 16 sensors.

Given the DMAIC cycle, we *defined* sensor sensitivity (m) as the variable to study with units of volts per Newton. The sensitivity *measurements* were performed at multiple input voltages as described in Table 1. From the 64 sensors at 19 different voltages, a total of 1216 sensitivities were collected. No lot control was considered for the 64 sensors; this is important to note because the manufacturer has stated different part-to-part repeatability for single-lot sensors [27]. Given the setup of Figure 2, sensitivity was obtained from a least-squares fitting process with general formula:

$$U_o = mF + b \tag{7}$$

where (b) was the y-intercept resulting from the fit. It must be clarified that b was not considered in this study; instead, we measured the output voltage at 0 N (U_o^{0N}). The reasoning behind this decision is provided in the next step. The *analysis* stage comprised an extensive evaluation of the experimental data in perspective of the theoretical foundations from Section 2. Before discussing the experimental data, we had to develop the following theoretical *claims*:

(i) Regardless of U, larger compressive forces increase the current density as the interparticle separation is reduced and the contact resistance is lowered; see (1)–(5). (ii) For null applied force, incremental U yields larger current density, thus U_o^{0N} is increased as well; see (2). However, we must recall that the constriction resistance is a voltage-independent phenomenon, thus changing U does not modify R_{con}; see (4). (iii) By taking the derivative of $J(U,s(F))$ with respect to F, we note that for the same force profile, the derivative increases for larger U. The rate of change between $J(U,s(F))$ and F determines sensor sensitivity as it relates the change in sensor current with the input force; this statement only applies for the tunneling resistance. Claim (iii) can be summarized as follows: larger U yields greater sensitivity, whereas reducing U diminishes sensor sensitivity.

(iv) It is clear from Figure 1 that both piezoresistive phenomena occur simultaneously under any applied stress. Nonetheless, the experimental data support the hypothesis that one phenomenon usually dominates over the other for a given sensor. For example: when the contact resistance dominates, the percolation paths are the main source of piezoresistivity; whereas quantum tunneling is less important. Under this scenario, U_o^{0N} is large but m is small. To understand this, we must compare (2) and (3) with (4) as follows: given the predominant percolation paths along the FSR, a large U_o^{0N} is naturally expected because these paths tend to easily transport electrons from one electrode to the other, however, when an external force is applied, the change in resistance is small as predicted by (4), i.e., the contact resistance changes with the inverse of the applied force. In contrast, when quantum tunneling dominates, the percolation paths are incomplete or nonexistent; in this case, (2) and (3) instead play a major role. Under this scenario, a small increment in s—caused by an external force—results in a dramatic change in current density due to the exponential dependence in (2) and (3); this can be understood as a large sensitivity. Nonetheless, when quantum tunneling dominates and the sensor is unloaded, U_o^{0N} is small because there are not percolation paths that connect both electrodes.

Experimental results supporting previous statements are next presented. Thereafter, *improve* and *control* stages are described.

4. Results

Figure 3 shows the experimental data for two specimens at different input voltages. The claims (i)–(iv) from Section 3 can be validated with the results reported in this figure, as follows: (i) larger forces increase U_o; (ii) for a given sensor, increasing U, also increases U_o^{0N}, and similarly, (iii) larger U increases m as well. Finally, (iv) if U_o^{0N} is large, then percolation dominates and m is usually small. The opposite case can be also observed in Figure 3; if U_o^{0N} is small, then quantum tunneling dominates, and m is generally large.

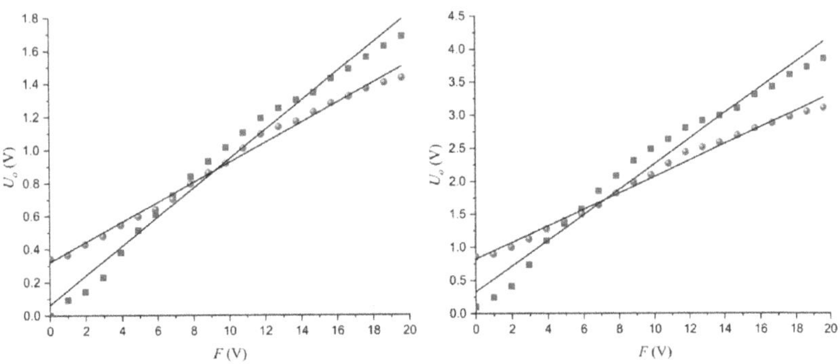

Figure 3. Plots of the output voltage (U_o) as a function of the applied force (F) for two different specimens at different input voltages (U). The fit to (7) has been overlaid in both cases. The sensor where percolation dominates is shown with red spherical markers; the sensor where quantum tunneling dominates is shown with blue square markers. Data taken at (**left**) U = 3 V and (**right**) U = 6 V.

From the experimental data resulting at each input voltage, we obtained sensor sensitivity for each specimen; then, we calculated the mean (μ) and standard deviation, σ, at each U using a probability fit to a Gaussian distribution. Previously, we applied Anderson-Darling tests to verify the null hypothesis to different probability distributions; the Gaussian distribution was the most likely distribution for each dataset.

As a part of the *improve* stage, the ratio of μ/σ as a function of U is shown in Figure 4; this chart lets us assess which input voltage naturally yields the highest part-to-part repeatability. The quotient μ/σ is known in literature as the inverse of the coefficient of variation. Note that at U = 3 V, the dispersion in sensitivity is the lowest. Hence, we can state that when operating Interlink sensors at 3 V, part-to-part repeatability is naturally maximized. We focus on this start point to further reduce σ. As previously mentioned, this study comprised the application of compensation techniques at the last stage of final application deployment, therefore, we could only trim U to target a desired sensitivity for a given sensor; see claim (iii) from the *Methods* section. In practice, it was possible to change U by means of a DAC; see Figure 2.

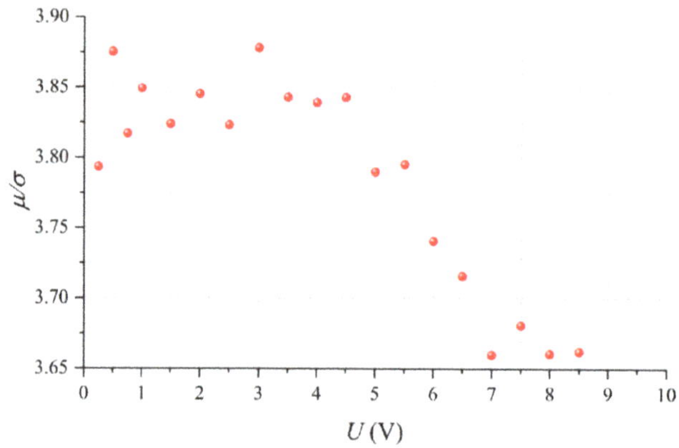

Figure 4. Inverse of the coefficient of variation (μ/σ) estimated at different input voltages (U) for the 48 sensors under study. Probability fitting to a Gaussian distribution was performed at each U to estimate μ and σ.

4.1. *Sensor Classification on the Basis of the Output Voltage at Null Force*

From the experimental data taken at U = 3 V, we plotted in Figure 5 each sensor sensitivity, m, with its corresponding output voltage at null force, U_o^{ON}. Later, the experimental data were fitted to a Gaussian probability distribution resulting in a mean value of μ_{3V} = 0.0735 V/N with standard deviation σ_{3V} = 0.019 V/N.

As predicted by claim (iv), we observed that sensors exhibiting a large U_o^{ON} tend to show a low m, and that similarly, sensors with a small U_o^{ON} tended to show a large m. In other words, U_o^{ON} is approximately related to sensor sensitivity. To the best of the authors' knowledge, this observation has not been published elsewhere in specialized literature, but it represents a powerful tool, as described in the next paragraph. For compensation purposes in the *improve* stage, we have segmented the dataset from Figure 5 into three regions according to U_o^{ON}: (A) sensors where quantum tunneling dominates that meet the criteria $U_o^{ON} < U_{th}^{low}$, (B) sensors where quantum tunneling and contact resistance have similar weight, and (C), sensors where contact resistance dominates that meet the criteria $U_o^{ON} > U_{th}^{high}$. The procedure for setting the threshold voltages (U_{th}^{low}, U_{th}^{high}) is described in the next section.

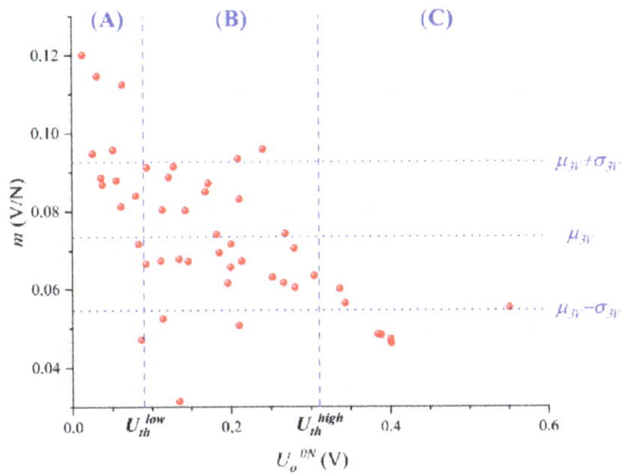

Figure 5. Sensitivity (m) as a function of the output voltage at null force ($U_o{}^{0N}$) for each specimen. The sensitivity data were fit to a Gaussian probability distribution resulting in the mean, μ_{3V} = 0.0735 V/N, and standard deviation, σ_{3V} = 0.019 V/N. Low threshold voltage ($U_{th}{}^{low}$) and high threshold voltage ($U_{th}{}^{high}$) were defined for classification purposes. Data were taken at U = 3 V for 48 sensors.

The Importance of $U_o{}^{0N}$ for assessing sensor's sensitivity is twofold. First, $U_o{}^{0N}$ can be measured at negligible cost after product manufacturing, i.e., a simple driving circuit is required with no mechanical setup, see Figure 2. Second, we can discard non-compliant sensors based on $U_o{}^{0N}$ measurements; this ultimately enhances part-to-part repeatability as described next: by keeping the sensors that meet the criteria $U_{th}{}^{high} < U_o{}^{0N} < U_{th}{}^{low}$, we retain the sensors from Region (B) and discard the sensors from Regions (A) and (C). In practice, this implies that 19 out of the 48 sensors are discarded and 29 are kept; it must be emphasized that the sensors from Region (B) represent our desired target, for this reason, we discard the noncompliant sensors belonging to Regions (A) and (C).

Subsequently, we perform a new fit to a Gaussian distribution from the 29 selected sensors that results in the mean μ_B = 0.0718 V/N with standard deviation σ_B = 0.0142 V/N. Note that μ_B slightly changed from the previously reported value of μ_{3V} = 0.0735 V/N, but that σ_B decreased 25% from its original value of σ_{3V} = 0.019V/N. From a Six Sigma standpoint, this represents a 1.5σ improvement. Finally, the quotient μ_B/σ_B can be recalculated with the 29 selected sensors as 5.05, which is greater than the value of 3.87 reported in Figure 4. This simple procedure is useful for enhancing part-to-part repeatability, but unfortunately, it occurs at the cost of discarding sensors, which unavoidably represents material waste. Therefore, in the next section, we introduce an alternative compensation technique for the sensors in Regions (A) and (C).

Before proceeding with the compensation technique, we must recall the random distribution of nanoparticles along the polymer matrix. This fact unavoidably causes some specimens from Region (B) to show either a sensitivity larger than $\mu_{3V} + \sigma_{3V}$ or lower than $\mu_{3V} - \sigma_{3V}$; the physical reasons for this behavior are manifold, but we can point out some of them: a given specimen may have a particle count below average; thus, sensitivity is lowered. Another possible reason is the random spacing of clusters that creates isolated conglomerates in the polymer matrix; these separated clusters are just too far from each other to enable quantum tunneling. See Figure 1 for a schematic representation of this.

4.2. Compensation Technique to Enhance Part-to-Part Repeatability

Following the *improve* stage; new driving voltages must be found for the sensors belonging to Regions (A) and (C), see Figure 5; this procedure is supported by claim (iii). By doing this, we can closely match each specimen's sensitivity with μ_{3V}, i.e., the target sensitivity.

For the sensors belonging to Region (A), we must find a new U that lowers the sensitivity of each specimen. In this case, the new U must be lower than 3 V. For the sensors belonging to Region (C), the opposite occurs; the new input voltage must be higher than 3 V to increase their sensitivity. As stated in claim (ii), a change in U also modifies $U_o{}^{0N}$; this is not a problem itself because $U_o{}^{0N}$ can be measured anytime in the final application.

In order to find a relationship among m, $U_o{}^{0N}$, and U, we plotted them together in Figure 6 for the sensors belonging to Region (A). Then, a least-squares fit was applied to find a numerical expression that relates these three variables. The best, yet simplest, function was found to be:

$$U = a \cdot \hat{U}_o{}^{0N} + b \cdot \hat{m} + c \tag{8}$$

where a, b, c are coefficients resulting from the fit. Normalization was performed for the variables $U_o{}^{0N}$ and m, where $\hat{U}_o{}^{0N} = U_o{}^{0N}/U$ and $\hat{m} = m/U$. By doing this, we avoided quadratic functions and used a simple 3D plane resulted instead. The same procedure was repeated for the sensors belonging to Region (C). Both fitting results are summarized in Table 2. The fit to (8) allowed us to determine the optimal threshold voltages for regions (A), (B), and (C). The threshold voltages, $U_{th}{}^{low}$ and $U_{th}{}^{high}$, were previously defined in Figure 5; they let us assess whether quantum tunneling or constriction resistance dominates for a given sensor. The procedure for determining $U_{th}{}^{low}$ and $U_{th}{}^{high}$ is described next.

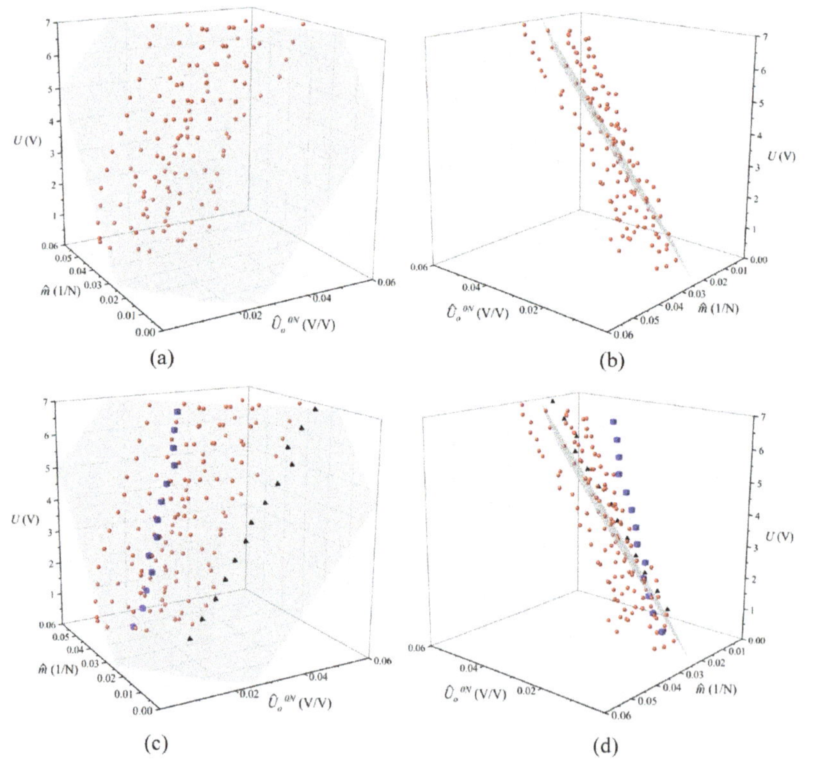

Figure 6. 3D plot of the normalized sensitivity ($\hat{m}$), the normalized output voltage at null force ($\hat{U}_o{}^{0N}$), and the input voltage (U). Data corresponding to the 12 sensors belonging to Region (A). (**a**,**b**) Isometric views of the 3D fit using Equation (8) as the model with parameters shown in Table 2. (**c**,**d**) Isometric views of the 3D fit with special markers. The blue cubes and black pyramid markers correspond to individual sensors that shift in different directions along the plane when U is changed.

Table 2. Parameters resulting from the fit. Parameters *a* through *c* were obtained from a least-squares fit to (8) with coefficient of correlation R^2. The mean sensitivity (μ) for each region (μ_A, μ_C) was measured at U = 3 V.

	a (V)	*b* (N·V)	*c* (V)	R^2	μ (V/N)
Region (A)	132.3	139.1	−3.36	0.67	$\mu_A = 0.091$
Region (C)	101	485.2	−18.2	0.94	$\mu_C = 0.052$

The higher $U_{th}{}^{low}$ is set, the more sensors are considered as a part of Region (A). However, by doing this, the coefficient of determination (R^2) resulting from the fit is reduced; this occurs because we are embracing sensors that fall out the criteria for this region, i.e., in Region (A) quantum tunneling dominates. The same procedure was repeated to obtain $U_{th}{}^{high}$, but in this case, $U_{th}{}^{high}$ has to be reduced in order to embrace more sensors in Region (C). In practice, $U_{th}{}^{low}$ and $U_{th}{}^{high}$ were found by an iterative process that aimed to obtain the largest $U_{th}{}^{low}$ that minimized R^2 in Region (A), and the smallest $U_{th}{}^{high}$ that minimized R^2 in Region (C). Finally, $U_{th}{}^{low}$ and $U_{th}{}^{high}$ resulted in 90 mV and 310 mV, respectively.

Compensating voltages for the sensors in Region (A) and (C) can also be found from (8). Nonetheless, we cannot directly replace $U_o{}^{ON}$ in (8) and set a targeted sensitivity, m, to obtain the new U. As previously mentioned, a change in U modifies both: $U_o{}^{ON}$ and m; see claims (ii), (iii). The 3D surface resulting from (8) can be understood as the plane where $U_o{}^{ON}$ and m are shifted when U is changed. Nonetheless, every sensor exhibits a different rate of change in $U_o{}^{ON}$ as a function of U; this is exemplified with the pyramid and cubic markers in Figure 6. Each marker corresponds to individual sensor data that can be well fitted in the 3D plane, but they move in different angles along the surface when U is changed. Therefore, we are limited to obtain an average rate of change for $U_o{}^{ON}$ as a function of U; this average comprises the 12 sensors belonging to Region (A). In the Conclusion section, the scenario is discussed when we consider an individual rate of change for each sensor.

Finally, given the mean sensitivity at 3 V for the sensors in Region (A), μ_A, the model from (8), the target sensitivity, μ_{3V}, and the average rate of change of $U_o{}^{ON}$ as a function of U, we computed the new U resulting in U_A = 2.5 V. Similarly, the same procedure was performed for the sensors in Region (C), but using μ_C and the average rate of change of $U_o{}^{ON}$ for these sensors; this resulted in the new voltage of U_C = 3.15 V.

Later, experimental sensor characterization was performed at the new voltages U_A and U_C; this was done for the sensors belonging to Regions (A) and (C), respectively. The experimental results are plotted in Figure 7. Nonetheless, the sensor data from Region (B) is the same plotted in Figure 5 since U remained unchanged for these sensors. For the whole dataset for Figure 7, a new fit to a Gaussian distribution was performed, resulting in μ_{comp} = 0.0727 V/N and σ_{comp}= 0.016 V/N. This represents a 15% reduction in the standard deviation when compared with σ_{3V}. Although this reduction in the standard deviation is lower than previously reported for the discarding method; it is demonstrated that changing U is an effective way to fine-tune the sensitivity of FSRs.

4.3. *Assessing the Compensation Technique from a Six Sigma Perspective*

As a part of the *control* stage of the DMAIC cycle, we tested an additional group of 16 Interlink FSR 402 sensors. These sensors were initially characterized at U = 3 V to obtain $U_o{}^{ON}$ and m. Figure 8 shows the summary of this characterization with an overlay of $U_{th}{}^{low}$, $U_{th}{}^{high}$, μ_{3V}, and σ_{3V} for classification purposes. An Anderson-Darling test to different probability distributions demonstrated that these data could be well fitted to a Gaussian distribution with $\mu_{control}$ = 0.0774 V/N and $\sigma_{control}$ = 0.0189 V/N. Note that the start point before compensation was quite similar to μ_{3V} and σ_{3V} from the *improve* stage.

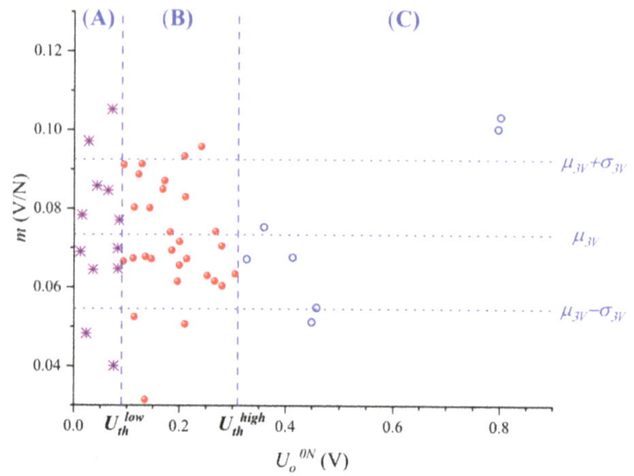

Figure 7. Sensitivity (m) as a function of the output voltage at null force ($U_o{}^{0N}$) for each specimen after compensation. Compensation was performed for the sensors belonging to Regions (A) and (C) using different input voltages (U_A, U_C). For comparison purposes, the y-axis scale was held unchanged from Figure 5. Different markers were used for the sensors of Regions (A) and (C) after compensation.

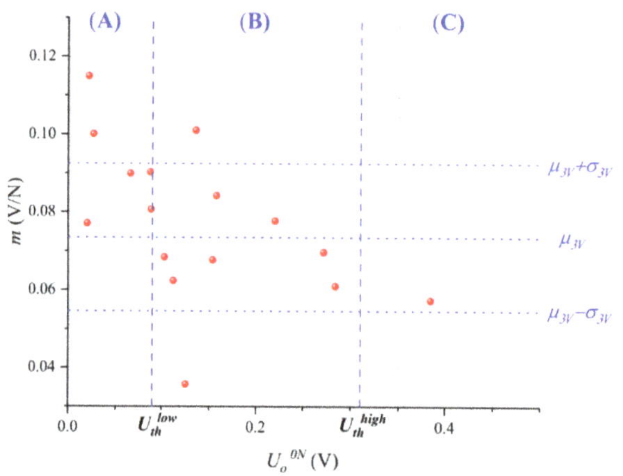

Figure 8. Sensitivity (m) as a function of the output voltage at null force ($U_o{}^{0N}$) for the sixteen specimens of the *control* stage. Data were taken at $U = 3$ V.

Later, we classified the sensors belonging to Regions (A), (B), and (C) on the basis of $U_{th}{}^{low}$ and $U_{th}{}^{high}$. Part-to-part repeatability was enhanced following two different approaches: first, by discarding the sensors from Regions (A) and (C) and keeping only those from Region (B), and second, by applying the compensation technique based on changing U.

From the first approach, 7 out the 16 sensors were discarded and 9 were kept. Then, we recalculated the mean and standard deviation, obtaining 0.0698 V/N and 0.0168 V/N, respectively; this represents an 11% reduction in the standard deviation and a Six Sigma improvement of 0.66σ. However, note that the mean sensitivity was altered from its original value.

From the second approach, new characterization was performed at the input voltages U_A and U_C for the sensors belonging to Regions (A) and (C), respectively. No compensation was performed for the sensors belonging to Region (B). The experimental results are shown in Figure 9. Later, we recalculated the mean and standard deviation, obtaining 0.0726 V/N and 0.0154 V/N, respectively; this represents an 18.5% reduction in the standard deviation amid a negligible variation for the mean sensitivity. This represents a 1.1σ improvement using the Six Sigma approach.

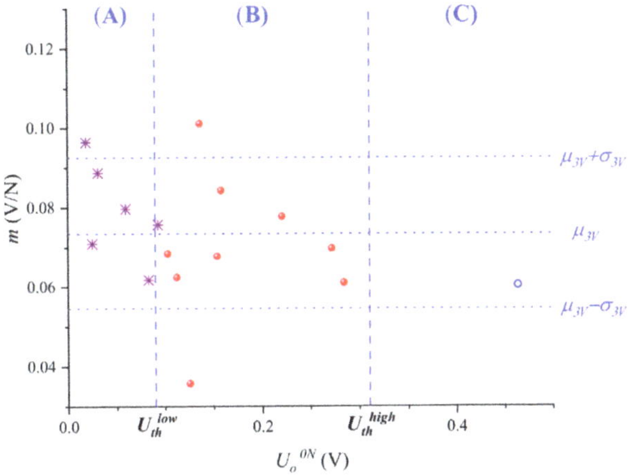

Figure 9. Sensitivity (m) as a function of the output voltage at null force (U_o^{0N}) for each specimen after compensation. Experimental data resulting from the *control* stage. Different markers were used for the sensors of Regions (A) and (C) after compensation.

4.4. Practical Considerations for the Implementation of the Proposed Methods

First of all, it must be emphasized that this research did not consider lot control for the specimens characterized. Therefore, if the manufacturing conditions remain unchanged over time (e.g., particle dimension, type of polymer, preparation conditions, and so on), the compensation techniques should work for any specimen of Interlink FSR 402. In regard to the feedback resistor, a change in R_f produces a linear variation in the magnitude of m, U_o^{0N}, U_{th}^{low} and U_{th}^{high}; this occurs because the amplifier is an inherently linear device. Therefore, if R_f is doubled or halved, the aforementioned magnitudes also change proportionally.

Nonetheless, a change in the driving voltage would impact m, U_o^{0N}, U_{th}^{low}, and U_{th}^{high} in a nonlinear fashion; this is expected because quantum tunneling is a nonlinear phenomenon. Besides, we must recall that during the *improve* stage, we found U = 3 V as the optimal voltage that naturally maximizes the quotient μ/σ; see Figure 4. Therefore, we set U = 3 V as the starting point to subsequently trim the input voltage in the *improve* and *control* stages. However, if we set a different starting point for U, we would deviate from the results reported here.

Finally, given the relatively low number of sensors considered in this research (64 specimens), only three regions were defined. By considering more regions, a more suitable compensating voltage could be applied for each specimen. The experimental data employed in this study can be found in [56].

5. Conclusions

Changing the input voltage is an effective way to trim the sensitivity of force-aensing resistors (FSRs). The output voltage at null force (U_o^{0N}) can give us a hint about the individual sensor's sensitivity without requiring individual characterization. To the best of authors' knowledge, this observation has not been reported elsewhere. With the information provided from U_o^{0N}, we could set new input voltages that let us reduce the dispersion in sensors' sensitivity and/or discard noncompliant sensors. Both methods were explored in this study, resulting in a reduction of 18.5% and 11% in the standard deviation of sensor sensitivity for each procedure, respectively. These results were obtained in the *control* stage of the DMAIC cycle for a bunch of 16 Interlink FSR 402 sensors. These percentages represent 1.1σ and 0.66σ improvements according to the Six Sigma methodology.

A promising technique to be explored in the authors' future work is based on individually defining new input voltages for each sensor; by doing this, we can more accurately match the sensitivity of each specimen with a target sensitivity, thus achieving a larger Six Sigma improvement. However, this procedure requires that we previously track U_0^{0N} for multiple input voltages and later incorporate such information in the final compensation formula.

Author Contributions: Conceptualization, L.P.-M. and A.O.G.-P.; hardware set-up, A.P.; figure design and data analysis, V.M.F.; validation, C.P. All authors have read and agreed to the published version of the manuscript.

Funding: This work was funded by Universidad Católica de Colombia through the Internal Call for Projects 2021, grant No CON0000456 and by Universidad Antonio Nariño through grant No PI/UAN-2021-701GIBIO.

Data Availability Statement: Experimental data can be found here [56].

Conflicts of Interest: The authors declare no conflict of interest.

Appendix A. Foundations of the Six Sigma Methodology

Six Sigma is a continuous improvement methodology that has been widely applied in the manufacturing and services sectors. Its implementation focuses on the application of the DMAIC cycle (Define, Measure, Analyze, Improve, and Control) [41]. This methodology is based on the reduction of the process variability regarding a Critical to Quality (CTQ) or critical variable for the process or the client. Process variability is assessed on the basis of the standard deviation (σ) for any process parameter, e.g., when building antennas on a printed circuit board (PCB), the width and length of the paths have a strong influence on the antennas' bandwidth and frequency response [57]; therefore, we can deploy the DMAIC cycle to reduce the dispersion of the width and length of paths from their nominal value.

Six Sigma requires the collection of data and the application of analytical and statistical tools to identify the causes of the variation and thus achieve a 6σ level that translates into 3.4 errors per million opportunities [41]. A 6σ level implies that standard deviation is reduced by a factor of 6 from the original assessment. The opportunity is understood as the probability of noncompliance or the probability of not meeting the required specifications. The DMAIC cycle implemented by the Six Sigma methodology is used to identify and solve problems related to the process; analyzing the current state of a problem of interest, find its root cause, propose improvement solutions, and keep them durable over time.

References

1. Yeasmin, R.; Duy, L.T.; Han, S.; Seo, H. Intrinsically Stretchable and Self-Healing Electroconductive Composites Based on Supramolecular Organic Polymer Embedded with Copper Microparticles. *Adv. Electron. Mater.* **2020**, *6*, 2000527. [CrossRef]
2. Wang, X.; Liu, X.; Schubert, D.W. Highly Sensitive Ultrathin Flexible Thermoplastic Polyurethane/Carbon Black Fibrous Film Strain Sensor with Adjustable Scaffold Networks. *Nano-Micro Lett.* **2021**, *13*, 64. [CrossRef] [PubMed]
3. Yang, H.; Yuan, L.; Yao, X.; Zheng, Z.; Fang, D. Monotonic Strain Sensing Behavior of Self-Assembled Carbon Nanotubes/Graphene Silicone Rubber Composites under Cyclic Loading. *Compos. Sci. Technol.* **2020**, *200*, 108474. [CrossRef]
4. Zhao, Y.; Ren, M.; Shang, Y.; Li, J.; Wang, S.; Zhai, W.; Zheng, G.; Dai, K.; Liu, C.; Shen, C. Ultra-Sensitive and Durable Strain Sensor with Sandwich Structure and Excellent Anti-Interference Ability for Wearable Electronic Skins. *Compos. Sci. Technol.* **2020**, *200*, 108448. [CrossRef]
5. Xiang, D.; Zhang, X.; Han, Z.; Zhang, Z.; Zhou, Z.; Harkin-Jones, E.; Zhang, J.; Luo, X.; Wang, P.; Zhao, C.; et al. 3D Printed High-Performance Flexible Strain Sensors Based on Carbon Nanotube and Graphene Nanoplatelet Filled Polymer Composites. *J. Mater. Sci.* **2020**, *55*, 15769–15786. [CrossRef]
6. Fekiri, C.; Kim, H.C.; Lee, I.H. 3d-Printable Carbon Nanotubes-Based Composite for Flexible Piezoresistive Sensors. *Materials* **2020**, *13*, 5482. [CrossRef]
7. Aikawa, S.; Zhao, Y.; Yan, J. Development of High-Sensitivity Electrically Conductive Composite Elements by Press Molding of Polymer and Carbon Nanofibers. *Micromachines* **2022**, *13*, 170. [CrossRef]

8. Li, F.-C.; Kong, Z.; Wu, J.-H.; Ji, X.-Y.; Liang, J.-J. Advances in Flexible Piezoresistive Pressure Sensor. *Wuli Xuebao Acta Phys. Sin.* **2021**, *70*, 100703. [CrossRef]
9. Kanoun, O.; Bouhamed, A.; Ramalingame, R.; Bautista-Quijano, J.R.; Rajendran, D.; Al-Hamry, A. Review on Conductive Polymer/CNTs Nanocomposites Based Flexible and Stretchable Strain and Pressure Sensors. *Sensors* **2021**, *21*, 341. [CrossRef]
10. Kim, K.; Shin, S.; Kong, K. An Air-Filled Pad With Elastomeric Pillar Array Designed for a Force-Sensing Insole. *IEEE Sens. J.* **2018**, *18*, 3968–3976. [CrossRef]
11. Chen, D.; Cai, Y.; Huang, M.-C. Customizable Pressure Sensor Array: Design and Evaluation. *IEEE Sens. J.* **2018**, *18*, 6337–6344. [CrossRef]
12. Aoyagi, S.; Suzuki, M.; Morita, T.; Takahashi, T.; Takise, H. Bellows Suction Cup Equipped With Force Sensing Ability by Direct Coating Thin-Film Resistor for Vacuum Type Robotic Hand. *IEEEASME Trans. Mechatron.* **2020**, *25*, 2501–2512. [CrossRef]
13. Liang, J.; Wu, J.; Huang, H.; Xu, W.; Li, B.; Xi, F. Soft Sensitive Skin for Safety Control of a Nursing Robot Using Proximity and Tactile Sensors. *IEEE Sens. J.* **2020**, *20*, 3822–3830. [CrossRef]
14. Zhang, H.; Ren, P.; Yang, F.; Chen, J.; Wang, C.; Zhou, Y.; Fu, J. Biomimetic Epidermal Sensors Assembled from Polydopamine-Modified Reduced Graphene Oxide/Polyvinyl Alcohol Hydrogels for the Real-Time Monitoring of Human Motions. *J. Mater. Chem. B* **2020**, *8*, 10549–10558. [CrossRef]
15. Sun, X.; Liu, T.; Zhou, J.; Yao, L.; Liang, S.; Zhao, M.; Liu, C.; Xue, N. Recent Applications of Different Microstructure Designs in High Performance Tactile Sensors: A Review. *IEEE Sens. J.* **2021**, *21*, 10291–10303. [CrossRef]
16. Boada, M.; Lazaro, A.; Villarino, R.; Gil, E.; Girbau, D. Battery-Less NFC Bicycle Tire Pressure Sensor Based on a Force-Sensing Resistor. *IEEE Access* **2021**, *9*, 103975–103987. [CrossRef]
17. Pang, G.; Deng, J.; Wang, F.; Zhang, J.; Pang, Z.; Yang, G. Development of Flexible Robot Skin for Safe and Natural Humanâ—Robot Collaboration. *Micromachines* **2018**, *9*, 576. [CrossRef]
18. Cavallo, A.; Beccatelli, M.; Favero, A.; Al Kayal, T.; Seletti, D.; Losi, P.; Soldani, G.; Coppedè, N. A Biocompatible Pressure Sensor Based on a 3D-Printed Scaffold Functionalized with PEDOT:PSS for Biomedical Applications. *Org. Electron.* **2021**, *96*, 106204. [CrossRef]
19. Fernandez, F.D.M.; Khadka, R.; Yim, J.-H. Highly Porous, Soft, and Flexible Vapor-Phase Polymerized Polypyrrole-Styrene-Ethylene-Butylene-Styrene Hybrid Scaffold as Ammonia and Strain Sensor. *RSC Adv.* **2020**, *10*, 22533–22541. [CrossRef]
20. Lo, L.-W.; Zhao, J.; Wan, H.; Wang, Y.; Chakrabartty, S.; Wang, C. An Inkjet-Printed PEDOT:PSS-Based Stretchable Conductor for Wearable Health Monitoring Device Applications. *ACS Appl. Mater. Interfaces* **2021**, *13*, 21693–21702. [CrossRef]
21. Ou, L.; Song, B.; Liang, H.; Liu, J.; Feng, X.; Deng, B.; Sun, T.; Shao, L. Toxicity of Graphene-Family Nanoparticles: A General Review of the Origins and Mechanisms. *Part. Fibre Toxicol.* **2016**, *13*, 57. [CrossRef] [PubMed]
22. Zhang, K.; Song, C.; Wang, Z.; Gao, C.; Wu, Y.; Liu, Y. A Stretchable and Self-Healable Organosilicon Conductive Nanocomposite for a Reliable and Sensitive Strain Sensor. *J. Mater. Chem. C* **2020**, *8*, 17277–17288. [CrossRef]
23. Hussain, I.; Ma, X.; Luo, Y.; Luo, Z. Fabrication and Characterization of Glycogen-Based Elastic, Self-Healable, and Conductive Hydrogels as a Wearable Strain-Sensor for Flexible e-Skin. *Polymer* **2020**, *210*, 122961. [CrossRef]
24. Ma, Z.; Li, H.; Jing, X.; Liu, Y.; Mi, H.-Y. Recent Advancements in Self-Healing Composite Elastomers for Flexible Strain Sensors: Materials, Healing Systems, and Features. *Sens. Actuators Phys.* **2021**, *329*, 112800. [CrossRef]
25. Paredes-Madrid, L.; Palacio, C.A.; Matute, A.; Parra Vargas, C.A. Underlying Physics of Conductive Polymer Composites and Force Sensing Resistors (FSRs) under Static Loading Conditions. *Sensors* **2017**, *17*, 2108. [CrossRef]
26. Paredes-Madrid, L.; Fonseca, J.; Matute, A.; Gutiãrrez Velãsquez, E.I.; Palacio, C.A. Self-Compensated Driving Circuit for Reducing Drift and Hysteresis in Force Sensing Resistors. *Electronics* **2018**, *7*, 146. [CrossRef]
27. Interlink Electronics FSR400 Series Datasheet. 2017. Available online: https://f.hubspotusercontent20.net/hubfs/3899023/Integration%20Guides/FSR%20X%20%26%20UX%20Integration%20Guide%20-%20Interlink%20Electronics.pdf (accessed on 26 April 2022).
28. Peratech Inc. QTC SP200 Series Datasheet. Single Point Sensors. 2015. Available online: https://www.peratech.com/assets/uploads/datasheets/Peratech-QTC-DataSheet-SP200-Series-Nov15.pdf (accessed on 26 April 2022).
29. Tekscan Inc. FlexiForce, Standard Force & Load Sensors Model A201. Datasheet. 2017. Available online: https://www.tekscan.com/sites/default/files/resources/FLX-Datasheet-A201-RevI.pdf (accessed on 26 April 2022).
30. Omega Ultra Low Profile, Tension and Compression Load Cells. 2021. Available online: https://www.omega.com/en-us/force-strain-measurement/load-cells/lchd/p/LCHD-7-5K (accessed on 26 April 2022).
31. Urban, S.; Ludersdorfer, M.; van der Smagt, P. Sensor Calibration and Hysteresis Compensation with Heteroscedastic Gaussian Processes. *IEEE Sens. J.* **2015**, *15*, 6498–6506. [CrossRef]
32. Nguyen, X.A.; Chauhan, S. Characterization of Flexible and Stretchable Sensors Using Neural Networks. *Meas. Sci. Technol.* **2021**, *32*, 075004. [CrossRef]
33. Albright, T.B.; Hobeck, J.D. High-Fidelity Stochastic Modeling of Carbon Black-Based Conductive Polymer Composites for Strain and Fatigue Sensing. *J. Mater. Sci.* **2021**, *56*, 6861–6877. [CrossRef]
34. Castellanos-Ramos, J.; Navas-Gonzalez, R.; Macicior, H.; Sikora, T.; Ochoteco, E.; Vidal-Verdu, F. Tactile Sensors Based on Conductive Polymers. *Microsys. Technol.* **2010**, *16*, 765–776. [CrossRef]
35. Meyer, D.; Maehling, P.; Varghese, T.; Lewis, J. Calibration Considerations for Six Sigma^{TM} Accuracy and Precision in Combustion Pressure Measurement. *SAE Int. J. Commer. Veh.* **2017**, *10*, 508–517. [CrossRef]

36. Milačič, M.; Booden, V.; Grimes, J.; Maier, B. Hydrogen Leak Detection Method Derived Using DCOV Methodology. *SAE Tech. Pap.* **2009**, *9*, 97–102. [CrossRef]
37. Yusof, N.S.B.; Sapuan, S.M.; Sultan, M.T.H.; Jawaid, M. Materials Selection of "Green" Natural Fibers in Polymer Composite Automotive Crash Box Using DMAIC Approach in Six Sigma Method. *J. Eng. Fibers Fabr.* **2020**, *15*, 1558925020920773. [CrossRef]
38. Prabhakaran, R.T.D.; Babu, B.J.C.; Agrawal, V.P. Quality Modeling and Analysis of Polymer Composite Products. *Polym. Compos.* **2006**, *27*, 329–340. [CrossRef]
39. Arvidsson, M.; Gremyr, I.; Johansson, P. Use and Knowledge of Robust Design Methodology: A Survey of Swedish Industry. *J. Eng. Des.* **2003**, *14*, 129–143. [CrossRef]
40. Lucero, B.; Linsey, J.; Turner, C.J. Frameworks for Organising Design Performance Metrics. *J. Eng. Des.* **2016**, *27*, 175–204. [CrossRef]
41. Pyzdek, T. *The Six Sigma Handbook, Revised and Expanded*; McGraw Hill: New York, NY, USA, 2002.
42. Dauzon, E.; Lin, Y.; Faber, H.; Yengel, E.; Sallenave, X.; Plesse, C.; Goubard, F.; Amassian, A.; Anthopoulos, T.D. Stretchable and Transparent Conductive PEDOT:PSS-Based Electrodes for Organic Photovoltaics and Strain Sensors Applications. *Adv. Funct. Mater.* **2020**, *30*, 2001251. [CrossRef]
43. Cvek, M.; Kutalkova, E.; Moucka, R.; Urbanek, P.; Sedlacik, M. Lightweight, Transparent Piezoresistive Sensors Conceptualized as Anisotropic Magnetorheological Elastomers: A Durability Study. *Int. J. Mech. Sci.* **2020**, *183*, 105816. [CrossRef]
44. Ding, S.; Han, B.; Dong, X.; Yu, X.; Ni, Y.; Zheng, Q.; Ou, J. Pressure-Sensitive Behaviors, Mechanisms and Model of Field Assisted Quantum Tunneling Composites. *Polymer* **2017**, *113*, 105–118. [CrossRef]
45. Kalantari, M.; Dargahi, J.; Kovecses, J.; Mardasi, M.G.; Nouri, S. A New Approach for Modeling Piezoresistive Force Sensors Based on Semiconductive Polymer Composites. *IEEE ASME Trans. Mechatr.* **2012**, *17*, 572–581. [CrossRef]
46. Li, C.; Thostenson, E.T.; Chou, T.-W. Dominant Role of Tunneling Resistance in the Electrical Conductivity of Carbon Nanotube–Based Composites. *Appl. Phys. Lett.* **2007**, *91*, 223114. [CrossRef]
47. Awasthi, S.; Gopinathan, P.S.; Rajanikanth, A.; Bansal, C. Current–Voltage Characteristics of Electrochemically Synthesized Multi-Layer Graphene with Polyaniline. *J. Sci. Adv. Mater. Dev.* **2018**, *3*, 37–43. [CrossRef]
48. Oskouyi, A.B.; Uttandaraman, S.; Mertiny, P. Current-Voltage Characteristics of Nanoplatelet-Based Conductive Nanocomposites. *Nanoscale Res. Lett.* **2014**, *9*, 369. [CrossRef] [PubMed]
49. Panozzo, F.; Zappalorto, M.; Quaresimin, M. Analytical Model for the Prediction of the Piezoresistive Behavior of {CNT} Modified Polymers. *Compos. Part B Eng.* **2017**, *109*, 53–63. [CrossRef]
50. Clayton, M.F.; Bilodeau, R.A.; Bowden, A.E.; Fullwood, D.T. Nanoparticle Orientation Distribution Analysis and Design for Polymeric Piezoresistive Sensors. *Sens. Actuators Phys.* **2020**, *303*, 111851. [CrossRef]
51. Simmons, J. Electrical Tunnel Effect between Dissimilar Electrodes Separated by a Thin Insulating Film. *J. Appl. Phys.* **1963**, *34*, 2581–2590. [CrossRef]
52. Zhang, X.-W.; Pan, Y.; Zheng, Q.; Yi, X.-S. Time Dependence of Piezoresistance for the Conductor-Filled Polymer Composites. *J. Pol. Sci. Part B Pol. Phys.* **2000**, *38*, 2739–2749. [CrossRef]
53. Esquinazi, P.; Barzola-Quiquia, J.; Dusari, S.; García, N. Length Dependence of the Resistance in Graphite: Influence of Ballistic Transport. *J. Appl. Phys.* **2012**, *111*, 033709. [CrossRef]
54. Mikrajuddin, A.; Shi, F.G.; Kim, H.K.; Okuyama, K. Size-Dependent Electrical Constriction Resistance for Contacts of Arbitrary Size: From Sharvin to Holm Limits. *Mat. Sci. Semicon. Proc.* **1999**, *2*, 321–327. [CrossRef]
55. Paredes-Madrid, L.; Matute, A.; Bareño, J.O.; Parra Vargas, C.A.; Gutierrez Velásquez, E.I. Underlying Physics of Conductive Polymer Composites and Force Sensing Resistors (FSRs). A Study on Creep Response and Dynamic Loading. *Materials* **2017**, *10*, 1334. [CrossRef]
56. Paredes-Madrid, L.; Garzon Posada, A.; Fontalvo, V.; Peña Puerto, A.; Palacio Gómez, C.A. Force Sensing Resistor Data for Enhancing Part-to-Part Repeatability Using a Six Sigma Approach. *IEEE Dataport.* **2021**, *2021*, 16. [CrossRef]
57. Emara, M.K.; Tomura, T.; Hirokawa, J.; Gupta, S. All-Dielectric Fabry–Pérot-Based Compound Huygens' Structure for Millimeter-Wave Beamforming. *IEEE Trans. Antennas Propag.* **2021**, *69*, 273–285. [CrossRef]

micromachines

Article

High-Performance Flexible Piezoresistive Sensor Based on $Ti_3C_2T_x$ MXene with a Honeycomb-like Structure for Human Activity Monitoring

Yue Su †, Kainan Ma †, Fang Yuan, Jun Tang, Ming Liu * and Xu Zhang *

Institute of Semiconductors, Chinese Academy of Sciences, Beijing 100864, China; yuesu@semi.ac.cn (Y.S.); makainan@semi.ac.cn (K.M.); fyuan@semi.ac.cn (F.Y.); junt@semi.ac.cn (J.T.)
* Correspondence: liuming@semi.ac.cn (M.L.); zhangxu@semi.ac.cn (X.Z.)
† These authors contributed equally to this work.

Abstract: Wearable and flexible pressure sensors have sparked great interest due to their unique capacity to conformally attach to the surface of the skin and quantify human activities into recordable electric signals. As a result, more and more research efforts are being devoted to developing high-sensitivity and cost-effective flexible sensors for monitoring an individual's state of activity. Herein, a high-performance flexible piezoresistive sensor was designed and fabricated by combing 2D transition metal carbides, nitrides, and carbonitrides (MXene) with a honeycomb-like structure formed by femtosecond filamentating pulses. The sensing mechanism is attributed to the change of the connecting conductive paths between the top interdigital electrodes and the bottom microstructured films coated with MXene. The obtained sensing device demonstrates high sensitivity of 0.61 kPa^{-1}, relatively short response time, and excellent reliability and stability. Benefiting from the aforementioned extraordinary sensing performance, the sensor can be used with success to monitor tiny physiological signals, detect large deformations during human movement, and distinguish finger gestures, thus demonstrating its broad prospects in physiological analysis systems, health monitoring systems, and human–machine interaction.

Keywords: flexible piezoresistive sensor; MXene; honeycomb-like structure; femtosecond filamentating pulses

Citation: Su, Y.; Ma, K.; Yuan, F.; Tang, J.; Liu, M.; Zhang, X. High-Performance Flexible Piezoresistive Sensor Based on $Ti_3C_2T_x$ MXene with a Honeycomb-like Structure for Human Activity Monitoring. *Micromachines* **2022**, *13*, 821. https://doi.org/10.3390/mi13060821

Academic Editor: Nam-Trung Nguyen

Received: 5 May 2022
Accepted: 23 May 2022
Published: 25 May 2022

1. Introduction

Flexible pressure sensors with excellent force-to-electric ability have aroused a huge surge of research interest due to their enormous potential in diverse applications, such as personal healthcare monitoring, artificial intelligence interaction, humanoid robotics, etc. [1–5]. Over the years, great efforts have been devoted to developing high-performance sensing devices to capture information feedback from various human activities, which is of great value for timely diagnosis and treatment of the disease [6]. Compared with capacitive [7,8], piezoelectric [9–11], and triboelectric [12] pressure sensors, piezoresistive-type sensors have been considered excellent candidates for next-generation sensors due to their merits of remarkable sensing performance, feasible manufacturing technique, excellent conformal contact, and low energy consumption [6,13,14]. Despite considerable advances recently, the low-cost, large-scale fabrication of piezoresistive pressure sensors integrating low-strain and large deformation detecting capabilities still faces enormous challenges.

Rational selection of sensing materials has been employed as an important factor to achieve the fabrication of high-performance and low-cost flexible pressure sensors. Various conductive materials, such as graphene [15], conductive polymer [16], metal nanowires [17], and carbon nanotubes (CNTs) [18], have been demonstrated as the active layers for flexible piezoresistive sensors due to their great conductivity and excellent mechanical properties. In addition, two-dimensional transition metal dichalcogenides such as $MoSe_2$ nanosheets [19] and WSe_2 nanosheets [20] can also be used as active material. The emerging two-dimensional transition metal carbides, nitrides, and carbonitrides (MXenes), based on their characteristics of adjustable interlayer spacing, excellent mechanical flexibility, and outstanding metallic conductivity, have great potential for developing next-generation high-performance piezoresistive sensors [21–23]. MXenes are typically obtained by selectively etching the A-layer from a ceramic material called the MAX phase and can be defined as $M_{n+1}X_nT_x$ (n = 1, 2, 3), where M denotes an early transition metal, X represents carbon or nitrogen, and T means surface functional groups [24]. In recent years, research on MXenes-based pressure sensors has been gradually carried out. Unfortunately, MXene-based sensors rapidly reach the deformation limit under external pressure due to the single 2D planar structure of MXene materials [25] and the unavoidable stacking during the preparation process [26], thereby limiting device performance. Therefore, there is a need to develop approaches to introduce microstructures into MXene-based sensors so that their performance in sensing devices can be greatly improved.

Up to now, non-planar microstructures with different morphologies have been introduced into MXene-based piezoresistive sensors to obtain higher sensitivity and shorter response time, such as the bioinspired microspines [6], wrinkle structure [1], and protrusion structures [23]. The preparation techniques involved in these microstructures are mostly based on the use of everyday objects with surface microstructures as molds or pre-stretching methods [1,6,23]. Compared with the expensive and complex traditional photolithography technique, these micro-nano processing approaches are simple, convenient, and low-cost, but it is difficult to meet the needs of free customization. Most recently, it was demonstrated that femtosecond laser filament micro-nano processing technology can provide a useful strategy to arbitrarily adjust the morphology of the flexible substrates composing the sensors, thereby developing a robust alternative to fabricating high-performance sensors in a simple, low-cost, and fast way [27–29]. The constructed flexible piezoresistive pressure sensors with filament-processed microstructure exhibit high sensitivity to pressure and vibration, which can be used with success to monitor various human physiological signals [28]. Moreover, with the assistance of the principal component analysis (PCA) algorithm and artificial neural networks, the fabricated sensor can unambiguously identify different phonations, and total recognition accuracy rate can reach 93.4% [28,29]. However, from the perspective of the practical application of flexible wearable smart devices, it is necessary to further improve the sensing performance of the device to sensitively monitor both subtle and large movements of the human body.

Herein, a highly-sensitive piezoresistive sensor was assembled by combing $Ti_3C_2T_x$ MXene nanosheets with honeycomb-like microarchitecture polydimethylsiloxane (PDMS) film fabricated by femtosecond laser pulses in the self-channeling regime. As stress increases, the honeycomb-like microstructures rapidly increase the connecting conductive paths between interface states so that the performance of the pressure sensor can be enhanced. The obtained MXene-based piezoresistive sensor with a honeycomb-like structure exhibits an outstanding sensitivity (0.61 kPa^{-1}), a short response time (160 ms), and excellent cycling stability, and can be mass-produced based on a simple preparation process. Benefiting from the above advantages, the fabricated sensor can monitor both tiny physiological signals, such as facial expressions and artery pulse, and large human movements, such as knee bending and elbow swing. Meanwhile, it can also work effectively in distinguishing various finger gestures, providing a convenient communication tool for people with language barriers. Therefore, our work provides a competitive route for large-scale preparation of low-cost, high-sensitivity piezoresistive sensors with broad prospects in health, sports, and communication applications.

2. Experiment Details

The fabrication of the MXene-based piezoresistive sensor with a honeycomb-like structure was carried out with the procedures illustrated in Figure 1.

Preparation of $Ti_3C_2T_x$ MXene. $Ti_3C_2T_x$ aqueous dispersions were synthesized by chemically exfoliating the Al layer of Ti_3AlC_2 by HCl + LiF etchant, as shown in Figure 1a. First, 2 g of LiF was slowly added into 40 mL of 9 M HCl, followed by continuous stirring at room temperature for 30 min. Subsequently, 2 g of MAX phase precursor (Ti_3AlC_2) powder was gradually added to the above solution for selective etching, which was stirred for 24 h at 35 °C. Then, the resulting MXene suspension was washed with deionized water and centrifuged repeatedly at 3500 rpm until the pH was about 5–6. The sediment was then dried in a vacuum oven and the resulting product was sonicated with ethanol for 1 h. Finally, the obtained powder was added into deionized water, and then sonicated and centrifuged at 3500 rpm for 3 min to collect the $Ti_3C_2T_x$ supernatant. In this experiment, the concentration of the $Ti_3C_2T_x$ solution was quantified to 5 mg/mL by further dilution or concentration by vacuum evaporation.

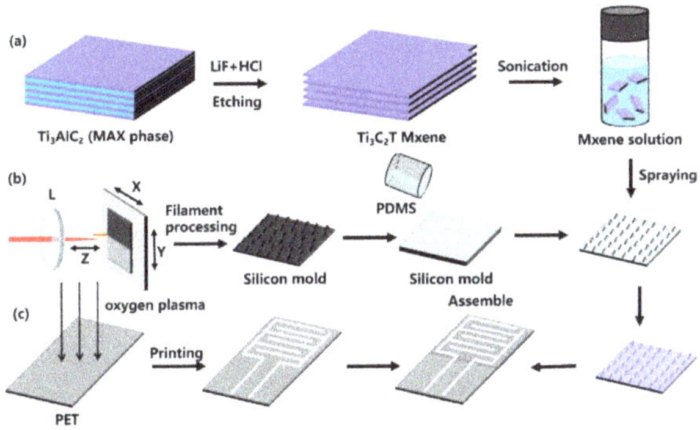

Figure 1. (**a**) Preparation of $Ti_3C_2T_x$ MXene. (**b**) Preparation of Micro-Patterned MXene/PDMS conductive film. (**c**) Assembly of the MXene-based piezoresistive sensor.

Preparation of Micro-Patterned MXene/PDMS conductive film. Figure 1b briefly shows the fabrication process of the MXene sensing thin film with a honeycomb-like structure. First, linearly-polarized laser pulses with a pulse width of 35 fs, a central wavelength of 800 nm, a repetition rate of 500 Hz, and a pulse energy of 2 mJ were generated based on a commercial Ti:sapphire laser system (Spectra-Physics, Spitfifire ACE) and a self-built optical system. After that, the laser pulse was loosely focused by a 25.4 mm-diameter fused silica lens (f = 100 cm) to form a 3 cm-long laser filament with an estimated clamped intensity of the filament core of about 150–200 TW/cm^2. Then, the mold with the honeycomb-like microarchitecture was prepared through the interaction between the single filament and a 2 mm-thick silicon wafer. The sample translation speed and the separation between adjacent processed lines correspond to 1.25 mm/s and 100 μm, respectively. The patterned flexible PDMS thin film was then achieved by replicating the reversed micropatterns from the fabricated silicon mold. Finally, a facile and scalable drip-coating method was employed to deposit the $Ti_3C_2T_x$ MXene nanosheets of high conductivity onto the microstructured PDMS thin film by van der Waals force.

Assembly of the MXene-based piezoresistive sensor. As shown in Figure 1c, the cleaned PET film was first exposed to oxygen plasma (2 min) for hydrophilic treatment. Then, the interdigitated electrode patterned silver conductive ink thin layers were printed on the PET substrate through an inkjet printer and thermally dried in an oven at 100 °C for 1 h. Subsequently, with the assistance of 3 M tape, the MXene sensing film and the interdigital electrode were assembled into the MXene-based piezoresistive sensor.

Performance Characterization. The morphology and structural analysis of the samples was performed using a field emission scanning electron microscope (SEM model Regulus8100 by Hitachi). In order to analyze the phase composition and crystal structure of the fabricated $Ti_3C_2T_x$ MXene, the XRD patterns were observed by a Rigaku D/MAX

2550 diffractometer at a scan angle of 4° to 60°. The sensing performance of the MXene-based piezoresistive sensor was measured with a homemade testing machine, which is composed of a force gauge and a data acquisition system (KeithleyDAQ6510).

3. Result and Discussion

In Figure 2a,a′, we show the differently-magnified SEM images of the filament-processed silicon mold surface at the ablation speed of 1.25 mm/s, corresponding to the number of 40 laser shots hitting the same position. It can be found that highly-repeatable honeycomb-like microarchitectures with an average size of about 15–20 μm are formed on the surface after the interaction between the filament and the silicon wafer. In addition, we also conducted microscopic analysis on the surface of MXene-based PDMS film, as shown in Figure 2b,b′. From the high magnification SEM image in Figure 2b′, it can be seen that the stacked $Ti_3C_2T_x$ can be densely coated on the microstructured PDMS film, which is attributed to the abundant hydrophilic groups, such as -H, -OH, generated on the surface of the MXene after chemical etching. In order to further investigate the crystalline structure of $Ti_3C_2T_x$ thin film, we measured its XRD spectra, as shown in Figure 3a. The strong diffraction peak (002) located at 6° indicates the formation of $Ti_3C_2T_x$ nanosheets with hexagonal crystal structure by etching. In addition, the (104) peak related to the Al element around 39° vanishes, revealing the successful extraction of Al-atoms from Ti_3AlC_2 and indicating that the etching degree was relatively complete during the preparation process.

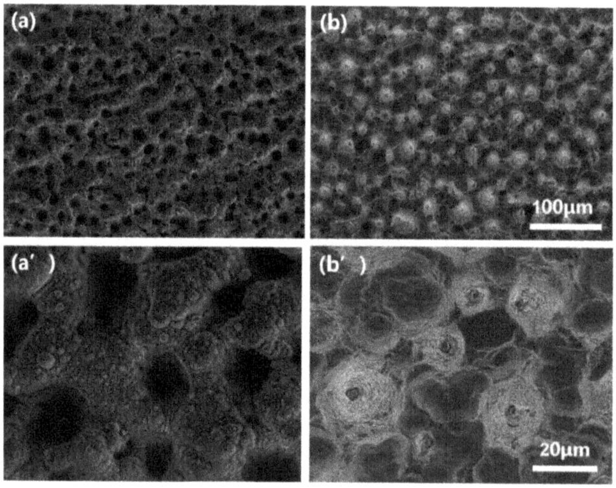

Figure 2. (**a**,**a′**) The top-view SEM images of the filament-processed silicon mold surface under low and high magnifications. (**b**,**b′**) Differently magnified SEM images of the fabricated MXene-based PDMS film surface.

Based on the above-mentioned filament-processed microstructure and MXene sensing material, we assembled and characterized the sensing performance of the MXene-based piezoresistive sensor with a honeycomb-like structure. The sensitivity, as a key parameter for evaluating device performance, can be defined as $S = (\Delta R/R_0)/\Delta P$, where ΔR is the resistance change before and after the applied pressure, and R_0 and ΔP are the initial value without applied pressure and the change of the applied pressure. The $\Delta R/R_0$ versus ΔP for the MXene-based piezoresistive sensor is displayed in Figure 3, from which it can be clearly seen that the fabricated sensor shows three linear regions. The fitted sensitivities are, respectively, 6.10×10^{-1} kPa^{-1} (comparable with those reported in previous literatures [30,31]), 2.60×10^{-2} kPa^{-1}, and 1.26×10^{-3} kPa^{-1}, for the range of 0–1.5 kPa, 1.5–8 kPa and 8–50 kPa. For this sensor, the superior sensing performance may be attributed to the specific morphology of the filament-processed microstructure and the extraordinary properties of $Ti_3C_2T_x$ nanosheets. In addition, we also compare the sensitivity and responsivity of the sensing devices constructed with different sensing materials, as shown in Figure 3b. As can be seen from Figure 3b, the MXene-based device exhibits higher responsivity, and its sensitivity under the pressure of 8 kPa is more than two times higher than that of the SWCNTs-based piezoresistive sensor (our previously prepared device) [26]. The above results indicate that the MXene-based sensor is more suitable for monitoring human activity signals and, consequently, may have the potential for health, sports, and communication applications.

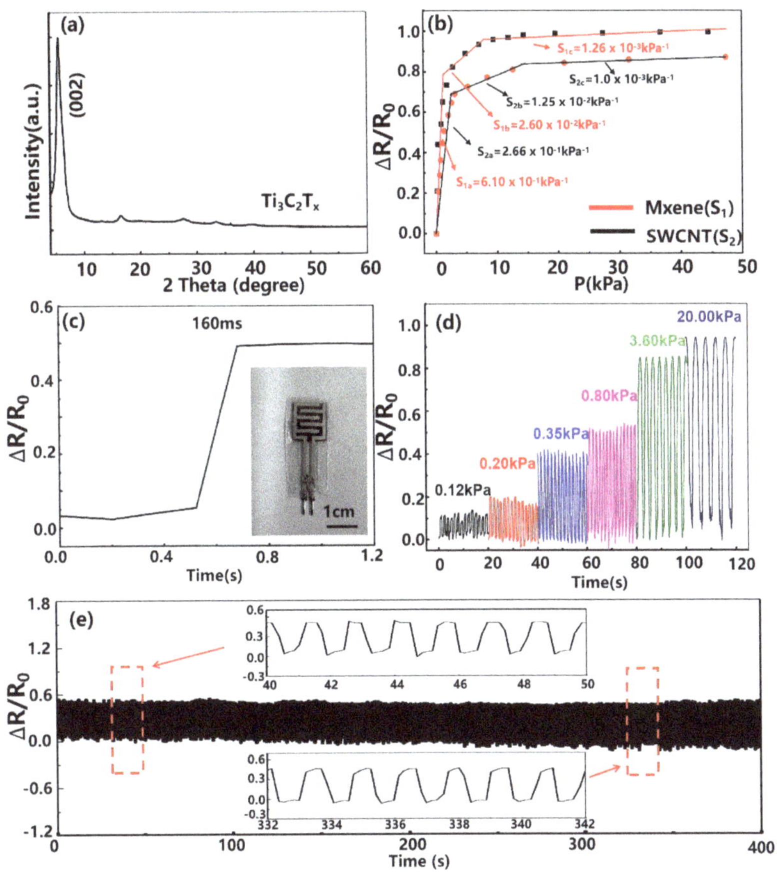

Figure 3. (**a**) XRD spectra of $Ti_3C_2T_x$ materials. (**b**) The resistance response versus pressure change for the filament-processed MXene-based piezoresistive sensor and the SWCNTs-based piezoresistive sensor. (**c**) Instant response of the MXene-based piezoresistive sensor. Inset: Photograph of the fabricated sensing device taken by a digital camera. (**d**) Relative resistance variation of the fabricated MXene-based piezoresistive sensor under serial pressures. (**e**) Real-time resistance response of the MXene-based piezoresistive sensor for loading/unloading cycles under 0.8 kPa, at a 0.64-Hz repetition rate.

The working mechanism of the MXene-based piezoresistive sensor and the existence of the three different sensitivities can be explained as follows. In the initial state, several conductive paths are formed between the MXene/PDMS conductive film and the interdigitated electrodes due to the presence of contact points. When external pressure is applied to the device, the contact area between the relatively large honeycomb-like architecture and the interdigitated electrodes increases, resulting in a significant increase in the number of conductive paths. As the pressure increases further, the parts that are already in contact provide support of a certain level, which slows down the increase rate of the conductive path. In the high-pressure regime, the honeycomb-like microstructures have been compressed to a flat shape, that is, at their maximum compression positions, and thus the increased pressure contributes only to a small increase in the conducting area of the interface state.

In addition, the fast response time of the device is also of great significance for practical applications, as it can ensure a timely response in the monitoring process of ultrafast pressure signals. As illustrated in Figure 3c, although the elastic deformation of the microstructured PDMS film will cause a certain response time delay, the fabricated device exhibits a short response time of 160 ms (less than the human body's response time of 400 ms), which is sufficient for human activity monitoring. In order to demonstrate the fabricated sensor more intuitively, a photograph of the fabricated sensing device is shown in the inset of Figure 3c. It can be seen that the prepared sensor has an overall size of 3.0×1.2 cm^2 with a thickness of 450 μm, and the sensing area occupies 1.00×1.25 cm^2. Figure 3d shows the stable cyclic performance of the fabricated MXene-based piezoresistive sensor under serial pressures (multiple cyclic loading under the applied pressures of 0.12 kPa, 0.2 kPa, 0.35 kPa, 0.8 kPa, 3.6 kPa, and 20 kPa). Moreover, we also measured the long-term cyclic stability of the fabricated MXene-based piezoresistive sensor, that is, loading–unloading cycling tests at a frequency of 0.64 Hz under a pressure of 0.8 kPa (Figure 3e). It was found that under the pressure of 0.8 kPa, the relative resistance response amplitude of the device is about 50.0%, which is consistent with the obtained sensitivity curve (Figure 3b). After long-term loading and unloading tests, the shape and intensity of the electrical signal change of this sensor remains almost the same with a slight attenuation, showing its excellent long-term repeatability and stability.

The outstanding performance of the fabricated MXene-based piezoresistive sensor with a honeycomb-like structure, with its high sensitivity, repeatability and excellent conformal contact, allows it to be useful as a wearable device for real-time tiny physiological signal and large-scale human motion monitoring. In human health monitoring, the detection of joint activities is of great significance for motion monitoring and gesture recognition. By attaching the fabricated sensor to different joint positions of the human body with the assistance of medical tape, we obtained the responsive signal waveforms for bending the elbow (Figure 4a), wrist (Figure 4b), finger (Figure 4c), and knee (Figure 4f), where each flexion of the joint maintains almost the same angle. It can be seen that the motion state can be clearly distinguished by comparing the shape and intensity of the graphs; that is, the responsivity tends to increase due to the increase of pressure during the bending process. The above results clearly demonstrate the potential of our fabricated sensor in the field of large-scale human motion monitoring.

To demonstrate the tiny deformation detection capability of our sensor, the device was attached to the subject's cheek to monitor changes in facial muscles as expressions change. Three main emotional facial expressions, namely anger, smiling, and laughter, were investigated, as shown in Figure 4d. In fact, human facial expressions are produced by several muscle groups. It was found that when the tester makes different facial expressions, the obtained waveforms have significant differences, mainly reflected in the change of relative resistance and the movement time of micro-muscle groups. The differences in characteristic peaks are mainly attributed to the deviation of facial skin or muscle movement when people perform different expressions. It was found that when the subject makes different facial expressions, the resistance change waveforms are significantly different, which is caused by deviations in the degree of facial skin or muscle movement. From Figure 4d, the laughing expression results in the largest change in the device resistance value, which may be attributed to the fact that happy laughter requires large facial muscle fluctuations and excitation of all muscle groups on the face. In addition, when the subject repeatedly makes the same facial expression, the response curves obtained by the device show consistent characteristics.

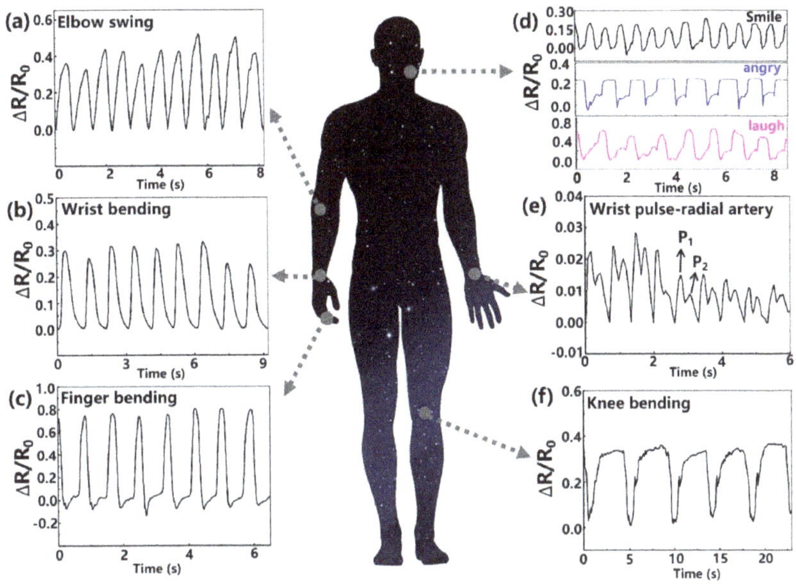

Figure 4. Applications of the MXene-based piezoresistive sensor for real-time monitoring of human activities. Resistance change waveforms of the wearable sensor in detecting human movements: (**a**) elbow swing, (**b**) wrist bending, (**c**) finger bending, (**d**) facial expressions, (**e**) wrist pulse, and (**f**) knee bending.

We have transformed the signals of the facial expressions into a frequency domain to make the difference among them more apparent, as shown in Figure 5. The main peaks of the three expressions are of different frequencies and amplitudes. The smile expression has a shorter-term muscle movement than laughter and anger, so its signal on the time domain is narrower than the other two, while it is obvious in the frequency domain that the frequency of its main peak is higher than the other two. Furthermore, though the anger and laughter are almost the same widths in the time domain, the angry expression muscle retention time is longer, which is indicated by the period of the flat top on their curves. Correspondingly, the difference between anger and laughter is much easier to detect in the frequency domain because, though the frequencies of the main peaks of the angry signal and the laughing signal are almost the same, the angry signal has higher harmonic components, which form a longer-term flat top. The facial expressions sampled with our sensor have significantly different features, so a threshold discriminant method based on signal processing or a classifier based on machine learning can be applied for pattern recognition of the expressions. Based on the aforementioned test, it is shown that our fabricated sensor can be used to recognize the delicate deformations of human facial expressions, which could contribute to observing the state of thinking of the human brain.

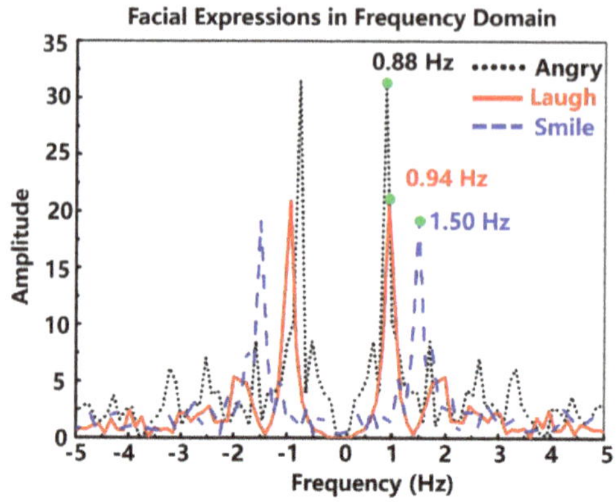

Figure 5. The signals of the facial expressions transformed into the frequency domain.

Moreover, we test the capability of the fabricated sensor to identify weak physiological signals. The sensor was tightly attached to the skin of one's wrist for non-invasive continuous monitoring of the radial artery blood pressure wave. As shown in Figure 4e, the obtained pulse signal is consistent with the typical radial artery characteristic waveform, containing two distinct peaks, namely, the systolic (P1) and diastolic (P2) peaks [32]. The signal processing system matched with the sensor to automatically extract the pulse rate is shown in Figure 6a. Firstly, a high pass filter with a cut-off frequency of 0.5 Hz is applied to remove the direct-current component. Then, a Fourier transformer module transforms the time-domain radial artery pulse signal to the frequency domain. After that, the main component extractor picks up the most significant frequency component, which indicates the frequency of the artery pulsing. As shown in Figure 6b, the most significant frequency component is at 1.5255 Hz, equivalent to 91.5 bpm, close to the actual pulse rate of the subject, 90 bpm. Since the commercial wearable bracelets allow measurement error within ±5 bpm in the resting state, our sensor is qualified for commercial applications. Additionally, some important medical information for health monitoring can also be obtained, such as the reflection index (R.I.) and the stiffness index (S.I.).

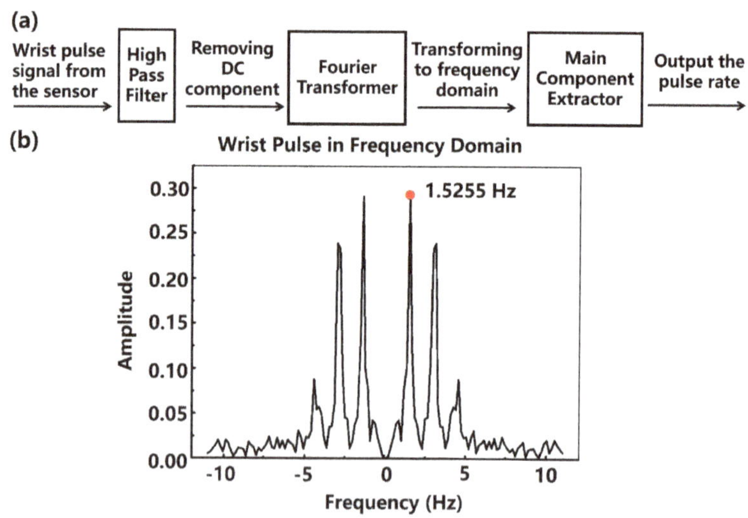

Figure 6. (**a**) The diagram of extracting the pulse rate from the sensor attached to the skin of the subject's wrist. (**b**) The Fourier-transformed wrist pulse signal.

Furthermore, the applications of the fabricated MXene-based piezoresistive sensor for exquisite and sophisticated finger gesture recognition are illustrated in Figure 7. As shown in Figure 7a, the gesture recognition glove is assembled by integrating the smart glove with our fabricated MXene-based piezoresistive sensor. Five independent sensing devices are stuck on the joint parts of the glove fingers to measure the bending and tilting angles of the five fingers. Figure 7b shows the schematic diagrams of Arabic numeral gestures in a daily scenario and the corresponding responsive output obtained. It was found that the numbers "0" to "9" can be unambiguously distinguished according to the intensity of the piezoresistive output, which mainly depends on the bending angle and number of the finger joints. From number "1" to number "5", the highly responsive piezoresistive output gradually decreases due to the reduction in the number of bent fingers. When the gesture matches the number "6", the index, middle, and ring fingers curl up, resulting in a reduction in the resistance of the corresponding sensors. When the gesture transitioned from the number "7" to the number "0", the amplitude of the piezoresistive response elevated gradually, which may be attributed to the increase in the magnitude and number of finger bends. Thus, our MXene-based sensor with a honeycomb-like structure can be used to register various sophisticated gestures, which has broad application prospects in sign-to-speech translation, virtual reality, and other fields.

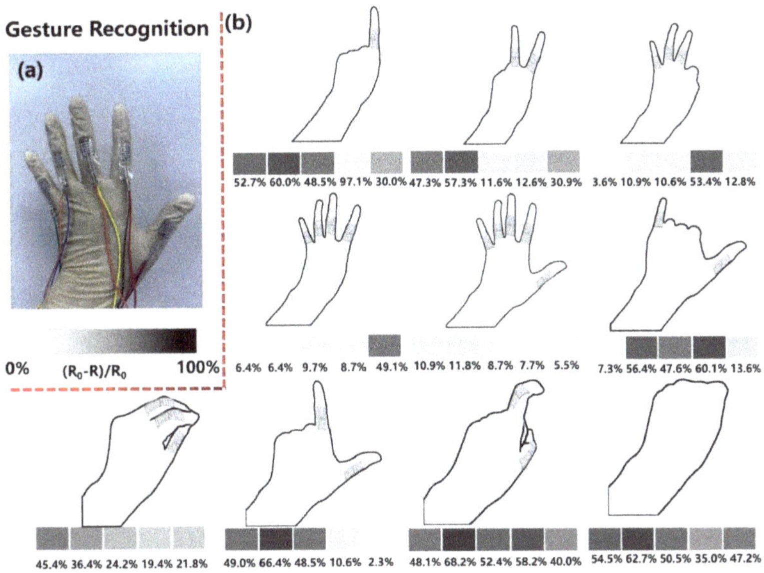

Figure 7. (**a**) The gesture recognition glove based on the fabricated piezoresistive sensor. The prepared five MXene-based piezoresistive sensing devices were assembled at the joints of the glove fingers. (**b**) Gestures of Arabic numerals and their corresponding responsive output.

4. Conclusions

In summary, we fabricated a sensitive flexible MXene-based piezoresistive sensor with a honeycomb-like structure through drip-coating MXene nanosheets combined with femtosecond filamentating pulses processing. With the specific morphology of the filament-processed microstructure and the extraordinary properties of $Ti_3C_2T_x$ nanosheets, the obtained MXene composite sensor features a high sensitivity of 0.61 kPa^{-1}, which is more than two times higher than that of our previously prepared SWCNTs-based piezoresistive sensing device [26]. Additionally, the device also demonstrates a short response time of 160 ms and excellent repeatability under loading/unloading cycles. The sensing mechanism of the fabricated device can be explained by the change in the conducting area of the interface state between the MXene/PDMS conductive film and the interdigitated electrodes. This extraordinary sensing performance attributes enabled us to realize the monitoring of sensing signals from tiny muscle movement to large-scale human motions, including artery pulse, facial expressions, and joint motion. Subsequently, we assembled a smart glove that was used with success for exquisite and sophisticated gesture recognition. Therefore, we hope that our fabricated sensing device can be used in next-generation wearable electronic devices for healthcare monitoring and human activity real-time detection.

Author Contributions: Conceptualization, X.Z., M.L. and F.Y.; methodology, F.Y., Y.S. and K.M.; software, K.M., Y.S. and J.T.; validation, Y.S., K.M. and J.T.; formal analysis, Y.S. and K.M.; investigation, Y.S., K.M., F.Y. and J.T.; resources, X.Z., M.L. and J.T.; data curation, Y.S., K.M. and F.Y.; writing—original draft preparation, Y.S. and K.M.; writing-review and editing, X.Z. and M.L.; visualization, Y.S., K.M., F.Y. and J.T.; project administration, M.L. and X.Z.; funding acquisition, M.L. and X.Z.; Supervision, F.Y., M.L. and X.Z. All authors have read and agreed to the published version of the manuscript.

Funding: This work has been supported by the Key Research Program of Frontier Science, Chinese Academy of Sciences (Grant No. QYZDY-SSW-JSC004).

Institutional Review Board Statement: Not applicable.

Informed Consent Statement: Ethical review and approval were waived for this study as it only involved in vitro monitoring by the first author. Informed consent was obtained from all subjects involved in the study.

Data Availability Statement: Not applicable.

Conflicts of Interest: The authors declare no conflict of interest.

References

1. Yan, J.F.; Ma, Y.N.; Li, X.X.; Zhang, C.K.; Cao, M.L.; Chen, W.; Luo, S.J.; Zhu, M.; Gao, Y.H. Flexible and high-sensitivity piezoresistive sensor based on MXene composite with wrinkle structure. *Ceram. Int.* **2020**, *46*, 23592–23598. [CrossRef]
2. Yamamoto, Y.; Harada, S.; Yamamoto, D.; Honda, W.; Arie, T.; Akita, S.; Takei, K. Printed multifunctional flexible device with an integrated motion sensor for health care monitoring. *Sci. Adv.* **2016**, *2*, e1601473. [CrossRef] [PubMed]
3. Yamada, T.; Hayamizu, Y.; Yamamoto, Y.; Yomogida, Y.; Izadi-Najafabadi, A.; Futaba, D.N.; Hata, K. A stretchable carbon nanotube strain sensor for human-motion detection. *Nat. Nanotechnol.* **2011**, *6*, 296–301. [CrossRef] [PubMed]
4. Yu, X.G.; Xie, Z.Q.; Yu, Y.; Lee, J.; Vazquez-Guardado, A.; Luan, H.W.; Ruben, J.; Ning, X.; Akhtar, A.; Li, D.F.; et al. Skin-integrated wireless haptic interfaces for virtual and augmented reality. *Nature* **2019**, *575*, 473. [CrossRef] [PubMed]
5. Zhu, M.L.; Sun, Z.D.; Zhang, Z.X.; Shi, Q.F.; He, T.Y.Y.; Liu, H.C.; Chen, T.; Lee, C.K. Haptic-feedback smart glove as a creative human-machine interface (HMI) for virtual/augmented reality applications. *Sci. Adv.* **2020**, *6*, eaaz8693. [CrossRef]
6. Cheng, Y.F.; Ma, Y.A.; Li, L.Y.; Zhu, M.; Yue, Y.; Liu, W.J.; Wang, L.F.; Jia, S.F.; Li, C.; Qi, T.Y.; et al. Bioinspired Microspines for a High-Performance Spray $Ti_3C_2T_x$ MXene-Based Piezoresistive Sensor. *Acs Nano* **2020**, *14*, 2145–2155. [CrossRef]
7. Lee, J.; Kwon, H.; Seo, J.; Shin, S.; Koo, J.H.; Pang, C.; Son, S.; Kim, J.H.; Jang, Y.H.; Kim, D.E.; et al. Conductive Fiber-Based Ultrasensitive Textile Pressure Sensor for Wearable Electronics. *Adv. Mater.* **2015**, *27*, 2433–2439. [CrossRef]
8. Luo, Y.S.; Shao, J.Y.; Chen, S.R.; Chen, X.L.; Tian, H.M.; Li, X.M.; Wang, L.; Wang, D.R.; Lu, B.H. Flexible Capacitive Pressure Sensor Enhanced by Tilted Micropillar Arrays. *Acs Appl. Mater. Interfaces* **2019**, *11*, 17796–17803. [CrossRef]
9. Yang, T.; Pan, H.; Tian, G.; Zhang, B.B.; Xiong, D.; Gao, Y.Y.; Yan, C.; Chu, X.; Chen, N.J.; Zhong, S.; et al. Hierarchically structured PVDF/ZnO core-shell nanofibers for self-powered physiological monitoring electronics. *Nano Energy* **2020**, *72*, 104706. [CrossRef]
10. Deng, W.L.; Yang, T.; Jin, L.; Yan, C.; Huang, H.C.; Chu, X.; Wang, Z.X.; Xiong, D.; Tian, G.; Gao, Y.Y.; et al. Cowpea-structured PVDF/ZnO nanofibers based flexible self-powered piezoelectric bending motion sensor towards remote control of gestures. *Nano Energy* **2019**, *55*, 516–525. [CrossRef]
11. Tian, G.; Xiong, D.; Su, Y.H.; Yang, T.; Gao, Y.Y.; Yan, C.; Deng, W.; Jin, L.; Zhang, H.T.; Fan, X.Q.; et al. Understanding the Potential Screening Effect through the Discretely Structured ZnO Nanorods Piezo Array. *Nano Lett.* **2020**, *20*, 4270–4277. [CrossRef] [PubMed]
12. Yan, C.; Gao, Y.Y.; Zhao, S.L.; Zhang, S.L.; Zhou, Y.H.; Deng, W.L.; Li, Z.W.; Jiang, G.; Jin, L.; Tian, G.; et al. A linear-to-rotary hybrid nanogenerator for high-performance wearable biomechanical energy harvesting. *Nano Energy* **2020**, *67*, 104235. [CrossRef]
13. Park, J.; Lee, Y.; Hong, J.; Ha, M.; Jung, Y.D.; Lim, H.; Kim, S.Y.; Ko, H. Giant Tunneling Piezoresistance of Composite Elastomers with Interlocked Microdome Arrays for Ultrasensitive and Multimodal Electronic Skins. *Acs Nano* **2014**, *8*, 4689–4697. [CrossRef] [PubMed]
14. Tien, N.T.; Jeon, S.; Kim, D.I.; Trung, T.Q.; Jang, M.; Hwang, B.U.; Byun, K.E.; Bae, J.; Lee, E.; Tok, J.B.H.; et al. A Flexible Bimodal Sensor Array for Simultaneous Sensing of Pressure and Temperature. *Adv. Mater.* **2014**, *26*, 796–804. [CrossRef] [PubMed]

15. Li, W.W.; Geng, X.M.; Guo, Y.F.; Rong, J.Z.; Gong, Y.P.; Wu, L.Q.; Zhang, X.M.; Li, P.; Xu, J.B.; Cheng, G.S.; et al. Reduced Graphene Oxide Electrically Contacted Graphene Sensor for Highly Sensitive Nitric Oxide Detection. *Acs Nano* **2011**, *5*, 6955–6961. [CrossRef]
16. Kannichankandy, D.; Pataniya, P.M.; Narayan, S.; Patel, V.; Sumesh, C.K.; Patel, K.D.; Solanki, G.K.; Pathak, V.M. Flexible piezo-resistive pressure sensor based on conducting PANI on paper substrate. *Synth. Met.* **2021**, *273*, 116697. [CrossRef]
17. Ge, J.; Sun, L.; Zhang, F.R.; Zhang, Y.; Shi, L.A.; Zhao, H.Y.; Zhu, H.W.; Jiang, H.L.; Yu, S.H. A Stretchable Electronic Fabric Artificial Skin with Pressure-, Lateral Strain-, and Flexion-Sensitive Properties. *Adv. Mater.* **2016**, *28*, 722–728. [CrossRef]
18. Yeom, C.; Chen, K.; Kiriya, D.; Yu, Z.B.; Cho, G.; Javey, A. Large-Area Compliant Tactile Sensors Using Printed Carbon Nanotube Active-Matrix Backplanes. *Adv. Mater.* **2015**, *27*, 1561. [CrossRef]
19. Pataniya, P.M.; Bhakhar, S.A.; Tannarana, M.; Zankat, C.; Patel, V.; Solanki, G.K.; Patel, K.D.; Jha, P.K.; Late, D.J.; Sumesh, C.K. Highly sensitive and flexible pressure sensor based on two-dimensional MoSe2 nanosheets for online wrist pulse monitoring. *J. Colloid Interface Sci.* **2021**, *584*, 495–504. [CrossRef]
20. Pataniya, P.M.; Sumesh, C.K.; Tannarana, M.; Zankat, C.K.; Solanki, G.K.; Patel, K.D.; Pathak, V.M. Flexible paper based piezo-resistive sensor functionalised by 2D-WSe2 nanosheets. *Nanotechnology* **2020**, *31*, 435503. [CrossRef]
21. Naguib, M.; Kurtoglu, M.; Presser, V.; Lu, J.; Niu, J.J.; Heon, M.; Hultman, L.; Gogotsi, Y.; Barsoum, M.W. Two-Dimensional Nanocrystals Produced by Exfoliation of Ti_3AlC_2. *Adv. Mater.* **2011**, *23*, 4248–4253. [CrossRef] [PubMed]
22. Anasori, B.; Lukatskaya, M.R.; Gogotsi, Y. 2D metal carbides and nitrides (MXenes) for energy storage. *Nat. Rev. Mater.* **2017**, *2*, 16098. [CrossRef]
23. Xiang, Y.C.; Fang, L.; Wu, F.; Zhang, S.F.; Ruan, H.B.; Luo, H.J.; Zhang, H.; Li, W.J.; Long, X.M.; Hu, B.S.; et al. 3D Crinkled Alk-Ti_3C_2 MXene Based Flexible Piezoresistive Sensors with Ultra-High Sensitivity and Ultra-Wide Pressure Range. *Adv. Mater. Technol.* **2021**, *6*, 2001157. [CrossRef]
24. Jin, Y.K.; Wen, B.; Gu, Z.X.; Jiang, X.T.; Shu, X.L.; Zeng, Z.P.; Zhang, Y.P.; Guo, Z.N.; Chen, Y.; Zheng, T.T.; et al. Deep-Learning-Enabled MXene-Based Artificial Throat: Toward Sound Detection and Speech Recognition. *Adv. Mater. Technol.* **2020**, *5*, 2000262. [CrossRef]
25. Ma, Y.N.; Liu, N.S.; Li, L.Y.; Hu, X.K.; Zou, Z.G.; Wang, J.B.; Luo, S.J.; Gao, Y.H. A highly flexible and sensitive piezoresistive sensor based on MXene with greatly changed interlayer distances. *Nat. Commun.* **2017**, *8*, 1207. [CrossRef]
26. Zhu, Y.C.; Rajoua, K.; Le Vot, S.; Fontaine, O.; Simon, P.; Favier, F. Modifications of MXene layers for supercapacitors. *Nano Energy* **2020**, *73*, 104734. [CrossRef]
27. Su, Y.; Zhang, W.; Chen, S.M.; Yao, D.W.; Zhang, X.; Chen, H.D.; Xu, H.L. Piezoresistive Electronic-Skin Sensors Produced with Self-Channeling Laser Microstructured Silicon Molds. *IEEE Trans. Electron Devices* **2021**, *68*, 786–792. [CrossRef]
28. Su, Y.; Zheng, L.S.; Yao, D.W.; Zhang, X.; Chen, H.D.; Xu, H.L. Robust physiological signal monitoring by a flexible piezoresistive sensor microstructured with filamentating laser pulses. *Sensors Actuators Phys.* **2021**, *331*, 112907. [CrossRef]
29. Su, Y.; Ma, K.A.; Zhang, X.; Liu, M. Neural Network-Enabled Flexible Pressure and Temperature Sensor with Honeycomb-like Architecture for Voice Recognition. *Sensors* **2022**, *22*, 759. [CrossRef]
30. dos Santos, A.; Pinela, N.; Alves, P.; Santos, R.; Fortunato, E.; Martins, R.; Aguas, H.; Igreja, R. Piezoresistive E-Skin Sensors Produced with Laser Engraved Molds. *Adv. Electron. Mater.* **2018**, *4*, 1800182. [CrossRef]
31. Xiong, Y.X.; Zhu, Y.; Liu, X.B.; Zhu, P.L.; Hu, Y.G.; Sun, R.; Wong, C.P. A flexible pressure sensor based on melamine foam capped by copper nanowires and reduced graphene oxide. *Mater. Today Commun.* **2020**, *24*, 100970. [CrossRef]
32. Choong, C.L.; Shim, M.B.; Lee, B.S.; Jeon, S.; Ko, D.S.; Kang, T.H.; Bae, J.; Lee, S.H.; Byun, K.E.; Im, J.; et al. Highly Stretchable Resistive Pressure Sensors Using a Conductive Elastomeric Composite on a Micropyramid Array. *Adv. Mater.* **2014**, *26*, 3451–3458. [CrossRef] [PubMed]

micromachines

MDPI

Article

Monolayer MoS_2-Based Flexible and Highly Sensitive Pressure Sensor with Wide Sensing Range

Dandan Xu [1,2,†], Ling Duan [1,†], Suyun Yan [1], Yong Wang [3], Ke Cao [1,*], Weidong Wang [1,2,*], Hongcheng Xu [1], Yuejiao Wang [1], Liangwei Hu [4] and Libo Gao [1,*]

1 School of Mechano-Electronic Engineering, Xidian University, Xi'an 710071, China; ddxu@stu.xidian.edu.cn (D.X.); charlotte_duan@163.com (L.D.); yansuyun1018@163.com (S.Y.); xuhongcheng@stu.xidian.edu.cn (H.X.); yuejiwang4-c@my.cityu.edu.hk (Y.W.)
2 CityU-Xidian Joint Laboratory of Micro/Nano-Manufacturing, Xi'an 710071, China
3 School of Advanced Materials and Nanotechnology, Xidian University, Xi'an 710171, China; yongwang@xidian.edu.cn
4 School of Aerospace Engineering, Xiamen University, Xiamen 361102, China; 19920211151512@stu.xmu.edu.cn
* Correspondence: caoke@xidian.edu.cn (K.C.); wangwd@mail.xidian.edu.cn (W.W.); lbgao@xidian.edu.cn (L.G.)
† These authors contribute equally to this work.

Abstract: Flexible pressure sensors play an important role in flexible robotics, human-machine interaction (HMI), and human physiological information. However, most of the reported flexible pressure sensors suffer from a highly nonlinear response and a significant decrease in sensitivity at high pressures. Herein, we propose a flexible novel iontronic pressure sensor based on monolayer molybdenum disulfide (MoS_2). Based on the unique structure and the excellent mechanical properties as well as the large intercalation capacitance of MoS_2, the prepared sensor holds an ultra-high sensitivity (S_{max} = 89.75 kPa^{-1}) and a wide sensing range (722.2 kPa). Further, the response time and relaxation time of the flexible sensor are only 3 ms, respectively, indicating that the device can respond to external pressure rapidly. In addition, it shows long-term cycling stability (over 5000 cycles with almost no degradation) at a high pressure of 138.9 kPa. Finally, it is demonstrated that the sensor can be used in physiological information monitoring and flexible robotics. It is anticipated that our prepared sensor provide a reliable approach to advance the theory and practicality of the flexible sensor electronics.

Keywords: flexible electronics; flexible sensor; wearable sensor; iontronic pressure sensor; MoS_2

Citation: Xu, D.; Duan, L.; Yan, S.; Wang, Y.; Cao, K.; Wang, W.; Xu, H.; Wang, Y.; Hu, L.; Gao, L. Monolayer MoS_2-Based Flexible and Highly Sensitive Pressure Sensor with Wide Sensing Range. *Micromachines* **2022**, *13*, 660. https://doi.org/10.3390/mi13050660

Academic Editor: Joost Lötters

Received: 21 March 2022
Accepted: 20 April 2022
Published: 22 April 2022

1. Introduction

Flexible pressure sensors are widely used in human-machine interaction (HMI) [1–3], electronic skin (E-Skin) [4–6], and human health monitoring [7–11]. In particular, capacitive pressure sensors have attracted a broad attention with their merits of low power, excellent stability and fast response time. But building flexible capacitive pressure sensors with high sensitivity and wide sensing range has been a challenge when considered the sensitivity of conventional capacitive pressure sensors is typically relatively low.

Iontronic pressure sensors have recently attracted a lot of attention [12–17]. This type of interfacial capacitive sensor based on electrical double layer (EDL) capacitance has the advantages of ultra-high capacitance ($\mu F\ cm^{-2}$) and high signal-to-noise ratio (SNR), which are not available with conventional flat plate capacitive sensor. The EDL performance depends on many items, among which the electrode materials play a most important role. Currently reported sensitive materials for preparing iontronic pressure sensor is categorized into metal-based electrode materials [18,19], carbon-based materials [7], metal oxide indium tin oxide (ITO) [20], and MXene [21]. Metal electrodes typically exhibit smaller specific surface area and reduced EDL capacitance. In the case of carbon materials, the electrodes are thicker, mechanically prone to fracture and have longer ion transport paths, while MXene faces problems such as oxidation. Therefore, the deep research on new materials for iontronic pressure sensor and the further study of their sensing mechanisms are important for the development of flexible iontronic sensors.

As a new two-dimensional material, molybdenum disulfide (MoS_2) have attracted a lot of attention in recent years, and it has been widely used in supercapacitors, microelectronic devices, and batteries because of its rich elemental composition, wide band gap (1.29−1.90 eV), high electron mobility, excellent thermal properties and good mechanical properties [22–24]. Therefore the MoS_2 has the potential to be used as an electrode material in iontronic pressure sensors. However, the study on the monolayer MoS_2-based flexible iontronic pressure sensor are rarely reported.

Therefore, in order to solve the problem of sensitivity and sensing range balance of the flexible pressure sensor, we propose a novel iontronic pressure sensor based on MoS_2 monolayer. The monolayer MoS_2/gold composite electrode was obtained by wet transfer strategy. And the iontronic pressure sensor was assembled by the composite electrode and microstructural ionic film. Thanks to the excellent mechanical-electrical properties of MoS_2 and unique layered structure, the prepared sensor shows an unprecedented ultra-high sensitivity (S_{max} = 89.75 kPa^{-1}, S_{min} = 10.24 kPa^{-1}) and a wide sensing range (722.2 kPa). The response time and relaxation time of the sensor sensor are both 3 ms, respectively. Moreover, the possible sensing mechanism of the sensor is proposed. Finally, the application of this novel iontronic pressure sensor is carefully demonstrated.

2. Materials and Methods

2.1. Preparation of MoS_2/Au Composite Electrode

The Au/PI (polyimide) was prepared by magnetron sputtering. A thin 100 nm gold layer was sputtered on the cleaned PI film. A monolayer MoS_2 grew on the Si/SiO_2 (Silion/Silicon dioxide) wafer (SiO_2 layer thickness, 300 nm) by chemical vapor deposition (CVD) method. A PMMA (polymethyl methacrylate) film was covered on the MoS_2 layer as a supporting protective layer by spin-coating method (3000 r min^{-1}, 10 s), and dried completely at 100 °C for 10 min. Then, the samples were placed in a KOH (potassium hydroxide) solution (1 mol L^{-1}) with a heating temperature of 90 °C to etch the growth substrate to separate the PMMA/MoS_2 from the Si/SiO_2 substrate. After the sample was completely separated from the substrate, PMMA/MoS_2 was transferred to deionized water and soaked several times to remove the residue KOH solution, and then PMMA/MoS_2 was transferred to the Au/PI electrode. Finally, the PMMA film was removed using acetone, and the electrode was cleaned with isopropyl alcohol and deionized water several times to remove the residual acetone.

2.2. Preparation of PVA/H_3PO_4 Film with Pyramid Microstructure

1 g of Polyvinyl Alcohol (PVA) was mixed with deionized water (10 mL) under serious stirring at 85 °C for one hour until completely dissolved to a transparent state. Then the 85% H_3PO_4 solution (2 mL) was added to the prepared PVA solution drop by drop, and then stirring continued for half an hour at 90 °C to make the uniform dispersion. Finally, the prepared gel electrolyte was evenly coated on the mask template with pyramidal microstructure and the thickness was controlled at 100 μm by spin coating. After the film was dried at 60 °C for one hour, the film can be peeled off from the substrate.

2.3. Fabrication of Flexible Pressure Sensor

The prepared gold electrode with a monolayer MoS_2 on PI substrate, the poly (ethylene terephthalate) PET spacer layer, the PVA/H_3PO_4 ionic film with upside-down pyramid microstructure are stacked in a "sandwich" configuration PET with a thickness of about 100 μm was employed as the spacer, and finally 3M tape was used to ensure the fully integrated encapsulation of the device.

2.4. Material and Structural Characterization

Field emission scanning electron microscopy (FESEM, Quanta 450, 20 kV) were used to characterize the structure and morphology information. Raman spectroscopy were measured by a Renishaw inVia Raman spectrometer (Renishaw, Britain, UK). For each test on MoS_2, three samples were prepared and five points on each sample were measured.

2.5. Electrical and Mechanical Characterization

For the electromechanical characterization, the variable capacitance can be recorded by LCR impedance analyzer (Tonghui, TH 2827A, Changzhou, China) and IM3536 LCR meter (HIOKI, Taiwan, China) linkage test. The CV curves also were shown on the electrochemical working station (CHI 760E, Chen Hua, Shanghai, China) under the applied pressure. For featuring cyclic stability, the device was tested on a mechanical testing machine (Zhi Qu, 990B, Dongguan, China) with the pulling speed of 5 mm min^{-1}.

3. Results and Discussion

MoS_2 possess excellent intercalating pseudocapacitive and EDL properties for charge storage. And it has been shown that MoS_2 has faster electron transfer capability than oxide and higher theoretical specific capacitance than graphene when used as an electrode material. Therefore, in this study, a monolayer MoS_2 is introduced as the electrode to improve the capacitance and further boost the sensitivity of the sensor. The monolayer MoS_2 was transferred by wet transfer method (Figure 1) on an Au/PI film. The assembled device is shown in Figure 1b. The size of the prepared sensor is 1×1 cm^2, which is small compared with a coin, demonstrating its portable and flexible feature. Noted that the microstructural ionic film was beneficial to further improve the sensitivity of the sensor, which has been demonstrated by previous reports [25].

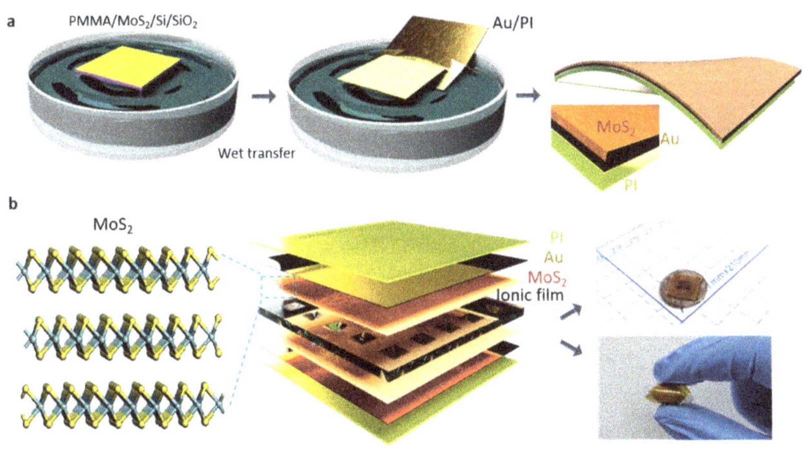

Figure 1. Preparation and schematic diagram of iontronic pressure sensor. (**a**) Wet transfer process of MoS_2 monolayer. (**b**) Digital optical image and flexible feature of iontronic pressure sensor and the corresponding diagram layout of each layer in the sensor.

The monolayer MoS_2 was transferred onto the Au/PI electrode by wet transfer method [26], and the optical microscope of the electrode is shown in Figure 2a. From which it can be observed that the large and continuous monolayer MoS_2 on the Au/PI film showed good integrity and only a very few cracks existed due to etching when the support layer was removed during the transfer process. The Raman results further certified the MoS_2 property. The number of layers of the sample can be determined according to the wave number difference between $E^1{}_{2g}$ and A_{1g}.from Raman spectra of monolayer MoS_2.

Wave number difference of bulk material is 26 cm^{-1}, while the wave number difference between A_{1g} and $E^1{}_{2g}$ in Figure 2b is 19 cm^{-1}. Hence, the preparative sample of MoS_2 can be regarded as monolayer with the thickness of 6.15 Å [27,28]. Further, SEM of the monolayer MoS_2 before and after the transfer was performed in Figure 2c,d. It can be observed that the monolayer MoS_2 on the Si/SiO_2 substrate is large area continuous growth without clear defect. Also, as shown in Figure 2d, the monolayer MoS_2 can still be observed as a continuous boundary which is similar to that before transfer to the target substrate by the wet transfer method. This further indicates that the monolayer MoS_2 obtained by the wet transfer method has a high integrity with few significant cracks, which lays the foundation for the preparation of the sensor electrode. Finally, SEM of the monolayer MoS_2 after 5000 cycle test was performed in Figure 2e,f.

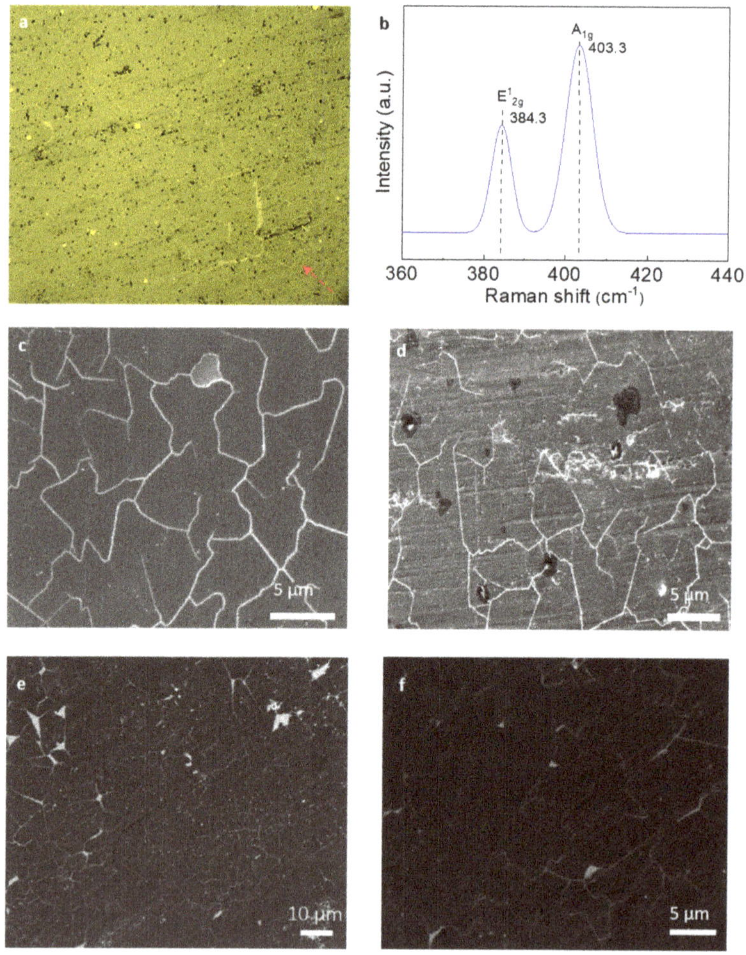

Figure 2. Characterization of the sensor electrodes for flexible sensors. Scanning electron microscopy (SEM) images of monolayer MoS_2. (**a**) Optical image of the monolayer MoS_2 on Au/PI film. (**b**) Raman data of the monolayer MoS_2. (**c**,**d**) SEM of the monolayer MoS_2 before and after transfer process. (**e**,**f**) SEM image after 5000 cycle stability test.

The digital optical image of the ionic membrane is shown in Figure 3a, which exhibits excellent flexibility. The microstructure of the inverted pyramid is clearly visible (Figure 3b).

The prepared PVA/H_3PO_4 ionic membrane was further characterized by SEM, as shown in Figure 3c. It can be clearly seen that the microstructure of the prepared ionic membrane is well established, exhibiting an inverted pyramidal structure with each size of about 100 μm. The cross-sectional structure of the ionic membrane is shown in Figure 3d, which shows that the membrane has a distinct pyramidal protrusion structure with a height of about 70 μm. The microstructure has been demonstrated to significantly improve the sensitivity of the flexible sensor. When the microstructure is compressed, due to the initial mechanically non-stationary state, it is easy to deform thus causing a rapid increase in the contact area between the electrode and the ionic membrane, thus increasing the capacitance and sensitivity of the sensor.

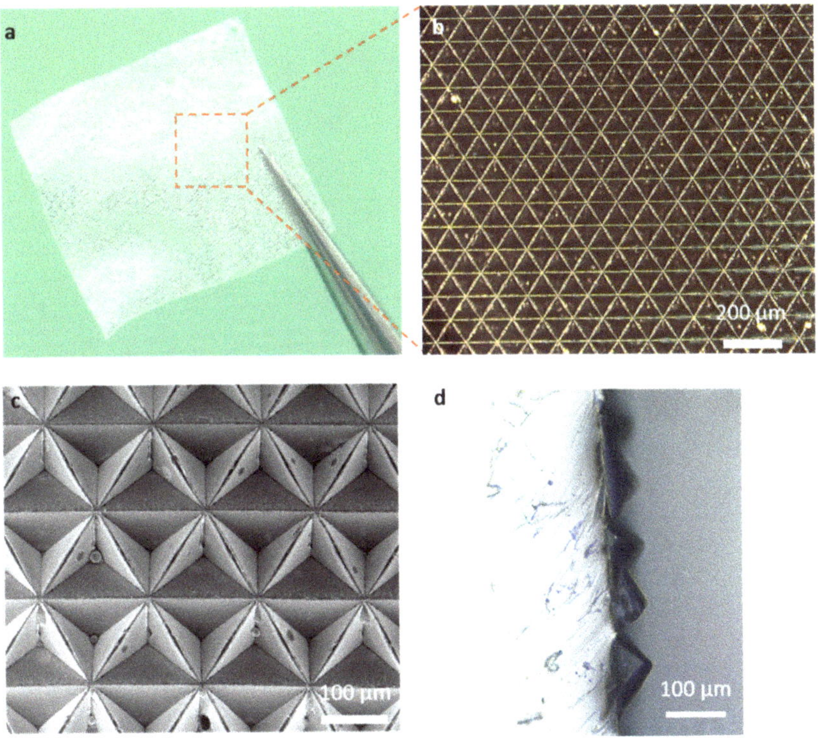

Figure 3. Characterization of the ionic film for flexible sensors. (**a**) Low and (**b**) high magnification of the optical image of the ionic film. (**c**) SEM image of the ionic film. (**d**) Cross-sectional view of the ionic film.

The performance of the sensor is investigated as shown in Figure 4. The maximum sensitivity (S_{max}) of the MoS_2-based iontronic pressure sensor is 89.75 kPa^{-1} up to 55.6 kPa, and the minimum sensitivity(S_{min}) is 10.24 kPa^{-1} in the range of 55.6 kPa to 722.2 kPa, which is superior to our previous work and other works [29–34]. Also, an iontronic pressure sensor with the same structure but without the MoS_2 layer was prepared as a control group. The results showed that its sensitivity was 0.031 kPa^{-1}, which demonstrated the important role of the MoS_2 layer. The increased capacitance is derived from the unique layer structure, huge specific surface area and pseudocapacitance characteristics.

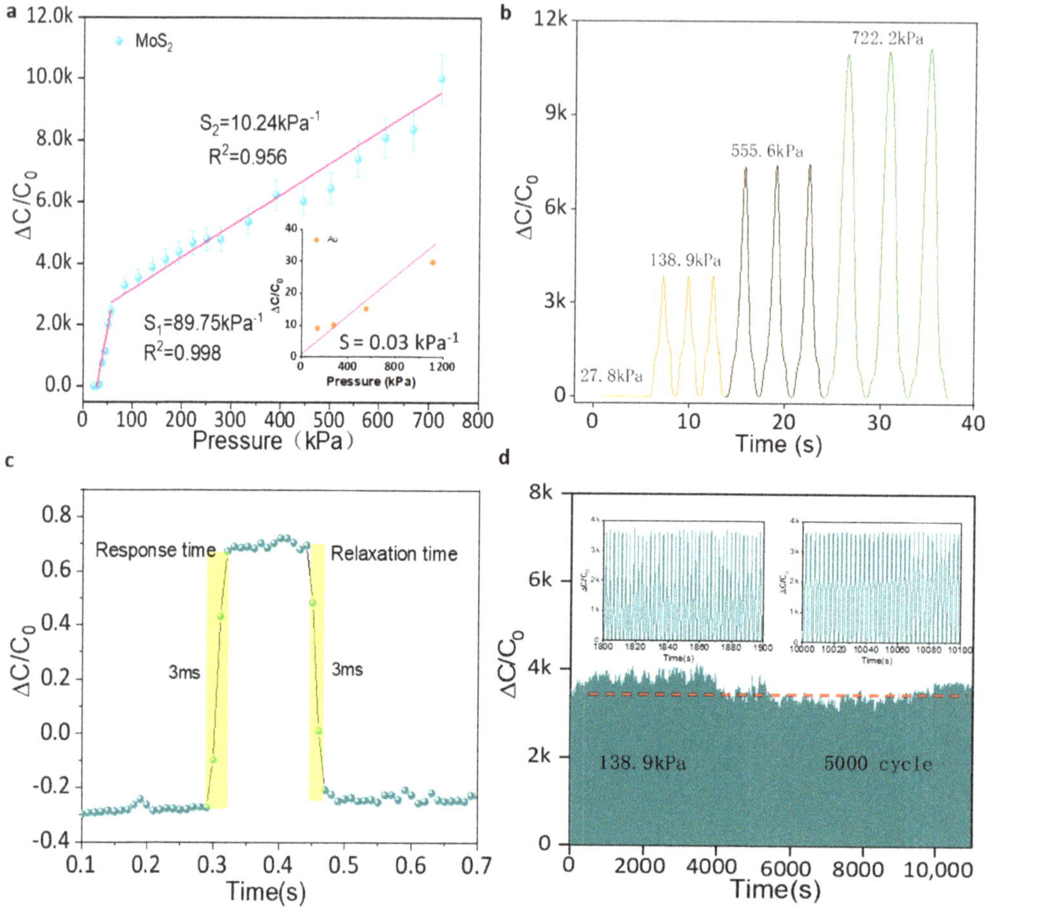

Figure 4. The performance of the flexible sensor. (**a**) Capacitance variation of the sensor under various pressure. (**b**) Continuous pressure test of the sensor. (**c**) Response and relaxation time of the flexible sensor. (**d**) Durability test of the sensor under applied pressure of 138.9 kPa for 5000 cycles.

The stability of the sensor subject to the same pressure is also one of the important factors. The sensor was tested under four different continuously pressure of 27.8 kPa, 138.9 kPa, 555.6 kPa, and 722.2 kPa respectively, as shown in Figure 4b. The signal of the sensor is constant for each test, and can be rapidly recovered after the force is removed. The stable and proportional electrical signal output at different pressures indicates that the sensor can maintain high stability with small signal drift. To evaluate the dynamic response speed of the pressure sensor, a weight of 1 g (equivalent pressure ~0.01 N) was gently placed on the pressure sensor followed by a quick release revealing a 3 ms response time and relaxation time in Figure 4c. This indicates that the prepared sensor can respond quickly to the applied external stimulus and has a good dynamic response. Even after continuous pressure of 138.9 kPa for 5000 cycles, as shown in Figure 4d, the sensor still maintained a stable signal output with no significant attenuation, demonstrating its long-term use application. Small cracks were found on the surface of monolayer MoS_2 but they did not affect final signal response in Figure 4d.

Also, the sensor was further tested at different frequencies (0.1, 0.2 and 1 Hz) at the same pressure (138.9 kPa). As shown in Figure 5a, the sensor maintained a stable signal at various frequency, demonstrating that the sensor's excellent dynamic response. Importantly, the CV curve of the sensor under 722.2 kPa was investigated to find the deep sensing mechanism. As shown in Figure 5b, the CV curve is not ideally rectangular, and shows a small peak in redox, which is further evidence of the pseudo-capacitance effect as described above.

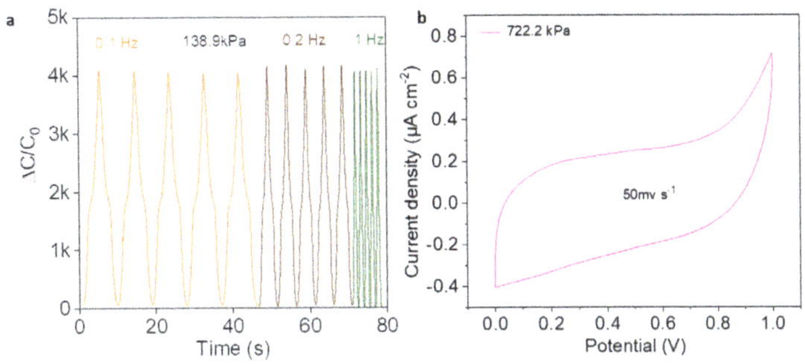

Figure 5. (**a**) Capacitance variation of the sensor under various frequencies. (**b**) CV curve of the sensor at a scan rate of 50 mV s^{-1}.

Based on the above, the superior performance of the sensor has been demonstrated and its sensing mechanism was revealed as shown in Figure 6. In this study, in addition to improving the sensitivity by MoS_2 layer, a spacer layer is also introduced between the ionic film and the electrode (Figure 6a). Initially, the capacitance is extremely small due to the presence of the spacer layer. When the sensor is subjected to pressure, the upper electrodes layer was bent due to stress and the distance between the two electrodes becomes narrow, resulting in a larger capacitance ($C_{PE,}$ Figure 6b). Further, when the force continues to be applied, the electrodes began contact with the ionic membrane (C_{EDL1}), causing a rapid increase in EDL capacitance. With further increase in force, the contact area between the microstructure of the ionic membrane and the lower electrode further increases and the capacitance further becomes larger (C_{EDL2}), leading to higher capacitance changes. In addition, it is obvious from the simulation that the deformation of the microstructure is mainly concentrated in the middle part of the microstructure rather than the tip of the structure, which may be due to the unique shape of the microstructure. It is also confirmed that the microstructure can improve the capacitance of the sensor.

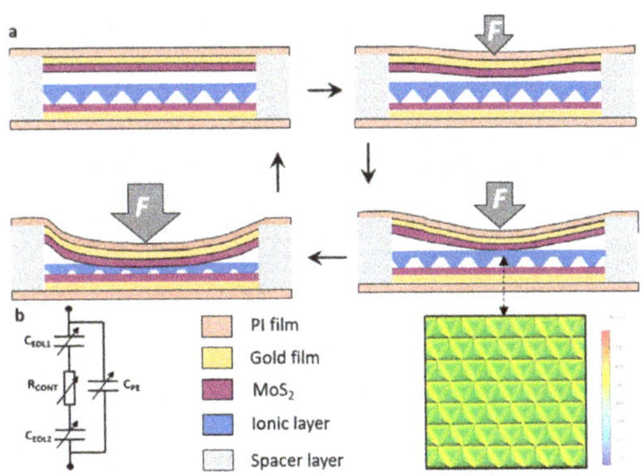

Figure 6. Schematic illustration of the working mechanism of the sensor. (**a**) Sensing mechanism of the iontronic pressure sensor and the stress concertation of the microstructure. (**b**) Corresponding capacitance variation of the sensor.

To investigate the application of the iontronic pressure sensor, the sensor was mounted on the human body to detect the movement. As shown in Figure 7a, with the rapid change of the finger to a small angle, the signal output of the sensor is consistent each time, indicating that the sensor can accurately recognize the bending of the finger. Subsequently, with a rapid change of the finger to a large angle (about 90°), the signal output amplitude increases significantly compared to the previous small angle change, suggesting that the sensor can accurately recognize different angle changes of the finger. In addition to the finger, the sensor can also detect wrist and arm bending as well as eye blinking (Figure 7b–d). These above demonstrated that the sensor has the ability to detect human motion and has important applications in the health rehabilitation of the human body. To simulate the scene of gas leak monitoring, an aurilave was used to imitate airflow. Under subtle signals, the sensor response was captured. Figure 7e has shown the variation of capacitance over time, which demonstrated that stable output can also be achieved for subtle signals. As summarized in Figure 7f [35–43], our Monolayer MoS_2-based iontronic pressure sensor shows an incomparably high sensitivity and an ultrabroad work range of pressure, out-performing existing capacitance pressure sensor.

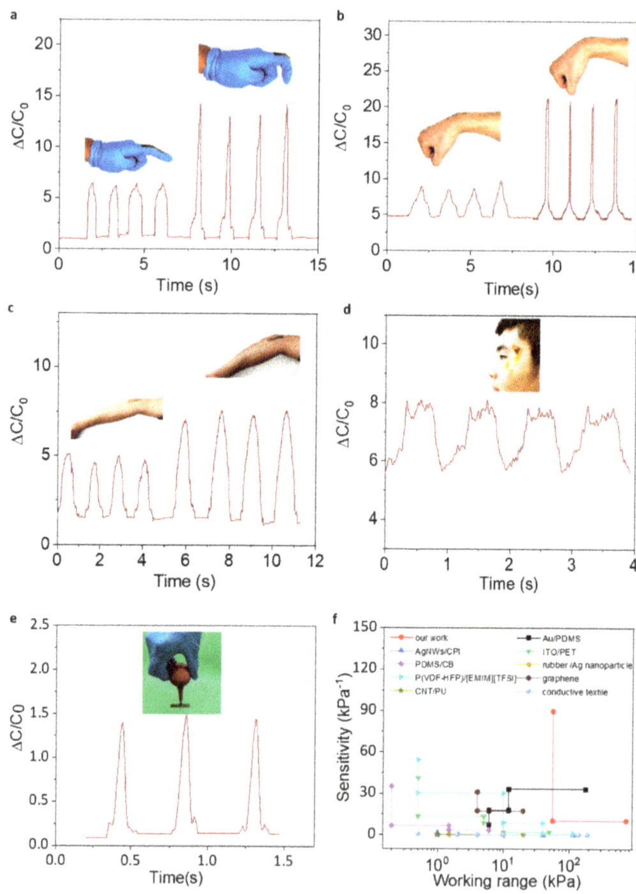

Figure 7. Practical application of the flexible iontronic sensor for montion monitoring. Capacitance variations of the sensor during (**a**) finger (**b**) wrist (**c**) arm bending and (**d**) eye blinking. (**e**) Capacitance response output of gas leak monitoring, and the image is shown in the inset. (**f**) Comparison of the sensitivity of our pressure sensor with existing capacitive sensors.

Further, we directly mounted the flexible sensor on the flexible manipulator and use the manipulator to grasp objects of different weights. In this study, pears and apples of different weights but similar shapes for comparison. As shown in Figure 8, the capacitance change trend of both is the similar, but the magnitude is different, which is due to the different magnitude of force applied by the manipulator when grasping two different weight objects. This successfully demonstrates that the prepared sensor has the ability to recognize the weight of the target and can be used as a tactile sensor to fix on the flexible manipulator to recognize the mass of the object.

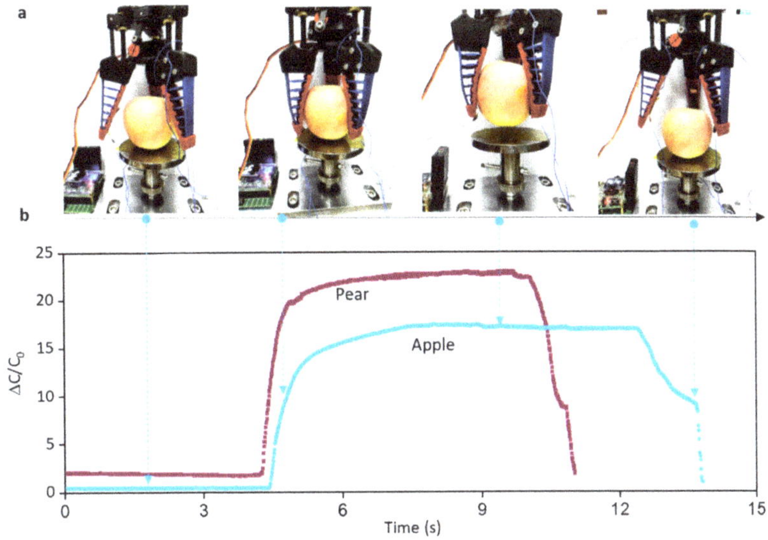

Figure 8. Application of the sensor on flexible robotics. (**a**) The process to grasp and release apple; (**b**) Corresponding steps for the soft manipulator to grasp and release the apple and pear sample.

4. Conclusions

In summary, a highly sensitive iontronic pressure sensor based on EDL sensing mechanism was designed and prepared. The electrodes based on a monolayer MoS_2 were successfully prepared by the wet transfer method, and the PVA/H_3PO_4 ionic membrane with inverted pyramid microstructure was beneficial to enhancing the sensitivity. The pressure sensor showed a high sensitivity of 89.75 kPa^{-1} in the sensing range of 722.2kPa. Additionally, the sensor exhibited a short response time and relaxation time. Through analysis, the unique structure-induced high capacitance, the rationally designed structure such as the microstructural ionic film and device configuration are all beneficial to enhancing the sensitivity and sensing range. Finally, it is verified that the sensor has a great potential for motion detection and flexible robotics applications. Our sensor design makes a novel strategy for flexible sensor electronics and will inspire more committee focus on 2D materials-based sensor.

The next research should focus on the following items.

Firstly, the effect of different layers of MoS_2 on the performance of the sensor and the mechanism of the effect should be investigated.

Secondly, the sensors should be further miniaturized to prepare higher integrated sensors such as the sensor arrays with a high sensitivity to achieve a large-scale application.

Author Contributions: Conceptualization, L.G. and K.C.; methodology, D.X., L.D. and Y.W. (Yong Wang); software, S.Y.; writing—original draft preparation, D.X., L.D.; writing—review and editing,

L.G., K.C., H.X., Y.W. (Yuejiao Wang) and L.H.; supervision, W.W. All authors have read and agreed to the published version of the manuscript.

Funding: This research was funded by the National Natural Science Foundation of China (No.61904141), the Fundamental Research Funds for the Central Universities (JB210407, JB210409), the Key Research and Development Program of Shaanxi (Program No. 2021GY-277), Innovation Fund of Xidian University, the Young Talent fund of University Association for Science and Technology in Shaanxi, China (No. 20210508), and the funding of Natural Science Foundation of Shaanxi Province (No. 2021JQ-181).

Informed Consent Statement: Informed consent was obtained from all subjects involved in the study.

Data Availability Statement: The data presented in this study are available on request from the corresponding author.

Conflicts of Interest: The authors declare no conflict of interest.

References

1. Li, X.; Wang, W.; Wu, L.; Zhao, H.; Wang, M.; Wang, Y.; Xu, H.; Liu, M.; Gao, L. Wearable, Self-Cleaning, Wireless Integrated Tactile Sensory System with Superior Sensitivity. *Sens. Actuators A Phys.* **2021**, *331*, 113027. [CrossRef]
2. Senthil Kumar, K.; Chen, P.-Y.; Ren, H. A Review of Printable Flexible and Stretchable Tactile Sensors. *Research* **2019**, *2019*, 3018568. [CrossRef] [PubMed]
3. Xiong, Y.; Shen, Y.; Tian, L.; Hu, Y.; Zhu, P.; Sun, R.; Wong, C.P. A Flexible, Ultra-Highly Sensitive and Stable Capacitive Pressure Sensor with Convex Microarrays for Motion and Health Monitoring. *Nano Energy* **2020**, *70*, 104436. [CrossRef]
4. Moses, O.A.; Gao, L.; Zhao, H.; Wang, Z.; Lawan Adam, M.; Sun, Z.; Liu, K.; Wang, J.; Lu, Y.; Yin, Z.; et al. 2D Materials Inks toward Smart Flexible Electronics. *Mater. Today* **2021**, *50*, 116–148. [CrossRef]
5. Xu, H.; Gao, L.; Wang, Y.; Cao, K.; Hu, X.; Wang, L.; Mu, M.; Liu, M.; Zhang, H.; Wang, W.; et al. Flexible Waterproof Piezoresistive Pressure Sensors with Wide Linear Working Range Based on Conductive Fabrics. *Nano-Micro Lett.* **2020**, *12*, 159. [CrossRef]
6. Zhao, X.F.; Hang, C.Z.; Wen, X.H.; Liu, M.Y.; Zhang, H.; Yang, F.; Ma, R.G.; Wang, J.C.; Zhang, D.W.; Lu, H.L. Ultrahigh-Sensitive Finlike Double-Sided E-Skin for Force Direction Detection. *ACS Appl. Mater. Interfaces* **2020**, *12*, 14136–14144. [CrossRef]
7. Xu, H.; Gao, L.; Zhao, H.; Huang, H.; Wang, Y.; Chen, G.; Qin, Y.; Zhao, N.; Xu, D.; Duan, L.; et al. Stretchable and Anti-Impact Iontronic Pressure Sensor with an Ultrabroad Linear Range for Biophysical Monitoring and Deep Learning-Aided Knee Rehabilitation. *Microsyst. Nanoeng.* **2021**, *7*, 92. [CrossRef]
8. Gao, L.; Fan, R.; Zhou, W.; Hu, X.; Cao, K.; Wang, W.; Lu, Y. Biomimetic and Radially Symmetric Graphene Aerogel for Flexible Electronics. *Adv. Electron. Mater.* **2019**, *5*, 1900353. [CrossRef]
9. Wang, C.; Hou, X.; Cui, M.; Yu, J.; Fan, X.; Qian, J.; He, J.; Geng, W.; Mu, J.; Chou, X. An Ultra-Sensitive and Wide Measuring Range Pressure Sensor with Paper-Based CNT Film/Interdigitated Structure. *Sci. China Mater.* **2020**, *63*, 403–412. [CrossRef]
10. Lee, Y.; Park, J.; Cho, S.; Shin, Y.E.; Lee, H.; Kim, J.; Myoung, J.; Cho, S.; Kang, S.; Baig, C.; et al. Flexible Ferroelectric Sensors with Ultrahigh Pressure Sensitivity and Linear Response over Exceptionally Broad Pressure Range. *ACS Nano* **2018**, *12*, 4045–4054. [CrossRef]
11. Peng, Y.; Wang, X.; Zhong, L.; Pang, K.; Chen, Y.; Wanga, M.; Liu, W. A Flexible Dual-Modal Sensing System for Synchronous Pressure and Inertial Monitoring of Finger Movement. *IEEE Sens. J.* **2021**, *21*, 10483–10490. [CrossRef]
12. Liu, Q.; Liu, Z.; Li, C.; Xie, K.; Zhu, P.; Shao, B.; Zhang, J.; Yang, J.; Zhang, J.; Wang, Q.; et al. Highly Transparent and Flexible Iontronic Pressure Sensors Based on an Opaque to Transparent Transition. *Adv. Sci.* **2020**, *7*, 2000348. [CrossRef] [PubMed]
13. Zhu, Z.; Li, R.; Pan, T. Imperceptible Epidermal–Iontronic Interface for Wearable Sensing. *Adv. Mater.* **2018**, *30*, 1705122. [CrossRef]
14. Xiao, Y.; Duan, Y.; Li, N.; Wu, L.; Meng, B.; Tan, F.; Lou, Y.; Wang, H.; Zhang, W.; Peng, Z. Multilayer Double-Sided Microstructured Flexible Iontronic Pressure Sensor with a Record-Wide Linear Working Range. *ACS Sens.* **2021**, *6*, 1785–1795. [CrossRef] [PubMed]
15. Nie, B.; Li, R.; Brandt, J.D.; Pan, T. Iontronic Microdroplet Array for Flexible Ultrasensitive Tactile Sensing. *Lab Chip* **2014**, *14*, 1107–1116. [CrossRef] [PubMed]
16. Kim, J.S.; Lee, S.C.; Hwang, J.; Lee, E.; Cho, K.; Kim, S.J.; Kim, D.H.; Lee, W.H. Enhanced Sensitivity of Iontronic Graphene Tactile Sensors Facilitated by Spreading of Ionic Liquid Pinned on Graphene Grid. *Adv. Funct. Mater.* **2020**, *30*, 1908993. [CrossRef]
17. Chang, Y.; Wang, L.; Li, R.; Zhang, Z.; Wang, Q.; Yang, J.; Guo, C.F.; Pan, T. First Decade of Interfacial Iontronic Sensing: From Droplet Sensors to Artificial Skins. *Adv. Mater.* **2021**, *33*, 2003464. [CrossRef]
18. Liu, Q.; Liu, Y.; Shi, J.; Liu, Z.; Wang, Q.; Guo, C.F. High-Porosity Foam-Based Iontronic Pressure Sensor with Superhigh Sensitivity of 9280 KPa−1. *Nano-Micro Lett.* **2022**, *14*, 21. [CrossRef]
19. Keum, K.; Eom, J.; Lee, J.H.; Heo, J.S.; Park, S.K.; Kim, Y.H. Fully-Integrated Wearable Pressure Sensor Array Enabled by Highly Sensitive Textile-Based Capacitive Ionotronic Devices. *Nano Energy* **2021**, *79*, 101631. [CrossRef]
20. Nie, B.; Li, R.; Cao, J.; Brandt, J.D.; Pan, T. Flexible Transparent Iontronic Film for Interfacial Capacitive Pressure Sensing. *Adv. Mater.* **2015**, *27*, 6055–6062. [CrossRef]
21. Gao, L.; Wang, M.; Wang, W.; Xu, H.; Wang, Y.; Zhao, H.; Cao, K.; Xu, D.; Li, L. Highly Sensitive Pseudocapacitive Iontronic Pressure Sensor with Broad Sensing Range. *Nano-Micro Lett.* **2021**, *13*, 140. [CrossRef] [PubMed]

22. Park, M.; Park, Y.J.; Chen, X.; Park, Y.K.; Kim, M.S.; Ahn, J.H. MoS_2-Based Tactile Sensor for Electronic Skin Applications. *Adv. Mater.* **2016**, *28*, 2556–2562. [CrossRef] [PubMed]
23. Qiu, D.; Chu, Y.; Zeng, H.; Xu, H.; Dan, G. Stretchable MoS 2 Electromechanical Sensors with Ultrahigh Sensitivity and Large Detection Range for Skin-on Monitoring. *ACS Appl. Mater. Interfaces* **2019**, *11*, 37035–37042. [CrossRef]
24. Veeralingam, S.; Sahatiya, P.; Badhulika, S. Papertronics: Hand-Written MoS on Paper Based Highly Sensitive and Recoverable Pressure and Strain Sensors. *IEEE Sens. J.* **2021**, *21*, 8943–8949. [CrossRef]
25. Ruth, S.R.A.; Feig, V.R.; Tran, H.; Bao, Z. Microengineering Pressure Sensor Active Layers for Improved Performance. *Adv. Funct. Mater.* **2020**, *30*, 2003491. [CrossRef]
26. Hong, M.; Yang, P.F.; Zhou, X.B.; Zhao, S.Q.; Xie, C.Y.; Shi, J.P.; Zhang, Z.P.; Fu, Q.; Zhang, Y.F. Decoupling the Interaction between Wet-Transferred MoS_2 and Graphite Substrate by an Interfacial Water Layer. *Adv. Mater. Interfaces* **2018**, *5*, 21. [CrossRef]
27. Li, H.; Zhang, Q.; Yap, C.C.R.; Tay, B.K.; Edwin, T.H.T.; Olivier, A.; Baillargeat, D. From Bulk to Monolayer MoS_2: Evolution of Raman Scattering. *Adv. Funct. Mater.* **2012**, *22*, 1385–1390. [CrossRef]
28. Li, T.; Guo, W.; Ma, L.; Li, W.; Yu, Z.; Han, Z.; Gao, S.; Liu, L.; Fan, D.; Wang, Z.; et al. Epitaxial growth of wafer-scale molybdenum disulfide semiconductor single crystals on sapphire. *Nat. Nanotechnol.* **2021**, *16*, 1201. [CrossRef]
29. Gao, L.; Cao, K.; Hu, X.; Xiao, R.; Gan, B.; Wang, W.; Lu, Y. Nano Electromechanical Approach for Flexible Piezoresistive Sensor. *Appl. Mater. Today* **2020**, *18*, 100475. [CrossRef]
30. Kang, M.; Kim, J.; Jang, B.; Chae, Y.; Kim, J.H.; Ahn, J.H. Graphene-Based Three-Dimensional Capacitive Touch Sensor for Wearable Electronics. *ACS Nano* **2017**, *11*, 7950–7957. [CrossRef]
31. Atalay, O.; Atalay, A.; Gafford, J.; Walsh, C. A Highly Sensitive Capacitive-Based Soft Pressure Sensor Based on a Conductive Fabric and a Microporous Dielectric Layer. *Adv. Mater. Technol.* **2018**, *3*, 1700237. [CrossRef]
32. Qin, R.; Hu, M.; Li, X.; Liang, T.; Tan, H.; Liu, J.; Shan, G. A New Strategy for the Fabrication of a Flexible and Highly Sensitive Capacitive Pressure Sensor. *Microsyst. Nanoeng.* **2021**, *7*, 100. [CrossRef] [PubMed]
33. Luo, Y.; Shao, J.; Chen, S.; Chen, X.; Tian, H.; Li, X.; Wang, L.; Wang, D.; Lu, B. Flexible Capacitive Pressure Sensor Enhanced by Tilted Micropillar Arrays. *ACS Appl. Mater. Interfaces* **2019**, *11*, 17796–17803. [CrossRef] [PubMed]
34. Zhou, Q.; Ji, B.; Wei, Y.; Hu, B.; Gao, Y.; Xu, Q.; Zhou, J.; Zhou, B. A Bio-Inspired Cilia Array as the Dielectric Layer for Flexible Capacitive Pressure Sensors with High Sensitivity and a Broad Detection Range. *J. Mater. Chem. A* **2019**, *7*, 27334–27346. [CrossRef]
35. Lee, J.; Kwon, H.; Seo, J.; Shin, S.; Koo, J.H.; Pang, C.; Son, S.; Kim, J.H.; Jang, Y.H.; Kim, D.E.; et al. Conductive fiber-based ultrasensitive textile pressure sensor for wearable electronics. *Adv. Mater.* **2015**, *27*, 2433–2439. [CrossRef]
36. Viry, L.; Levi, A.; Totaro, M.; Mondini, A.; Mattoli, V.; Mazzolai, B.; Beccai, L. Flexible three-axial force sensor for soft and highly sensitive artificial touch. *Adv. Mater.* **2014**, *26*, 2659–2664. [CrossRef]
37. Boutry, C.M.; Negre, M.; Jorda, M.; Vardoulis, O.; Chortos, A.; Khatib, O.; Bao, Z.N. A hierarchically patterned, bioinspired e-skin able to detect the direction of applied pressure for robotics. *Sci. Robot.* **2018**, *3*, 24. [CrossRef]
38. Qiu, Z.; Wan, Y.; Zhou, W.; Yang, J.; Yang, J.; Huang, J.; Zhang, J.; Liu, Q.; Huang, S.; Bai, N.; et al. Ionic Skin with Biomimetic Dielectric Layer Templated from Calathea Zebrine Leaf. *Adv. Funct. Mater.* **2018**, *28*, 37. [CrossRef]
39. Lu, P.; Wang, L.; Zhu, P.; Huang, J.; Wang, Y.; Bai, N.; Wang, Y.; Li, G.; Yang, J.; Xie, K.; et al. Iontronic pressure sensor with high sensitivity and linear response over a wide pressure range based on soft micropillared electrodes. *Sci. Bull.* **2021**, *66*, 1091–1100. [CrossRef]
40. Cho, S.H.; Lee, S.W.; Yu, S.; Kim, H.; Chang, S.; Kang, D.; Hwang, I.; Kang, H.S.; Jeong, B.; Kim, E.H.; et al. Micropatterned Pyramidal Ionic Gels for Sensing Broad-Range Pressures with High Sensitivity. *ACS Appl. Mater. Interfaces* **2017**, *9*, 10128–10135. [CrossRef]
41. Wan, Y.; Qiu, Z.; Huang, J.; Yang, J.; Wang, Q.; Lu, P.; Yang, J.; Zhang, J.; Huang, S.; Wu, Z.; et al. Natural Plant Materials as Dielectric Layer for Highly Sensitive Flexible Electronic Skin. *Small* **2018**, *14*, e1801657. [CrossRef] [PubMed]
42. Pruvost, M.; Smit, W.J.; Monteux, C.; Poulin, P.; Colin, A. Polymeric foams for flexible and highly sensitive low-pressure capacitive sensors. *NPJ Flex. Electron.* **2019**, *3*, 7. [CrossRef]
43. Chhetry, A.; Kim, J.; Yoon, H.; Park, J.Y. Ultrasensitive Interfacial Capacitive Pressure Sensor Based on a Randomly Distributed Microstructured Iontronic Film for Wearable Applications. *ACS Appl. Mater. Interfaces* **2019**, *11*, 3438–3449. [CrossRef] [PubMed]

micromachines

MDPI

Article

Automatic Classification Framework of Tongue Feature Based on Convolutional Neural Networks

Jiawei Li, Zhidong Zhang *, Xiaolong Zhu, Yunlong Zhao, Yuhang Ma, Junbin Zang, Bo Li, Xiyuan Cao and Chenyang Xue

Key Laboratory of Instrumentation Science & Dynamic Measurement, North University of China, Taiyuan 030051, China; jiaweili0123@163.com (J.L.); zxldtc112233@163.com (X.Z.); mailzyl@163.com (Y.Z.); myh_694@163.com (Y.M.); zangjunbin@nuc.edu.cn (J.Z.); lb@nuc.edu.cn (B.L.); caoxiyuan@nuc.edu.cn (X.C.); xuechenyang@nuc.edu.cn (C.X.)

* Correspondence: zdzhang@nuc.edu.cn

Abstract: Tongue diagnosis is an important part of the diagnostic process in traditional Chinese medicine (TCM). It primarily relies on the expertise and experience of TCM practitioners in identifying tongue features, which are subjective and unstable. We proposed a tongue feature classification framework based on convolutional neural networks to reduce the differences in diagnoses among TCM practitioners. Initially, we used our self-designed instrument to capture 482 tongue photos and created 11 data sets based on different features. Then, the tongue segmentation task was completed using an upgraded facial landmark detection method and UNET. Finally, we used ResNet34 as the backbone to extract features from the tongue photos and classify them. Experimental results show that our framework has excellent results with an overall accuracy of over 86 percent and is particularly sensitive to the corresponding feature regions, and thus it could assist TCM practitioners in making more accurate diagnoses.

Keywords: TCM tongue diagnosis; deep learning; convolutional neural network; tongue segmentation; image classification

Citation: Li, J.; Zhang, Z.; Zhu, X.; Zhao, Y.; Ma, Y.; Zang, J.; Li, B.; Cao, X.; Xue, C. Automatic Classification Framework of Tongue Feature Based on Convolutional Neural Networks. *Micromachines* **2022**, *13*, 501. https://doi.org/10.3390/mi13040501

Academic Editor: Bihan Wen

Received: 1 March 2022
Accepted: 22 March 2022
Published: 24 March 2022

1. Introduction

Traditional Chinese medicine (TCM) has received increasing attention and acknowledgment from medical experts since the World Health Organization (WHO) included it in the latest global medical guidelines [1,2]. Tongue diagnosis is a key criterion for TCM diagnosis; it is a noninvasive and convenient approach to assess human health [3,4]. However, tongue features are varied and intertwined, and subtle differences in features may correspond to completely different diseases. TCM practitioners rely on their expertise and previous experience to identify tongue features, which are susceptible to environmental factors, resulting in unstable and inaccurate diagnoses. Therefore, building a system that can objectively classify tongue features is urgently required.

Generally, tongue feature classification refers to identifying the types of different tongue features and assisting TCM practitioners to confirm diagnoses. Accurate classification is difficult due to the minor differences within the inner class. In addition, this classification mainly extracts and classifies the feature vector of the tongue body images, rather than the entire tongue regions. However, the tongue image captured by the camera has other information including lips, teeth, and even facial images. Image segmentation should be used to eliminate useless information, which can increase the accuracy of the classification and reduce the amount of calculation. Therefore, objectification and automated tongue diagnosis are mainly composed of two tasks; tongue image segmentation and tongue feature classification. Segmenting the tongue from the image is a prerequisite for tongue feature classification [5].

For tongue segmentation, traditional methods are generally used, including the region growing method, threshold method, watershed transformation method, edge detection, and snake model [6–11]. Nevertheless, the color around the tongue is similar to the color of the tongue's body, and the edge contour is relatively fuzzy. The application of traditional methods to achieve the effect of tongue image segmentation is not ideal. In recent years, more studies on using semantic segmentation based on deep convolutional neural networks (CNN) for tongue segmentation have been conducted, and the effect is better than some traditional image segmentation methods [12,13]. However, most training data sets used only contain normal tongue images, thereby reducing the clinical practicability and robustness of the segmentation algorithm.

In the research of tongue classification, many scholars digitize tongue features and classify them according to particular threshold standards [14,15]. Nevertheless, the accuracy of recognition is unsatisfactory. A number of scholars have also developed classification models for various tongue features using CNN. They use convolution operations to extract high-level semantic features and support vector machine (SVM), softmax, or other classifiers for classification. These approaches have been shown to be effective in the recognition of various features of tongue texture, tongue coating, and tongue shape [3,16–18]. These studies, however, only focus on a few features and do not systematically obtain all tongue features for analysis based on the theory of TCM tongue diagnosis; thus, the results lack clinical significance.

In this paper, we use the CNN approach to deal with these limitations. The main contribution and innovation of this paper is the optimization of the usability, efficiency, and interpretability of the tongue diagnosis classification method. First, our method is to classify 11 tongue features according to TCM theory, which can identify more comprehensive tongue features and is consistent with TCM theory to assist practitioners in making more standard diagnoses. Therefore, a clinical and standard data set with 11 tongue features was constructed due to the lack of a relevant open-source data set. In contrast, most previous studies have classified one or a few features, which are not applicable to clinical use. Second, we propose a tongue region extraction method, which can convert the original image of the tongue into a smaller pixel image before segmentation and classification to improve the efficiency and accuracy of segmentation and classification. Finally, we use GradCAM to visualize the decisions of the classification network, enabling the user to know where anomalous features appear and to better understand the basis of the classification. The proposed method is described in detail in Section 2, and the experimental results are presented in Section 3. Section 4 concludes this study.

2. Related Works

The main challenge in the domain of tongue diagnosis classification systems is to digitize the diagnostic experience of TCM practitioners in order to allow models to make as clinically meaningful and accurate diagnoses as TCM practitioners do. Such models are generally composed of two parts: tongue segmentation and classification of tongue features.

In the field of tongue segmentation, Li et al. [19] transformed the tongue image into HSI color space and used thresholding to obtain the initial region of the tongue body, and then used the tongue root and upper lip gap region to remove the irrelevant region to achieve the final segmentation. Shi et al. [20] combined the geometrical Snake model with the parameterized GVFSnake model to establish the C2G2FSnake tongue segmentation model. They introduced color space information as control information to update the external force parameters to control the accuracy and velocity of the curve, and then applied the gradient vector flow field (GVF) to obtain the results. However, these methods based on traditional image processing techniques will occasionally judge similarly colored regions around the tongue as tongue pixels as well, and the segmentation edge is rough. The segmentation effect of the traditional method is not ideal. In [12,13,21,22], semantic segmentation methods such as FCN, SegNet, and ResNet were applied to tongue segmentation. These methods using CNN for feature extraction can precisely distinguish the difference between tongue pixels and irrelevant pixels, and the contour of the segmentation result is smooth. Also, these methods do not require manual assistance at all. However, the data sets referenced by these methods are normal tongue images and lack clinicality. Therefore, this paper uses semantic segmentation based on CNN for the tongue segmentation and uses clinical tongue images as the data set for training, which is more clinically relevant.

In the field of tongue feature classification, Zhang et al. [14] extracted the tongue measurements, distances, areas, and their proportions from the tongue image and classified the tongue shape using a decision tree. Bo et al. [15] corrected the tongue deflection by applying three geometric criteria and then classified tongue shapes by analytic hierarchy process according to seven geometric features defined by various measurements of length, area, and angle of the tongue. These methods are used to convert specific tongue features into digital quantities by corresponding image processing, and then classifiers such as decision tree or SVM are used to achieve feature classification. However, the classification accuracy of these methods is not ideal, and it can only classify a few specific features, which is not very practical. Li et al. [3] proposed a method for the recognition of tooth marks. The method first generates suspicious regions based on concavity information, then extracts the depth features within the regions by CNN, and finally uses multiple-instance SVM to complete the final classification. Tang et al. [16] selected suspected rotten-greasy tongue coating patches firstly, and then used ResNet to extract features of each patch to complete the classification of the curdy or greasy coating by MI-SVM. These tongue feature classification methods based on deep learning can automatically extract more detailed and relevant features by CNN for different tongue features with higher accuracy. However, these classification methods currently classify only a few tongue features and cannot effectively assist physicians in clinical diagnosis. In addition, interpretability is very important for medical diagnosis results, but these methods are to draw direct conclusions without explainable process. Therefore, we built a classification model for 11 tongue features using CNN and visualized the decisions for the classification using GradCAM.

3. Methods

A tongue feature classification method is proposed, and the entire framework detail is shown in Figure 1. To extract the tongue region from the original input images, the recognition algorithm with 68 facial landmarks in the dlib library is upgraded. Then, the tongue images are segmented by the UNET [23]. Lastly, we build 11 tongue classification models by using the Residual Network (ResNet) [24] according to the different features. In addition, the Grad-CAM [25] is used to provide visual explanations for model decisions.

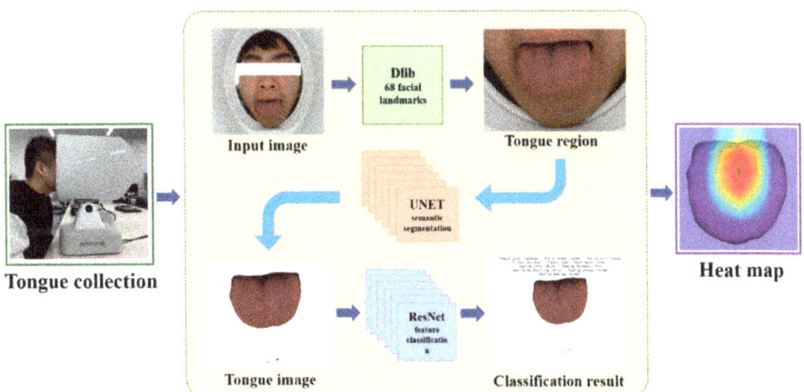

Figure 1. Overview of our framework.

3.1. Data Sets

The tongue images in the data set were collected with consistent conditions and high resolution to ensure the objectivity and effectiveness of the classification. The tongue images of the data set, which contains 482 photos with the pixel size of 3264 × 2448, were collected by a custom designed tongue diagnosis instrument (Figure 2). The tongue image was then classified by the expert TCM practitioners from Shanxi University of Chinese Medicine's Affiliated Hospital who have received extensive training and have normal or corrected vision. Eleven datasets were created based on different features, including six tongue body features and five tongue coating features (Table 1).

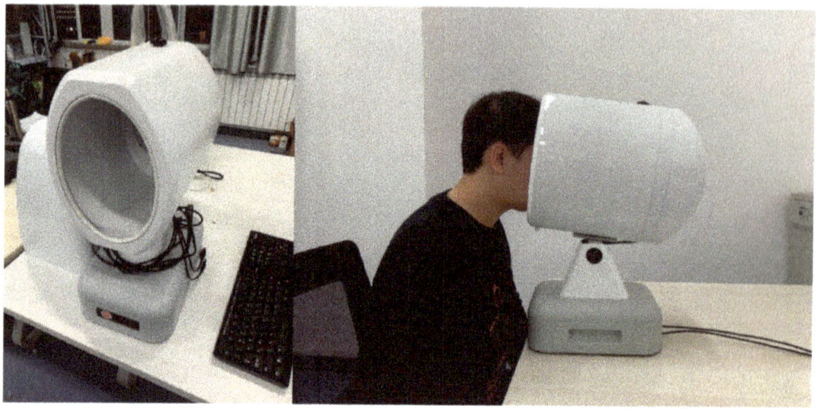

Figure 2. Tongue image acquisition with the tongue diagnosis instrument.

Table 1. Tongue Features.

Features	Inner-Class
Tongue color	Pale tongue, Light red tongue, Light cyanosed tongue, Red tongue, Deep red tongue, Cyanosed tongue, Ashen tongue, Red tongue borders and tip
Rough and tender tongue	Normal, Rough tongue, Tender tongue
Puffy and thin tongue	Normal, Puffy tongue, Swollen tongue, Thin tongue
Spots and prickles tongue	Normal, Spots and prickles tongue
Fissured tongue	Normal, Fissured tongue
Tooth-marked tongue	Normal, Tooth-marked tongue
Tongue coating color	White coating, Yellow coating, Grayish black coating
Thin and thick coating	Thin coating, Thick coating
Moist and dry coating	Moist coating, Slippery coating, Dry coating
Curdy and greasy coating	Normal, Greasy coating, Curdy coating
Peeled coating	Normal, Peeled coating

3.2. Extraction of the Tongue Region

The original image captured by the tongue diagnostic instrument is a photograph containing the entire face as well as the edges of the instrument. The extraction of the tongue region can improve the efficiency of tongue segmentation by accurately labeling the tongue segmentation and reducing the amount of image segmentation computation. However, no detector for images of sticking out tongues is available at present. In this stage, we upgraded the facial landmark recognition algorithm in the dlib library to extract the tongue region. The steps were as follows:

Step 1: Load the original image into the facial detector and obtain a matrix containing pixel coordinate information of all 68 facial landmarks. Extract the coordinates of each landmark using traversal and label them with sequential numbers. Then, record the coordinates of the 49th, 55th, and 9th of the 68 landmarks, which are the landmarks on the sides of the mouth and the bottom of the chin respectively.

Step 2: Calculate the bounding box of the tongue area according to the coordinate difference ratio in the result of step 1. The height, width, and center point coordinates of the bounding box are calculated as follows:

$$\text{height} = y_3 - \frac{y_1 + y_2}{2} + \frac{x_2 - x_1}{2} \tag{1}$$

$$\text{width} = 2 \times (x_2 - x_1) \tag{2}$$

$$\begin{bmatrix} x_{center} \\ y_{center} \end{bmatrix} = \begin{bmatrix} \frac{x_1+x_2}{2} \\ y_3 + \frac{x_2-x_1}{4} - \frac{\text{height}}{2} \end{bmatrix} \tag{3}$$

where (x_1, y_1) and (x_2, y_2) are the coordinates of the landmarks on the sides of the mouth, and (x_3, y_3) are the coordinates of the landmark on the bottom of the chin.

Step 3: Cut out the tongue region from the original image (Figure 3b) according to the bounding box.

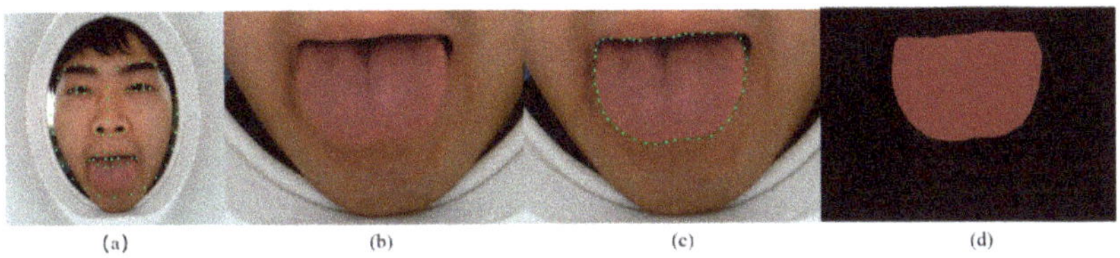

Figure 3. (**a**) Original image with 68 facial landmarks. (**b**) Tongue region image. (**c**) Edge annotation of tongue. (**d**) Tongue contour image.

We also built a data set for tongue segmentation after completing the task of extracting the tongue region of the original images. We manually labeled the edges of the tongue with the labelme software (Figure 3c) and generated corresponding contour images (Figure 3d). Tongue feature annotations were included with each contour image, and the data set was divided into the training set, validation set, and test set for the training of the segmentation model.

3.3. Segmentation of Tongue Images

As stated in Section 1, tongue segmentation can improve the effect of tongue feature classification by eliminating invalid information from the images. In this stage, we selected the UNET [23] based on deep CNN to complete the task of tongue segmentation. This method judges the category of each pixel in the image to obtain the tongue contour.

1. Structure: The structure of UNET can be divided into two parts (Figure 4). The first half of the UNET is the backbone feature extraction network. VGG16 [26] was chosen for feature extraction, which is a stack of convolution and maximum pooling operations. A feature layer with a new scale can be acquired after each pooling, resulting in five feature layers with distinct scales. The up-sampling part takes up the second half. The five feature layers were merged through the up-convolution method to produce an effective feature layer that contains all the features. The category of each pixel can be predicted according to the last obtained effective feature layer.

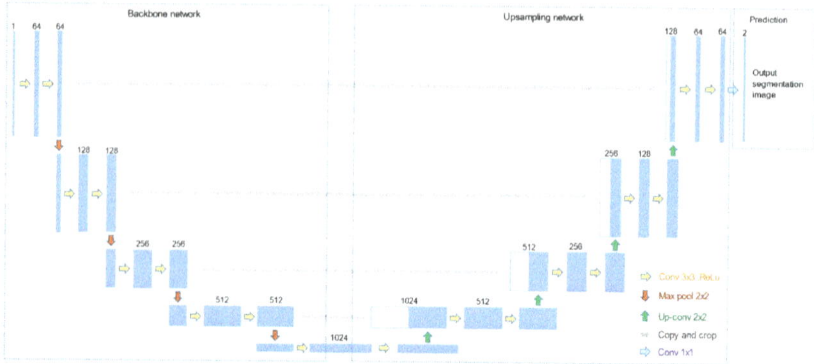

Figure 4. Structure of UNET.

2. Training: The data set for tongue segmentation was divided into the training set, validation set, and test set according to 8:1:1. The data were enhanced prior to training by random rotation and horizontal flipping of the image, as well as normalization. The loss included cross-entropy loss and dice loss. Adam algorithm was applied for optimization. Then, we used the official weight of the UNET network in the ImageNet data set as the initial weight for transfer learning. A total of 160 rounds were used to train the network. The weights of the backbone network were frozen in the first 80 rounds for rough training, and the learning rate was 1×10^{-4}. The global network was trained with a learning rate of 1×10^{-5} in the last 80 rounds for fine training.
3. Image Processing: The segmented contour images were processed by grayscale. Through observation, the generated gray image had a single gray level, with black pixels in the outer circle. Therefore, the grayscale image could be used as a mask to perform AND operate on the original tongue region image to realize the separation of the tongue. Then, we appended corresponding labels to the segmented tongue images to create classification data sets, which were also divided into the training, validation, and test sets for the training of the classification models.

3.4. Classification

We used the ResNet-34 [24] networks to classify tongue features.

1. Structure: Residual Network (ResNet-34) is a deep CNN with 34 layers, including 16 residual blocks, each with two layers (Figure 5). The last layer is an FC layer for tongue feature classification. The residual network increases the depth of the network through the connection of multiple residual blocks, while also avoiding the problem of gradient disappearance or gradient explosion.

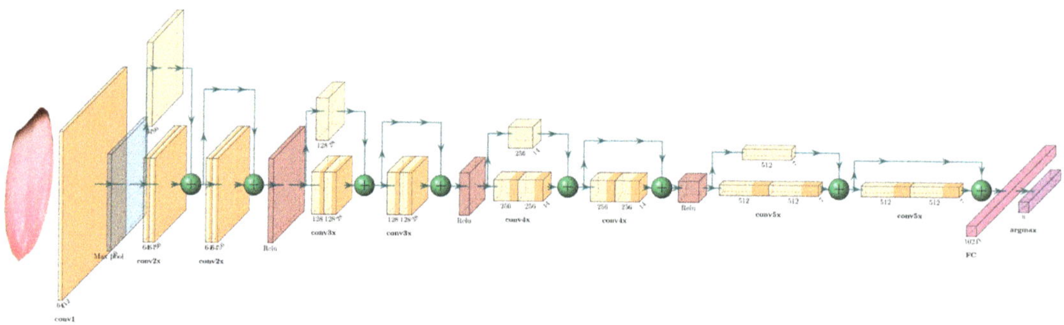

Figure 5. ResNet-34 Structure.

2. Training: The data sets for tongue feature classification are divided into the training set, validation set, and test set according to 6:2:2. The data are enhanced and normalized in the same way as the segmentation network preprocessing. The official model trained by ImageNet is used for initialization. The training is terminated after 160 rounds at a learning rate of 1×10^{-4}. The models are trained separately for 11 different tongue feature data sets. At the end of the network, the output layer is adjusted accordingly to the number of internal categories of the different features, and the final classification judgment is made using the argmax function.

3.5. Feature Visualization

In this stage, we used Grad-CAM [21] to produce visual explanations for decisions from tongue feature classification models.

TCM practitioners perform tongue diagnoses by identifying corresponding features in various positions on the tongue. Therefore, whether the corresponding feature area contributes more in tongue feature classification models should be determined. The Grad-CAM algorithm is a method for visualizing the feature maps in CNNs. The algorithm assigns importance values to each neuron through the gradient information of the back propagation of the network, and then generates a heat map by linear weighted summation to obtain the regions that the model focuses on. The tongue feature classification model can be divided into two parts: feature extraction and classifier. The classifier only classifies based on the extracted features, while the last convolutional layer of the feature extraction part is the one with the richest semantic information of tongue features. Therefore, the feature map of this layer is used for visualization. The specific realization of visualization of the corresponding regions of tongue features is as follows:

Step 1: Calculating the gradient of the classification score of the tongue feature to the last convolutional layer in conv5_x of ResNet-34. Then, similar to global average pooling, each pixel value is averaged in each channel dimension to obtain the neuron importance weight. The formula is as follows:

$$\alpha_k^c = \frac{1}{Z}\sum_i \sum_j \frac{\partial y^c}{\partial A_{ij}^k} \tag{4}$$

where Z represents the number of pixels in the feature map, and A_{ij}^k represents the pixel value at position (i, j) of the k-th feature map.

Step 2: the neuron importance weight is multiplied by each channel, added, and finally rectified by ReLU to obtain the heat map. The formula is as follows:

$$L_{\mathrm{Grad-CAM}}^c = \mathrm{ReLU}\left(\sum_k \alpha_k^c A^k\right) \tag{5}$$

4. Results

In this section, we present the experimental results of the proposed method. The metrics pixel accuracy (PA) and mean intersection over union (MIoU) are used to evaluate the effect of tongue image segmentation, and the accuracy (Acc) and F1-Score are adopted to evaluate the final classification results. The calculation formulas of PA, MIoU, Acc, and F1-Score are as follows:

$$\mathrm{PA} = \frac{\sum_{i=0}^k p_{ii}}{\sum_{i=0}^k \sum_{j=0}^k p_{ij}} \tag{6}$$

$$\mathrm{MIoU} = \frac{1}{k+1}\sum_{i=0}^{k} \frac{p_{ii}}{\sum_{j=0}^k p_{ij} + \sum_{j=0}^k p_{ji} - p_{ii}} \tag{7}$$

$$\mathrm{Acc} = \frac{\mathrm{TP}+\mathrm{TN}}{\mathrm{TP}+\mathrm{FN}+\mathrm{FP}+\mathrm{TN}} \tag{8}$$

$$\text{F1}-\text{score}=\frac{1}{c}\sum_{k=1}^{c}\frac{2p_k\times r_k}{p_k+r_k} \tag{9}$$

where k represents the number of categories other than the background, and p_{ij} represents the number of pixels belonging to the i category, predicted to be the j category pixels. Therefore, p_{ii} represents true positive (TP), and p_{ij} and p_{ji} are false positive (FP) and false negative (FN) respectively; p_k represents precision, and r_k represents recall rate.

4.1. Results of Tongue Image Segmentation

We have compared the GrabCut [27] algorithm, which is based on edge detection of image processing, with the UNET segmentation model after the same processing to verify the feasibility of the segmentation model. The effect comparison is shown in Figure 6, and the values of PA and MIoU corresponding to the segmentation results are shown in Table 2. The GrabCut algorithm has more isolated noises in the segmentation. On the contrary, the UNET algorithm based on deep CNN has a better tongue image segmentation effect.

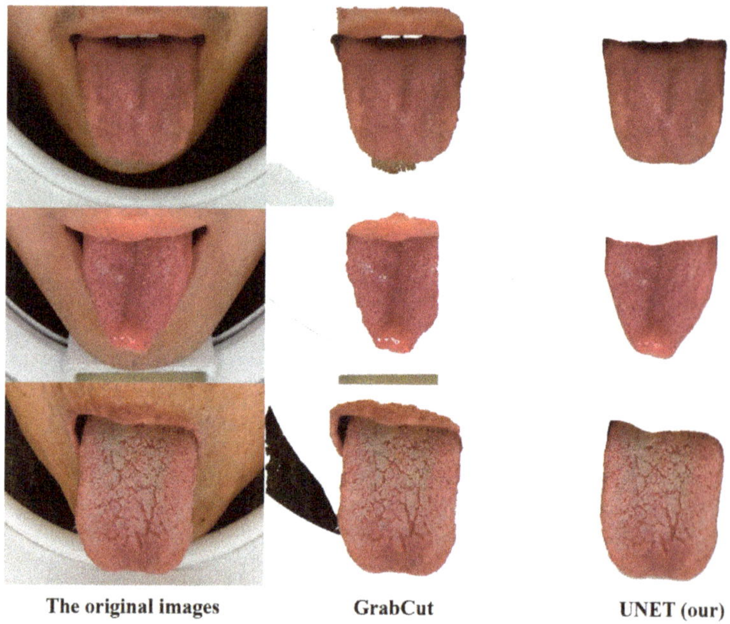

Figure 6. Segmentation effect comparison.

Table 2. GrabCut and UNET segmentation results.

Method	PA	MIoU
GrabCut	79.96%	66.26%
UNET	98.54%	97.14%

4.2. Results of Tongue Feature Classification

To verify the feasibility of the classification models, we test them on the test data sets. Figure 7 visualizes the result of tongue image classification. Table 3 shows the specific result parameters of the models for identifying the different categories of tongue features. The data show that the recognition effect is ideal, with an overall recognition accuracy of 86.14% and an overall F1-Score of 80.06%.

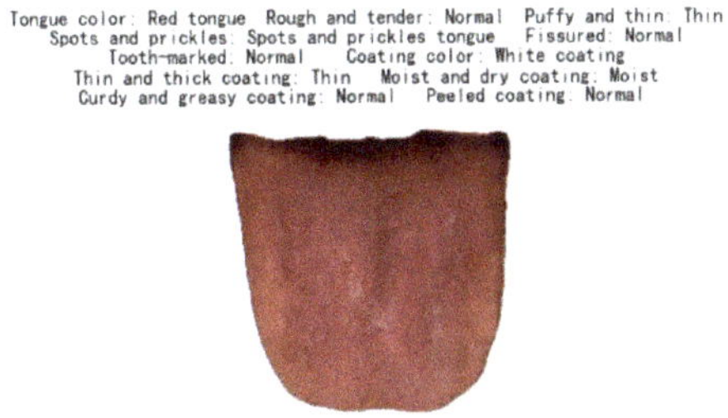

Figure 7. Tongue feature recognition results.

Table 3. Tongue feature recognition result parameters.

Feature	Acc	F1-Score
Tongue color	62.4%	55.2%
Rough and tender tongue	91.6%	83.6%
Puffy and thin tongue	86.3%	74.4%
Spots and prickles tongue	83.3%	76.5%
Fissured tongue	87.5%	82.9%
Tooth-marked tongue	86.7%	84.0%
Tongue coating color	87.5%	86.5%
Thin and thick coating	89.5%	89.2%
Moist and dry coating	87.4%	67.0%
Curdy and greasy coating	86.3%	87.2%
Peeled coating	98.9%	94.2%

4.3. Visualization of the Indicator Regions of Tongue Feature Classification

The Grad-CAM algorithm is used to process the final convolution layer of the classification model. Figure 8 shows the results of part of the feature classification model. The red area in the heat map has a greater impact on recognition, whereas the blue area has less impact. The observation shows that the classification model is more sensitive to the corresponding characteristic regions, confirming that it has high practical clinical value.

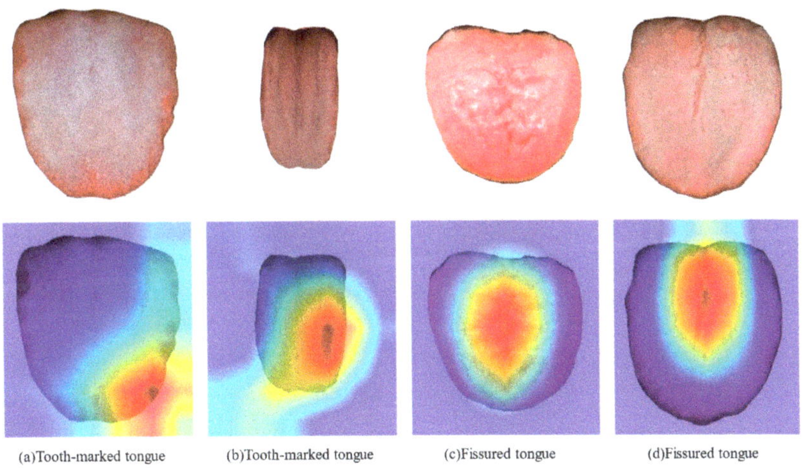

Figure 8. Heat map for different feature recognition.

5. Conclusions

We have proposed a tongue feature classification method based on CNN. The proposed method has three stages. First, the upgraded dlib facial 68 landmark algorithm is used to extract the tongue region. Second, the UNET network is used to accurately segment the tongue image. Finally, the ResNet network is used as the backbone feature extraction network to achieve the final classification of the characteristics of tongue texture and tongue coating and visualize the decisions using GradCAM. The experiment shows that the method has good accuracy for tongue feature recognition, with an overall accuracy rate of over 86%, and it is more sensitive to the corresponding feature regions. Compared with other approaches, our method identifies more comprehensive features and is more in line with TCM theory, which is suitable for clinical tongue diagnosis. When patients register at the hospital or have remote treatment, the approach can complete the classification of tongue features in advance, saving the time of TCM practitioners and thus allowing more patients to receive treatment. In addition, the GradCAM algorithm is applied to obtain a visible reference position of the network's decisions, which can help users to know the location of their tongue abnormalities and help novice practitioners to gain experience. However, the approach is not applicable to children with small faces or those who have difficulty extending their tongues because the tongue diagnostic instrument cannot perform accurate tongue acquisition in these individuals.

In addition, the proposed method can be found to have some deviations in the evaluation indexes when performing the classification of tongue color features. We believe there are two reasons for the phenomenon. One is because there is little variation in the inter-class features of tongue color, the confidence levels of these classes are similar and not high enough in the final fully connected layer of the network. The method selects the class with the highest confidence level by the argmax function, which leads to insufficient accuracy. Another is that the number of samples within the class of the data set for tongue color features is unevenly distributed, making the classification of these features less accurate. Therefore, the method still needs to be improved for the network model and data sets afterwards.

Author Contributions: Conceptualization, Z.Z. and J.L.; methodology, J.L. and X.Z.; software, J.L. and X.Z.; validation, Y.Z., Y.M., and J.L.; investigation, J.Z.; resources, X.C.; data curation, B.L. and J.L.; writing—original draft preparation, J.L.; writing—review and editing, Z.Z. and J.L.; visualization, J.L.; supervision, Z.Z.; project administration, C.X. and Z.Z. All authors have read and agreed to the published version of the manuscript.

Funding: This work was funded by the National Natural Science Foundation of China (62001430); the Shanxi Province Traditional Chinese Medicine Strong Province Special Project (204-110457); the Scientific & Technological Innovation Programs of Higher Education Institutions in Shanxi (2020L0275).

Data Availability Statement: The data presented in this study are available on request from the corresponding author. The data are not publicly available due to this data being supplied by Key Laboratory of Instrumentation Science & Dynamic Measurement (North University of China), Ministry of Education, and so cannot be made freely available.

Acknowledgments: The authors gratefully acknowledge the medical technical supports by Jiuzhang Men and his team at the Shanxi University of Chinese Medicine's Affiliated Hospital.

Conflicts of Interest: The authors declare no conflict of interest.

Ethical Statement: All subjects gave their informed consent for inclusion before they participated in the study. The study was conducted in accordance with the Declaration of Helsinki, and the protocol was approved by the Biological and Medical Ethics Committee of North University of China (110457).

References

1. Cyranoski, D. Why Chinese medicine is heading for clinics around the world. *Nature* **2018**, *561*, 448–450. [CrossRef] [PubMed]
2. Xu, Q.; Tang, W.; Teng, F.; Peng, W.; Zhang, Y.; Li, W.; Wen, C.; Guo, J. Intelligent Syndrome Differentiation of Traditional Chinese Medicine by ANN: A Case Study of Chronic Obstructive Pulmonary Disease. *IEEE Access* **2019**, *7*, 76167–76175. [CrossRef]
3. Li, X.; Zhang, Y.; Cui, Q.; Yi, X.; Zhang, Y. Tooth-Marked Tongue Recognition Using Multiple Instance Learning and CNN Features. *IEEE Trans. Cybern.* **2018**, *49*, 380–387. [CrossRef] [PubMed]
4. Zhou, H.; Hu, G.; Zhang, X. Constitution Identification of Tongue Image Based on CNN. In Proceedings of the 2018 11th International Congress on Image and Signal Processing, BioMedical Engineering and Informatics, CISP-BMEI 2018, Beijing, China, 13–15 October 2018; IEEE: Beijing, China, 2019; pp. 1–5.
5. Jiang, L.; Xu, W.; Chen, J. Digital imaging system for physiological analysis by tongue colour inspection. In Proceedings of the 2008 3rd IEEE Conference on Industrial Electronics and Applications, Singapore, 3–5 June 2008; IEEE: Singapore, 2008; pp. 1833–1836.
6. Li, W.; Hu, S.; Wang, S.; Xu, S. Towards the objectification of tongue diagnosis: Automatic segmentation of tongue image. In Proceedings of the 2009 35th Annual Conference of IEEE Industrial Electronics, Porto, Portugal, 3–5 November 2009; IEEE: Porto, Portugal, 2010; pp. 2121–2124.
7. Ning, J.; Zhang, D.; Wu, C.; Feng, Y. Automatic tongue image segmentation based on gradient vector flow and region merging. *Neural Comput. Appl.* **2012**, *21*, 1819–1826. [CrossRef]
8. Zhu, M.; Du, J. A Novel Approach for Color Tongue Image Extraction Based on Random Walk Algorithm. *Appl. Mech. Mater.* **2013**, *462–463*, 338–342. [CrossRef]
9. Li, C.; Wang, D.; Liu, Y.; Gao, X.; Shang, H. A novel automatic tongue image segmentation algorithm: Color enhancement method based on L*a*b* color space. In Proceedings of the 2015 IEEE International Conference on Bioinformatics and Biomedicine, BIBM 2015, Washington, DC, USA, 9–12 November 2015; IEEE: Washington, DC, USA, 2015; pp. 990–993.
10. Xu, W.; Kanawong, R.; Xu, D.; Li, S.; Ma, T.; Zhang, G.; Duan, Y. An automatic tongue detection and segmentation framework for computer-aided tongue image analysis. In Proceedings of the 2011 IEEE 13th International Conference on e-Health Networking, Applications and Services, Columbia, MO, USA, 13–15 June 2011; IEEE: Columbia, MO, USA, 2011; pp. 189–192.
11. Wu, J.; Zhang, Y.; Bai, J. Tongue Area Extraction in Tongue Diagnosis of Traditional Chinese Medicine. In Proceedings of the 2005 IEEE Engineering in Medicine and Biology 27th Annual Conference, Shanghai, China, 17–18 January 2006; IEEE: Shanghai, China, 2006; pp. 4955–4957.
12. Lin, B.; Xie, J.; Li, C.; Qu, Y. Deeptongue: Tongue Segmentation via Resnet. In Proceedings of the 2018 IEEE International Conference on Acoustics, Speech and Signal Processing, ICASSP 2018, Calgary, AB, Canada, 15–20 April 2018; IEEE: Calgary, AB, Canada, 2018; pp. 1035–1039.
13. Xue, Y.; Li, X.; Wu, P.; Li, J.; Wang, L.; Tong, W. Automated Tongue Segmentation in Chinese Medicine Based on Deep Learning. In Proceedings of the 25th International Conference on Neural Information Processing, ICONIP 2018, Siem Reap, Cambodia, 13–16 December 2018; Cheng, L., Leung, A., Ozawa, S., Eds.; Springer: Cham, Switzerland; Siem Reap, Cambodia, 2018; pp. 542–553.
14. Zhang, B.; Zhang, H. Significant Geometry Features in Tongue Image Analysis. *Evid.-Based Complement. Alternat. Med.* **2015**, *2015*, 897580. [CrossRef] [PubMed]
15. Huang, B.; Wu, J.; Zhang, D.; Li, N. Tongue shape classification by geometric features. *Inform. Sci.* **2010**, *180*, 312–324. [CrossRef]
16. Tang, Y.; Sun, Y.; Chiang, J.; Li, X. Research on Multiple-Instance Learning for Tongue Coating Classification. *IEEE Access* **2021**, *9*, 66361–66370. [CrossRef]
17. Xu, Q.; Zeng, Y.; Tang, W.; Peng, W.; Xia, T.; Li, Z.; Teng, F.; Li, W.; Guo, J. Multi-Task Joint Learning Model for Segmenting and Classifying Tongue Images Using a Deep Neural Network. *IEEE J. Biomed. Health* **2020**, *24*, 2481–2489. [CrossRef] [PubMed]
18. Wang, X.; Liu, J.; Wu, C.; Liu, J.; Li, Q.; Chen, Y.; Wang, X.; Chen, X.; Pang, X.; Chang, B.; et al. Artificial intelligence in tongue diagnosis: Using deep convolutional neural network for recognizing unhealthy tongue with tooth-mark. *Comput. Struct. Biotechnol. J.* **2020**, *18*, 973–980. [CrossRef] [PubMed]
19. Li, Z.; Yu, Z.; Liu, W.; Zhang, Z. Tongue Image Segmentation via Color Decomposition and Thresholding. In Proceedings of the 2017 4th International Conference on Information Science and Control Engineering, ICISCE 2017, Changsha, China, 21–23 July 2017; IEEE: Changsha, China, 2017; pp. 752–755.
20. Shi, M.; Li, G.; Li, F. C^2G^2FSnake: Automatic tongue image segmentation utilizing prior knowledge. *Sci. China Inf. Sci.* **2013**, *56*, 1–14. [CrossRef]
21. Qu, P.; Zhang, H.; Zhuo, L.; Zhang, J.; Chen, G. Automatic Tongue Image Segmentation for Traditional Chinese Medicine Using Deep Neural Network. In Proceedings of the Intelligent Computing Theories and Application, ICIC 2017, Liverpool, UK, 20 July 2017; Springer: Cham, Switzerland; Liverpool, UK, 2017; pp. 247–259.
22. Zhou, C.; Fan, H.; Li, C. TongueNet: Accurate Localization and Segmentation for Tongue Images using Deep Neural Networks. *IEEE Access* **2019**, *7*, 1. [CrossRef]
23. Ronneberger, O.; Fischer, P.; Brox, T. U-Net: Convolutional Networks for Biomedical Image Segmentation. *arXiv* **2015**, arXiv:1505.04597.
24. He, K.; Zhang, X.; Ren, S.; Sun, J. Deep Residual Learning for Image Recognition. *arXiv* **2015**, arXiv:1512.03385.

25. Selvaraju, R.R.; Cogswell, M.; Das, A.; Vedantam, R.; Parikh, D.; Batra, D. Grad-CAM: Visual Explanations from Deep Networks via Gradient-based Localization. *Int. J. Comput. Vis.* **2020**, *128*, 336–359. [CrossRef]
26. Karen, S.; Andrew, Z. Very Deep Convolutional Networks for Large-Scale Image Recognition. *arXiv* **2014**, arXiv:1409.1556.
27. Rother, C.; Kolmogorov, V.; Blake, A. GrabCut: Interactive foreground extraction using iterated graph cut. *ACM Trans. Graphic.* **2004**, *23*, 309–314. [CrossRef]

MDPI
St. Alban-Anlage 66
4052 Basel
Switzerland
Tel. +41 61 683 77 34
Fax +41 61 302 89 18
www.mdpi.com

Micromachines Editorial Office
E-mail: micromachines@mdpi.com
www.mdpi.com/journal/micromachines